Lecture Notes in Computer Science 6467

Commenced Publication in 1973
Founding and Former Series Editors:
Gerhard Goos, Juris Hartmanis, and Jan van Leeuwen

W0246069

Editorial Board

David Hutchison
Lancaster University, UK

Takeo Kanade
Carnegie Mellon University, Pittsburgh, PA, USA

Josef Kittler
University of Surrey, Guildford, UK

Jon M. Kleinberg
Cornell University, Ithaca, NY, USA

Alfred Kobsa
University of California, Irvine, CA, USA

Friedemann Mattern
ETH Zurich, Switzerland

John C. Mitchell
Stanford University, CA, USA

Moni Naor
Weizmann Institute of Science, Rehovot, Israel

Oscar Nierstrasz
University of Bern, Switzerland

C. Pandu Rangan
Indian Institute of Technology, Madras, India

Bernhard Steffen
TU Dortmund University, Germany

Madhu Sudan
Microsoft Research, Cambridge, MA, USA

Demetri Terzopoulos
University of California, Los Angeles, CA, USA

Doug Tygar
University of California, Berkeley, CA, USA

Gerhard Weikum
Max Planck Institute for Informatics, Saarbruecken, Germany

Swee-Huay Heng   Rebecca N. Wright
Bok-Min Goi (Eds.)

Cryptology and Network Security

9th International Conference, CANS 2010
Kuala Lumpur, Malaysia, December 12-14, 2010
Proceedings

 Springer

Volume Editors

Swee-Huay Heng
Multimedia University
Faculty of Information Science and Technology
Jalan Ayer Keroh Lama, 75450 Melaka, Malaysia
E-mail: shheng@mmu.edu.my

Rebecca N. Wright
Rutgers University
Department of Computer Science
96 Frelinghuysen Road, Piscataway, NJ, 08854, USA
E-mail: rebecca.wright@rutgers.edu

Bok-Min Goi
Universiti Tunku Abdul Rahman
Faculty of Engineering and Science
Kuala Lumpur Campus, Jalan Genting Klang, 53300 Kuala Lumpur, Malaysia
E-mail: goibm@utar.edu.my

Library of Congress Control Number: 2010940115

CR Subject Classification (1998): E.3, C.2, K.6.5, D.4.6, G.2.1, E.4

LNCS Sublibrary: SL 4 – Security and Cryptology

ISSN   0302-9743
ISBN-10   3-642-17618-6 Springer Berlin Heidelberg New York
ISBN-13   978-3-642-17618-0 Springer Berlin Heidelberg New York

springer.com

© Springer-Verlag Berlin Heidelberg 2010
Printed in Germany

Typesetting: Camera-ready by author, data conversion by Scientific Publishing Services, Chennai, India
Printed on acid-free paper   06/3180

Preface

The 9th International Conference on Cryptology and Network Security (CANS 2010) was held in Kuala Lumpur, Malaysia during December 12–14, 2010. The conference was co-organized by the Multimedia University (MMU), Malaysia, and Universiti Tunku Abdul Rahman (UTAR), Malaysia.

The conference received 64 submissions from 22 countries, out of which 21 were accepted after a careful and thorough review process. These proceedings also contain abstracts for two invited talks. All submissions were reviewed by at least three members of the Program Committee; those authored or co-authored by Program Committee members were reviewed by at least five reviewers. Program Committee members were allowed to use external reviewers to assist with their reviews, but remained responsible for the contents of the review and representing papers during the discussion and decision making. The review phase was followed by a 10-day discussion phase in which each paper with at least one supporting review was discussed, additional experts were consulted where needed, and final decisions were made.

We thank the Program Committee for their hard work in selecting the program. We also thank the external reviewers who assisted with reviewing and the CANS Steering Committee for their help. We thank Shai Halevi for use of his Web-Submission-and-Review software that was used for the electronic submission and review of the submitted papers, and we thank the International Association for Cryptologic Research (IACR) for Web hosting of the software.

The main goal of the conference is to promote research on all aspects of network security, as well as to build a bridge between research on cryptography and network security. The current edition continues to fulfill this goal with the broad range of areas covered by the high-quality accepted papers, including both theoretical and practical analysis of cryptographic primitives and protocols; secure systems and mechanisms for networked applications such as cloud computing, electronic voting, and mobile computing; and advances in anonymous credentials and their use. In addition, the conference featured two invited speakers to present their cutting-edge research: Kaoru Kurosawa of Ibaraki University, Japan, on "Cryptography for Unconditionally Secure Message Transmission in Networks" and Ahmad-Reza Sadeghi of Technical University Darmstadt, Germany, on "Cryptography Meets Hardware: Selected Topics of Hardware-based Cryptography"; these talks contributed greatly to the conference program.

Finally, we thank the local Organizing Committee for their dedication and commitment in organizing the conference. We also thank all authors who submitted papers, whether or not they were accepted, and all conference attendees for their contribution to a lively and energetic conference.

December 2010   Swee-Huay Heng
                                                                Rebecca N. Wright
                                                                   Bok-Min Goi

CANS 2010

The 9th International Conference on Cryptology and Network Security

Kuala Lumpur, Malaysia
December 12–14, 2010
Jointly organized by
Multimedia University, Malaysia
and
Universiti Tunku Abdul Rahman, Malaysia

General Chair

Bok-Min Goi   Universiti Tunku Abdul Rahman, Malaysia

Program Co-chairs

Swee-Huay Heng   Multimedia University, Malaysia
Rebecca N. Wright   Rutgers University, USA

Program Committee

Michel Abdalla   Ecole Normale Superieure, France
William Arbaugh   University of Maryland, USA
John Black   University of Colorado at Boulder, USA
Carlo Blundo   University of Salerno, Italy
Xavier Boyen   University of Liege, Belgium
Melissa Chase   Microsoft Research, USA
Sherman S.M. Chow   New York University, USA
Giovanni Di Crescenzo   Telcordia Technologies, USA
Rosario Gennaro   IBM Research, USA
Bok-Min Goi   Universiti Tunku Abdul Rahman, Malaysia
Matthew Green   Johns Hopkins University, USA
Tetsu Iwata   Nagoya University, Japan
Aaron D. Jaggard   Rutgers University, USA
Trevor Jim   AT&T Labs Research, USA
Charanjit Jutla   IBM Research, USA
Seny Kamara   Microsoft, USA
Jonathan Katz   University of Maryland, USA

Khoongming Khoo	DSO National Laboratories, Singapore
Aggelos Kiayias	University of Athens, Greece
Kwangjo Kim	KAIST, Korea
Vladimir Kolesnikov	Bell Labs, Alcatel-Lucent, USA
Lee Kai (Joseph) Liu	Institute for Infocomm Research, Singapore
Javier Lopez	University of Malaga, Spain
Mark Manulis	Technische Universität Darmstadt, Germany
Adam O'Neill	University of Texas at Austin, USA
Wakaha Ogata	Tokyo Institute of Technology, Japan
Raphael C.-W. Phan	Loughborough University, UK
Josef Pieprzyk	Macquarie University, Australia
David Pointcheval	CNRS & ENS, France
Georgios Portokalidis	Columbia University, USA
C. Pandu Rangan	Indian Institute of Technology, India
Rei Safavi-Naini	University of Calgary, Canada
Tomas Sander	Hewlett-Packard Labs, USA
Nitesh Saxena	Polytechnic Institute of NYU, USA
William Skeith	City College of New York, USA
Jessica Staddon	Google, USA
Angelos Stavrou	George Mason University, USA
Douglas Stinson	University of Waterloo, Canada
Ivan Visconti	University of Salerno, Italy
Poorvi Vora	George Washington University, USA
Xiaoyun Wang	Shandong University, China
Hoeteck Wee	Queens College, CUNY, USA
Susanne Wetzel	Stevens Institute of Technology, USA
Danfeng Yao	Virginia Tech, USA
Sung-Ming Yen	National Central University, Taiwan

Steering Committee

Yvo Desmedt	University College London, UK
Matthew Franklin	University of California at Davis, USA
Juan A. Garay	AT&T Labs - Research, USA
Yi Mu	University of Wollongong, Australia
David Pointcheval	CNRS & ENS, France
Huaxiong Wang	National Technological University, Singapore

Organizing Committee

Local Organizing Chair	Victor Hock-Kim Tan, UTAR
Secretariat	Ji-Jian Chin, MMU
Publicity	Yong-Haur Tay, UTAR
Finance	Thian-Song Ong, MMU

Webmaster   Goh Kah-Ong Michael, MMU
Logistics   Tommy Tong-Yuen Chai, UTAR
                     Zan-Kai Chong, UTAR
                     Swee-Eng Khor, MMU
                     Priya a/p Kulampurath Govindan Nair, UTAR
                     Huo-Chong Ling, MMU
                     Syh-Yuan Tan, UTAR
                     Yar-Ling Tan, UTAR
                     Wei-Chuen Yau, MMU

External Reviewers

Shweta Agrawal	Jian Guo	Alessandra Scafuro
Hadi Ahmadi	Tzipora Halevi	Elaine Shi
Werner Backes	Islam Hegazy	Jaechul Sung
Angelo De Caro	Vincenzo Iovino	Syh-Yuan Tan
David Cash	Srinivas Krishnan	Jheng-Hong Tu
Chien-Ning Chen	Virendra Kumar	Ashraful Tuhin
Ashish Choudhary	Wei-Chih Lien	Jonathan Voris
Jiali Choy	Yasuda Masaya	Douglas Wikstrom
Cheng-Kang Chu	Marine Minier	Huihui Yap
Dario Fiore	Sai Teja Peddinti	Aileen Zhang
Pierre-Alain Fouque	Nashad Safav	Tongjie Zhang

Table of Contents

Public Key Cryptography

Invited Talk II

Secure Mechanisms

Cryptographic Protocols

Anonymous Credentials

Cryptanalysis of Reduced-Round MIBS Block Cipher

Asli Bay, Jorge Nakahara Jr.*, and Serge Vaudenay

EPFL, Switzerland
{asli.bay,jorge.nakahara,serge.vaudenay}@epfl.ch

Abstract. This paper presents the first independent and systematic linear, differential and impossible-differential (ID) cryptanalyses of MIBS, a lightweight block cipher aimed at constrained devices such as RFID tags and sensor networks. Our contributions include linear attacks on up to 18-round MIBS, and the first ciphertext-only attacks on 13-round MIBS. Our differential analysis reaches 14 rounds, and our impossible-differential attack reaches 12 rounds. These attacks do not threaten the full 32-round MIBS, but significantly reduce its margin of security by more than 50%. One fact that attracted our attention is the striking similarity of the round function of MIBS with that of the Camellia block cipher. We actually used this fact in our ID attacks. We hope further similarities will help build better attacks for Camellia as well.

Keywords: cryptanalysis, lightweight block ciphers, RFID tags, sensor networks.

1 Introduction

This paper describes the first independent and systematic linear, differential and impossible-differential cryptanalyses on reduced-round variants of the MIBS block cipher. MIBS is a lightweight cipher, with a Feistel structure, aimed at ubiquitous but constrained environments, such as RFID tags and sensor networks [6]. MIBS operates on 64-bit blocks, uses keys of 64 or 80 bits and iterates 32 rounds. There is a striking similarity between the round functions of MIBS and Camellia ciphers [1]. This feature was actually exploited in our impossible-differential analysis of MIBS in Sect.5. Our results are summarized in Table 6.

Previous cryptanalytic results on MIBS, presented by its designers, concerned differential and linear relations on up to 4-round MIBS. Nonetheless, no full attacks were ever detailed. We provide better distinguishers and attacks on up to 18 rounds, effectively reducing the margin of security of MIBS by more than 50% as originally predicted by its designers.

This paper is organized as follows: Sect. 2 describes the main components of MIBS relevant for the attacks in this paper; Sect. 3 details linear relations

---

* This work was supported by the National Competence Center in Research on Mobile Information and Communication Systems (NCCR-MICS), a center of the Swiss National Science Foundation under grant number 5005-67322.

S.-H. Heng, R.N. Wright, and B.-M. Goi (Eds.): CANS 2010, LNCS 6467, pp. 1–19, 2010.

and attacks on reduced-round versions of MIBS; Sect. 4 presents differential characteristics and attacks; Sect. 5 presents impossible-differential distinguishers and attacks; Sect. 6 concludes this paper.

2   A Brief Description of MIBS

MIBS is a block cipher following a Feistel Network design [6]. MIBS operates on 64-bit blocks, uses keys of 64 or 80 bits, and iterates 32 rounds for both key sizes. All internal operations in MIBS are nibble-wise, that is, on 4-bit words. The round function F of MIBS has an SPN structure composed of an xor layer with a round subkey, an S layer of 4×4-bit S-boxes, and a linear transformation layer (with branch number 5), in this order.

For our attack purposes, the linear transformation (P layer) is most relevant. Let $(y_1, y_2, y_3, y_4, y_5, y_6, y_7, y_8)$ denote the input to this layer. Its output, $(y_1', y_2', y_3', y_4', y_5', y_6', y_7', y_8')$, can be described as

$$y_1' = y_1 \oplus y_2 \oplus y_4 \oplus y_5 \oplus y_7 \oplus y_8; \; y_2' = y_2 \oplus y_3 \oplus y_4 \oplus y_5 \oplus y_6 \oplus y_7;$$
$$y_3' = y_1 \oplus y_2 \oplus y_3 \oplus y_5 \oplus y_6 \oplus y_8; \quad y_4' = y_2 \oplus y_3 \oplus y_4 \oplus y_7 \oplus y_8;$$
$$y_5' = y_1 \oplus y_3 \oplus y_4 \oplus y_5 \oplus y_8; \quad y_6' = y_1 \oplus y_2 \oplus y_4 \oplus y_5 \oplus y_6;$$
$$y_7' = y_1 \oplus y_1 \oplus y_3 \oplus y_6 \oplus y_7; \; y_8' = y_1 \oplus y_3 \oplus y_4 \oplus y_6 \oplus y_7 \oplus y_8, \qquad (1)$$

where $\oplus$ denotes exclusive or.

The input text block to the i-th round is denoted (L_{i-1}, R_{i-1}), with $L_i, R_i \in \{0, 1\}^{32}$, and $(R_{i-1} \oplus F(K_i, L_{i-1}), L_{i-1})$ denotes the round output. (L_0, R_0) denotes a plaintext block.

The key schedule of MIBS is adapted from the key schedule of PRESENT [4]. There are two versions of key schedule of MIBS, both generating 32-bit round subkeys K_i for $1 \leq i \leq 32$, from 64-bit and 80-bit user keys, respectively. Let $state^i$ denote the ith round key state; $state^0$ denote the user key. The 80-bit version of key schedule of MIBS, with bit numbering in right-to-left order from 1 to 80, is as follows:

for $i = 1$ to $i = 32$,

$state^i = state^{i-1} \ggg 19$,

$state^i = S[state^i[80 \sim 77]] \| S[state^i[76 \sim 73]] \| state^i[72 \sim 1]$,

$state^i = state^i[80 \sim 20] \| state^i[19 \sim 15] \oplus$ Counter $\| state^i[14 \sim 1]$,

$K_i = state^i[80 \sim 49]$.

where '$\ggg$' means bitwise right-rotation, '$\|$' means string concatenation, and '$\sim$' indicates a sequence of bit positions. We refer to [6] for further details about MIBS components.

3   Linear Cryptanalysis

In [6], the designers claimed security of MIBS against linear cryptanalysis by providing a 4-round linear relation with 7 active S-boxes, and overall bias 2^{-8}.

Table 1. A 4-round linear relation for MIBS

Round i	ΓL_{i-1}	ΓR_{i-1}	Number of active S-boxes	Bias
1	00600600_x	02202220_x	1	2^{-2}
2	02202220_x	00660600_x	2	2^{-3}
3	00660600_x	00202200_x	2	2^{-3}
4	00202200_x	60666600_x	1	2^{-2}
5	60666600_x	00002200_x	-	-

They assumed that this relation was iterative (although it was not) and claimed resistance of the full 32-round MIBS to linear attacks.

Firstly, we derived the linear approximation table (LAT) for[1] the 4×4 S-box of MIBS. See Table 7 in the appendix. We note that this S-box is linearly 4-uniform[2] (an analogous concept to that used in DC, Sect. 4). Thus, the highest bias is 2^{-2}. We have found a better 4-round linear relation, described in Table 1, with only six active S-boxes and bias 2^{-7}. The last pair of bit masks in Table 1 stand for the output masks after the swapping of half blocks in a round.

We denote the input mask to the i-th round as $(\Gamma L_{i-1}, \Gamma R_{i-1})$. The $(i+1)$-th round input mask is the i-th round output mask. Values subscripted by 'x' are in hexadecimal base.

3.1 Searching for Linear Relations for MIBS

For a systematic linear analysis of MIBS, we automated the search procedure by creating a program to look for linear relations of MIBS according to the following criteria:

- focus on iterative linear relations, preferably;
- maximize the overall bias by minimizing the number of active S-boxes;
- use the fact that the S-box is linearly 4-uniform (Table 7);
- use the fact that the branch number of the P permutation in the F function of MIBS is 5 (which is claimed to be optimal)

Taking into account these criteria, the best result of our search is the 16-round linear relation with 30 active S-boxes and bias 2^{-31} in Table 2. From the LAT of MIBS, Table 7, there are six possible instantiations of this linear relation, that is, $(w, z) \in \{(2_x, 6_x), (6_x, 2_x), (4_x, e_x), (e_x, 4_x), (8_x, d_x), (d_x, 8_x)\}$, where we exploited the symmetry $w \overset{S-box}{\rightarrow} z$ and $z \overset{S-box}{\rightarrow} w$ (both with the same bias 2^{-2}). The last line of Table 2 accounts for the swapping between half blocks. The first 15 rounds of this distinguisher corresponds to the best 15-round linear relation (with 28 active S-boxes, and bias 2^{-29}) that will be used in a key-recovery attack in Sect. 3.2.

---

[1] The LAT of an S-box stands for a table containing an exhaustive enumeration of all linear approximations of the given S-box.

[2] It means that the largest entry in the LAT has value 4.

Table 2. A 16-round linear relation for MIBS

Round i	ΓL_{i-1}	ΓR_{i-1}	Number of active S-boxes	Bias
1	w000w0w0$_x$	00000000$_x$	0	2^{-1}
2	00000000$_x$	w000w0w0$_x$	2	2^{-3}
3	w000w0w0$_x$	z0000z00$_x$	3	2^{-4}
4	z0000z00$_x$	w000ww0w$_x$	2	2^{-3}
5	w000ww0w$_x$	z000zz0z$_x$	2	2^{-3}
6	z000zz0z$_x$	w000w0w00$_x$	3	2^{-4}
7	w0000w00$_x$	z000z0z0$_x$	2	2^{-3}
8	z000z0z0$_x$	00000000$_x$	0	2^{-1}
9	00000000$_x$	z000z0z0$_x$	2	2^{-3}
10	z000z0z0$_x$	w0000w00$_x$	3	2^{-4}
11	w0000w00$_x$	z000zz0z$_x$	2	2^{-3}
12	z000zz0z$_x$	w000ww0w$_x$	2	2^{-3}
13	w000ww0w$_x$	z00000z00$_x$	3	2^{-4}
14	z0000z00$_x$	w000w0w0$_x$	2	2^{-3}
15	w000w0w0$_x$	00000000$_x$	0	2^{-1}
16	00000000$_x$	w000w0w0$_x$	2	2^{-3}
17	w000w0w0$_x$	z0000z00$_x$	-	-

3.2   17-Round Multiple Linear Attack

We perform a key-recovery attack on 17-round MIBS by considering the first fifteen rounds of the linear distinguisher in Table 2, placed between rounds 2 and 16. We recover subkey bits from the first and last rounds.

The main relation for this 17-round attack is

$$(R_0 \oplus F(K_1, L_0)) \cdot \texttt{w000w0w0}_x \oplus (L_{17} \oplus F(K_{17}, R_{17})) \cdot \texttt{w000w0w0}_x = 0, \quad (2)$$

where w is one of the values indicated in Sect. 3.1. Due to the low branch number of the P layer (see Sect. 2), only two subkey nibbles need to be guessed in both $F(K_1, L_0)$ and $F(K_{17}, R_0)$. See Fig. 1. Following [3], we use four variations of (2) for four values of w that lead to linearly independent relations: $w \in \{2_x, 4_x, 8_x, d_x\}$. According to [3], the combined bias of these multiple linear relations is $\sqrt{4 \cdot (2^{-29})^2} = 2^{-28}$. The data complexity is $4/(2^{-28})^2 = 2^{58}$ KP.

The attack procedure follows [7]:

– Take 2^{58} known plaintexts and request the corresponding ciphertexts encrypted under the unknown secret key K.
– for $w \in \{2_x, 4_x, 8_x, d_x\}$ keep independent counters for each possible value of subkey bits which correspond to active S-boxes: S_1 and S_6 in both rounds 1 and 17.
– For each possible key, check that $(R_0 \oplus F(K_1, L_0)) \cdot \texttt{w000w0w0}_x \oplus (L_{17} \oplus F(K_{17}, R_{17})) \cdot \texttt{w000w0w0}_x = 0$ holds, where, for instance, $w = 6$:

  For each key candidate K_i, let T_i^w be the number of plaintexts such that $(R_0 \oplus F(K_{1,1}\|K_{1,6}, L_0)) \cdot \texttt{w000w0w0}_x \oplus (L_{17} \oplus F(K_{17,1}\|K_{17,6}, R_{17}) \oplus \texttt{w000w0w0}_x = 0$ for each w. Let $T_{\max}^w$ be the maximal value and $T_{\min}^w$ be the minimal value of all T_i^w's, then

  • If $|T_{\max}^w - N/2| > |T_{\min}^w - N/2|$ then adopt the key candidate corresponding to $T_{\max}^w$

- If $|T_{\min}^w - N/2| > |T_{\max}^w - N/2|$ then adopt the key candidate corresponding to $T_{\min}^w$, where $N = 2^{58}$ in this attack.
- the correct subkey is simultaneously suggested by the counters $T_{\max}^w$ or $T_{\min}^w$ corresponding to all four values of w.

According to the key schedule of MIBS, there is no overlapping between the subkeys $K_{1,1}$, $K_{1,6}$, $K_{17,1}$, $K_{17,6}$. Thus, the time complexity becomes $\frac{2^{16}}{2 \cdot 17} \cdot 2^{58} \approx 2^{69}$ 17-round MIBS encryptions because partial decryption of two nibbles in the first round and two other nibbles in the 17th round costs about half a round. The memory complexity is the 2^{58} blocks. The remaining 64 key bits can be recovered by exhaustive search without affecting the overall attack complexity. Following [9], the success probability of this attack, p_S, is computed assuming $N \cdot |p - 1/2|^2 = 4$, and $a = 8$

$$p_S = \Phi(2 \cdot \sqrt{N} \cdot |p - 1/2| - \Phi^{-1}(1 - 2^{-a-1})) \approx 0.9794$$

where Φ is the cumulative distribution function of the standard normal distribution.

3.3 Ciphertext-Only Attack

Assuming the input plaintext is coded as ASCII text, we can perform a ciphertext-only attack. In this setting though, the codebook size is reduced to $2^{64-8} = 2^{56}$, since the most significant bit of every byte is zero. We use the first 13 rounds of (Table 2), which imply the following linear relation: $L_0 \cdot \texttt{80008080}_\mathbf{x} \oplus L_{17} \cdot \texttt{e0000e00}_\mathbf{x} \oplus R_{17} \cdot \texttt{80008080}_\mathbf{x} = 0$, with bias 2^{-27}. We perform a distinguish-from-random attack, using $2 \cdot (2^{-27})^{-2} = 2^{55}$ CO, and equivalent number of encryptions. The memory complexity is negligible. According to [7], assuming Matsui's algorithm 1, the success probability of this distinguishing attack is about 97.7%.

3.4 18-Round Linear Attack

We can use the full 16-round relation in Table 2 with bias 2^{-31} for a key-recovery attack on 18-round MIBS. The attack procedure is similar to the one in Sect. 3.2, but this time we recover $K_{1,1}$, $K_{1,6}$, $K_{18,6}$, $K_{18,7}$, $K_{18,8}$. We found no overlapping in these subkeys, so we recover 20 subkey bits in total. The linear relations for this attack is

$$(R_0 \oplus R_{18} \oplus F(K_1, L_0)) \cdot \texttt{w000w0w0}_\mathbf{x} \oplus (L_{18} \oplus F(K_{18}, R_{18})) \cdot \texttt{z0000z00}_\mathbf{x} = 0, \quad (3)$$

For each pair (w, z) in Sect. 3.1 we have an independent linear relation. Following [3], the combined bias of these multiple linear relations is $\sqrt{6 \cdot (2^{-31})^2} = 2^{-29.7}$. The data complexity is $3/(2^{-29.7})^2 = 2^{60.98}$ KP.

The time complexity is $2^{20} \cdot 2^{60.98} \cdot 5/8 \cdot 1/18 \approx 2^{76.13}$ 18-round computations, since partial decryption of two nibble in the first round, and three nibbles in the 18th round costs less than one-round computation. Memory complexity is the same as data complexity. According to [9], the success probability of this attack is 72.14%.

4   Differential Cryptanalysis

Differential cryptanalysis (DC) was originally proposed by Biham and Shamir in [2]. In [6], the designers claim security of MIBS against DC by providing a 4-round characteristic with six active S-boxes, and probability 2^{-15}. They assumed that this characteristic was iterative (although it is not) and claimed resistance of the full 32-round MIBS to DC.

4.1   Searching for Differential Characteristics of MIBS

We have computed the difference distribution table (DDT) for[3] the 4×4 S-box of MIBS. See Table 8 in the appendix. We note that this S-box is differentially 4-uniform[4]. So, the highest probability for any difference propagation across this S-box is 2^{-2}.

For a systematic differential analysis of MIBS, we automated the search for differential characteristics by creating a program to look for differential characteristics for MIBS according to the following criteria:

(a) focus on iterative characteristics, preferably;
(b) maximize the overall probability by minimizing the number of active S-boxes;
(c) use the fact that the S-box is differentially 4-uniform (Table 8) [8];
(d) use the fact that the branch number of the P permutation in the F function of MIBS is 5

Using these criteria, we have found two 12-round differential characteristics, both with probability 2^{-56}. These characteristics have 28 active S-boxes in total, and for each S-box we chose the largest entries in the DDT. One characteristic is detailed in Table 3. The other characteristic is obtained from Table 3 by turning it upside-down (due to the symmetry of the Feistel Network scheme).

4.2   13-Round Differential Attack

We perform a key-recovery attack on 13-round MIBS by placing the 12-round characteristic in Table 3 in rounds 1 up to 12. We recover 24 subkey bits from the 13th round. The attack procedure is as follows:

(a) take $c \cdot 2^{56}$ pairs of plaintext blocks P_i and P_j which satisfy $P_i \oplus P_j =$ (EEE0E0EE$_x$, 50500550$_x$) and obtain their corresponding ciphertexts $C_i = (L_{13}^i, R_{13}^i)$ and $C_j = (L_{13}^j, R_{13}^j)$;
(b) keep counters for each possible value of six subkey nibbles of K_{13} corresponding to the six E_x nibble differences in the right half of the ciphertext, namely $K_{13,1}$, $K_{13,2}$, $K_{13,3}$, $K_{13,5}$, $K_{13,7}$ and $K_{13,8}$;
(c) keep only those text pairs for which the right half of the ciphertext difference equals EEE0E0EE$_x$;

---

[3] The DDT of an S-box stands for a table containing an exhaustive enumeration of all pairs of input/output differences for the given S-box.

[4] It means that the largest entry in the DDT has value 4.

Table 3. A 12-round differential characteristic for MIBS

Round i	ΔL_{i-1}	ΔR_{i-1}	Number of active S-boxes	Probability
1	EEE0E0EE$_x$	50500550$_x$	6	2^{-12}
2	00000050$_x$	EEE0E0EE$_x$	1	2^{-2}
3	00EEE000$_x$	00000050$_x$	3	2^{-6}
4	05005000$_x$	00EEE000$_x$	2	2^{-4}
5	00E000E0$_x$	05005000$_x$	2	2^{-4}
6	55500000$_x$	00E000E0$_x$	3	2^{-6}
7	00000000$_x$	55500000$_x$	0	1
8	55500000$_x$	00000000$_x$	3	2^{-6}
9	00E000E0$_x$	55500000$_x$	2	2^{-4}
10	05005000$_x$	00E000E0$_x$	2	2^{-4}
11	00EEE000$_x$	05005000$_x$	3	2^{-6}
12	00000050$_x$	00EEE000$_x$	1	2^{-2}
13	EEE0E0EE$_x$	00000050$_x$	-	-

(d) for each plaintext pair with indices i, j, compute $P^{-1}(L_{13}^i \oplus L_{13}^j \oplus 00000050_x)$, and compare with the output difference of the S-box layer inside $F(K_{13}, R_{13}^i)$ $\oplus F(K_{13}, R_{13}^j)$; discard the pairs that do not match one of the seven possible output differences of the S-box, according to the DDT (Table 8) with input difference E_x; from the input difference to the 13th round, increment counters corresponding to each suggested 24 subkey bits by the input difference EEE0E0EE$_x$, and $P^{-1}(L_{13}^i \oplus L_{13}^j \oplus 00000050_x)$;

Following [2], we estimate the signal-to-noise ratio (SNR), as $2^{24} \cdot 2^{-56}/(1 \cdot 2^{-32} \cdot (7/15)^6 \cdot (2^{-4})^2) = 2^{14}$, since $p = 2^{-56}$, $k = 24$, $\alpha = 1$ (we expect one subkey on average to be suggested in step (d)), $\beta = 2^{-32} \cdot (7/15)^6 \cdot (2^{-4})^2$, since $\Delta R_{13} =$ EEE0E0EE$_x$ gives a 32-bit condition, every output difference to an S-box whose input difference is E_x can have only seven possible nonzero output differences, and the two S-boxes with input difference 0 can only have 0 output difference. We estimate about $c = 32$ right pairs to uniquely determine the correct subkey values. This means 2^{61} CP. Step (c) imposes a 32-bit condition on the pairs. So, about $2^{61}/2^{32} = 2^{29}$ pairs survive. In step (d), the complexity corresponds to $2 \cdot 2^{29}$ one-round computations. This corresponds to about $2^{30}/13 \approx 2^{26.3}$ 13-round computations. The memory complexity corresponds to 2^{24} counters. If the user key has 64 bits, the remaining 40 key bits requires 2^{40} 13-round computations; if the key is 80-bit long, then the remaining 56 key bits requires 2^{56} 13-round computations.

According to [9], the success probability p_S of this attack, for SNR $= 2^{14}$, $a = 7$ (i.e. assuming we expect the correct 24-bit subkey to be ranked among the 7 highest counters), $N = 2^{61}$ CP, $p = 2^{-56}$, is

$$p_S = \Phi(\frac{\sqrt{p \cdot N \cdot SNR} - \Phi^{-1}(1 - 2^{-a})}{\sqrt{SNR + 1}}) \approx 0.9999$$

4.3   14-Round Differential Attack

For 14-round MIBS, we studied a key-recovery attack by placing the 12-round characteristic in Table 3 between rounds 2 and 13. We recover subkey bits from K_1 and K_{14} at the same time. The attack procedure is as follows:

(a) consider m structures of plaintexts, such that R_0 contains all possible 32-bit values, but in the L_0, half of the text contain arbitrary 32-bit values, and half of them contain $L_0 \oplus \text{50500550}_x$. Each structure, thus, contain $2^{32} \cdot 2^{32} = 2^{64}$ pairs with which difference $(\text{50500550}_x, \Delta R_0)$, where ΔR_0 is a nonzero 32-bit difference;

(b) keep only those text pairs for which the right half of the ciphertext difference equals EEE0E0EE_x;

(c) prepare counters for each possible value of four subkey nibbles of K_1 corresponding to the four 5_x nibble differences in the left half of the plaintext, namely $K_{1,1}$, $K_{1,3}$, $K_{1,6}$ and $K_{1,7}$, and each of the six nibbles of K_{14} corresponding to the six E_x nibble differences in the right half of the ciphertext; this corresponds to 40 subkey bits;

(d) for each pair of plaintext with indices i, j, compute $P^{-1}(R_0^i \oplus R_0^j \oplus \text{EEE0E0EE}_x)$, and compare it with the output difference of the S-box layer inside $F(K_1, L_0^i) \oplus F(K_1, L_0^j)$; discard the pairs that do not match one of the seven possible output differences of the S-box layer, according to the DDT (Table 8) with input difference 5_x; also, the S-boxes with input difference 0 can only have 0 output difference; from the input difference to the 1st round, increment counters corresponding to each suggested 16 subkey bits by the input difference 50500550_x, and $P^{-1}(R_0^i \oplus R_0^j \oplus \text{EEE0E0EE}_x)$;

(e) analogously, compute $P^{-1}(L_{14}^i \oplus L_{14}^j \oplus \text{00000050}_x)$, and compare it with the output difference of the S-box layer inside $F(K_{14}, R_{14}^i) \oplus F(K_{14}, R_{14}^j)$; discard the pairs that do not match one of the seven possible output differences of the S-box, according to the DDT (Table 8) with input difference E_x; also, the S-boxes with input difference 0 can only have 0 output difference; from the input difference to the 14th round, increment counters corresponding to each suggested 24 subkey bits by the input difference EEE0E0EE_x, and $P^{-1}(L_{14}^i \oplus L_{14}^j \oplus \text{00000050}_x)$;

Following [2], we estimate the signal-to-noise ratio (SNR), as $2^{40} \cdot 2^{-56} / (1 \cdot 2^{-32} \cdot (7/15)^4 \cdot (2^{-4})^4 \cdot (7/15)^6 \cdot (2^{-4})^2) = 2^{50}$, since $p = 2^{-56}$, $k = 40$, $\alpha = 1$ (we expect one subkey on average to be suggested in steps (d) and (e)), $\beta = 2^{-32} \cdot (7/15)^4 \cdot (2^{-4})^4 \cdot (7/15)^6 \cdot (2^{-4})^2$, since $\Delta R_{14} = \text{EEE0E0EE}_x$ gives a 32-bit condition, every output difference to an S-box whose input difference is 5_x or E_x can have only seven possible nonzero output differences, and the S-boxes with input difference 0 can only have 0 output difference. We estimate about $m = 128$ structures to determine the correct subkey values. This means $2^{7+33} = 2^{40}$ CP. Step (c) imposes a 32-bit condition on the pairs. So, about $2^{7+64}/2^{32} = 2^{39}$ pairs survive. In step (d), the complexity corresponds to $2 \cdot 2^{39}$ one-round computations. The same holds in step (e). This corresponds to about $2^{41}/14 \approx 2^{37.2}$ 14-round computations. The memory complexity corresponds to 2^{40} counters. If the user key has 64 bits, the remaining 24 key bits requires 2^{24} 14-round computations; if the key is 80-bit long, then the remaining 40 key bits requires 2^{40} 14-round computations.

According to [9], the success probability p_S of this attack, for SNR $= 2^{50}$, $a = 8$ (i.e. assuming we expect the correct 40-bit subkey to be ranked among the 8 highest counters), $N = 2^{40}$ CP, $p = 2^{-56}$, is

$$p_S = \Phi(\frac{\sqrt{p \cdot N \cdot SNR} - \Phi^{-1}(1 - 2^{-a})}{\sqrt{SNR + 1}}) \approx 0.5015$$

5   Impossible-Differential Cryptanalysis

There is a striking similarity between the round functions of MIBS and Camellia [1] block ciphers. Therefore, inspired by the impossible differential attack on Camellia, proposed by Wu et al. in [11], we have constructed a similar 8-round impossible differential for MIBS, as the one built for Camellia proposed in [10]. Then, we use this 8-round impossible differential to attack 12-round MIBS.

We have found the following 8-round impossible differential for MIBS:

$$(00000000_\mathbf{x}, 000000s0_\mathbf{x}) \overset{8r}{\nrightarrow} (0000h000_\mathbf{x}, 00000000_\mathbf{x}). \quad (4)$$

where u and v are nonzero nibble differences, and the broken arrow indicates that the difference in the left hand side does not cause the difference in the right hand side.

We have also found another 8-round impossible differential distinguisher for MIBS: $(00000000_\mathbf{x}, 00s00000_\mathbf{x}) \overset{8r}{\nrightarrow} (0000000h_\mathbf{x}, 00000000_\mathbf{x})$.

5.1   Some Properties of MIBS for 80-Bit User Key

We have exploited two properties of MIBS to use in the attack:

Property 1. Let $K_i = (K_{i,1}, K_{i,2}, \ldots, K_{i,8})$ denote the i-th round subkey, where $K_{i,1}$ is the most significant nibble. Then, K_1 and K_2 share 13 bits in common: $K_1[1 \sim 13] = K_2[20 \sim 32]$ or $K_{1,1}\|K_{1,2}\|K_{1,3}\|K_{1,4}[1] = K_{2,5}[4]\|K_{2,6}\|K_{2,7}\|K_{2,8}$ where values inside square brackets index bit positions.

Property 2. (similar to [5]) For any 32-bit strings X, X^*, if there exists a nonzero nibble difference s such that $P^{-1}(X \oplus X^* \oplus 000000s0_\mathbf{x})$ is of the form $??0?00??_\mathbf{x}$, then s is unique (? denotes any nibble value). The same holds for a nonzero nibble difference h.

Proof. Suppose there are two nibble differences s and w that satisfy this property. Then, P is a linear transformation relative to xor, $P^{-1}(X \oplus X^* \oplus 000000s0_\mathbf{x})$ $\oplus P^{-1}(X \oplus X^* \oplus 000000w0_\mathbf{x}) = P^{-1}(000000s0_\mathbf{x}) \oplus P^{-1}(000000w0_\mathbf{x})$ and has the form $??0?00??_\mathbf{x}$. But, $P^{-1}(000000s0_\mathbf{x}) \oplus P^{-1}(000000w0_\mathbf{x}) = ss0ss0ss_\mathbf{x} \oplus$ $ww0ww0ww_\mathbf{x}$. From the fifth nibble position, it follows that $s \oplus w = 0$, which is a contradiction.

5.2   Construction of 8-Round Impossible Differential Distinguisher

This 8-round impossible differential characteristic (4) is constructed by concatenating two 3-round differentials, and putting two connection rounds in between the two differentials. See Fig. 4. The first 3-round differential, depicted in Table 4, is built as follows: let the input difference to the first round be $(\Delta L_0, \Delta R_0) = (00000000_x, 000000s0_x)$ where s is a non-zero nibble difference and after the first round, the input difference to the second round will be $(\Delta L_1, \Delta R_1) = (000000s0_x, 00000000_x)$. Then in the second round, the input difference $000000s0_x$ to the S layer leads to the output difference $000000t0_x$, where t is a nonzero nibble difference. After applying the P layer, the output difference of the F-function will be $tt0t00tt_x$. The input difference to the third round is $(\Delta L_2, \Delta R_2) = (tt0t00tt_x, 000000s0_x)$. Afterwards, the difference $\Delta L_2 = tt0t00tt_x$ becomes $t_1t_20t_400t_7t_8$ after the S layer where t_1, t_2, t_4, t_7 and t_8 are non-zero nibble differences. Then, it evolves to $(c_1c_2c_3c_4c_5c_6c_7c_8)$, c_i are nonzero nibble differences, after the application of the P layer, and the output difference of the third round turns out to be $(\Delta L_3, \Delta R_3) = (c_1c_2c_3c_4c_5c_6c_7c_8 \oplus 000000s0_x, tt0t00tt_x)$. This completes the first differential.

Table 4. The first 3-round truncated differential for MIBS (in encryption direction)

Round i	ΔL_{i-1}	ΔR_{i-1}
1	00000000_x	$000000s0_x$
2	$000000s0_x$	00000000_x
3	$tt0t00tt_x$	$000000s0_x$
4	$c_1c_2c_3c_4c_5c_6c_7c_8 \oplus 000000s0_x$	$tt0t00tt_x$

The second 3-round differential in Table 5 is constructed as follows: let the output difference of round 8 be $(\Delta L_8, \Delta R_8) = (0000h000_x, 00000000_x)$ and if this difference is rolled back through round 8, then the output difference of round 7 becomes $(\Delta L_7, \Delta R_7) = (00000000_x, 0000h000_x)$. The difference $\Delta L_7 = 0000h000_x$ will be $0000w000_x$ after the application of the S layer in round 7 and the difference evolves to $www0ww00_x$ after the P layer where w denotes a nonzero nibble. Then, the output difference of round six becomes $(\Delta L_6, \Delta R_6) = (www0ww00_x, 0000h000_x)$ becomes $w_1w_2w_30w_5w_600$, where w_i are *nonzero nibble differences*, after the S layer and we get the input difference of round six as $(\Delta L_5, \Delta R_5) = (e_1e_2e_3e_4e_5e_6e_7e_8 \oplus 0000h000_x, www0ww00_x)$. This completes the second 3-round differential.

Concatenating these two 3-round differentials, we obtain an 8-round impossible differential distinguisher. One can see in Fig. 4, the input and output differences of the F-function in round 5 are $(e_1e_2e_3e_4e_5e_6e_7e_8) \oplus 0000h000_x$ and $(c_1c_2c_3c_4c_5c_6c_7c_8) \oplus 000000s0_x \oplus www0ww00_x = (c_1 \oplus w, c_2 \oplus w, c_3 \oplus w, c_4, c_5 \oplus w, c_6 \oplus w, c_7 \oplus s, c_8)$, respectively. Since the output difference of the S layer has

Table 5. The second 3-round truncated differential for MIBS (in decryption direction)

Round i	ΔL_{i-1}	ΔR_{i-1}
8	0000h000$_x$	00000000$_x$
7	00000000$_x$	0000h000$_x$
6	0000h000$_x$	www0ww00$_x$
5	www0ww00$_x$	$e_1e_2e_3e_4e_5e_6e_7e_8 \oplus$0000h000$_x$

to be equal to input difference of the P layer, that is, $S[(e_1e_2e_3e_4e_5e_6e_7e_8) \oplus$
0000h000$_x$] $= P^{-1}(c_1 \oplus w, c_2 \oplus w, c_3 \oplus w, c_4, c_5 \oplus w, c_6 \oplus w, c_7 \oplus s, c_8)$, we have:
$P^{-1}(c_1 \oplus w, c_2 \oplus w, c_3 \oplus w, c_4, c_5 \oplus w, c_6 \oplus w, c_7 \oplus s, c_8) = P^{-1}(c_1c_2c_3c_4c_5c_6c_7c_8) \oplus$
$P^{-1}(000000s0_x) \oplus P^{-1}(\text{www0ww00}_x) = (t_1t_20t_400t_7t_8) \oplus \text{ss0ss0ss}_x \oplus \text{0000w000}_x$
$= (t_1 \oplus s, t_2 \oplus s, 0, t_4 \oplus s, s \oplus w, 0, t_7 \oplus s, t_8 \oplus s)$.

We can see that the output difference of the third and sixth S-boxes are zero in
round five, which implies the input differences of these S-boxes are zero, too since
they are bijective. Therefore, $e_3 = e_6 = 0$ where $e_3 = w_1 \oplus w_2 \oplus w_3 \oplus w_5 \oplus w_6$,
$e_6 = w_1 \oplus w_2 \oplus w_5 \oplus w_6$. But, if $e_3 = w_1 \oplus w_2 \oplus w_3 \oplus w_5 \oplus w_6 = 0$ and
$e_6 = w_1 \oplus w_2 \oplus w_5 \oplus w_6 = 0$, then this leads to $w_3 = 0$ which contradicts the
assumption that w_3 is nonzero.

5.3   12-Round Impossible Differential Attack on MIBS with 80-Bit User Key

Fig. 3 depicts our 12-round impossible differential attack. We start in round
1 and end in round 12. But it can be constructed anywhere between rounds
1 and 32 due to the key schedule of MIBS for 80-bit user keys. From Fig. 3,
the required plaintexts for the attack have the form $(\Delta L_0, \Delta R_0) = (\text{uu0u00uu}_x,$
$P(\text{??0?00??}_x) \oplus 000000?0_x)$ where 'u' and '?' are nonzero nibble differences.

This attack is different from the conventional impossible differential attack in
a way that we exploit the equality of some subkey bits to eliminate wrong key
guesses by using the impossible differential. Instead of eliminating pairs round
by round, we can make a different analysis to reduced the time complexity of
the attack: the ciphertext pairs which satisfy the impossible differential should
have the output difference of round 10: $\Delta L_{10} = (\text{0000h000}_x, \text{00000000}_x)$, where
h is a nonzero nibble. When the S-box of MIBS is analyzed, one can see that
the number of nonzero entries of each row of the DDT is at most 2^3, that is
each nonzero input difference to the S-box causes at most 2^3 nonzero output
differences. Therefore, the nonzero nibble h can take $2^4 - 1 = 15$ different values
and in round 11, the output differences of the S-box, which corresponds to h,
has at most $15 \cdot 2^3$ possible nonzero output differences. Then in Round 12, five
nonzero nibbles at positions $(1, 2, 3, 5, 6)$ have at most $(2^3)^5$ nonzero output
differences which result in at most $15 \cdot 2^3 \cdot (2^3)^5 \approx 2^{22}$ possible output differences
after the S layer.

The attack procedure is as follows:

Data Collection

Choose 2^m structures of plaintexts of each structure is of the form:

$$\Delta L_0 \qquad\qquad = (uua_3ua_5a_6uu)$$
$$\Delta R_0 = P(x_1x_2b_3x_4b_5b_6x_7x_8) \oplus (c_1c_2c_3c_4c_5c_6yc_8)$$

where (a_i, b_j, c_j) are constants and (u, x_i, y) takes all possible nonzero values. So, each structure has $(2^4)^7 = 2^{28}$ plaintexts which constitute $\frac{1}{2} \cdot 2^{28} \cdot 2^{28} = 2^{55}$ plaintext pairs. Since we take 2^m structures, there are 2^{55+m} plaintexts pairs in total.

Data Filtering and Key Elimination

- The analysis that we made above shows that the probability of a random pair passes the test is $2^{-42} = 2^{22} \cdot 2^{-64}$, therefore after this filtering step $2^{55+m} \cdot 2^{22} \cdot 2^{-64} = 2^{13+m}$ pairs remain.
- For each remaining pair $((L_0, R_0), (L_{12}, R_{12}))$ and $((L_0^*, R_0^*), (L_{12}^*, R_{12}^*))$, do the following steps:
    1. By Property 2, there is only one nibble u which satisfies $P^{-1}(L_0 \oplus L_0^* \oplus 000000u0_x)$, and it has the form ??0?00??$_x$. Therefore, for each pair of plaintexts compute $P^{-1}(L_0 \oplus L_0^* \oplus 000000u0_x)$ to find the unique value of u by trying all possible values of u. It is analogous to find the unique value of h.
    2. Afterwards, in rounds 1 and 12, since the input and output differences of the S-boxes are known, the subkey nibbles $(K_{1,1}, K_{1,2}, K_{1,4}, K_{1,7}, K_{1,8})$ and $(K_{12,1}, K_{12,2}, K_{12,3}, K_{12,5}, K_{12,6})$ are suggested with the help of the DDT.
    3. Guess further 24 subkey bits (6 nibbles) of rounds 1 and 12, namely, $(K_{1,3}, K_{1,5}, K_{1,6}, K_{12,4}, K_{12,7}, K_{12,8})$, then do the followings:
        (a) For every remaining pair, encrypt plaintexts through the first round, and decrypt their corresponding ciphertexts through the last round to obtain intermediate values (L_1, L_1^*) and (R_{11}, R_{11}^*), respectively.
        (b) Compute the suggested bits of the subkey nibbles $K_{2,7}$ and $K_{11,5}$ using the values L_1, L_1^*, u and $P^{-1}(L_0 \oplus L_0^*)$ for round 2 and R_{11}, R_{11}^*, h and $P^{-1}(R_{12} \oplus R_{12}^*)$ for round 12.
        (c) By Property 1, check the subkey nibbles satisfying the following relation $K_{2,7} = K_{1,2}[2 \sim 4]||K_{1,3}[1]$. This equality implies a 4-bit condition on pairs and any pair which satisfies the equality eliminates one wrong 68-bit subkey value: $(K_{1,1}, K_{1,2}, K_{1,4}, K_{1,7}, K_{1,8}, K_{12,1}, K_{12,2}, K_{12,3}, K_{12,5}, K_{12,6}, K_{1,3}, K_{1,5}, K_{1,6}, K_{12,4}, K_{12,7}, K_{12,8}, K_{11,5})$. Each pair eliminates 2^{-4} of all subkey guesses, so after the first pair the number of remaining keys is $2^{68}(1-2^{-4})$. Since we have 2^{13+m} pairs, there are $2^{68}(1-2^{-4})^{2^{13+m}}$ wrong subkeys. For $m = 0$, no wrong subkeys survive.

5.4   Complexity Analysis

Data Complexity: We set $m = 0$, because it is enough to take just one structure of plaintexts. So, the data complexity the attack is 2^{28} chosen plaintexts (CP).

Memory Access: In data filtering step, we have to have access to all 2^{22} output differences $(\Delta L_{12}, \text{ggg0gg00}_x)$ stored in a hash table to identify the useful pairs. Therefore, this step needs $2^{22} \cdot 2^{28} = 2^{50}$ memory access (MA). We approximate the cost of one round MIBS encryption to be equivalent to one memory access. Thus, 2^{50} memory accesses cost about $2^{46.42}$ 12-round MIBS encryptions.

Time complexity:

- Step 1 needs at most two one-round MIBS encryption per remaining pair. Therefore, the time complexity of this step is at most $2^{13} \cdot \dfrac{2}{12} \approx 2^{10.41}$ 12-round MIBS encryptions. We do not need to try all possible 15 values of a and h, since the computation is less than two round encryptions.
- The time complexity of Step 2 is less than $2^{13} \cdot \dfrac{2}{12} \approx 2^{10.41}$ 12-round MIBS encryptions. Because, for five active S-boxes, we use the DDT to find the suggested keys, which costs less than one round encryption.
- In Step 3(a), since we guess 24 bits of subkeys; the time complexity of this step is at most $2 \cdot 2^{13} \cdot 2^{24} \cdot \dfrac{2}{12} \approx 2^{35.42}$ 12-round MIBS encryptions. In Step 3(b), the time complexity is less than $2^{13} \cdot \dfrac{2}{12} \approx 2^{10.41}$ 12-round MIBS encryptions. Note that the complexity of checking 4-bit equality in subkeys is negligible.

Memory Complexity: The storage of all chosen plaintexts and their corresponding ciphertexts is $2^{28} \cdot 2 = 2^{29}$ blocks. In step 1, we have to store 2^{22} possible output differences which need 2^{22} blocks of memory. In the key elimination step, since we have 2^{68} bits subkey guess, we need $2^{68} \cdot 2^{-6} = 2^{62}$ blocks of memory.

To conclude, the time complexity of the attack is dominated by the data filtering step, which is $2^{46.42}$ 12-round MIBS encryptions. The memory and the data complexities are 2^{62} blocks and 2^{28} CP, respectively. The attack recovers 68 bits of the 80-bit secret key; the remaining 12-bit of the secret key can be found by exhaustive search.

6   Conclusions

This paper described the first independent and systematic linear, differential and impossible-differential analyses of reduced-round versions of the MIBS block cipher [6]. Actually, we presented the best known-plaintext attack so far on up to 18-round MIBS, and the first ciphertext-only attack on 13-round MIBS. These attacks do not threaten the full 32-round MIBS, but reduce by more than 50% its margin of security.

Table 6 summarizes the complexities of all attacks on reduced-round MIBS described in this paper.

Table 6. Attack complexities on reduced-round MIBS block cipher

#Rnds	Time	Data	Memory	Key Size (bits)	Source	Comments	Success Prob.
12	$2^{46.42}$	2^{28} CP	2^{62}	80	Sect. 5	ID, key-recovery	—
13	2^{40}	2^{61} CP	2^{24}	64	Sect. 4.2	DC, key-recovery	99.9%
13	2^{55}	2^{55} CO	—	64 or 80	Sect. 3.3	LC,distinguishing	97.7%
13	2^{56}	2^{61} CP	2^{24}	80	Sect. 4.2	DC, key-recovery	99.9%
14	$2^{37.2}$	2^{40} CP	2^{40}	64	Sect. 4.3	DC, key-recovery	50.15%
14	2^{40}	2^{40} CP	2^{40}	80	Sect. 4.3	DC, key-recovery	50.15%
17	2^{69}	2^{58} KP	2^{58}	80	Sect. 3.2	LC, key-recovery	97.94%
18	$2^{76.13}$	$2^{60.98}$ KP	$2^{60.98}$	80	Sect. 3.4	LC, key-recovery	72.14%

time complexity is number of reduced-round encryptions;
LC: Linear Cryptanalysis; DC: Differential Cryptanalysis; ID: Imposs. Differential
CP: Chosen Plaintext; KP: Known Plaintext; CO: Ciphertext Only.

References

1. Aoki, K., Ichikawa, T., Kanda, M., Matsui, M., Moriai, S., Nakajima, J., Tokita, T.: Camellia: A 128-bit block cipher suitable for multiple platforms - design and analysis. In: Stinson, D.R., Tavares, S. (eds.) SAC 2000. LNCS, vol. 2012, pp. 39–56. Springer, Heidelberg (2001)
2. Biham, E., Shamir, A.: Differential Cryptanalysis of DES-like Cryptosystems. Journal of Cryptology 4(1), 3–72 (1991)
3. Biryukov, A., De Cannière, C., Quisquater, M.: On Multiple Linear Approximations. In: Franklin, M. (ed.) CRYPTO 2004. LNCS, vol. 3152, pp. 1–22. Springer, Heidelberg (2004)
4. Bogdanov, A., Knudsen, L.R., Leander, G., Paar, C., Poschman, A., Robshaw, M.J.B., Seurin, Y., Vikkelsoe, C.: PRESENT: An ultra-lightweight block cipher. In: Paillier, P., Verbauwhede, I. (eds.) CHES 2007. LNCS, vol. 4727, pp. 450–466. Springer, Heidelberg (2007)
5. Lu, J., Kim, J.-S., Keller, N., Dunkelman, O.: Improving the efficiency of impossible differential cryptanalysis of reduced round camellia and MISTY1. In: Malkin, T.G. (ed.) CT-RSA 2008. LNCS, vol. 4964, pp. 370–386. Springer, Heidelberg (2008)
6. Izadi, M.I., Sadeghiyan, B., Sadeghian, S.S., Khanooki, H.A.: MIBS: a new lightweight Block Cipher. In: Garay, J.A., Miyaji, A., Otsuka, A. (eds.) CANS 2009. LNCS, vol. 5888, pp. 334–348. Springer, Heidelberg (2009)
7. Matsui, M.: Linear Cryptanalysis Method for DES Cipher. In: Helleseth, T. (ed.) EUROCRYPT 1993. LNCS, vol. 765, pp. 386–397. Springer, Heidelberg (1994)
8. Nyberg, K.: Differentially Uniform Mappings for Cryptography. In: Helleseth, T. (ed.) EUROCRYPT 1993. LNCS, vol. 765, pp. 55–64. Springer, Heidelberg (1994)
9. Selçuk, A.A.: On Probability of Success in Linear and Differential Cryptanalysis. Journal of Cryptology 1(21), 1–19 (2008)
10. Wu, W., Zhang, W., Feng, D.: Impossible differential cryptanalysis of reduced-round ARIA and Camellia. Journal of Computer Science and Technology 22(3), 449–456 (2007)
11. Wu, W., Zhang, L., Zhang, W.: Improved Impossible-Differential Cryptanalysis of Reduced-Round Camellia. In: Avanzi, R., Keliher, L., Sica, F. (eds.) SAC 2008. LNCS, vol. 5381, pp. 442–456. Springer, Heidelberg (2009)

A   Appendix - Figures and Tables

Table 7. Linear Approximation Table (LAT) of the S-box of MIBS

	0_x	1_x	2_x	3_x	4_x	5_x	6_x	7_x	8_x	9_x	A_X	B_x	C_x	D_x	E_x	F_x
0_x	8	0	0	0	0	0	0	0	0	0	0	0	0	0	0	0
1_x	0	-2	0	2	0	-2	-4	-2	2	0	-2	0	2	0	2	-4
2_x	0	0	-2	-2	-2	2	-4	0	0	4	2	-2	-2	-2	0	0
3_x	0	2	2	0	2	0	0	2	-2	4	0	2	0	2	-2	-4
4_x	0	-2	-2	4	-2	0	0	2	0	-2	2	0	-2	0	-4	-2
5_x	0	0	-2	2	2	-2	0	0	2	2	0	4	-4	0	2	2
6_x	0	-2	4	2	0	-2	0	-2	0	2	4	-2	0	2	0	2
7_x	0	4	0	0	0	-4	0	0	-2	-2	2	-2	-2	-2	2	-2
8_x	0	2	2	4	0	2	-2	0	-2	0	0	2	2	-4	0	2
9_x	0	0	2	-2	-4	-4	-2	2	0	0	-2	2	0	0	-2	2
A_x	0	-2	0	-2	-2	0	2	-4	-2	0	2	4	0	-2	0	-2
B_x	0	0	4	0	-2	2	2	2	2	4	0	0	0	-2	-2	2
C_x	0	0	0	0	2	-2	2	-2	2	2	-2	-2	0	-4	-4	0
D_x	0	2	0	-2	2	0	-2	0	4	-2	4	2	2	0	-2	0
E_x	0	4	-2	2	-4	0	2	-2	2	2	0	0	2	2	0	0
F_x	0	2	2	0	0	2	-2	-4	0	-2	-2	0	-4	2	-2	0

Table 8. (xor) Difference Distribution Table (DDT) of the S-box of MIBS

	0_x	1_x	2_x	3_x	4_x	5_x	6_x	7_x	8_x	9_x	A_x	B_x	C_x	D_x	E_x	F_x
0_x	16	0	0	0	0	0	0	0	0	0	0	0	0	0	0	0
1_x	0	0	0	0	2	0	0	2	2	2	0	4	2	0	2	0
2_x	0	2	0	2	0	0	0	4	0	0	2	2	2	0	0	2
3_x	0	0	2	0	0	2	2	2	0	0	0	2	4	2	0	0
4_x	0	0	0	2	0	2	2	2	2	4	0	0	0	0	0	2
5_x	0	0	2	2	2	0	0	2	0	0	0	0	0	2	4	2
6_x	0	0	2	0	0	2	0	0	4	0	2	0	2	0	2	2
7_x	0	2	2	2	4	2	0	0	0	2	0	0	2	0	0	0
8_x	0	0	0	0	2	0	2	0	0	2	2	0	2	2	0	4
9_x	0	4	0	0	2	2	0	0	2	0	0	2	0	2	0	2
A_x	0	2	0	4	0	0	2	0	2	0	0	0	2	2	2	0
B_x	0	0	2	2	2	0	2	0	2	0	4	2	0	0	0	0
C_x	0	2	2	0	0	0	4	0	0	2	0	2	0	0	2	2
D_x	0	2	4	0	0	0	0	2	2	2	2	0	0	2	0	0
E_x	0	2	0	0	2	4	2	2	0	0	2	0	0	0	2	0
F_x	0	0	0	2	0	2	0	0	0	2	2	2	0	4	2	0

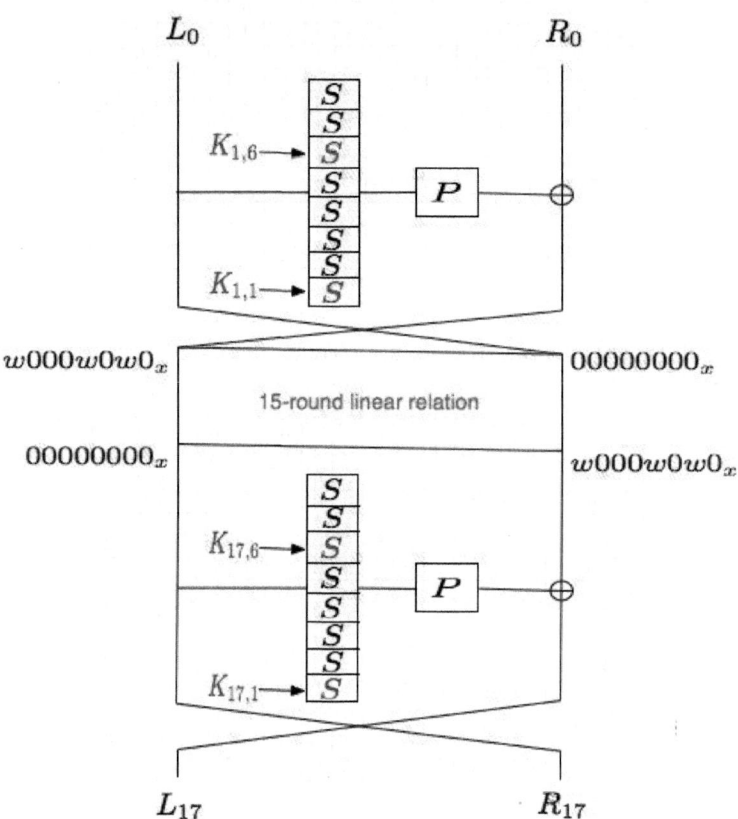

Fig. 1. Linear attack on 17-round MIBS

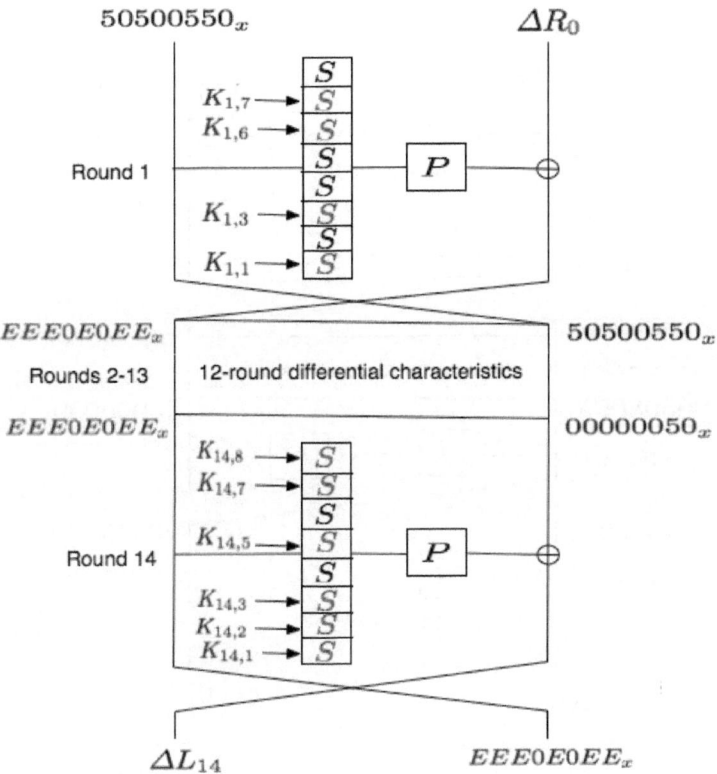

Fig. 2. Differential attack on 14-round MIBS

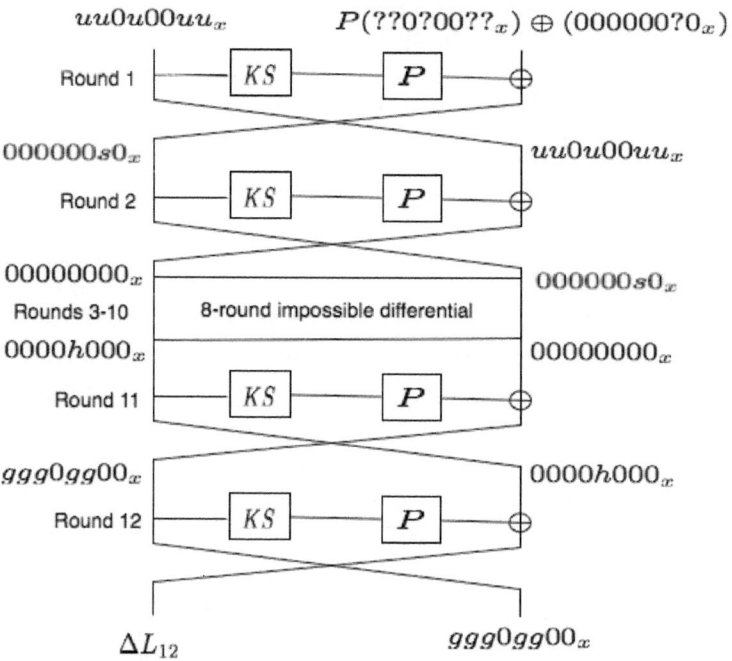

Fig. 3. Impossible differential attack on 12-round MIBS

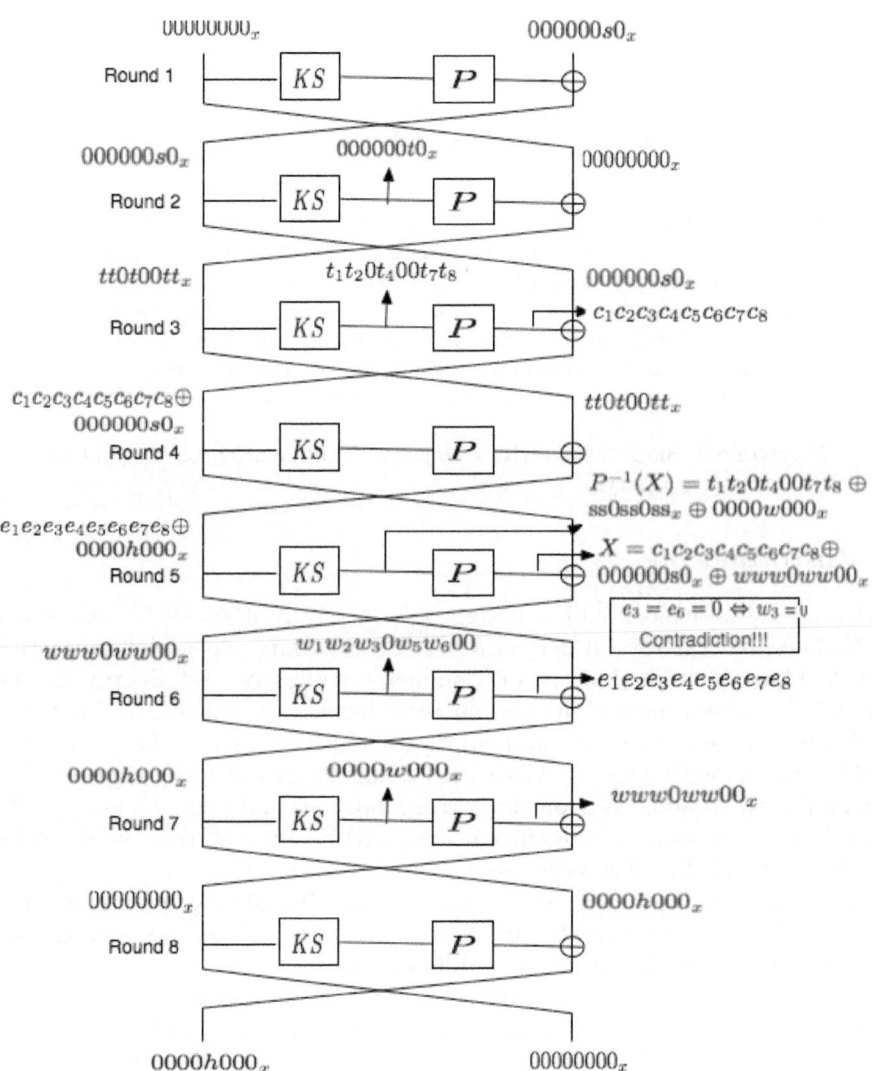

Fig. 4. 8-round impossible differential of MIBS

Impossible Differential Cryptanalysis of ARIA Reduced to 7 Rounds

Chenghang Du and Jiazhe Chen

Key Lab of Cryptologic Technology and Information Security,
Ministry of Education, Shandong University, Jinan, 250100, P.R. China
{chenghangdu,jiazhechen}@mail.sdu.edu.cn

Abstract. This paper studies the security of the block cipher ARIA against impossible differential cryptanalysis. We find a new impossible differential property of ARIA, and propose an attack against ARIA-256 reduced to 7 rounds based on this property, while previous attacks can only attack ARIA up to 6 rounds. Our new attack needs 2^{125} chosen plaintexts and 2^{238} 7-round encryptions. This is the best result for impossible differential cryptanalysis of ARIA known so far.

Keywords: Block cipher, ARIA, Impossible Differential, Data complexity, Time complexity.

1 Introduction

ARIA [12,15] is a block cipher designed by a group of South Korean experts in 2003. ARIA was established as a Korean Standard block cipher algorithm (KS X 1213) by the Ministry of Commerce, Industry and Energy in 2004. ARIA is a general-purpose involution SPN block cipher algorithm, optimized for lightweight environments and hardware implementation. The interface of ARIA is the same as AES [7]. ARIA has 128-bit block size with 128/192/256-bit key, and in the original version the corresponding round numbers are 10/12/14 respectively [12], while in the current one, ARIA v1.0 [15], the round numbers are altered to 12/14/16 respectively.

The designers, Daesung Kwon et al., gave the initial cryptanalysis of ARIA [12]. It contained differential and linear cryptanalysis [4,14], truncated differential cryptanalysis [10], impossible differential cryptanalysis [1], square attack [3,11], higher order differential cryptanalysis [10], interpolation attack [9], and so on. Later in 2004, Alex Biryukov et al. performed a security evaluation of ARIA in which they focused on dedicated linear cryptanalysis and truncated differential cryptanalysis [5], and found attack on ARIA up to 7 rounds. But they didn't evaluate the security against impossible differential cryptanalysis which is an important attacking method of block cipher. Wenli Wu et al. found a non-trivial 4-round impossible differential path in the first place, which led to an attack on 6-round ARIA requiring about 2^{121} chosen plaintexts and about 2^{112} encryptions [17]. Then Shenhua Li proposed an improved impossible differential attack, which needed 2^{96} 6-round encryptions, and reduced the chosen plaintexts number to 2^{120} [13].

S.-H. Heng, R.N. Wright, and B.-M. Goi (Eds.): CANS 2010, LNCS 6467, pp. 20–30, 2010.
© Springer-Verlag Berlin Heidelberg 2010

Impossible differential cryptanalysis is a kind of technique that uses differentials with probability 0 to get rid of the wrong keys, in order to obtain the right key. These differentials are called impossible differentials. Since its appearance, researchers discovered that it can be used to analyze many block ciphers, such as AES, and get some good results [2,3,6,8,16].

In this paper, we propose a new impossible differential path, which leads to the attack of ARIA-256 reduced to 7 rounds. We use the "early-abort technique" introduced in [4,17] to reduce the time complexity of our attack. The data complexity is 2^{125}, while the time complexity is less than 2^{238} in our attack.

We organize our paper as follows. Section 2 gives a description of ARIA. A 4-round impossible differential path of ARIA is described in Section 3. In Section 4 we present our impossible differential attack on 7-round ARIA-256. And we conclude our paper in Section 5.

2 Description of ARIA

ARIA is a 128-bit SPN structure block cipher. ARIA-256 supports 256-bit key length, and the corresponding round number is 16. Each round consists of the following three parts:

Round Key Addition(AK): This is done by XORing the 128-bit round key $k_i, 1 \le i \le 17$. The round key is derived from the master key (MK) through the key schedule. The detail of the key schedule is in [15].

Substitution Layer(SL): Applying the non-linear 8×8-bit S-boxes in parallel on each byte of the state. ARIA uses 2 S-boxes S_1, S_2 and their inverses S_1^{-1}, S_2^{-1}. Each S-box is defined to be an affine transformation of the inversion function over $GF(2^8)$.

$$S: GF(2^8) \longrightarrow GF(2^8), \ S_1: \ x \longrightarrow Q \cdot x^{-1} \oplus q,$$

where

$$Q = \begin{pmatrix} 1 & 0 & 0 & 0 & 1 & 1 & 1 & 1 \\ 1 & 1 & 0 & 0 & 0 & 1 & 1 & 1 \\ 1 & 1 & 1 & 0 & 0 & 0 & 1 & 1 \\ 1 & 1 & 1 & 1 & 0 & 0 & 0 & 0 \\ 1 & 1 & 1 & 1 & 1 & 0 & 0 & 0 \\ 0 & 1 & 1 & 1 & 1 & 1 & 0 & 0 \\ 0 & 0 & 1 & 1 & 1 & 1 & 1 & 0 \\ 0 & 0 & 0 & 1 & 1 & 1 & 1 & 1 \end{pmatrix} \quad and \quad q = \begin{pmatrix} 1 \\ 1 \\ 0 \\ 0 \\ 0 \\ 1 \\ 1 \\ 0 \end{pmatrix}.$$

and

$$S_2: \ x \longrightarrow T \cdot x^{247} \oplus t,$$

where

$$T = \begin{pmatrix} 0 & 1 & 0 & 1 & 1 & 1 & 1 & 0 \\ 0 & 0 & 1 & 1 & 1 & 1 & 0 & 1 \\ 1 & 1 & 0 & 1 & 0 & 1 & 1 & 1 \\ 1 & 0 & 0 & 1 & 1 & 1 & 1 & 0 \\ 0 & 0 & 1 & 0 & 1 & 1 & 0 & 0 \\ 1 & 0 & 0 & 0 & 0 & 0 & 0 & 1 \\ 0 & 1 & 0 & 1 & 1 & 1 & 0 & 1 \\ 1 & 1 & 0 & 1 & 0 & 0 & 1 & 1 \end{pmatrix} \quad and \ \ t = \begin{pmatrix} 0 \\ 1 \\ 0 \\ 0 \\ 0 \\ 1 \\ 1 \\ 1 \end{pmatrix}.$$

There are two types of substitution layers to be used so as to make the cipher involution.

$$LS_o = (S_1, S_2, S_1^{-1}, S_2^{-1}, S_1, S_2, S_1^{-1}, S_2^{-1}, S_1, S_2, S_1^{-1}, S_2^{-1}, S_1, S_2, S_1^{-1}, S_2^{-1}),$$
$$LS_e = (S_1^{-1}, S_2^{-1}, S_1, S_2, S_1^{-1}, S_2^{-1}, S_1, S_2, S_1^{-1}, S_2^{-1}, S_1, S_2, S_1^{-1}, S_2^{-1}, S_1, S_2).$$

LS_o is for the odd rounds, while LS_e is for the even rounds.

Diffusion Layer(DL): A 16×16 involution binary matrix with branch number 8 was selected to improve the diffusion effect. It's a simple linear map in which the 128-bit plaintexts are treated as byte matrices of size 4×4.

The 128-bit plaintext includes 16 bytes with every byte numbered as the following:

0	4	8	12
1	5	9	13
2	6	10	14
3	7	11	15

The diffusion layer is given by $DL : X \rightarrow Y, \ \ Y = AX$
where

$$X = (x_0, x_1, x_2, x_3, x_4, x_5, x_6, x_7, x_8, x_9, x_{10}, x_{11}, x_{12}, x_{13}, x_{14}, x_{15})^T,$$
$$Y = (y_0, y_1, y_2, y_3, y_4, y_5, y_6, y_7, y_8, y_9, y_{10}, y_{11}, y_{12}, y_{13}, y_{14}, y_{15})^T,$$

$$A = \begin{pmatrix} 0 & 0 & 0 & 1 & 1 & 0 & 1 & 0 & 1 & 1 & 0 & 0 & 0 & 1 & 1 & 0 \\ 0 & 0 & 1 & 0 & 0 & 1 & 0 & 1 & 1 & 1 & 0 & 0 & 1 & 0 & 0 & 1 \\ 0 & 1 & 0 & 0 & 1 & 0 & 1 & 0 & 0 & 0 & 1 & 1 & 1 & 0 & 0 & 1 \\ 1 & 0 & 0 & 0 & 0 & 1 & 0 & 1 & 0 & 0 & 1 & 1 & 0 & 1 & 1 & 0 \\ 1 & 0 & 1 & 0 & 0 & 1 & 0 & 0 & 1 & 0 & 0 & 1 & 0 & 0 & 1 & 1 \\ 0 & 1 & 0 & 1 & 1 & 0 & 0 & 0 & 0 & 1 & 1 & 0 & 0 & 0 & 1 & 1 \\ 1 & 0 & 1 & 0 & 0 & 0 & 0 & 1 & 0 & 1 & 1 & 0 & 1 & 1 & 0 & 0 \\ 0 & 1 & 0 & 1 & 0 & 0 & 1 & 0 & 1 & 0 & 0 & 1 & 1 & 1 & 0 & 0 \\ 1 & 1 & 0 & 0 & 1 & 0 & 0 & 1 & 0 & 0 & 1 & 0 & 0 & 1 & 0 & 1 \\ 1 & 1 & 0 & 0 & 0 & 1 & 1 & 0 & 0 & 0 & 0 & 1 & 1 & 0 & 1 & 0 \\ 0 & 0 & 1 & 1 & 0 & 1 & 1 & 0 & 1 & 0 & 0 & 0 & 0 & 1 & 0 & 1 \\ 0 & 0 & 1 & 1 & 1 & 0 & 0 & 1 & 0 & 1 & 0 & 0 & 1 & 0 & 1 & 0 \\ 0 & 1 & 1 & 0 & 0 & 0 & 1 & 1 & 0 & 1 & 0 & 1 & 1 & 0 & 0 & 0 \\ 1 & 0 & 0 & 1 & 0 & 0 & 1 & 1 & 1 & 0 & 1 & 0 & 0 & 1 & 0 & 0 \\ 1 & 0 & 0 & 1 & 1 & 1 & 0 & 0 & 0 & 1 & 0 & 1 & 0 & 0 & 1 & 0 \\ 0 & 1 & 1 & 0 & 1 & 1 & 0 & 0 & 1 & 0 & 1 & 0 & 0 & 0 & 0 & 1 \end{pmatrix}.$$

DL is an involution. So we have $DL^{-1} = DL$.

3   4-Round Impossible Differentials of ARIA

Several 4-round impossible differentials of the ARIA were presented in [13,17]. In this section, we propose some new impossible differential paths of 4-round ARIA.

We use X_m^I and X_m^O to denote the input and output of round m, while X_m^S denotes the intermediate value after the application of SL of round m. $X_{m,n}$ denotes the n-th byte of X_m, while R_m denotes the m-th round. We analyze the 4-round impossible differential of R_3 to R_6.

One new impossible differential path states that, given a pair of X_3^I which is equal in all bytes except the 3rd byte, then after 4 rounds encryption the ciphertext differences ΔX_6^O can't be like this $(j, 0, j, 0, 0, 0, 0, 0, j, 0, 0, j, 0, 0, 0, 0)$, i.e., the ciphertext pair has nonzero equal difference at bytes $(0, 2, 8, 11)$, and no difference at the other bytes.

We expressed the property like this:

$$(0, 0, c, 0, 0, 0, 0, 0, 0, 0, 0, 0, 0, 0, 0, 0) \nRightarrow (j, 0, j, 0, 0, 0, 0, 0, j, 0, 0, j, 0, 0, 0, 0) \quad (1)$$

where c and j denote any nonzero value.

The path is illustrated in Fig.1.

Proof: To start with the first 2 rounds, suppose the difference of inputs satisfies the left part of (1). The first 2-round differential is obtained as follows:

The input difference $\Delta X_3^I = (0, 0, c, 0, 0, 0, 0, 0, 0, 0, 0, 0, 0, 0, 0, 0)$ is preserved through the AK operation of R_3. This difference is in a single byte, so the difference after the SL of R_3 is still in a single byte, i.e., $\Delta X_3^S = (0, 0, d, 0, 0, 0, 0, 0, 0, 0,$

24   C. Du and J. Chen

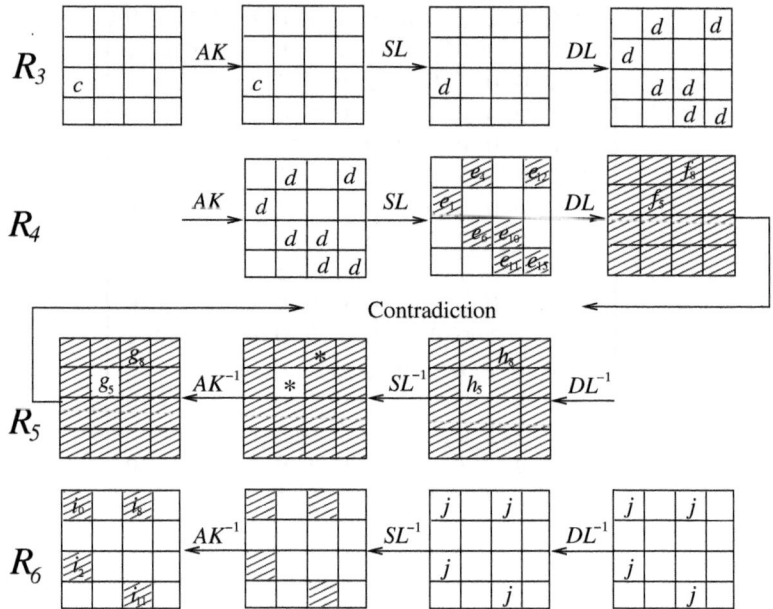

Fig. 1. 4-round impossible differential path of ARIA

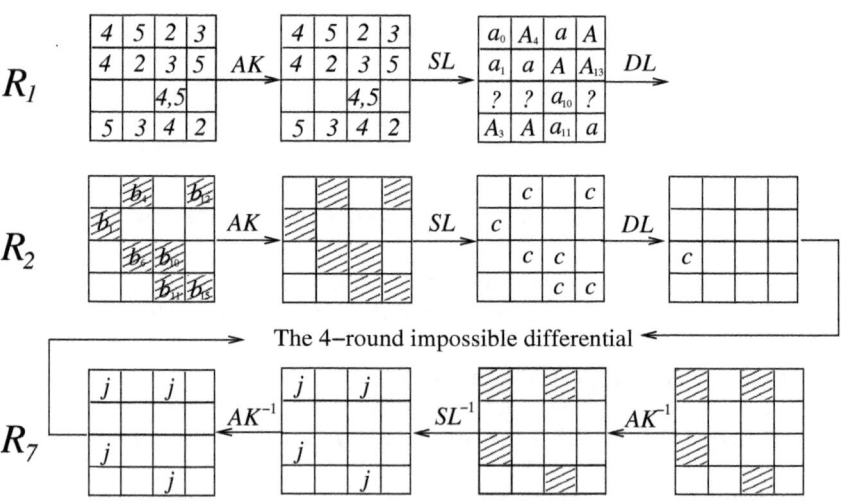

Fig. 2. Impossible Differential Cryptanalysis of 7-round ARIA

$0, 0, 0, 0, 0, 0)$, where d is an unknown nonzero byte. And the DL of R_3 makes
the differential become $\Delta X_3^O = (0, d, 0, 0, d, 0, d, 0, 0, 0, d, d, d, 0, 0, d)$.

After AK and SL of R_4, the difference is $\Delta X_4^S = (0, e_1, 0, 0, e_4, 0, e_6, 0, 0, 0, e_{10}, e_{11}, e_{12}, 0, 0, e_{15})$.

Finally, after the DL of R_4, the difference evolves into $\Delta X_4^O = (f_0, f_1, f_2, f_3, f_4,$ $f_5, f_6, f_7, f_8, f_9, f_{10}, f_{11}, f_{12}, f_{13}, f_{14}, f_{15})$, where we have $f_5 = f_8 = e_1 \oplus e_4 \oplus e_{10} \oplus$ e_{15}. Hence, $\Delta X_3^I = (0, 0, c, 0, 0, 0, 0, 0, 0, 0, 0, 0, 0, 0, 0, 0)$ evolves with probability one into ΔX_4^O, which has same value in bytes 5 and 8.

Now, we investigate how the inverse of the last 2 rounds works on the right part of (1).

The second differential ends after R_6 with difference $\Delta X_6^O = (j, 0, j, 0, 0, 0, 0,$ $0, j, 0, 0, j, 0, 0, 0, 0)$.

After rolling back this difference through DL, we get $\Delta X_6^S = (j, 0, j, 0, 0, 0, 0,$ $0, j, 0, 0, j, 0, 0, 0, 0)$. Then after the transformation SL^{-1} and AK^{-1}, the difference is evolved into $\Delta X_6^I = (i_0, 0, i_2, 0, 0, 0, 0, 0, i_8, 0, 0, i_{11}, 0, 0, 0, 0)$ where i_0, i_2, i_8, i_{11} are unknown nonzero byte values.

After DL^{-1} of R_5, the difference is changed to $\Delta X_5^S = (h_0, h_1, h_2, h_3, h_4, h_5,$ $h_6, h_7, h_8, h_9, h_{10}, h_{11}, h_{12}, h_{13}, h_{14}, h_{15})$. Here $h_5 = i_1 \oplus i_3 \oplus i_4 \oplus i_9 \oplus i_{10} \oplus i_{14} \oplus i_{15} =$ $0 \oplus 0 \oplus 0 \oplus 0 \oplus 0 \oplus 0 \oplus 0 = 0$, and $h_8 = i_0 \neq 0$.

Therefore, when rolling back this difference through SL and AK of R_5, we get $\Delta X_5^I = (g_0, g_1, g_2, g_3, g_4, g_5, g_6, g_7, g_8, g_9, g_{10}, g_{11}, g_{12}, g_{13}, g_{14}, g_{15})$. And we know $g_5 = SL^{-1}(X_5^S) \oplus SL^{-1}(X_5^S \oplus h_5) = 0$, and $g_8 = SL^{-1}(X_5^S) \oplus SL^{-1}(X_5^S \oplus h_8) \neq 0$.

So we have $g_5 \neq g_8$. And also this property stands with probability one.

This differential contradicts the first differential with probability one, which has $f_5 = f_8$.

This contradiction is emphasized in Fig.1.

Some other impossible differential paths like (1) can also be found either. It's just the position has been altered. For instance,

$$(0, 0, 0, c, 0, 0, 0, 0, 0, 0, 0, 0, 0, 0, 0, 0) \not\Rightarrow (0, j, 0, 0, j, j, 0, 0, j, 0, 0, 0, 0, 0, 0, 0),$$
$$(c, 0, 0, 0, 0, 0, 0, 0, 0, 0, 0, 0, 0, 0, 0, 0) \not\Rightarrow (0, 0, 0, 0, j, 0, 0, 0, 0, 0, j, j, 0, j, 0, 0),$$
$$(c, 0, 0, 0, 0, 0, 0, 0, 0, 0, 0, 0, 0, 0, 0, 0) \not\Rightarrow (0, j, 0, 0, j, j, 0, 0, j, 0, 0, 0, 0, 0, 0, 0),$$
$$(0, c, 0, 0, 0, 0, 0, 0, 0, 0, 0, 0, 0, 0, 0, 0) \not\Rightarrow (0, j, 0, j, 0, 0, 0, 0, 0, j, j, 0, 0, 0, 0, 0),$$
$$(0, c, 0, 0, 0, 0, 0, 0, 0, 0, 0, 0, 0, 0, 0, 0) \not\Rightarrow (0, 0, 0, 0, 0, 0, 0, 0, 0, j, j, 0, j, j, 0, 0).$$

4   7-Round Impossible Differential Attack on ARIA-256

In this section, we present an impossible differential cryptanalysis of ARIA-256 reduced to 7 rounds, using the 4-round impossible differential path, which was described in previous section, with additional two rounds at the beginning and one round at the end as shown in Fig.2. Best previous impossible differential attacks could only apply to ARIA reduced to 6-round. Note that the last round of ARIA doesn't have the diffusion layer, but an additional AK.

4.1   Four Equations

We discover some amazing properties of DL transformation, which lead to this cryptanalysis. As illustrated in Fig.2, 4 significant equations of bytes in ΔX_1^S

are found, which make ΔX_1^S evolve into ΔX_2^I with 9 bytes $(0, 2, 3, 5, 7, 8, 9, 13, 14)$ equaling to zero with probability $p = 2^{-24}$, while in the random case the probability is $p = 2^{-72}$. The four equations are:

$$\Delta X_{1,5}^S = \Delta X_{1,8}^S = \Delta X_{1,15}^S \tag{2}$$

$$\Delta X_{1,7}^S = \Delta X_{1,9}^S = \Delta X_{1,12}^S \tag{3}$$

$$\Delta X_{1,0}^S \oplus \Delta X_{1,1}^S \oplus \Delta X_{1,10}^S \oplus \Delta X_{1,11}^S = 0 \tag{4}$$

$$\Delta X_{1,3}^S \oplus \Delta X_{1,4}^S \oplus \Delta X_{1,10}^S \oplus \Delta X_{1,13}^S = 0 \tag{5}$$

In Fig.2, as we can see, in R_1 we use $a, A, a_i, A_j, (i, j \in \{0, 1, ..., 15\})$ to present $\Delta X_{1,k}^S, (k \in \{0, 1, ..., 15\})$, i.e., $a_1 = \Delta X_{1,1}^S$, $A_4 = \Delta X_{1,4}^S$, $a = \Delta X_{1,5}^S = \Delta X_{1,8}^S = \Delta X_{1,15}^S$, etc., and each number in the states before and after AK of R_1 corresponds with an equation. So the 4 equations become:

$$\Delta X_{1,5}^S = \Delta X_{1,8}^S = \Delta X_{1,15}^S = a \tag{2}$$

$$\Delta X_{1,7}^S = \Delta X_{1,9}^S = \Delta X_{1,12}^S = A \tag{3}$$

$$a_0 \oplus a_1 \oplus a_{10} \oplus a_{11} = 0 \tag{4}$$

$$A_3 \oplus A_4 \oplus A_{10} \oplus A_{13} = 0 \tag{5}$$

Now we prove that $p = 2^{-24}$.

Proof: We use structure here to make our argument much easier and more explicit. A structure is defined as a set of 2^{56} differential values of plaintexts which equal to zero in all but 7 bytes $(1, 4, 6, 10, 11, 12, 15)$.

Randomly choose ΔX_2^I from this structure, and through DL^{-1} transformation, we get all the values of $\Delta X_{1,i}^S, 0 \leq i \leq 15$ as in Fig.2. And we have $\Delta X_{1,5}^S = \Delta X_{2,1}^I \oplus \Delta X_{2,4}^I \oplus \Delta X_{2,10}^I \oplus \Delta X_{2,15}^I$, $\Delta X_{1,8}^S = \Delta X_{2,1}^I \oplus \Delta X_{2,4}^I \oplus \Delta X_{2,10}^I \oplus \Delta X_{2,15}^I$ and $\Delta X_{1,15}^S = \Delta X_{2,1}^I \oplus \Delta X_{2,4}^I \oplus \Delta X_{2,10}^I \oplus \Delta X_{2,15}^I$. Therefore, no matter what the values of $\Delta X_{2,1}^I$, $\Delta X_{2,4}^I$, $\Delta X_{2,10}^I$, $\Delta X_{2,15}^I$ are, equation (2) stands with probability one.

Likewise, equation (3) always holds with probability one.

At the same time, we get $a_0 = \Delta X_{2,4}^I \oplus \Delta X_{2,6}^I, a_1 = \Delta X_{2,4}^I \oplus \Delta X_{2,12}^I, a_{10} = \Delta X_{2,6}^I \oplus \Delta X_{2,15}^I, a_{11} = \Delta X_{2,4}^I \oplus \Delta X_{2,12}^I, A_3 = \Delta X_{2,10}^I \oplus \Delta X_{2,11}^I, A_4 = \Delta X_{2,11}^I \oplus \Delta X_{2,15}^I, A_{13} = \Delta X_{2,6}^I \oplus \Delta X_{2,10}^I$.

It's easy to verify that equations (4) and (5) stand with probability one as well.

It suggests that, if we demand the values of ΔX_2^I are all in the structure defined above, the corresponding values of ΔX_1^S must fulfill all the equations $(2) - (5)$. So we can eliminate all the values of ΔX_1^S that can't satisfy all the equations without deleting a right one. We use p_2, p_3, p_4, p_5 to denote the probability of equations $(2) - (5)$ respectively. We can easily find out that $p_2 = 2^{-16}$, $p_3 = 2^{-16}$, $p_4 = 2^{-8}$, $p_5 = 2^{-8}$. So the number of ΔX_1^S is narrowed down to:

$$N = 2^{128} \times \prod_{i=2}^{5} p_i = 2^{80}.$$

Since DL is a linear transformation, and there are 2^{56} values of ΔX_1^I in the structure, the number of corresponding values of ΔX_1^S which make ΔX_2^I be elements of the structure is also 2^{56}. Spontaneously, any ΔX_1^S which satisfies all the 4 equations evolves ΔX_2^I into the structure with probability $p = \frac{2^{56}}{2^{80}} = 2^{-24}$. □

4.2   The Procedure of 7-Round Attack on ARIA-256

The procedure of this attack is as follows. We use $k_{m,n}$ to denote the n-th byte of k_m.

Step 1. Randomly select 2^{125} plaintexts, and such plaintexts proposes $2^{125} \times 2^{125} \times \frac{1}{2} = 2^{249}$ pairs.

Step 2. Select pairs whose ciphertext pairs have zero difference at the twelve bytes $(1, 3, 4, 5, 6, 7, 9, 10, 12, 13, 14, 15)$. The expected number of such pairs is $2^{249} \times 2^{-96} = 2^{153}$.

Step 3. Guess the 4-byte value $(k_{8,0}, k_{8,2}, k_{8,8}, k_{8,11})$ of the last round key k_8. For each ciphertext pair (C, C'), compute $\Delta X_7^I = SL^{-1}(C \oplus k_8) \oplus SL^{-1}(C' \oplus k_8)$, and choose pairs whose difference ΔX_7^I are same at the 4 bytes $\Delta X_{7,0}^I, \Delta X_{7,2}^I, \Delta X_{7,8}^I, \Delta X_{7,11}^I$. The expected number of the remaining pairs is $2^{153} \times 2^{-24} = 2^{129}$.

Step 4. Next guess all 16 bytes of k_1. But we don't guess all the 16 bytes values at once, we separate them into 5 parts, using the "four equations" presented in the previous subsection.

Step 4.1 Guess 3-byte value $(k_{1,5}, k_{1,8}, k_{1,15})$ of the first round key k_1, and for those plaintext pairs (P, P') with such ciphertext pairs, compute $\Delta X_1^S = SL(P \oplus k_1) \oplus SL(P' \oplus k_1)$ at the above 3 bytes. Choose pairs whose difference ΔX_1^S are same at these 3 bytes. The expected number of such pairs is $2^{129} \times 2^{-16} = 2^{113}$.

Step 4.2 Guess 3-byte value $(k_{1,7}, k_{1,9}, k_{1,12})$ of k_1, and for the remaining pairs (P, P') compute like above, $\Delta X_1^S = SL(P \oplus k_1) \oplus SL(P' \oplus k_1)$ at the 3 bytes $(7, 9, 12)$. And discard those pairs which have different values at bytes $(7, 9, 12)$. The number of the remaining pairs is $2^{113} \times 2^{-16} = 2^{97}$.

Step 4.3 Guess 4-byte value $(k_{1,0}, k_{1,1}, k_{1,10}, k_{1,11})$ of k_1, and compute $\Delta X_1^S = SL(P \oplus k_1) \oplus SL(P' \oplus k_1)$ at the 4 bytes $(0, 1, 10, 11)$. Choose pairs which satisfy the equation : $\Delta X_{1,0}^S \oplus \Delta X_{1,1}^S \oplus \Delta X_{1,10}^S \oplus \Delta X_{1,11}^S = 0$. So there are $2^{97} \times 2^{-8} = 2^{89}$ pairs left.

Step 4.4 Guess 3-byte value $(k_{1,3}, k_{1,4}, k_{1,13})$ of k_1, and compute $\Delta X_1^S = SL(P \oplus k_1) \oplus SL(P' \oplus k_1)$ at the 3 bytes $(3, 4, 13)$. Get rid of pairs which don't satisfy the equation : $\Delta X_{1,3}^S \oplus \Delta X_{1,4}^S \oplus \Delta X_{1,10}^S \oplus \Delta X_{1,13}^S = 0$. The number of the remaining pairs is $2^{89} \times 2^{-8} = 2^{81}$.

Step 4.5 Guess the last 3-byte value $(k_{1,2}, k_{1,6}, k_{1,14})$ of k_1, and compute $\Delta X_1^S = SL(P \oplus k_1) \oplus SL(P' \oplus k_1)$ at the those 3 bytes like above.

Step 4.6 For all 16 bytes values of ΔX_1^S, compute $\Delta X_2^I = DL(\Delta X_1^S)$, pick up pairs whose difference ΔX_2^I are zero at 9 bytes $(0, 2, 3, 5, 7, 8, 9, 13, 14)$. The probability is $p = 2^{-24}$. So the number of the remaining pairs is $2^{81} \times 2^{-24} = 2^{57}$.

Step 5. Guess 7-byte value at $(k_{2,1}, k_{2,4}, k_{2,6}, k_{2,10}, k_{2,11}, k_{2,12}, k_{2,15})$ of k_2, and compute $\Delta X_2^S = SL(X_2^I \oplus k_1) \oplus SL(X_2^{II} \oplus k_1)$ at the 7 bytes $(1, 4, 6, 10, 11, 12, 15)$. Choose pairs whose difference ΔX_2^S are same at the 7 bytes $(1, 4, 6, 10, 11, 12, 15)$. The probability is 2^{-48}.

Step 6. Since such a difference is impossible, every value of k_2 which satisfies the difference is wrong value. After we analyze 2^{57} pairs, there are only $2^{56} \times (1 - 2^{-48})^{2^{57}} \approx 2^{-662.3}$ wrong value of k_2 left.

Unless the assumptions on k_8 and k_1 are both correct, it is expected that we can get rid of the whole 56-bit values of k_2 for each 160-bit value of (k_8, k_1), since the number of remaining wrong value of (k_8, k_1, k_2) is about $2^{32} \times 2^{128} \times 2^{-662} = 2^{-512} \approx 0$ [2]. Hence if there remains a value of k_2, we can assume the value of (k_8, k_1, k_2) is right.

4.3   Time Complexity

Next we analyze the time complexity of our attack.

In Step 3, if we compute all the values of those 4 bytes at once, the time complexity of this step will be $2 \times (2^{153} \times 2^{32} \times \frac{4}{16}) = 2^{184}$. But actually we only need 3×2^{168}. Because we can first compute $\Delta X_{7,0}^I$ and $\Delta X_{7,2}^I$, and check if they are equal. And get rid of the pairs which don't. For the rest pairs, continue to compute $\Delta X_{7,8}^I$, and compare with the value of $\Delta X_{7,0}^I$. If they are equal, remain the corresponding pairs, and so on. This is what is called the "early-abort technique". Since we only need to compute 4 bytes here. Thus this step requires $2 \times (2^{153} \times 2^{16} + 2^{145} \times 2^{24} + 2^{137} \times 2^{32}) \times \frac{4}{16} = 3 \times 2^{168}$ one round operations.

And we also use the "early-abort technique" in all the rest steps.

In Step 4.1, because we just compute 3 bytes of plaintext pairs, and only AK and SL are operated, so we consider it as $\frac{3}{16} \times \frac{2}{3}$ one round operations. So just like Step 3, this step requires $2^{32} \times 2 \times (2^{129} \times 2^{16} + 2^{121} \times 2^{24}) \times \frac{3}{16} \times \frac{2}{3} = 2^{176}$ one round operations.

Similarly, Step 4.2 needs $2^{32} \times 2 \times 2^{24} \times (2^{113} \times 2^{16} + 2^{105} \times 2^{24}) \times \frac{3}{16} \times \frac{2}{3} = 2^{184}$ one round operations.

In Step 4.3, we encrypt the 4 bytes $(0, 1, 10, 11)$ of plaintext pairs, and also only AK and SL are operated. So this step demands $2^{32} \times 2 \times 2^{48} \times (2^{97} \times 2^{32}) \times \frac{4}{16} \times \frac{2}{3} = \frac{1}{3} \times 2^{209}$ one round operations.

Just like Step 4.3, Step 4.4 requires $2^{32} \times 2 \times 2^{80} \times (2^{89} \times 2^{24}) \times \frac{3}{16} \times \frac{2}{3} = 2^{223}$ one round operations.

Step 4.5 needs $2^{32} \times 2 \times 2^{104} \times (2^{81} \times 2^{24}) \times \frac{3}{16} \times \frac{2}{3} = 2^{239}$ one round operations.

In Step 4.6, as we considered in Step 4.1, because only DL is operated, we consider it as $\frac{1}{3}$ one round operations.

So here we require $2^{32} \times 2^{128} \times 2 \times 2^{81} \times \frac{1}{3} = \frac{1}{3} \times 2^{242}$ one round operations.

And in Step 5, like Step 3, we demand $2^{32} \times 2^{128} \times 2 \times (2^{57} \times 2^{16} + 2^{49} \times 2^{24} + 2^{41} \times 2^{32} + 2^{33} \times 2^{40} + 2^{25} \times 2^{48} + 2^{17} \times 2^{56}) \times \frac{7}{16} = 21 \times 2^{231}$ one round operations.

Therefore, the total time complexity is $(3 \times 2^{168} + 2^{176} + 2^{184} + \frac{1}{3} \times 2^{209} + 2^{223} + \frac{1}{3} \times 2^{242} + 21 \times 2^{231}) \times \frac{1}{7} = 2^{238}$ encryptions of ARIA-256 reduced to 7 rounds.

Consequently, our attack requires about 2^{125} chosen plaintexts and less than 2^{238} encryptions of 7-round ARIA-256.

5 Conclusion

In this paper, we present a new impossible differential attack against ARIA-256 reduced to 7 rounds. This attack requires 2^{125} chosen plaintexts and 2^{238} encryptions. Our result is the best impossible differential cryptanalysis result on ARIA as far as we know to date. In Table 1, we compare the new attack with the previous impossible differential attacks.

Table 1. Comparison of impossible differential cryptanalysis of ARIA variants

Variant	Number of Rounds	Chosen Plaintexts	Time Complexity	Source
ARIA-128	6	2^{121}	2^{112}	Ref.[17]
ARIA-128	6	2^{120}	2^{96}	Ref.[13]
ARIA-256	7	2^{125}	2^{238}	This paper

Acknowledgments

The authors would like to thank Professor Xiaoyun Wang for her valuable instructions and suggestions. The authors also thank Chengliang Tian and Keting Jia for their useful help. This research is supported by the National 973 Program of China (Grant No.2007CB807902) and the National Natural Science Foundation of China (Grant No.60910118).

References

1. Biham, E., Biryukov, A., Shamir, A.: Cryptanalysis of Skipjack Reduced to 31 Rounds Using Impossible Differentials. In: Stern, J. (ed.) EUROCRYPT 1999. LNCS, vol. 1592, pp. 12–23. Springer, Heidelberg (1999)
2. Biham, E., Dunkelman, O., Keller, N.: Related-Key Impossible Differential Attacks on 8-round AES-192. In: Pointcheval, D. (ed.) CT-RSA 2006. LNCS, vol. 3860, pp. 21–33. Springer, Heidelberg (2006)
3. Biham, E., Keller, N.: Cryptanalysis of Reduced Variants of Rijndael. In: The Third AES Candidate Conference (2000)
4. Biham, E., Shamir, A.: Differential Cryptanalysis of DES-like Cryptosystems. Journal of Cryptology 4(1), 3–72 (1991)
5. Biryukov, A., De Canniere, C., Lano, J., Ors, S.B., Preneel, B.: Security and Performance Analysis of Aria. Version 1.2 (Janaury 7, 2004)

6. Cheon, J.H., Kim, M., Kim, K., et al.: Improved Impossible Differential Cryptanalysis of Rijndael and Crypton. In: Kim, K. (ed.) ICISC 2001. LNCS, vol. 2288, pp. 39–49. Springer, Heidelberg (2002)
7. Daemen, J., Rijmen, V.: The Design of Rijndael. In: Information Security and Cryptography. Springer, Heidelberg (2002)
8. Jakimoski, G., Desmedt, Y.: Related-Key Differential Cryptanalysis of 192-bit key AES Variants. In: Matsui, M., Zuccherato, R. (eds.) SAC 2003. LNCS, vol. 3006, pp. 208–221. Springer, Heidelberg (2004)
9. Jakobsen, T., Knudsen, L.R.: The Interpolation Attack against Block Ciphers. In: Biham, E. (ed.) FSE 1997. LNCS, vol. 1267, pp. 28–40. Springer, Heidelberg (1997)
10. Knudsen, L.R.: Truncated and Higher Order Differentials. In: Preneel, B. (ed.) FSE 1994. LNCS, vol. 1008, pp. 196–211. Springer, Heidelberg (1995)
11. Knudsen, L.R., Wagner, D.: Integral Cryptanalysis (extended abstract). In: Daemen, J., Rijmen, V. (eds.) FSE 2002. LNCS, vol. 2365, pp. 629–632. Springer, Heidelberg (2002)
12. Kwon, D., Kim, J., Park, S., et al.: New Block Cipher: ARIA. In: Lim, J.-I., Lee, D.-H. (eds.) ICISC 2003. LNCS, vol. 2971, pp. 432–445. Springer, Heidelberg (2004)
13. Li, S., Song, C.: Improved Impossible Differential Cryptanalysis of ARIA. In: ISA 2008, pp. 129–132. IEEE Computer Society, Los Alamitos (April 2008)
14. Matsui, M.: Linear Cryptanalysis Method for DES Cipher. In: Helleseth, T. (ed.) EUROCRYPT 1993. LNCS, vol. 765, pp. 386–397. Springer, Heidelberg (1994)
15. National Security Research Institute: Specification of ARIA, Version 1.0 (January 2005), http://www.nsri.re.kr/ARIA/doc/ARIAspecification-e.pdf
16. Phan, R.C.: Impossible Differential Cryptanalysis of 7-round AES. Inf. Process. Lett. 91(1), 33–38 (2004)
17. Wu, W., Zhang, W., Feng, D.: Impossible Differential Cryptanalysis of Reduced-Round ARIA and Camellia. Journal of Computer Science and Technology 22(3), 449–456 (2007)

An Algorithm Based Concurrent Error Detection Scheme for AES

Chang N. Zhang, Qian Yu, and Xiao Wei Liu

Department of Computer Science, University of Regina, Canada
{zhang,yu209,liu273}@cs.uregina.ca

Abstract. With the wide-spread practical applications of AES, not only high performance, but also strong reliability is desirable to all the cryptosystem. In this paper, a lightweight concurrent AES error detection scheme which is based on the algorithm based fault tolerant (ABFT) technique is proposed. Two versions of scheme are presented to satisfy different application requirements. The first general version scheme can detect single error for the whole AES process with high efficiency. Another run-time version scheme is used to immediately end the error round with no time delay and no computation wasted on the rest rounds for propagating errors. Utilizing the ready-made arithmetic units in AES, single error can be detected by the sender and prevent the misdirected information from sending out. The results of the hardware FPGA implementation and simulation show that the proposed scheme can be integrated both on software and hardware without making many changes to the original AES implementation.

Keywords: Advanced Encryption Standard, fault tolerance, error detection, ABFT.

1 Introduction

In 2001, through the evaluation of several essential criteria among candidate algorithm nominations, National Institute of Standards and Technology (NIST) finally issued Advanced Encryption Standard (AES) as a replacement for triple DES (3DES), which is also well know as Rijndael. As a symmetric block cipher, AES is proved to not only have comparable security strength, but also achieve significant efficiency improvement for implementation on software or hardware.

However, with the wide-spread of the AES applications, differential types of faults may be bought in. Several efforts were devoted into fault tolerance of the transformations and rounds in AES algorithm. Guido Bertoni et al presented a fault model for AES and analyzed the behavior of the AES algorithm in the presence of faults [1]. They also proposed a fault detection technique for a hardware implementation of the AES algorithm which is based on the parity codes [2]. Moreover, they developed an analytical error model for the parity-based EDC for the AES encryption algorithm and is capable of locating single-bit transient and permanent faults [4, 5]. Later, the same group further described the complete error model extended to include the Key Schedule (KS) part and presented the results of the software

S.-H. Heng, R.N. Wright, and B.-M. Goi (Eds.): CANS 2010, LNCS 6467, pp. 31–42, 2010.

simulations of the model [7]. L. Berveglieri et al proposed an extension to an existing
AES architecture to provide error detection and fault tolerance [3]. Kaijie Wu et al
presented a low-cost concurrent checking method for the AES encryption algorithm by
using parity checking which can detect faults during normal operation and deliberately
injected faults [6]. Mark Karpovsky et al presented a method of protecting a hardware
implementation of the AES against a side-channel attack known as Differential Fault
Analysis attack [8]. Chih-Hsu Yen et al proposed several error-detection schemes for
AES which are based on the (n+1, n) cyclic redundancy check over GF (2^8) [9]. Luca
Breveglieri et al presented an operation-centered approach to the incorporation of fault
detection into cryptographic device implementation through the use of Error Detection
Codes [10]. Ramesh Karri et al presented a fault-tolerant architecture for symmetric
block ciphers which is based on a hardware pipeline for encryption and decryption [11].
P. Maistri et al presented the results of a validation campaign on an AES core protected
with some error detection mechanisms [15]. Mojtaba Valinataj et al combined and
reinforced the parity prediction scheme with a partially distributed TMR to achieve
more reliability against multiple simultaneous errors [16].

Other efforts focus on relevant fault detection field. L. Berveglieri et al presented
suggestions for providing fault detection capabilities in recent block ciphers and came
to the conclusion that the detection capability of any code depends on the type of the
code, the frequency of checkpoints and the level of redundancy [14]. Ramesh Karri et al
presented a technique to concurrently detect errors in block ciphers as well as a new
encoding strategy [17].

As a summary, the parity bit check coding technique has been introduced and widely
applied to the basic operations of AES. As a result, the parity bit needs to be generated
and checked for every individual AES operation which brings in considerable time and
hardware overhead. The algorithm based tolerant (ABFT) technique is a general
concept for designing efficient fault tolerant schemes based on structures of the
algorithms [18-20]. Compare with other fault tolerant schemes, the ABFT makes use of
the computational nature of the targeted algorithm and poses a conceptual way to better
create a fault tolerant version by altering the algorithm computation so that its output
contains extra information for error detection and correction. It has relatively low
overhead and no additional arithmetical logic unit is required. Due to the essentiality of
the AES computational nature, we believe that the algorithm based fault tolerant
(ABFT) technique can be more suitable for AES computations. By utilizing the basic
arithmetic units in AES, errors can be accurately detected with lighter overheads.

This paper is organized as follows. Section 2 briefly reviews the AES algorithm and
introduces the notations for further discussion. Section 3 presents two proposed fault
detection scheme for rounds and whole process. Section 4 shows the implementation
and simulation results of the general version scheme. The last Section is a conclusion.

2 The AES Algorithm and Relevant Notations

Originally, the Rijndael proposal for the AES algorithm has three alternatives of block
length and key length, which are 128, 192, or 256 bits respectively. For the sake of
simplicity, we only adopt the AES parameters of 128-bit key size, 128-bit plaintext

block size, 10 rounds, 128-bit round key size, and 44-word expanded key size, which are the most commonly-used parameter set. Due to the similarity between encryption and decryption parts, just encryption is introduced as an illustration of our schemes. There is one initial round followed by 9 four-step rounds and ended by a tenth final round.

2.1 The Initial Round

The first initial round only performs the AddRoundKey operation. "m" stands for the total ordinal number of each round. "e" is the results for every round. "P" and "K" are defined as plaintext and round key respectively. For the initial round 0, we have $m = 0$, the plaintext state P and round 0th key state K^0 are represented as,

$$P = \begin{bmatrix} P_{00} & P_{01} & P_{02} & P_{03} \\ P_{10} & P_{11} & P_{12} & P_{13} \\ P_{20} & P_{21} & P_{22} & P_{23} \\ P_{30} & P_{31} & P_{32} & P_{33} \end{bmatrix};$$

$$K^0 = \begin{bmatrix} K^0_{00} & K^0_{01} & K^0_{02} & K^0_{03} \\ K^0_{10} & K^0_{11} & K^0_{12} & K^0_{13} \\ K^0_{20} & K^0_{21} & K^0_{22} & K^0_{23} \\ K^0_{30} & K^0_{31} & K^0_{32} & K^0_{33} \end{bmatrix};$$

According to the operation, the equation for round 0 can be derived as follows:

$$\begin{bmatrix} e^0_{0j} \\ e^0_{1j} \\ e^0_{2j} \\ e^0_{3j} \end{bmatrix} = \begin{bmatrix} P_{0j} \\ P_{1j} \\ P_{2j} \\ P_{3j} \end{bmatrix} \oplus \begin{bmatrix} K^0_{0j} \\ K^0_{1j} \\ K^0_{2j} \\ K^0_{3j} \end{bmatrix};$$

where $0 \le j \le 3$ and e^0_{ij} is the result for round 0, $0 \le i \le 3$.

2.2 The 9 Rounds

The 9 rounds have the same sub-operations for each round, which are SubBytes, ShiftRows, MixColumns and AddRoundKey according to the order. For these 9 rounds, $1 \le m \le 9$, "a" is the input of round m, "b" "r" and "c" stands for the intermediate results for SubBytes, ShiftRows and MixColumns operations respectively, and e^m also represents the results of round m.

Input of Round m:

$$a_{ij}^m = e_{ij}^{m-1},$$

where $1 \le m \le 9$, $0 \le i \le 3$, $0 \le j \le 3$;

SubBytes Operation:

$$b_{ij}^m = S[a_{ij}^m],$$

where S stands for the substitution by the S-box and $1 \le m \le 9$, $0 \le i \le 3$, $0 \le j \le 3$.

ShiftRows Operation:

$$\begin{bmatrix} r_{0j}^m \\ r_{1j}^0 \\ r_{2j}^0 \\ r_{3j}^0 \end{bmatrix} = \begin{bmatrix} b_{0j}^m \\ b_{1,j+1}^m \\ b_{2,j+2}^m \\ b_{3,j+3}^m \end{bmatrix},$$

where $0 \le j \le 3$ and $1 \le m \le 9$.

MixColumns Operation:

$$\begin{bmatrix} c_{0j}^m \\ c_{1j}^m \\ c_{2j}^m \\ c_{3j}^m \end{bmatrix} = \begin{bmatrix} 02 & 03 & 01 & 01 \\ 01 & 02 & 03 & 01 \\ 01 & 01 & 02 & 03 \\ 03 & 01 & 01 & 02 \end{bmatrix} \cdot \begin{bmatrix} r_{0j}^m \\ r_{1j}^0 \\ r_{2j}^0 \\ r_{3j}^0 \end{bmatrix} \text{ in GF } (2^8),$$

where $0 \le j \le 3$ and $1 \le m \le 9$.

AddRoundKey Operation:

$$\begin{bmatrix} e_{0j}^m \\ e_{1j}^m \\ e_{2j}^m \\ e_{3j}^m \end{bmatrix} = \begin{bmatrix} c_{0j}^m \\ c_{1j}^m \\ c_{2j}^m \\ c_{3j}^m \end{bmatrix} \oplus \begin{bmatrix} K_{0j}^m \\ K_{1j}^m \\ K_{2j}^m \\ K_{3j}^m \end{bmatrix},$$

where $0 \le j \le 3$ and $1 \le m \le 9$. Note that $c_{ij}^m \oplus K_{ij}^m = c_{ij}^m + K_{ij}^m$ in GF (2^8).

As a conclusion, the equation for each round can be represented as:

$$E_j^m = \begin{bmatrix} e_{0j}^m \\ e_{1j}^m \\ e_{2j}^m \\ e_{3j}^m \end{bmatrix} = \begin{bmatrix} 02 & 03 & 01 & 01 \\ 01 & 02 & 03 & 01 \\ 01 & 01 & 02 & 03 \\ 03 & 01 & 01 & 02 \end{bmatrix} \cdot \begin{bmatrix} S[e_{0j}^m] \\ S[e_{1,j+1}^m] \\ S[_{2,j+2}^m] \\ S[e_{3,j+3}^m] \end{bmatrix} + \begin{bmatrix} K_{0j}^m \\ K_{1j}^m \\ K_{2j}^m \\ K_{3j}^m \end{bmatrix} \text{ in GF } (2^8), \qquad (1)$$

where $0 \le j \le 3$ and $1 \le m \le 9$.

2.3 The Final Round

The final round has no MixColumns operation, and the three operations and their execution order is the same as the previous 9 rounds. In this 10th round, we have $m = 10$ and the following equation:

$$
\begin{bmatrix} e_{0j}^{10} \\ e_{1j}^{10} \\ e_{2j}^{10} \\ e_{3j}^{10} \end{bmatrix} = \begin{bmatrix} S[e_{0j}^{9}] \\ S[e_{1,j+1}^{9}] \\ S[_{2,j+2}^{9}] \\ S[e_{3,j+3}^{9}] \end{bmatrix} + \begin{bmatrix} K_{0j}^{10} \\ K_{1j}^{10} \\ K_{2j}^{10} \\ K_{3j}^{10} \end{bmatrix} \text{ in GF } (2^8),
$$

where $0 \le j \le 3$.

3 The Algorithm Based Error Detection Schemes for AES

For round 0 to 10, let $K = \begin{bmatrix} K_{ij} \end{bmatrix} = [\sum_{m=0}^{10} K_{ij}^{m}]$, and $E = \begin{bmatrix} e_{ij} \end{bmatrix} = \begin{bmatrix} \sum_{m=0}^{10} e_{ij}^{m} \end{bmatrix}$, both in GF (2^8), where $0 \le i, j \le 3$, we have:

$$
E_j = \begin{bmatrix} e_{0j} \\ e_{1j} \\ e_{2j} \\ e_{3j} \end{bmatrix} = \begin{bmatrix} P_{0j} \\ P_{1j} \\ P_{2j} \\ P_{3j} \end{bmatrix} + \begin{bmatrix} 02 & 03 & 01 & 01 \\ 01 & 02 & 03 & 01 \\ 01 & 01 & 02 & 03 \\ 03 & 01 & 01 & 02 \end{bmatrix} \cdot \begin{bmatrix} \sum_{m=o}^{8} S[e_{0j}^{m}] \\ \sum_{m=0}^{8} S[e_{1,j+1}^{m}] \\ \sum_{m=0}^{8} S[e_{2,j+2}^{m}] \\ \sum_{m=0}^{8} S[e_{3,j+3}^{m}] \end{bmatrix} + \begin{bmatrix} S[e_{0j}^{9}] \\ S[e_{1,j+1}^{9}] \\ S[_{2,j+2}^{9}] \\ S[e_{3,j+3}^{9}] \end{bmatrix} + \begin{bmatrix} K_{0j} \\ K_{1j} \\ K_{2j} \\ K_{3j} \end{bmatrix}
$$

in GF (2^8),   (2)

where $0 \le j \le 3$.

Assuming that only a single error may occur during the whole AES process, and the SubByte operation is implemented by some fault tolerant look-up table (e.g. Hamming Code) so that it is error-free for the SubByte operation.

Due to the linear computational nature of the equation (2), the algorithm based fault tolerant (ABFT) technique can be applied to pre-compute some known parameters as check-sums. By storing certain intermediate results, this equation can be used to detect error. We propose two versions of the error detection scheme. The general version can detect an error for the whole total 11 rounds AES process. Another run-time version can detect error and immediately stop the round in which the error exists so that it can prevent the error from propagating to the following rounds.

3.1 The General Version of Concurrent Error Detection Scheme

Firstly, the plaintext state and the round key state have to be XORed together,

$$PK = \left[PK_{ij} \right],$$ (3)

where $PK_{ij} = \sum_{m=0}^{10} K_{ij}^m \oplus P_{ij}$, and $0 \le i, j \le 3$;

Secondly, while the rounds are going, two intermediate results of each rounds need to be stored, which are the final results of each round and the substitution results after SubBytes operation of each round. And then they are pre-computed in the following ways.

$$E = \left[e_{ij} \right], \text{ where } e_{ij} = \sum_{m=0}^{10} e_{ij}^m, \text{ and } 0 \le i, j \le 3;$$

$$S_{0j} = \sum_{m=0}^{8} S\left[e_{0j}^m \right], \; S_{1j} = \sum_{m=0}^{8} S\left[e_{1,j+1}^m \right], \; S_{2j} = \sum_{m=0}^{8} S\left[e_{2,j+2}^m \right], \; S_{3j} = \sum_{m=0}^{8} S\left[e_{3,j+3}^m \right],$$

where $0 \le j \le 3$.

Thirdly, according to above notations and the equation (2), we can obtain the following equation:

$$E' = \begin{bmatrix} 02 & 03 & 01 & 01 \\ 01 & 02 & 03 & 01 \\ 01 & 01 & 02 & 03 \\ 03 & 01 & 01 & 02 \end{bmatrix} \cdot \begin{bmatrix} S_{0j} \\ S_{1j} \\ S_{2j} \\ S_{3j} \end{bmatrix} + \begin{bmatrix} S[e_{0j}^9] \\ S[e_{1,j+1}^9] \\ S[_{2,j+2}^9] \\ S[e_{3,j+3}^9] \end{bmatrix} + \begin{bmatrix} PK_{0j} \\ PK_{1j} \\ PK_{2j} \\ PK_{3j} \end{bmatrix} \text{ in GF } (2^8)$$ (4)

where $0 \le j \le 3$.

At last, in order to know whether error occurs, the only work is to compare the results of above equation E' with the stored values E. If they equal then there is no error, otherwise, single error occurs in some round. By knowing the existence of errors, the sender can be informed of the false encryption result at once. In that case, error can be blocked. The general version of concurrent error detection scheme is shown in Figure 1.

If we count one round with four basic operations as a time unit, the proposed general version ABFT scheme requires only the final two basic operations, MixColumns and AddRoundKey, to perform multiplication and addition. In that case, the overhead is about 1/20 of the total AES processing time. If the AES is implemented by pipeline, and each pipeline unit performs one of the four basic operations, then the proposed ABFT scheme only affects the pipeline latency and needs some extra storage for the intermediate results, but still maintain the same throughput.

3.2 The Run-Time Concurrent Error Detection Scheme

The goal of the run-time error detection version is to find an error and immediately stop the process.

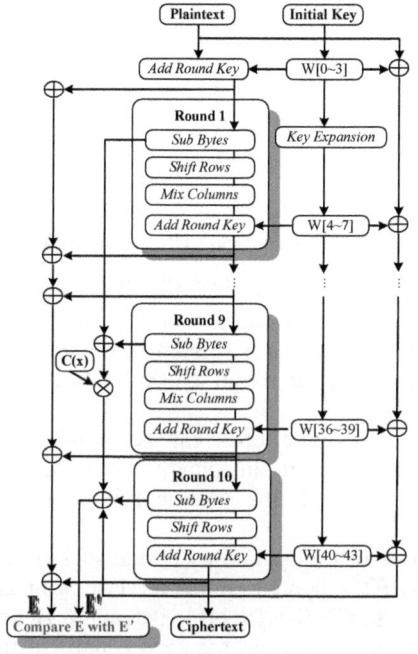 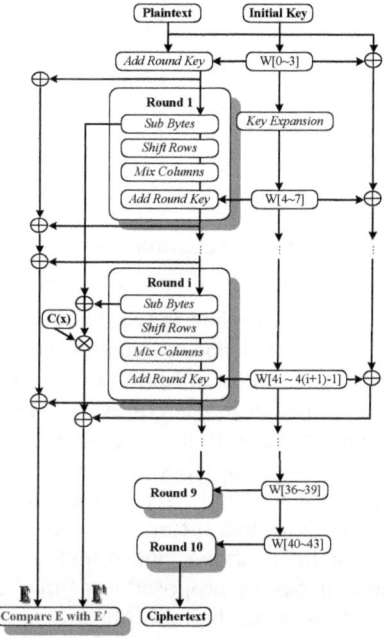

Fig. 1. The general version of error detection scheme for AES

Fig. 2. The run-time version of error detection scheme for AES

Similar to the previous discussion, we can pre-compute the PK according to equation (3). Then the intermediate results of SubBytes and the results of each round are stored and organized in the same way as the general version, however, the computational method varies. The flow chart of the run-time error detection version is shown in Figure 2.

Let "h" represent the number of round which has just finished and use the following notation:

$$E^h = \left[e_{ij}^h \right], \text{ where } e_{ij}^h = \sum_{m=0}^{h} e_{ij}^m \text{ , and } 0 \le i, j \le 3 ; \tag{5}$$

$$S_{0j}^h = \sum_{m=0}^{h} S\left[e_{0j}^m \right], S_{1j}^h = \sum_{m=0}^{h} S\left[e_{1,j+1}^m \right], S_{2j}^h = \sum_{m=0}^{h} S\left[e_{2,j+2}^m \right], S_{3j}^h = \sum_{m=0}^{h} S\left[e_{3,j+3}^m \right],$$

where $0 \le j \le 3$;

$$PK_{ij}^h = \sum_{m=0}^{h} K_{ij}^m \oplus P_{ij} ,$$

where $0 \le i, j \le 3$

Also, there is some modification based on equation (1), that is,

$$
E_h^{'} = \begin{bmatrix} e_{0j}^h \\ e_{1j}^h \\ e_{2j}^h \\ e_{3j}^h \end{bmatrix} = \begin{bmatrix} 02 & 03 & 01 & 01 \\ 01 & 02 & 03 & 01 \\ 01 & 01 & 02 & 03 \\ 03 & 01 & 01 & 02 \end{bmatrix} \cdot \begin{bmatrix} S_{0j}^h \\ S_{1j}^h \\ S_{2j}^h \\ S_{3j}^h \end{bmatrix} + \begin{bmatrix} PK_{0j}^h \\ PK_{1j}^h \\ PK_{2j}^h \\ PK_{3j}^h \end{bmatrix} \text{ in GF } (2^8), \tag{6}
$$

where $0 \le j \le 3$.

By using this equation, we can detect the errors in each round by comparing the value E^h given by equation (5) with the value $E_h^{'}$ of the equation (6), $h = 1, 2, \cdots, 9$. An error exists when they do not equal. If the error happened in round 10 (h=10), we still can use the check equation (4) indicated in the general version to compute the check value. In order to find errors in real-time during the round, check scheme has to be performed at the end of every round.

Using the run-time concurrent error detection scheme, error can be found before the whole AES process finishes. Once error occurs in certain round, there is no need to do the rest rounds. By performing the run-time check, the work and time spent on computing useless information can be saved.

As defined above, one round time with four basic operations is considered as a time unit. So the proposed run-time version ABFT scheme requires the last two basic operations to do the multiplication and addition in GF (2^8) for error detection. Hence, the overhead is about 1/2 of the total AES processing time for each round. In this way, if sub-operations are implemented by pipeline units, only pipeline latency is affected.

4   Hardware Implementation and Simulation Results

As a symmetric encryption algorithm standard, AES has become one of the most important crypto-algorithms implemented on a variety of platforms, such as Field Programmable Gate Array (FPGA) and Application Specific Integrated Circuit (ASIC). Among them, there are two most basic and commonly adopted architectures, which are rolling architecture and unrolling architecture [21]. The rolling architecture uses a feedback structure where the data are iteratively transformed by the round functions. This approach has the advantage of small area but the disadvantage of a low throughput. The unrolling architecture pipelines the eleven rounds and inserts registers between every two rounds. This kind of architecture achieves a high throughput, but compromises the area that is approximately 10 times larger than the rolling architecture.

For each round of the AES encryption and decryption operations, a different round key is required. In general, two methods exist to generate the round keys for the eleven rounds [22]. The first way is to pre-compute the round keys and store them into a register or memory for all the incoming plain texts in one session. However, this register/ memory based method requires a large memory or register for key storage. Another way is to generate the round keys is in an on-the-fly fashion, which allows the key expansion scheduling running concurrently with the data encryption/ decryption rounds even if the initial key is changed.

The proposed general version of the algorithm based concurrent error detection AES scheme has been implemented according to the rolling architecture and shared memory is chosen as the way to generate round keys.

The Xilinx SpartanIII XC3S400 FPGA device is used to prototype the proposed scheme. Simulation is done by Modelsim PE version. Xilinx ISE synthesizes and implements the design. Very-High-Speed Integrated Circuit Hardware Description Language (VHDL) is chosen as the description language and top-level source type in Xilinx ISE.

The ABFT AES architecture is partitioned into four different modules performing distinct functions, each of which synchronously cooperates with other modules by using linked signals. Assembling these components together, a ciphertext state and an error detection signal are obtained after each encryption process, which feed back the encrypted ciphertext and error detection result.

The program control module takes charge of the procedure and sends out time-sequential commands to the AES core module. The AES core is responsible for all the sub-operations in the rolling architecture. These sub-operations include add round key, substitute bytes, shift rows, mix columns, error detection computation and result state comparisons. This module acts as an intermediary between the program control module, S-box module, and key-ram module. The S-box module is used for S-box table lookup for SubByte operation. It is a 16*16 ROM and each element is an 8-bit data. The key-ram module stores the 44 word round keys for the 11 rounds. In this implementation, round keys are pre-generated before the encryption and are stored in the key RAM in advance. A 16-bit RAM bus is used to transfer the two bytes round keys to AES core module.

Table 1 gives the specification of the proposed scheme generated by the Xilinx ISE. The comparison is performed between the original AES encryption and the proposed scheme.

Table 1. Comparison between the original AES and the proposed scheme

Logic Utilization	Original	ABFT AES	Overhead
# of Slices	943	1254	32.9%
# of Slice Flip Flops	289	433	49.8%
# of 4 Input LUTs	1802	2376	31.9%
Clock Period (ns) (Clock Frequency MHz)	17.494 (57.162)	17.760 (56.306)	1.52%

We use Modelsim PE edition to simulate the proposed scheme. Figure 3 shows the simulated wave graph of our proposed scheme. The implementation details and input parameters are set in the following way: the clock frequency is 50MHz with no particular constraint specification. The plaintext is chosen to be "00 11 22 33 44 55 66 77 88 99 aa bb cc dd ee ff" (in hexadecimal), and the initial key is "f6 cc 34 cd c5 55 c5 41 82 54 26 02 03 ad 3e cd". In this graph, the left column lists out the input, output and intermediate parameters in our implementation. Next to this column is the detailed data of each parameter. And in the right most view, you can find the variation of every signal in the process of ABFT AES. The output signal (encrypted ciphertext) turns out to be

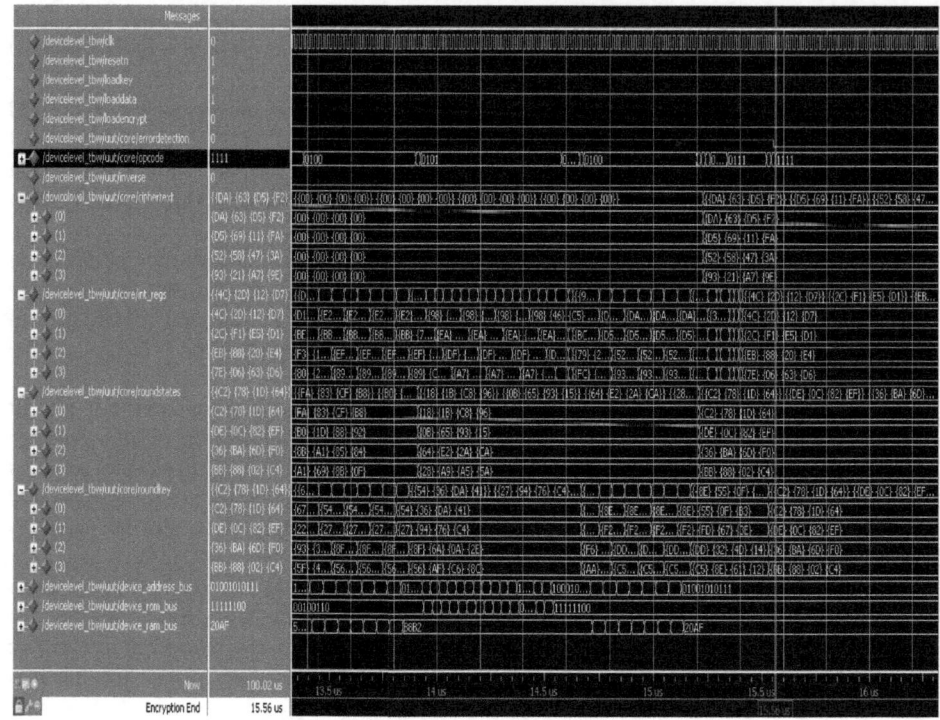

Fig. 3. Simulation result of ABFT AES scheme

what we expected, which is "da d5 52 93 63 69 58 21 d5 11 47 a7 f2 fa 3a 9e". Essentially the point we want to show in this figure is that the logic of our design has been verified on FPGA and accordingly, the waves prove the proposed scheme can achieve AES encryption with fault tolerance in a parallel way.

Since there is no united way to evaluate various implementations of AES algorithm across various platforms, and since every group employs different technology libraries, chooses different tools and even sets up different constrained parameters to test their designs, it is not quite comparable between these schemes merely because of their presented implementation results.

The proposed scheme has a reasonable hardware overhead compared to the existing schemes. Furthermore, the relative overhead will be much less if the unrolling architecture is implemented.

5   Conclusion

In this paper, a lightweight concurrent error detection scheme is proposed. This error detection scheme is based on the ABFT technique and the computational nature of the AES algorithm. Utilizing the ready-made arithmetic units in the original design, single error can be efficiently detected by the sender. In this way, useless computation and false crypto code can prevent propagation. According to the practical requirements, two versions of the scheme are presented. The general version deals with the whole AES process and the error detection procedure occurs at the end of all rounds. The run-time version performs error detection for every round. Hence, it is capable of terminating the error round immediately. Compared to other fault tolerant schemes for AES, the proposed scheme only brings in overheads of computational time spent on calculating the detection equations, as well as additional memory or register for storing intermediate results. Moreover, without doing much modification to the AES architecture, this scheme can be integrated both on software and hardware in an easy way. The rolling architecture is chosen to implement the general version on Xilinx FPGA board. The simulation result shows that our scheme has a reasonable hardware overhead compared to the existing schemes and the relative overhead will be much less if the unrolling architecture is to be implemented.

References

1. Bertoni, G., Breveglieri, L., Koren, I., Maistri, P., Piuri, V.: On the propagation of faults and their detection in a hardware implementation of the advanced encryption standard. In: 13th IEEE International Conference on Application-Specific Systems, Architectures and Processors (ASAP 2002), p. 303 (2002)
2. Bertoni, G., Breveglieri, L., Koren, I., Maistri, P., Piuri, V.: A parity code based fault detection for an implementation of the advanced encryption standard. In: 17th IEEE International Symposium on Defect and Fault Tolerance in VLSI Systems (DFT 2002), p. 51 (2002)
3. Breveglieri, L., Koren, I., Maistri, P.: Incorporating error detection and online reconfiguration into a regular architecture for the advanced encryption standard. In: 20th IEEE International Symposium on Defect and Fault Tolerance in VLSI Systems (DFT 2005), pp. 72–80 (2005)
4. Bertoni, G., Breveglieri, L., Koren, I., Maistri, P., Piuri, V.: Detecting and locating faults in VLSI implementations of the advanced encryption standard. In: 18th IEEE International Symposium on Defect and Fault Tolerance in VLSI Systems (DFT 2003), p. 105 (2003)
5. Bertoni, G., Breveglieri, L., Koren, I., Maistri, P., Piuri, V.: Error analysis and detection procedures for a hardware implementation of the advanced encryption standard. IEEE Transactions on Computers, 492–505 (April 2003)
6. Bertoni, G., Breveglieri, L., Koren, I., Maistri, P., Piuri, V.: Low cost concurrent error detection for the advanced encryption standard. IEEE Transactions on Computers, 492–505 (2003)
7. Bertoni, G., Breveglieri, L., Koren, I., Maistri, P., Piuri, V.: An efficient hardware-based fault diagnosis scheme for AES: performances and cost. In: 19th IEEE International Symposium on Defect and Fault Tolerance in VLSI Systems (DFT 2004), pp. 130–138 (2004)
8. Karpovsky, M., Kulikowski, K.J., Taubin, A.: Robust protection against fault-injection attacks on smart cards implementing the advanced encryption standard. In: International Conference on Dependable Systems and Networks (DSN 2004), p. 93 (2004)
9. Yen, C.-H., Wu, B.-F.: Simple error detection methods for hardware implementation of Advanced Encryption Standard. IEEE Transactions on Computers, 720–731 (2006)
10. Breveglieri, L., Koren, I., Maistri, P.: An operation-centered approach to fault detection in symmetric cryptography ciphers. IEEE Transactions on Computers, 635–649 (2007)

11. Karri, R., Wu, K., Mishra, P., Kim, Y.: A fault tolerant architecture for symmetric block ciphers. In: IEEE International Symposium on Defect and Fault Tolerance in VLSI Systems (DFT 2001), p. 0427 (2001)
12. Kermani, M.M., Reyhani-Masoleh, A.: Parity-based fault detection architecture of S-box for Advanced Encryption Standard. In: 21st IEEE International Symposium on Defect and Fault-Tolerance in VLSI Systems (DFT 2006), pp. 572–580 (2006)
13. Mozaffari-Kermani, M., Reyhani-Masoleh, A.: A structure-independent approach for fault detection hardware implementations of the Advanced Encryption Standard. In: Workshop on Fault Diagnosis and Tolerance in Cryptography (FDTC 2007), pp. 47–53 (2007)
14. Breveglieri, L., Koren, I., Maistri, P.: Detection faults in four symmetric key block ciphers. In: 15th IEEE International Conference on Application-Specific Systems, Architectures and Processors (ASAP 2004), pp. 258–268 (2004)
15. Maistri, P., Vanhauwaert, P., Leveugle, R.: Evaluation of register-level protection techniques for the Advanced Encryption Standard by multi-level fault injections. In: 22nd IEEE International Symposium on Defect and Fault-Tolerance in VLSI Systems (DFT 2007), pp. 499–507 (2007)
16. Valinataj, M., Safari, S.: Fault tolerant arithmetic operations with multiple error detection and correction. In: 22nd IEEE International Symposium on Defect and Fault-Tolerance in VLSI Systems (DFT 2007), pp. 188–196 (2007)
17. Karri, R., Kuznetsov, G., Goessel, M.: Concurrent error detection in block ciphers. In: International Test Conference 2003 (ITC 2003), p. 919 (2003)
18. Patel, J.H., Fung, L.Y.: Concurrent error detection in ALU's by recomputing with shifted operands. IEEE Trans. Comput. C-31, 589–595 (1982)
19. Gulati, R.K., Reddy, S.M.: Concurrent error detection in VLSI array structures. In: Proc. IEEE Internet, Conf. on Computer Design, pp. 488–491 (1986)
20. Kuhn, R.H.: Yield enchancement by fault-tolerant systolic arrays in VLSI and modern signal processing, pp. 178–184. Prentice-Hall, Englewood Cliffs (1985)
21. Qin, H., Sasao, T., Iguchi, Y.: An FPGA design of AES encryption circuit with 128-bit keys. In: Great Lakes Symposium on VLSI, Proceedings of the 15th ACM Great Lakes Symposium on VLSI, Chicago, Illinois, USA, pp. 147–151 (2005)
22. Guürkaynak, F.K., Burg, A., Felber, N., Fichtner, W., Gasser, D., Hug, F., Kaeslin, H.: A 2 Gb/s balanced AES crypto-chip implementation. In: Great Lakes Symposium on VLSI, Proceedings of the 14th ACM Great Lakes Symposium on VLSI, Boston, MA, USA, pp. 39–40 (2004)

Cryptography for Unconditionally Secure Message Transmission in Networks (Invited Talk)

Kaoru Kurosawa

Ibaraki University, Japan
kurosawa@mx.ibaraki.ac.jp

We consider the model of unconditionally secure (r-round, n-channel) message transmission schemes which was introduced by Dolev et al. [1]. In this model, there are n channels between a sender and a receiver, and an infinitely powerful adversary **A** may corrupt (observe and forge) the messages sent through t out of n channels. The sender wishes to send a secret s to the receiver in r-round without sharing any key with the receiver.

We say that a message transmission scheme is perfectly secure if it satisfies perfect privacy and perfect reliability. The perfect privacy means that the adversary **A** learns no information on s, and the perfect reliability means that the receiver can output s correctly. We say that it is *almost* secure if it satisfies perfect privacy and *almost* reliability.

In this talk, we survey some protocols and bounds for the above problem. We also describe some new results for general adversary structures [4,5]. In particular, we introduce the general error decodable secret sharing scheme and show its application to 1-round perfectly secure message transmission schemes [4].

References

1. Dolev, D., Dwork, C., Waarts, O., Yung, M.: Perfectly Secure Message Transmission. J. ACM 40(1), 17–47 (1993)
2. Kurosawa, K., Suzuki, K.: Almost Secure (1-round, n-channel) Message Transmission Scheme. In: Desmedt, Y. (ed.) ICITS 2007. LNCS, vol. 4883, pp. 99–112. Springer, Heidelberg (2009); Also appeared in Cryptology ePrint Archive: Report 2007/076 (2007)
3. Kurosawa, K., Suzuki, K.: Truly Efficient 2-Round Perfectly Secure Message Transmission Scheme. IEEE Transactions on Information Theory 55(11), 5223–5232 (2009)
4. Kurosawa, K.: General Error Decodable Secret Sharing Scheme and Its Application. Cryptology ePrint Archive: Report 2009/263 (2009)
5. Kurosawa, K.: Round-Efficient Perfectly Secure Message Transmission Scheme against General Adversary. Cryptology ePrint Archive: Report 2010/450 (2010)

S.-H. Heng, R.N. Wright, and B.-M. Goi (Eds.): CANS 2010, LNCS 6467, p. 43, 2010.
© Springer-Verlag Berlin Heidelberg 2010

Performance and Security Aspects of Client-Side SSL/TLS Processing on Mobile Devices

Johann Großschädl and Ilya Kizhvatov

University of Luxembourg,
Laboratory of Algorithmics, Cryptology and Security (LACS),
6, rue Richard Coudenhove-Kalergi, L–1359 Luxembourg, Luxembourg
{johann.groszschaedl,ilya.kizhvatov}@uni.lu

Abstract. The SSL/TLS protocol is the de-facto standard for secure Internet communications, and supported by virtually all modern e-mail clients and Web browsers. With more and more PDAs and cell phones providing wireless e-mail and Web access, there is an increasing demand for establishing secure SSL/TLS connections on devices that are relatively constrained in terms of computational resources. In addition, the cryptographic primitives executed on the client side need to be protected against side-channel analysis since, for example, an attacker may be able to monitor electromagnetic emanations from a mobile device. Using an RSA-based cipher suite has the advantage that all modular exponentiations on the client side are carried out with public exponents, which is uncritical regarding performance and side-channel leakage. However, the current migration to AES-equivalent security levels makes a good case for using an Elliptic Curve Cryptography (ECC)-based cipher suite. We show in this paper that, for high security levels, ECC-based cipher suites outperform their RSA counterparts on the client side, even though they require the integration of diverse countermeasures against side-channel attacks. Furthermore, we propose a new countermeasure to protect the symmetric encryption of messages (i.e. "bulk data") against Differential Power Analysis (DPA) attacks. This new countermeasure, which we call *Inter-Block Shuffling (IBS)*, is based on an "interleaved" encryption of a number of data blocks using a non-feedback mode of operation (such as counter mode), and randomizes the order in which the individual rounds of the individual blocks are executed. Our experimental results indicate that IBS is a viable countermeasure as it provides good DPA-protection at the expense of a slight degradation in performance.

1 Introduction

In the past, research in network security was conducted under the assumption that the endpoints of a communication channel are secure; an adversary could only attack the communication itself. A typical attack in this scenario started with eavesdropping on network traffic, followed by the modification, injection, or replay of messages with the goal to compromise the security of (parts of) the network [19]. However, with the current paradigm shift to more and more cell

S.-H. Heng, R.N. Wright, and B.-M. Goi (Eds.): CANS 2010, LNCS 6467, pp. 44–61, 2010.
© Springer-Verlag Berlin Heidelberg 2010

phones, PDAs, and other mobile or embedded devices being used to access the Internet, this adversary model must be adapted to incorporate attacks on the communication endpoints themselves too. For example, an adversary can try to obtain the secret key(s) used to encrypt the communication by analyzing side-channel information (e.g. power consumption or EM emanations) leaking from a device [22,9]. Recent research [29] shows that EM analysis is possible from a distance as far as 50 cm[1]. In the worst case, an EM attack on a mobile phone or PDA may even be conducted without the owner of the device being able to notice it [30]. Therefore, secure networking does not only require sophisticated protocols, but also a secure implementation of these protocols and the involved cryptographic algorithms [23]. In particular, the cryptographic algorithms have to be protected against all known forms of side-channel attack.

The "de-facto" standard for secure communication over an insecure, open network like the Internet is the Secure Sockets Layer (SSL) protocol [8] and its successor, the Transport Layer Security (TLS) protocol [6]. Both use a combination of public-key and secret-key cryptographic techniques to guarantee the confidentiality, integrity, and authenticity of data transfer between two parties (typically a client and a server). The SSL protocol is composed of two layers and includes a number of sub-protocols. At the lower level is the SSL Record Protocol, which specifies the format of data transmission between client and server, including encryption and integrity checking [8]. It encapsulates several higher-level protocols, one of which is the SSL Handshake Protocol. The main tasks of the handshake protocol are the negotiation of a set of cryptographic algorithms, the authentication of the server (and, optionally, of the client[2]), as well as the establishment of a *pre-master secret* via asymmetric (i.e. public-key) techniques [8]. Both the client and the server derive a master secret from this pre-master secret, which is then used by the record protocol to generate shared keys for symmetric encryption and message authentication.

1.1 Efficient and Secure Implementation of the Handshake Protocol

The SSL/TLS protocol is "algorithm-independent" (or "algorithm-agile") in the sense that it supports different algorithms for one and the same cryptographic operation, and allows the communicating parties to make a choice among them [8]. At the beginning of the handshake phase, the client and the server negotiate a *cipher suite*, which is a well-defined set of algorithms for authentication, key agreement, symmetric encryption, and integrity checking. Both SSL and TLS specify the use of RSA or DSA for authentication, and RSA or Diffie-Hellman

---

[1] Note that an attacker does not necessarily need to have direct physical access to the target device in order to monitor EM emanations. Consequently, he does not need to have the device under his possession to mount a side-channel attack; it suffices to place an EM probe in the vicinity of the device.

[2] Most Internet applications use SSL only for server-side authentication, which means that the server is authenticated to the client, but not vice versa. Client authentication is typically done at the application layer (and not the SSL layer), e.g. by entering a password and sending it to the server over a secure SSL connection.

for key establishment. In 2006, the TLS protocol was revised to support Elliptic Curve Cryptography (ECC) [2,13], and since then, cipher suites using ECDH for key exchange and ECDSA for authentication can be negotiated during the handshake phase [3]. The results from [11] and [12] clearly show that SSL/TLS servers reach significantly better performance and throughput figures when the handshakes are carried out with ECC instead of RSA. On the client side, however, the situation is not that clear since the cryptographic operations executed during the handshake seem to favor RSA cipher suites over their ECC-based counterparts. When using an RSA cipher suite, all modular exponentiations on the client side are performed with public exponents, which are typically small [8]. In the case of an ECC-based cipher suite, however, the client has to execute two scalar multiplications for ephemeral ECDH key exchange, and at least one double-scalar multiplication to validate the server's certificate[3], which is quite costly. In addition, the scalar multiplications for ECDH key exchange need to be protected against Simple Power Analysis (SPA) attacks, whereas RSA-based key transport (or, more precisely, the encryption of a random number using the public RSA key from the server's certificate) is rather uncritical with respect to side-channel leakage from the client.

It is widely presumed that, due to efficiency reasons, RSA cipher suites are better suited for SSL/TLS handshake processing on resource-restricted clients than ECC-based cipher suites. For example, Gupta et al. compared in [11] the handshake time of OpenSSL 0.9.6b using a 1024-bit RSA cipher suite versus a 163-bit ECC cipher suite, and found the former outperforming the latter by 30% when executed on a PDA operating as client. VeriSign, a major international Certification Authority (CA), prefers RSA cipher suites over their ECC-based counterparts for mobile clients since, as mentioned in [41], "very few platforms have problems with RSA." However, the ongoing migration to AES-equivalent security levels (e.g. 256-bit ECC, 3072-bit RSA) makes a good case to reassess the "ECC vs. RSA" question for client-side SSL processing. Surprisingly, the relative performance of ECC and RSA-based cipher suites on the client side has not yet been studied for security levels beyond 193 and 2048 bits, respectively (at least we are not aware of such a study). With the present paper we intend to fill this gap and demonstrate that a handshake with a cipher suite based on 256-bit ECC is roughly 30% faster than a handshake with 3072-bit RSA, while ECC wins big over RSA at higher security levels. To support these claims, we provide a detailed performance analysis of a "lightweight" SSL implementation into which we integrated a public-key crypto library optimized for client-side SSL processing on mobile devices. We also show that the protection of ECDH key exchange against side-channel attacks has almost no impact on the overall handshake time.

---

[3] Instead of sending a single certificate to the client, the server may also send a chain of two or more certificates linking the server's certificate to a trusted certification authority (CA). However, throughout this paper we assume that the certificate chain consists of just one certificate, and hence a single signature verification operation is sufficient to check the validity of the certificate.

1.2 Efficient and Secure Implementation of the Record Protocol

Besides the ECDH key exchange, also the symmetric encryption of application data (i.e. "bulk data") using a block cipher such as the AES may leak sensitive information through power or EM side channels, which can be exploited by an adversary to mount a Differential Power Analysis (DPA) attack [22]. Numerous countermeasures against DPA attacks on the AES have been proposed in the past 10 years; from a high-level point of view they can be broadly categorized into Hiding and Masking [25]. Typical examples of the hiding countermeasure to protect a software implementation of the AES include the random insertion of dummy instructions/operations/rounds and the shuffling of operations such as S-box look-ups. The goal of hiding is to randomize the power consumption by performing all leaking operations at different moments of time in each execution. Masking, on the other hand, conceals every key-dependent intermediate result with a random value, the so-called mask, in order to break the correlation between the "real" (i.e. unmasked) intermediate result and the power consumption. However, masking in software is extremely costly in terms of execution time, whereas hiding provides only a marginal protection against DPA attacks [25]. Therefore, these countermeasures are not very well suited for an SSL/TLS client since the amount of data to be encrypted can be fairly large, and hence a significant performance degradation is less acceptable than, for example, for a smart card application that encrypts just a few 128-bit blocks of data.

In order to solve this problem, we introduce *Inter-Block Shuffling (IBS)*, a new countermeasure to protect the AES (and other round-based block ciphers) against DPA attacks. IBS belongs to the category of "hiding" countermeasures and encrypts/decrypts several 128-bit blocks of data in a randomly interleaved fashion. It can be applied whenever large amounts of data are to be encrypted or decrypted, which is often the case when transmitting emails or HTML files over an SSL connection. The SSL record protocol specifies a payload of up to 2^{14} bytes, which corresponds to 1,024 blocks of 128 bits [8]. A straightforward encryption of this payload starts with the first block, then continues with the second block, and so on, until the last block has been processed. However, when using IBS, the individual rounds of the blocks are executed "interleaved" and in random order. More precisely, the encryption starts with the first round of a randomly chosen block, followed by the first round of another randomly chosen block, and so on, until the first round of each block has been performed. Then the encryption of the up to 1024 blocks continues with the second round (again the blocks are processed in random order), followed by the remaining rounds until all rounds of all 1024 blocks have been executed. Of course, IBS can only be used with a non-feedback mode of operation such as the Counter Mode or the Galois/Counter Mode [31,36]. Contrary to IBS, the shuffling countermeasures sketched in the previous paragraph randomize the sequence of operations within *one* block, hence they can be referred to as "intra-block shuffling." Our experimental results show that IBS is significantly more effective than other software countermeasures (in particular intra-block shuffling) as it achieves a high degree of DPA-resistance at the expense of a small performance degradation.

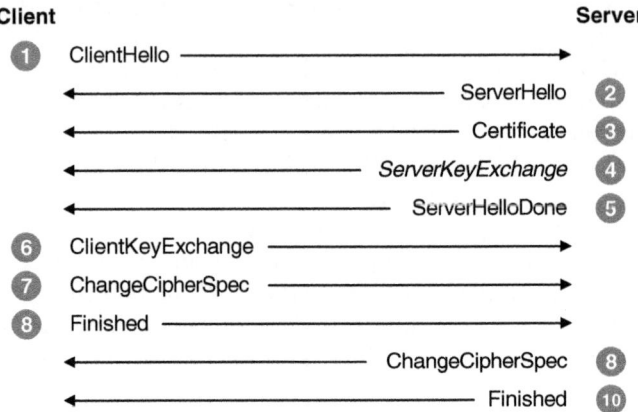

Fig. 1. SSL handshake with server authentication [37]

2   Handshake Protocol

The Secure Sockets Layer (SSL) protocol and its successor, the Transport Layer
Security (TLS) protocol, are standardized protocol suites for enabling secure
communication between a client and a server over an insecure network [6]. The
main focus in the design of these protocols lay in modularity, extensibility, and
transparency. Both SSL and TLS use a combination of asymmetric (i.e. public-
key) and symmetric (i.e. secret-key) cryptographic techniques to authenticate
the communicating parties and encrypt the data being transferred. The actual
algorithms to be used for authentication and encryption are negotiated during
the handshake phase of the protocol. SSL/TLS supports traditional public-key
cryptosystems (i.e. RSA, DSA, Diffie-Hellman) as well as ECC schemes such as
ECDSA and ECDH.

2.1   Handshake with Server Authentication

The SSL protocol contains several sub-protocols, one of which is the handshake
protocol. After agreeing upon a *cipher suite*[4], which defines the cryptographic
primitives to be used and their domain parameters, the server is authenticated
to the client and a pre-master secret is established using public-key techniques
[8]. Figure 1 shows an overview of the messages exchanged during this process
(see [37] for a detailed description). When using an RSA-based cipher suite, the
pre-master secret is established through key transport: The client generates a
random number and sends it in RSA-encrypted form to the server. On the other
hand, when using an ECC cipher suite, the pre-master secret is established via
ECDH key exchange.

---

[4] A cipher suite is a pre-defined combination of three cryptographic algorithms: A key
exchange/authentication algorithm, an encryption algorithm, and a MAC algorithm.

In the *ClientHello* message the client sends its supported cipher suites to the server, who confirms the selected suite in its own *ServerHello* message. Then, the server transmits its certificate and an optional *ServerKeyExchange* message to the client. The latter is only sent if the public key contained in the certificate is not sufficient to establish a shared pre-master secret, which is, for example, the case when the certificate is authorized for signing only and used in combination with ephemeral Diffie-Hellman (or ephemeral ECDH) for key agreement. In this scenario, the server generates another public key, signs it using the private key associated with the certificate, and embeds both the key and the signature into the *ServerKeyExchange* message. The *ServerHelloDone* message concludes the hello phase of the handshake; after sending this message, the server waits for a response from the client.

The client first checks the validity of the certificate (or chain of certificates), which requires, among other things[5], the verification of the signature(s) in the certificate. Then, the client extracts the public key from the certificate; the type and intended use of this key depends on the negotiated cipher suite [3,6]. If the server has sent a *ServerKeyExchange* message, the client verifies the signature (using the public key from the certificate) and retrieves the public key contained in this message. Now the client possesses all keys required to agree upon a Pre-Master Secret (PMS) with the server. Depending on the cipher suite, the PMS is established through either key transport or key exchange. In the former case (i.e. key transport), the client generates the PMS, which is simply a 46-byte random number, and encrypts it using the public key[6] from the certificate. The client sends the encrypted PMS to the server as part of the *ClientKeyExchange* message. On the other hand, if the PMS is established via key exchange, the client generates a Diffie-Hellman (or ECDH) key pair and includes the public part of this key pair in the *ClientKeyExchange* message. The client performs a conventional Diffie-Hellman (resp. ECDH) computation taking the secret key of the freshly generated key pair and the public key[7] of the server as input; the result is a shared key, which is used as PMS. Following the *ClientKeyExchange* message, the client sends a *ChangeCipherSpec* message to the server to indicate that subsequent messages will be encrypted. Finally, the *Finished* message is the first one protected with the just negotiated algorithms and keys; it allows the server to verify that the negotiation was successful.

Upon receipt of the *ClientKeyExchange* message, the server derives the PMS either directly through encryption of this message and extraction of the 46-byte random number (if the negotiated cipher suite is based on RSA key transport)

---

[5] In particular, the client has to check the validity period and revocation status of the certificate, and that it was issued by a trusted Certification Authority (CA).

[6] If key agreement is realized through key transport, the public key contained in the server's certificate must be an RSA key authorized for encryption. The encryption of the PMS not only ensures confidentiality, but also allows the client to verify that the server possesses the secret key corresponding to the certified public key.

[7] Depending on the cipher suite, the server's public key is contained in either the certificate (if the cipher suite is based on static Diffie-Hellman/ECDH keys) or the *ServerKeyExchange* message (in the case of ephemeral Diffie-Hellman/ECDH).

or through a Diffie-Hellman/ECDH computation using the public key embedded in this message and its own secret key as input (if the cipher suite is based on Diffie-Hellman or ECDH key exchange). The server derives a master secret from the PMS, which, in turn, is used by the record protocol to generate secret keys for bulk encryption. Successful decryption of the client's *Finished* message confirms that the server has computed the PMS correctly. Then, the server sends a *ChangeCipherSpec* and a *Finished* message to the client; the former signals the transition to encrypted communication, whereas the latter allows the client to verify the correctness of the server's PMS computation.

A handshake based on an ECC cipher suite as specified in [3] is executed in a similar way as one based on a cipher suite that uses RSA or DSA for authentication and Diffie-Hellman for key agreement. All ECC cipher suites require the server to possess an ECDSA-signed certificate, i.e. the client has to perform ECDSA instead of RSA or DSA verifications. The PMS is established through either static or ephemeral ECDH key exchange. Static ECDH implies that the certificate contains an ECDH-capable key and, hence, the server does not send a *ServerKeyExchange* message. In the other case (i.e. ephemeral ECDH), the server's certificate contains an ECDSA-capable public key, and the server sends its ephemeral ECDH key in the *ServerKeyExchange* message, which is signed under the private key corresponding the certified public key.

ECC-based cipher suites differ significantly from their RSA-based counterparts with respect to the computational load they impose on the client. If an RSA-based cipher suite has been negotiated, the client needs to carry out two modular exponentiations: one to verify the RSA signature of the certificate, and the second to encrypt the PMS. However, both exponentiations are performed with public exponents, which are usually small, e.g. $2^{16} + 1$. Unfortunately, the situation is less favorable for the client when an ECC-based cipher suite is used [11]. ECDH key exchange requires the client to execute two scalar multiplications (one to generate a key pair, the second to obtain the shared key), whereas ECDSA verification requires a double-scalar multiplication [13], all of which are relatively costly in relation to exponentiation with a small exponent.

2.2 Performance Evaluation

The implementation results from [11] show that the overall handshake time on the client side is significantly increased when using an ECC-based cipher suite instead of an RSA cipher suite, at least for low security levels (163-bit ECC and 1024-bit RSA). To our knowledge, the relative performance of ECC-based and RSA cipher suites has not yet been evaluated for higher (e.g. AES-equivalent) security levels. We intend to fill this gap by presenting a detailed performance analysis of a "lightweight" SSL/TLS implementation supporting both ECC and RSA cipher suites. In the following, we briefly summarize our implementation of the public-key primitives (i.e. RSA, ECDSA, ECDH), including the integrated countermeasures against side-channel attacks.

We used MatrixSSL [34], a lightweight SSL implementation for mobile and embedded devices, as starting point of our performance analysis. MatrixSSL is optimized towards small code size and low memory footprint, but nonetheless provides both server and client functionality. The source code is written in pure ANSI C and available under the GNU General Public License. MatrixSSL, in its original form, contains cipher suites based on traditional public-key algorithms (i.e. RSA, DSA, and Diffie-Hellman), but not the ECC cipher suites defined in [3]. Therefore, we replaced the entire public-key part of the MatrixSSL crypto library by our own library called MiniPKC. MiniPKC supports all the public-key cryptosystems needed in SSL, including ECC schemes over both prime and binary extension fields. Our priority in the design of MiniPKC lay on small code size and low memory footprint rather than on pure performance. MiniPKC is a generic public-key library in the sense that allows for arithmetic on operands of arbitrary size. It supports arbitrary curves and fields for ECC, but contains performance-optimized implementations for "standardized" domain parameters [13]. We aimed to protect MiniPKC against all possible forms of side-channel attack; in the context of client-side handshake processing, this boils down to a protection against Simple Power Analysis (SPA) [23].

Implementation Details and SPA Countermeasures. As mentioned previously, MiniPKC supports RSA, DSA, Diffie-Hellman, as well as ECDSA and ECDH on elliptic curves over arbitrary prime and binary extension fields. The modular multiplication (resp. squaring) operation is implemented on basis of the CIOS method for Montgomery reduction as described in [21]. The square-and-multiply algorithm is used for modular exponentiation if the exponent is small (e.g. a public exponent in RSA), whereas the m-ary method with $m = 2^4$ comes into operation for large exponents (i.e. four bits of the exponent are processed at a time).

The arithmetic in $\mathbb{F}_p$ is modular arithmetic, i.e. addition and multiplication modulo the prime p. MiniPKC uses the C functions implementing Montgomery multiplication and squaring not only for exponentiation (e.g. RSA), but also to perform $\mathbb{F}_p$-arithmetic such as needed for ECC. Our optimized implementation of the arithmetic for standardized fields contains dedicated modular reduction functions for generalized-Mersenne (GM) primes [13]. In addition, we unrolled the inner loops of certain arithmetic operations. MiniPKC represents points on an elliptic curve over $\mathbb{F}_p$ using the mixed Jacobian-affine coordinates described in [13, Section 3.2.2] and performs scalar multiplications via a window method with a window size of four. Double-scalar multiplications, such as carried out in ECDSA verification, are realized according to Algorithm 3.48 in [13].

MiniPKC also supports ECC on elliptic curves over $\mathbb{F}_{2^m}$. The multiplication of two binary polynomials is accomplished using the left-to-right comb method (Algorithm 2.36 in [13]) in combination with Karatsuba's technique [17]. On the other hand, the square of a binary polynomial is computed in linear time with the help of a small look-up table. MiniPKC contains a generic reduction routine for arbitrary irreducible polynomials, similar to the one in OpenSSL [33]. The scalar multiplication of a point on an elliptic curve over $\mathbb{F}_{2^m}$ is implemented on basis of the projective version of the López-Dahab algorithm [13, p. 103].

Table 1. Execution times (in msec) of cryptographic operations and SSL handshakes using RSA and ECC-based cipher suites of different cryptographic strength

Bits of security	RSA-based cipher suite				ECC-based cipher suite			
	Key size	Key establ.	Sign. verif.	Hand-shake	Key size	Key establ.	Sign. verif.	Hand-shake
80	1024	6.7	6.7	32.7	160	14.0	8.3	41.7
112	2048	26.3	26.3	72.6	224	37.6	18.1	75.2
128	3072	58.1	58.1	134.2	256	46.1	29.4	94.8
192	7680	404.7	404.7	829.4	384	155.8	99.2	275.0
256	15360	1605.5	1605.5	3235.6	512	265.3	169.9	455.3

Koschuch et al. [23] analyzed different cipher suites regarding side-channel leakage from the server and the client. On the client side, RSA cipher suites do not need to be protected against side-channel attacks, whereas the ECC-based cipher suites from [3] require countermeasures against Simple Power Analysis (SPA). MiniPKC is SPA-resistant as its implementation of the field and curve arithmetic does not contain any key- or data-dependent branches or load/store operations. Consequently, a scalar multiplication executes always exactly the same sequence of operations and instructions, irrespective of the scalar and the base point. To achieve this, it is important to avoid conditional subtractions in Montgomery multiplication and other operations such as addition in $\mathbb{F}_p$. Moreover, "irregularities" in the execution of the window method must be avoided (e.g. special consideration of zero-digits in the scalar k), which can be achieved by representing k with a digit set that does not contain zero [13].

Experimental Results. We integrated MiniPKC along with the ECC-based cipher suites from [3] into MatrixSSL version 1.7 [34], which increased the code size (i.e. the size of the binary executable) from 110 kB to approximately 150 kB. For comparison, the code size of OpenSSL [33] is more than 2 MB [23]. The memory (i.e. RAM) footprint of MatrixSSL, when operating as client, is merely 10 kB. Table 1 summarizes the execution time of key establishment, signature verification, and a full handshake for both RSA and ECC-based cipher suites of different security levels. We measured the timings on a Compaq iPAQ h3600 PDA featuring a 200 MHz StrongARM SA-1100 processor. The iPAQ PDA ran MatrixSSL with ECC support and operated as client in our experiments. It was connected to a PC via the USB port of its cradle and initiated SSL handshakes with an OpenSSL server running there.

Each row in Table 1 lists the timings of an RSA and an ECC cipher suite of comparable cryptographic strength according to the recommendations of the NIST [32] (e.g. 3072-bit RSA is comparable to 256-bit ECC). The leftmost cell of each row indicates the number of bits of security provided by the algorithms and key sizes listed in that row (e.g. 3072-bit RSA and 256-bit ECC provide 128 bits of security, which means they are comparable to e.g. AES-128). In the case of ECC, the execution time of key establishment is the time the client needs to perform two scalar multiplications. The overall handshake time was measured on the client as the time that elapsed from sending the *ClientHello* message to receiving and checking the *Finished* message from the server. In summary, the timings in Table 1 show performance advantages for RSA over ECC when the security level is low (i.e. 1024-bit RSA, 160-bit ECC), which confirms the results of Gupta et al. [11]. However, the relative performance of RSA and ECC-based cipher suites turns into the opposite when increasing the security level, mainly because the key length of ECC scales linearly with that of symmetric ciphers (e.g. AES), while RSA keys go up sub-exponentially [32]. A handshake with a 256-bit ECC cipher suite is roughly 30% faster than an RSA-based handshake of comparable cryptographic strength. On the other hand, ECC cipher suites win big over RSA at security levels corresponding to 192 and 256-bit AES; in the latter case, the ECC-based handshake outperforms its RSA counterpart by a factor of more than seven.

When using an ECC cipher suite, the cryptographic operations (i.e. ECDH key exchange, ECDSA verification) constitute between 53.5% and 95.6% of the handshake time. The ECC-timings in Table 1 were obtained with cipher suites based on NIST-approved elliptic curves over $\mathbb{F}_p$; using a binary extension field of roughly the same order as underlying algebraic structure increased the execution time of ECDH and ECDSA, as well as the full handshake time, by some 20–30%. Making the $\mathbb{F}_p$-arithmetic and the window method for scalar multiplication resistant against SPA attacks incurred a small performance degradation of about 10% compared to a "straightforward" implementation. This result is somewhat in contrast with the findings of Koschuch et al. [23], who reported a 50% jump in the handshake time due to SPA countermeasures. However, this gap is caused by different algorithms for scalar multiplication: Koschuch et al's SPA-resistant implementation employs the Montgomery ladder [13], which is significantly slower than the window method with a window size of four.

3   Record Protocol

The SSL/TLS record protocol is layered above the Transport Control Protocol (TCP) but below other SSL/TLS sub-protocols such as the handshake protocol [6,8]. It provides private (i.e. encrypted) and reliable (i.e. integrity-checked) communication between a client and a server. Among the supported algorithms for the so-called "bulk encryption" are the block ciphers AES and 3DES, and the stream cipher RC4. The integrity of the messages is protected via a keyed MAC calculated using a cryptographic hash function (e.g. MD5 or SHA-1) in combination with a secret key. All secret keys and initialization vectors needed for bulk encryption and message authentication are derived from the so-called *master secret*, which, in turn, is generated from the pre-master secret that was negotiated during the handshake phase of the protocol. The record protocol is also responsible for the fragmentation of the messages into chunks of up to 16 kB, and the encapsulation of these chunks with appropriate headers to create so-called *records*, which are forwarded to the transport protocol.

To date, the security of embedded SSL implementations against side-channel attacks was considered only for the public-key part, i.e. the handshake protocol

[23]. Besides ECDH key exchange, also the symmetric encryption of application data (i.e. the records) using a block cipher such as the AES may leak sensitive information through power or EM side channels. This leakage can be exploited by an adversary to conduct a side-channel attack, Differential Power Analysis (DPA) [22] being the most practical one. In this section we propose and analyze a new countermeasure to protect the record protocol (i.e. the secret-key portion of SSL) against power analysis and EM attacks.

3.1 Motivation for a New Countermeasure

Numerous countermeasures against DPA attacks on block ciphers have been proposed in the past 10 years. From a high-level point of view, they can be broadly categorized into hiding and masking [25]. The goal of hiding is to randomize the power consumption by performing all leaking operations at different moments of time in each execution. Masking, on the other hand, tries to conceal every key-dependent intermediate result with a random value, the so-called mask, in order to break the correlation between the "real" (i.e. unmasked) intermediate result and the power consumption. In general, masking offers good protection at the expense of very large performance (or memory) overhead, whereas hiding usually provides less protection but does so at lower implementation cost.

Effect of Timing Disarrangement. Hiding can be implemented in software in two flavors: by shuffling the order of the operations or by introducing random delays through dummy operations. Both result in timing disarrangement of the target operation in a side channel trace. A theoretical analysis in [5,24] shows that, if the moment when the target operation occurs is uniformly distributed across k time instants, the number of side-channel traces needed for a successful DPA attack grows in k^2 when the attack is performed straightforwardly, or linearly with k in case integration and windowing techniques (practically verified also in [38]) are employed. The latter figure can be used to roughly estimate the effect of timing disarrangement countermeasures.

Efficiency and Limitations of Shuffling. Shuffling is relatively attractive to implementers since it introduces almost no performance overhead. However, the disarrangement resulting from shuffling alone is often not sufficiently large. To illustrate this, let us consider a natural choice for AES: shuffling the order of the Sbox operations within a round. This introduces a disarrangement over 16 time instances, thus the DPA attack with windowing and integration will require 16 times more traces compared to the unprotected implementation. Attacking an unprotected implementation on an embedded micro-controller requires some 100 traces [25]. Therefore, several thousands of traces will be required to attack the implementation protected with the described shuffling, which is still a feasible amount. In practice, shuffling is strengthened by inserting dummy operations to increase k and is combined with masking [15,39,35]. So the problem is evident: shuffling is efficient when k is large; increasing k is, however, not possible due to the nature of the algorithms being protected. The countermeasure we propose in the next subsection allows one to overcome this limitation.

	Block 1	Block 2	Block 3	Block 4	Block 5
Round 10	10	20	30	40	50
⋮	⋮	⋮	⋮	⋮	⋮
Round 3	3	13	23	33	43
Round 2	2	12	22	32	42
Round 1	1	11	21	31	41

	Block 1	Block 2	Block 3	Block 4	Block 5
Round 10	47	50	48	49	46
⋮	⋮	⋮	⋮	⋮	⋮
Round 3	12	15	13	11	14
Round 2	7	10	6	8	9
Round 1	4	3	1	5	2

Fig. 2. Conventional encryption (top) versus encryption using IBS (bottom)

Galois/Counter Mode. Our novel countermeasure is designed for non-feed-back modes of operation of block ciphers, in particular the counter mode and Galois/Counter mode (GCM) [7], which provides authenticated encryption. The GCM has not shown any security flaws [28], is standardized by the NIST, and included in TLS cipher suites [36]. Hence, it is very likely that AES-GCM will become a mode of choice in embedded implementations of TLS. Note that the HMAC scheme with CBC mode of operation specified in TLS was found to be vulnerable to side-channel attacks [27], which is not the case for the MAC part of GCM. An efficient implementation of AES-GCM resistant to timing attacks was recently introduced by Käsper et al. [18]. Even though their implementation does not consider resistance against DPA attacks, the techniques described in [18] can be combined with our new countermeasure against DPA.

3.2 Our New Countermeasure: Inter-Block Shuffling

In this subsection we introduce the Inter-Block Shuffling (IBS) countermeasure to protect the AES (and other round-based block ciphers) against DPA. IBS is applicable in the context of TLS when the counter mode [31] or GCM is used.

Let us consider AES-GCM in TLS [36]. The TLS record protocol specifies a payload of up to 2^{14} bytes, which corresponds to 1024 blocks of 128 bits [6] forming a record. A straightforward encryption of this payload starts with the first block, continues with the second block, and so on, until the last block has been processed. This is illustrated on the top of Figure 2 for a small example in which five blocks are encrypted, whereby for each block 10 rounds are performed (i.e. 50 rounds altogether). On the other hand, when using IBS, the individual rounds of the blocks are executed "interleaved" and in a random order. More precisely, the encryption starts with the first round of a randomly chosen block (which is Block 3 in the example shown on the bottom of Figure 2), followed by the first round of another randomly chosen block (Block 5), and so on, until the first round of each block has been performed. Then, the encryption of the up to

1024 blocks continues with the second round (again the blocks are processed in random order), followed by the remaining rounds until all rounds of all blocks have been executed. Contrary to IBS, the traditional shuffling countermeasures mentioned in Subsection 3.1 randomize the sequence of operations within one block, hence they can be referred to as "intra-block shuffling."

IBS breaks the correlation between the traces corresponding to encryption rounds and the inputs/outputs of block encryptions. This thwarts side-channel attacks which require knowledge of the inputs or outputs corresponding to a specific trace, in particular DPA. The computations of the GHASH function in the MAC part of the GCM are chained and, consequently, should be computed without randomization. However, this does not introduce an opportunity for a key-recovery DPA attack. In the sequel, we show that our IBS countermeasure can be implemented efficiently and provides good security against side-channel attacks.

Implementation Efficiency. The overhead of our IBS is relatively small. In a concrete implementation using IBS, one can pick the values of the counter used for encryption from random positions following the Fisher-Yates algorithm [20] for obtaining a random shuffle. For example, a fast PRNG providing uniform pseudo-random integers would suffice [26]. As the counter is incremental [7], the counter values can be produced on-the-fly knowing the block number. One has to keep all the ciphertexts buffered for the MAC part; however, 16 kB of RAM (in the case of AES) should not be a problem on a state-of-the-art PDA or cell phone. If less memory is available, inter-block shuffling can be performed within the smaller group of blocks. So, IBS provides a trade-off between security and memory requirements.

Encrypting (resp. decrypting) a 128-bit block of data using the AES takes between 639 and 1,605 clock cycles on an iPAQ PDA featuring a StrongARM processor. These performance figures are based on the implementation reported in [1], whereby the exact cycle count depends on the size of the look-up tables being used. When applying our IBS countermeasure, the 128-bit State is written to memory after each round, and the State of a different (randomly selected) block is loaded. Assuming 128-bit keys and, hence, 10 rounds per block, a total of 9 additional load/store operations of 128-bit States have to be carried out in relation to an unprotected implementation. Since a 128-bit AES State consists of four 32-bit words, these extra load/store operations take 72 clock cycles on a StrongARM processor, assuming that they hit the data cache. Our practical results indicate that IBS increases the AES encryption time by between 4.5 and 11.3%, depending on the concrete implementation (not taking into account the time needed to generate random numbers and update the counter value). One random integer (of size 10 bits for 1024 blocks in a TLS record) per encryption round is required.

The overhead of IBS is much smaller than that of intra-block shuffling since the latter requires additional dummy blocks of instructions, several random integers per round, and combination with masking [15] to achieve a comparable security level. For example, it was shown in [38] that advanced DPA attacks on an AES implementation protected by these techniques require 500 times more traces than for an unprotected implementation, while the performance overhead is more than 100%. On the other hand, with our IBS we can achieve the same security against DPA with less than 25% (a rough estimate) overhead.

3.3   Security Analysis of AES-GCM with IBS

Starting already from a small number of blocks (i.e. 16), IBS introduces more timing disarrangement than the intra-block shuffling. For a maximum number of blocks, which is typically reached when transmitting E-mails or HTML files over a TLS connection, a DPA attack with windowing and integration requires (at least) 1000 times more traces than for an unprotected implementation. This complexity estimation is quite pessimistic since for integration one would have to acquire very long traces such that the additional cost of trace processing is significant.

It was recently demonstrated that the exact knowledge of the incremental counter values in counter mode and GCM is not necessary to mount a successful first-order DPA attack [16]. However, this attack still requires the adversary to know the correspondence between the traces and the inputs (or outputs), which is *not* the case for the IBS. Hence, IBS is immune to this attack.

An attack that could be mounted against our IBS is the recently introduced unknown plaintext template attack [14]. For this attack to work, the adversary must 1) be able to profile an implementation [25] and 2) should know the times when corresponding key bytes are processed, which are stronger assumptions than for DPA. If these are considered to be feasible, IBS should be combined with intra-block shuffling to thwart the unknown plaintext template attack.

Problem with Counter Structure. There is a problem due to the structure of the 16-byte counter block in GCM, which is defined in [7,36] as follows.

$$\underbrace{\texttt{ID}}_{\text{4 bytes}} \;||\; \underbrace{\texttt{SEQNUM}}_{\text{8 bytes}} \;||\; \underbrace{\texttt{BLOCKCTR}}_{\text{4 bytes}}$$

Here ID is the server or client identifier that is a part of the key material and kept secret. SEQNUM is the value that should be unique for each TLS record and is transmitted in clear; it may be the record sequence number. ID and SEQNUM form the 12-byte GCM nonce. BLOCKCTR is the 32-bit counter initialized to 1 at the beginning of the TLS record and incremented by 1 for each successive 128-bit block.

We can observe that the bytes of SEQNUM do not change within a single TLS record, so IBS has no effect on these bytes. At the same time, SEQNUM changes from one TLS record to the other, which enables a conventional DPA attack in the case of AES. When SEQNUM is simply a sequence counter, only few of the low-order bytes of SEQNUM will change, and the corresponding key bytes can be efficiently recovered via DPA. The key bytes corresponding to the fixed known bytes of SEQNUM can be recovered with the collision attack techniques in [4]. The same holds for the recovery of the two higher order bytes of `BLOCKCTR` that will be zero in all blocks since the number of blocks in a TLS record is $\leq 2^{10}$.

Because of the fixed parts in the counter block, 10 bytes of the AES key can be recovered despite IBS. For AES-192 and AES-256, the security margin is still sufficiently large, but for AES-128 only 48 unknown bits are left in the key. In the following, we show that a simple modification of the counter can completely thwart this attack.

Repairing the Counter. To thwart the described attack, the counter should be updated in such a way that, for different blocks within a TLS record, all of the counter bytes are different. This can be achieved by applying a bijective transformation with good diffusion properties; in other words, we want to make each bit of the output dependent on all the bits of the input, but the relations can be linear. An LFSR would perfectly suit these requirements. We suggest to generate the values of the counter as described in [7,36], but before encryption feed each counter block into a 128-bit LFSR with maximum period, clock the LFSR for 128 steps to propagate the difference, and take the resulting state as the output. The clocking of the LFSR can be implemented efficiently in a word-oriented way as in [42]. Of course, the suggested modification of the counter is problematic in practice since GCM is already a well-established standard and implemented in several products. Nonetheless, our results demonstrate that an algorithmic- or protocol-level countermeasure can be very efficient (see e.g. [10] for another example). On the other hand, protecting a given block without any modification of the usage scheme (i.e. applying masking) is very costly.

In a classical scenario, using a plain incremental counter in a counter mode of operation is considered secure. In a side-channel scenario, however, it appears that the plain counter does not allow one to implement some countermeasures like, for example, the presented IBS. Moreover, it has been shown in [40] that a plain incremental counter allows for fault attacks, and that using an LFSR to update the counter prevents these attacks. We conclude that standardizing an LFSR-based counter update for counter modes of operation is desirable because it allows for efficient prevention of implementation attacks.

4   Conclusions

In this paper, we studied the interplay between network security and applied cryptography (i.e. resistance against side-channel attacks) by taking client-side SSL/TLS processing on mobile devices as example. We conducted a detailed performance analysis of the handshake protocol for different cipher suites and found that a handshake using a 256-bit ECC cipher suite is roughly 30% faster than an RSA-based handshake of comparable cryptographic strength, whereas ECC wins big over RSA at higher security levels. This result in favor of ECC was found despite the fact that ECDH key exchange requires countermeasures against SPA attacks, which increased the handshake time of our SSL stack by just up to 10%. We also introduced IBS, a novel countermeasure to protect the bulk encryption carried out by the record protocol against DPA attacks. IBS is algorithm-independent (i.e. works with any round-based cipher), and provides reliable protection against DPA attacks at the expense of a slight performance degradation (between 4.5 and 11.3%). In summary, our results shows that good countermeasures against side-channel attacks do not need to be costly, i.e. it is possible to achieve high security without sacrificing performance.

References

1. Atasu, K., Breveglieri, L., Macchetti, M.: Efficient AES implementations for ARM based platforms. In: Proceedings of the 19th ACM Symposium on Applied Computing (SAC 2004), pp. 841–845. ACM Press, New York (2004)
2. Blake, I.F., Seroussi, G., Smart, N.P.: Elliptic Curves in Cryptography. Cambridge University Press, Cambridge (1999)
3. Blake-Wilson, S., Bolyard, N., Gupta, V., Hawk, C., Möller, B.: Elliptic Curve Cryptography (ECC) Cipher Suites for Transport Layer Security (TLS). Internet Engineering Task Force, Network Working Group, RFC 4492 (May 2006)
4. Bogdanov, A., Kizhvatov, I., Pyshkin, A.: Algebraic methods in side-channel collision attacks and practical collision detection. In: Chowdhury, D.R., Rijmen, V., Das, A. (eds.) INDOCRYPT 2008. LNCS, vol. 5365, pp. 251–265. Springer, Heidelberg (2008)
5. Clavier, C., Coron, J.-S., Dabbous, N.: Differential power analysis in the presence of hardware countermeasures. In: Koç, Ç.K., Paar, C. (eds.) CHES 2000. LNCS, vol. 1965, pp. 252–263. Springer, Heidelberg (2000)
6. Dierks, T., Rescorla, E.K.: The transport layer security (TLS) protocol version 1.2. Internet Engineering Task Force, Network Working Group, RFC 5246 (August 2008)
7. Dworkin, M.: Recommendation for block cipher modes of operation: Galois/Counter mode and GMAC. NIST Special Publication 800-38D (November 2007),
   http://csrc.nist.gov/publications/nistpubs/800-38D/SP-800-38D.pdf
8. Freier, A.O., Karlton, P., Kocher, P.C.: The SSL Protocol Version 3.0. Internet Draft (November 1996), http://wp.netscape.com/eng/ssl3/draft302.txt
9. Gebotys, C.H., Ho, S.C., Tiu, C.C.: EM analysis of Rijndael and ECC on a wireless Java-based PDA. In: Rao, J.R., Sunar, B. (eds.) CHES 2005. LNCS, vol. 3659, pp. 250–264. Springer, Heidelberg (2005)
10. Guajardo, J., Mennink, B.: Towards side-channel resistant block cipher usage or can we encrypt without side-channel countermeasures? Cryptology ePrint Archive, Report 2010/015 (2010), http://eprint.iacr.org/
11. Gupta, V., Gupta, S., Chang Shantz, S., Stebila, D.: Performance analysis of elliptic curve cryptography for SSL. In: Proceedings of the 3rd ACM Workshop on Wireless Security (WiSe 2002), pp. 87–94. ACM Press, New York (2002)
12. Gupta, V., Stebila, D., Fung, S., Chang Shantz, S., Gura, N., Eberle, H.: Speeding up secure Web transactions using elliptic curve cryptography. In: Proceedings of the 11th Annual Network and Distributed System Security Symposium (NDSS 2004), pp. 231–239. Internet Society, San Diego (2004)
13. Hankerson, D.R., Menezes, A.J., Vanstone, S.A.: Guide to Elliptic Curve Cryptography. Springer, Heidelberg (2004)

14. Hanley, N., Tunstall, M., Marnane, W.P.: Unknown plaintext template attacks. In: Youm, H.Y., Yung, M. (eds.) WISA 2009. LNCS, vol. 5932, pp. 148–162. Springer, Heidelberg (2009)
15. Herbst, C., Oswald, E., Mangard, S.: An AES smart card implementation resistant to power analysis attacks. In: Zhou, J., Yung, M., Bao, F. (eds.) ACNS 2006. LNCS, vol. 3989, pp. 239–252. Springer, Heidelberg (2006)
16. Jaffe, J.: A first-order DPA attack against AES in counter mode with unknown initial counter. In: Paillier, P., Verbauwhede, I. (eds.) CHES 2007. LNCS, vol. 4727, pp. 1–13. Springer, Heidelberg (2007)
17. Karatsuba, A.A., Ofman, Y.P.: Multiplication of multidigit numbers on automata. Soviet Physics - Doklady 7(7), 595–596 (1963)
18. Käsper, E., Schwabe, P.: Faster and timing-attack resistant AES-GCM. In: Clavier, C., Gaj, K. (eds.) CHES 2009. LNCS, vol. 5747, pp. 1–17. Springer, Heidelberg (2009)
19. Kaufman, C., Perlman, R., Speciner, M.: Network Security: Private Communication in a Public World. Prentice Hall, Englewood Cliffs (2002)
20. Knuth, D.E.: Seminumerical Algorithms, 3rd edn. The Art of Computer Programming, vol. 2. Addison-Wesley, Reading (1998)
21. Koç, Ç.K., Acar, T., Kaliski, B.S.: Analyzing and comparing Montgomery multiplication algorithms. IEEE Micro. 16(3), 26–33 (1996)
22. Kocher, P.C., Jaffe, J., Jun, B.: Differential power analysis. In: Wiener, M.J. (ed.) CRYPTO 1999. LNCS, vol. 1666, pp. 388–397. Springer, Heidelberg (1999)
23. Koschuch, M., Großschädl, J., Payer, U., Hudler, M., Krüger, M.: Workload characterization of a lightweight SSL implementation resistant to side-channel attacks. In: Franklin, M.K., Hui, L.C., Wong, D.S. (eds.) CANS 2008. LNCS, vol. 5339, pp. 349–365. Springer, Heidelberg (2008)
24. Mangard, S.: Hardware countermeasures against DPA – A statistical analysis of their effectiveness. In: Okamoto, T. (ed.) CT-RSA 2004. LNCS, vol. 2964, pp. 222–235. Springer, Heidelberg (2004)
25. Mangard, S., Oswald, E., Popp, T.: Power Analysis Attacks: Revealing the Secrets of Smart Cards. Springer, Heidelberg (2007)
26. Marsaglia, G.: Xorshift RNGs. Journal of Statistical Software 8(14), 1–6 (2003)
27. McEvoy, R., Tunstall, M., Murphy, C.C., Marnane, W.P.: Differential power analysis of HMAC based on SHA-2, and countermeasures. In: Kim, S., Yung, M., Lee, H.-W. (eds.) WISA 2007. LNCS, vol. 4867, pp. 317–332. Springer, Heidelberg (2007)
28. McGrew, D.A., Viega, J.: The security and performance of the Galois/Counter Mode (GCM) of operation. In: Canteaut, A., Viswanathan, K. (eds.) INDOCRYPT 2004. LNCS, vol. 3348, pp. 343–355. Springer, Heidelberg (2004)
29. Meynard, O., Guilley, S., Danger, J.-L., Sauvage, L.: Far correlation-based EMA with a precharacterized leakage model. In: Proceedings of the 13th Conference on Design, Automation and Test in Europe (DATE 2010), pp. 977–980. IEEE Computer Society Press, Los Alamitos (2010)
30. Mills, E.: Leaking crypto keys from mobile devices. CNET News (October 2009), http://news.cnet.com/8301-27080_3-10379115-245.html
31. Modadugu, N., Rescorla, E.K.: AES Counter Mode Cipher Suites for TLS and DTLS. Internet draft (June 2006), http://tools.ietf.org/pdf/draft-ietf-tls-ctr-01.pdf
32. National Institute of Standards and Technology (NIST). Recommendation for Key Management – Part 1: General (Revised). Special Publication 800-57 (March 2007), http://csrc.nist.gov/publications/PubsSPs.html

33. OpenSSL Project. OpenSSL 0.9.7k (September 2006), `http://www.openssl.org`
34. PeerSec Networks, Inc. MatrixSSL 1.7.1 (September 2005),
    `http://www.matrixssl.org`
35. Rivain, M., Prouff, E., Doget, J.: Higher-order masking and shuffling for software
    implementations of block ciphers. In: Clavier, C., Gaj, K. (eds.) CHES 2009. LNCS,
    vol. 5747, pp. 171–188. Springer, Heidelberg (2009)
36. Salowey, J.A., Choudhury, A.K., McGrew, D.A.: AES Galois Counter Mode (GCM)
    Cipher Suites for TLS. Internet Engineering Task Force, Network Working Group,
    RFC 5288 (August 2008)
37. Thomas, S.A.: SSL and TLS Essentials: Securing the Web. John Wiley & Sons,
    Inc., Chichester (2000)
38. Tillich, S., Herbst, C.: Attacking state-of-the-art software countermeasures – A case
    study for AES. In: Oswald, E., Rohatgi, P. (eds.) CHES 2008. LNCS, vol. 5154,
    pp. 228–243. Springer, Heidelberg (2008)
39. Tillich, S., Herbst, C., Mangard, S.: Protecting AES software implementations on
    32-bit platforms against power analysis. In: Katz, J., Yung, M. (eds.) ACNS 2007.
    LNCS, vol. 4521, pp. 141–157. Springer, Heidelberg (2007)
40. Tirtea, R., Deconinck, G.: Specifications overview for counter mode of operation.
    Security aspects in case of faults. In: Proceedings of the 12th IEEE Mediter-
    ranean Electrotechnical Conference (MELECON 2004), vol. 2, pp. 769–773. IEEE,
    Los Alamitos (2004)
41. VeriSign, Inc. Secure Wireless E-Commerce with PKI from VeriSign. White paper
    (January 2000), `https://www.verisign.com/server/rsc/wp/wap/index.html`
42. Zhang, M., Carroll, C., Chan, A.: The software-oriented stream cipher SSC2. In:
    Schneier, B. (ed.) FSE 2000. LNCS, vol. 1978, pp. 31–48. Springer, Heidelberg
    (2000)

A Practical Cryptographic Denial of Service Attack against 802.11i TKIP and CCMP

Martin Eian

Department of Telematics
Norwegian University of Science and Technology
martin.eian@item.ntnu.no

Abstract. This paper proposes a highly efficient cryptographic denial of service attack against 802.11 networks using 802.11i TKIP and CCMP. The attacker captures one frame, then modifies and transmits it twice to disrupt network access for 60 seconds. We analyze, implement and experimentally validate the attack. We also propose a robust solution and recommendations for network administrators.

1 Introduction

IEEE 802.11 is a standard for wireless local area networks[1] [1]. The 802.11i amendment to the standard specifies the robust security network (RSN) [2]. An RSN supports two security mechanisms, the temporal key integrity protocol (TKIP) and counter mode with cipher block chaining message authentication code protocol (CCMP).

TKIP was designed to be backward compatible with existing hardware, which put computational constraints on the message integrity code (MIC) algorithm. The TKIP MIC is vulnerable to attacks due to these constraints. Countermeasures were thus introduced to detect and respond to attacks. If two TKIP MIC failures are detected within 60 seconds, all security associations using TKIP are terminated and the negotiation of new security associations using TKIP is disabled for 60 seconds.

The intended long term security mechanism for 802.11 networks, CCMP, has strong confidentiality and integrity protection. CCMP does not use countermeasures to compensate for vulnerabilities. The most common default configuration for 802.11 access points (APs) using 802.11i is to support both TKIP and CCMP. This provides backward compatibility, as well as a stronger security mechanism for clients that support it.

802.11 has been extensively used during the last decade in computers, mobile phones, wireless security cameras and vehicular communication systems. 802.11 networks are thus attractive targets for adversaries that seek to disrupt communications through the use of denial of service (DoS) attacks. An attacker could use physical layer jamming to disrupt a wireless network. In a typical jamming

---

[1] In this paper, "network" is a synonym for a "basic service set" (BSS) in 802.11.

S.-H. Heng, R.N. Wright, and B.-M. Goi (Eds.): CANS 2010, LNCS 6467, pp. 62–75, 2010.

attack the attacker transmits continuously. A distributed intrusion detection system can locate the attacker by measuring the received signal strength on multiple sensors. More sophisticated and efficient attacks target the 802.11 medium access control (MAC) layer. A common MAC layer attack is the deauthentication attack from Bellardo and Savage [3]. Transmitting a deauthentication frame, which takes less than 100 microseconds, disrupts a network using 802.11i for approximately 1 second [4]. The deauthentication attack is far more efficient than physical layer jamming. In general, the higher the efficiency of the attack, the more difficult it is to locate the attacker. Vulnerabilities that can be exploited by highly efficient DoS attacks should be found and amended.

The motivation of this work is to make 802.11 more resilient to DoS attacks by finding and amending the abovementioned vulnerabilities. This paper makes five principal contributions. First, we analyze the 802.11 standard and discover a highly efficient cryptographic DoS attack. Second, we show that the attack also works against clients using CCMP as the pairwise cipher in networks that support both TKIP and CCMP. Third, we demonstrate that the attack works even if 802.11e quality of service (QoS) support is disabled in the AP. Fourth, we implement the attack and experimentally validate the analytical results. Fifth, we propose a robust solution to the vulnerability and temporary measures to limit the exposure to the vulnerability.

A more general lesson from this work is the connection between the cryptographic protocol design and network availability. Information and network security was traditionally categorized as confidentiality, integrity and availability. In the case of TKIP, availability was intentionally put at risk to improve the integrity of the protocol. Later changes to the protocol resulted in a severe DoS vulnerability, and design flaws even put clients using newer security mechanisms at risk. The use of formal methods, models and tools to analyze the confidentiality and integrity properties of cryptographic protocols is a well developed field of research, but this is not the case for availability. The development of formal methods, models and tools for the analysis of availability in cryptographic protocols might help future protocol designers construct more robust protocols.

The rest of this paper is structured as follows. Section 2 reviews related work. In Section 3, we present relevant parts of the 802.11 standard and a vulnerability analysis. Section 4 provides the attack implementation. Section 5 presents the experimental setup, and Section 6 contains the results. In Section 7, we discuss the results, propose a solution, and provide recommendations for wireless network administrators. Section 8 concludes the work.

2 Related Work

Researchers have discovered several DoS vulnerabilities in the 802.11 standard. An early paper on this topic by Bellardo and Savage demonstrated that DoS attacks were practical [3]. In the years after the publication of this paper, such attacks have become much easier to carry out due to readily available software such as the aircrack-ng tool suite [5] and the driver support for 802.11 monitor mode and frame injection.

One of the most widely implemented DoS attacks against 802.11 is the deauthentication attack [3], which disconnects a client[2] by transmitting one deauthentication frame. Figure 4 in Appendix A illustrates the attack. When 802.11i is used, eight frames must be exchanged between the access point (AP) and the client before it is reconnected. Aime et al. performed measurements of the efficiency of the deauthentication attack, and concluded that transmitting one frame per second was sufficient to completely block the wireless channel in a network using 802.11i [4].

Smith mentioned that there are a number of challenges associated with deliberately invoking the TKIP countermeasures [6, Ch. 6]. He concluded that other DoS attacks against 802.11 were likely easier to mount.

Glass and Muthukkumarasamy published experimental results of a DoS attack in 2007 [7]. The TKIP countermeasures were invoked using a man-in-the-middle technique. They showed that it is possible to mount such an attack in a laboratory environment, but difficult to consistently establish the attacker as a man-in-the-middle between the client and AP. The reason for this difficulty is that the attacker has to compete with the legitimate AP. Since 802.11 uses a wireless broadcast medium, the client will receive messages from both the attacker and the AP, and might choose to connect to the AP rather than the attacker.

Beck and Tews published the first partial key recovery attack against TKIP in 2009 [8]. One of their key observations was that the QoS mechanisms introduced in 802.11e [9] made replay attacks against TKIP possible.

Halvorsen et al. proposed that the attack from Beck and Tews could be used as a cryptographic DoS attack [10] . The attacker has to transmit 129 frames on average to cause the network to shut down for 60 seconds. This attack is less efficient than the deauthentication attack, since the attacker has to transmit more than two frames to cause one second of disruption. The authors assumed that the 802.11e QoS features had to be enabled in the AP for the attack to work.

Könings et al. published two new DoS attacks against 802.11 in 2009 [11]. The attacks exploit the channel switch and channel assessment mechanisms of 802.11h [12]. Their paper also provides an overview and classification of previous DoS attacks against 802.11. With regards to DoS attacks invoking TKIP countermeasures, they only mention the paper by Glass and Muthukkumarasamy [7].

3   Vulnerability Analysis

Some background material from 802.11 is required to analyze the TKIP DoS vulnerability. Only the most relevant parts are covered in this paper, see 802.11-2007 [1] for more details.

3.1   TKIP and 802.11e

TKIP provides confidentiality and integrity for 802.11 networks by the use of the stream cipher RC4 and the message integrity code (MIC) Michael. The input

---

[2] In this paper, "client" is a synonym for a "non-AP station" (STA) in 802.11.

values to the MIC are the plaintext data, destination address, source address and QoS priority. TKIP generates a new RC4 key for each frame, using a key mixing function. The input values to the key mixing function are the temporal key currently in use, the transmitter's address and the TKIP sequence counter (TSC), a 48-bit monotonically increasing counter. To construct a frame, the MIC is appended to the data, then an integrity check value (ICV) is computed over the data and MIC. TKIP uses a 32-bit cyclic redundancy check (CRC-32) to compute the ICV. Finally, the data, MIC and ICV are encrypted by computing a bitwise XOR with the key stream generated by RC4. Figure 5 in Appendix B illustrates the structure of a TKIP frame.

The TSC is used to prevent replay attacks. If a frame is received with a TSC value that is equal to or less than the previous value seen, then this frame is discarded. This posed a problem when the QoS mechanisms in 802.11e were introduced. With 802.11e, frames may be transmitted out of order due to different priorities. For example, a frame carrying voice traffic may be transmitted before a frame carrying data from a file transfer, even though the voice frame has a higher TSC. To avoid legitimate frames being dropped, 802.11e introduced a separate TSC for each QoS priority at the receiver. The QoS priority is an integer value stored in the QoS Control field of the medium access control (MAC) header in 802.11, as illustrated in Figure 1. 802.11e defines 8 priority classes (0-7).

The wireless multimedia (WMM) specification from the Wi-Fi Alliance, based on 802.11e, defines 4 priority classes (0-3). The QoS Control field is only present in QoS frames. To determine if a frame has a QoS Control field, the receiver inspects the frame type and subtype in the Frame Control field of the MAC header.

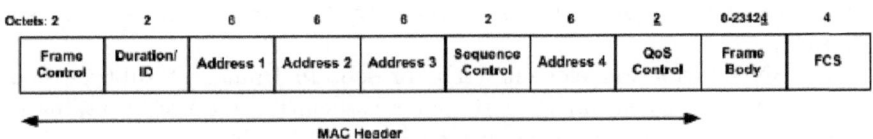

Figure 7-1—MAC frame format

Fig. 1. The 802.11 MAC frame [1]

The MIC used in TKIP is vulnerable to forgery attacks with a complexity of $O(2^{20})$ [13]. To compensate for this, TKIP uses countermeasures to detect and respond to attacks. All MIC failures at the AP and clients are recorded. If a client experiences a MIC failure, it sends an integrity protected failure report to the AP. If two or more MIC failures or failure reports are observed within 60 seconds, countermeasures are invoked. The countermeasures are to terminate all security associations using TKIP, and to refuse any new security associations using TKIP for 60 seconds. An attacker is thus limited to one MIC forgery attempt per minute.

The designers of TKIP tried to make it difficult to deliberately invoke the countermeasures. The TSC and ICV are checked before the MIC. If either fail, then the frame is discarded and the MIC is not checked. If an attacker changes the TSC, then the encryption key of the RC4 cipher is also changed, so the encrypted ICV would decrypt incorrectly. However, the changes made by 802.11e makes it possible to perform countermeasures based DoS attacks against TKIP. The QoS priority is one of the input values to the MIC, but not to the key mixing function or ICV. If the QoS priority is changed, then the MIC is invalid, but the ICV remains valid. The TSC of the new frame will be checked against the TSC for the new priority. Since a transmitter uses a single, monotonically increasing TSC counter, it is highly probable that the receiver will accept the TSC of a frame that has its QoS priority modified before it is retransmitted. To perform the DoS attack, the attacker captures traffic. When a QoS TKIP frame is observed, the attacker modifies the QoS priority and retransmits the frame twice. The receiver then invokes countermeasures, resulting in at least 60 seconds of downtime.

3.2   TKIP and CCMP

As mentioned in the introduction, the most common AP configuration is to allow both TKIP and CCMP. 802.11i specifies two temporal security associations between an AP and a client. The pairwise transient key security association (PTKSA) protects unicast traffic using a pairwise transient key (PTK). The group transient key security association (GTKSA) protects broadcast and multicast traffic from the AP to clients using a group transient key (GTK). When a client transmits a broadcast or multicast frame, it is protected by the PTK. The frame is decrypted by the AP, encrypted with the GTK, and transmitted to the wireless network. According to Section 9.2.7 of 802.11, clients should discard broadcast and multicast frames that have their own address as the source address [1].

The RSN information element (IE), present in frames of subtype beacon, probe response, association request and reassociation request, contains information about the security mechanisms supported by the transmitter. Figure 2 illustrates the RSN IE. The RSN IE supports multiple pairwise cipher suites, but only one group cipher suite. A network that supports both TKIP and CCMP has to use TKIP as the group cipher suite for all clients.

	Element ID	Length	Version	Group Cipher Suite	Pairwise Cipher Suite Count	Pairwise Cipher Suite List	AKM Suite Count	AKM Suite List	RSN Capabilities	PMKID Count	PMKI DList
Octets:	1	1	2	4	2	4-m	2	4-n	2	2	16-s

Figure 7-72—RSN information element format

Fig. 2. The RSN information element [1]

Knowing that CCMP clients use TKIP as the group cipher suite in networks that support both TKIP and CCMP, one might ask what happens if MIC failures occur on broadcast or multicast frames. Section 8.3.2.4 of 802.11 provides the answer [1]:

> The number of MIC failures is accrued independent of the particular key context. Any single MIC failure, whether detected by the Supplicant or the Authenticator and whether resulting from a group MIC key failure or a pairwise MIC key failure, shall be treated as cause for a MIC failure event.
>
> [...]
>
> If less than 60 s have passed since the most recent previous MIC failure, delete the PTKSA and GTKSA. Deauthenticate from the AP and wait for 60 s before (re)establishing a TKIP association with the same AP. A TKIP association is any IEEE 802.11 association that uses TKIP for its pairwise or group cipher suite.

To make the DoS attack work against clients using CCMP, the attacker waits for a broadcast or multicast frame from the AP. The attack is then carried out as described in Subsection 3.1. This attack works even if none of the clients use TKIP as the pairwise cipher suite. A side effect of such an attack is that the AP also invokes countermeasures due to the MIC failure reports received from the clients.

One might ask what happens to new security associations using CCMP as the pairwise cipher when the AP invokes countermeasures. Section 8.3.2.4.1 of 802.11 specifies the AP behavior as follows [1]:

> If less than 60 s have passed since the most recent previous MIC failure, the Authenticator shall deauthenticate and delete all PTKSAs for all STAs using TKIP. If the current GTKSA uses TKIP, that GTKSA shall be discarded, and a new GTKSA constructed, but not used for 60 s. The Authenticator shall refuse the construction of new PTKSAs using TKIP as one or more of the ciphers for 60 s. At the end of this period, the MIC failure counter and timer shall be reset, and creation of PTKSAs accepted as usual.

The statement "PTKSAs using TKIP as one or more of the ciphers" contradicts the definition of a PTKSA, which contains only one cipher. If the term "PTKSA" in this context is interpreted to mean "SA", then the statement is consistent with the rest of the standard. Such an interpretation implies that clients using CCMP as the pairwise cipher and TKIP as the group cipher will not be allowed to connect to the AP while countermeasures are in effect. As will be shown in Section 6, the experimental results support this interpretation.

3.3   Networks without 802.11e QoS Support

Since the attack relies on 802.11e QoS support, one might ask what happens if QoS support is disabled in the AP. 802.11 is vague on this point, but Section 6.1.1.2 provides a partial answer [1]:

At QoS STAs associated in a QoS BSS, MSDUs with a priority of Contention are considered equivalent to MSDUs with TID 0, and those with a priority of ContentionFree are delivered using the contention-free delivery if a point coordinator (PC) is present in the AP. If a PC is not present, MSDUs with a priority of ContentionFree shall be delivered using an UP of 0. At STAs associated in a non-QoS BSS, all MSDUs with an integer priority are considered equivalent to MSDUs with a priority of Contention.

The last sentence implies that clients should accept QoS frames even if associated in a network that does not support QoS. The key word in this sentence is "equivalent", which is open to interpretation. If it means that the integer priority is used as input to the TKIP MIC, then the attack works even if QoS is disabled in the AP, as long as the priority is not equal to 0. As will be shown in Section 6, the experimental results support this interpretation. To convert a regular data frame to a QoS frame, an attacker has to flip one bit in the Frame Control field and insert a two byte long QoS Control field.

3.4 Analysis Summary

Invoking the TKIP countermeasures is easy due to the modifications in 802.11e. An attacker captures a broadcast or multicast frame from the AP, then modifies the QoS priority and retransmits the frame twice. Figure 3 in Appendix A illustrates the attack. The attacker does not need to prevent the reception of the original frame at the clients. The attack is a simple retransmission of a modified frame, and does not use a man-in-the-middle technique. Since the frame is a broadcast frame, all clients except the one that transmitted the frame will invoke countermeasures. The AP will also invoke countermeasures due to the MIC failure reports from the clients. If the frame is not a QoS frame, the attacker flips one bit in the MAC header and inserts the QoS Control field. The attack also works against clients that use CCMP as the pairwise cipher. Furthermore, the attack only relies on QoS support in the clients. Disabling QoS support in the AP does not prevent the DoS attack.

4   Implementation

The aircrack-ng [5] tool suite was used as a framework for the vulnerability assessment tool implementation. The implementation depends on a wireless network interface card with driver support for 802.11 monitor mode and frame injection. Network interface cards with the Atheros AR2413 and AR5001X+ chipsets were used as the attacker in the experiments. The driver used was the Linux ath5k driver. A network interface card with the Intel 3945ABG chipset was tested, but not usable as the attacker because it replaced the source MAC address of all transmitted frames with the MAC address of the network interface card. The implementation listens for a TKIP frame using 802.11 monitor mode, then modifies and retransmits the frame to invoke the TKIP countermeasures. The source code for the modification and retransmission of frames is included in Appendix C.

Run-time configuration of several attack parameters is supported to make the vulnerability assessment tool more flexible. The default behavior is to listen for TKIP frames from the AP, and then perform the attack. The run-time options for the vulnerability assessment tool, tkipdos-ng, are included in Appendix D.

To give wireless network administrators the opportunity to test the vulnerability of their networks, tkipdos-ng will be made available as free software licensed under the GNU General Public License version 2.

5   Experimental Validation

Based on the analysis in Section 3, several vulnerability tests were constructed to validate the theoretical analysis on a wide array of different products. The hypotheses and experimental design are detailed in this section. The hypotheses to be tested were as follows:

1. Converting a non-QoS TKIP frame into a QoS TKIP frame with priority 1, 2 or 3 will cause a MIC failure in clients with QoS support even if QoS is disabled in the AP
2. Clients using CCMP as the pairwise cipher and TKIP as the group cipher will invoke countermeasures if they experience two TKIP MIC failures within a 60 second time period
3. Clients using CCMP as the pairwise cipher and TKIP as the group cipher will not connect to an AP if TKIP countermeasures in the client are currently active for that AP
4. Clients using CCMP as the pairwise cipher and TKIP as the group cipher will not be able to establish a connection to an AP with active countermeasures

To test the hypotheses, two experiments were designed. The first experiment tests hypotheses 1, 2 and 3. An AP is configured with support for both TKIP and CCMP. QoS and TKIP countermeasures are disabled in the AP. Two clients connect to the AP using CCMP as the pairwise cipher suite. The attacker listens for broadcast or multicast TKIP frames from the AP. When such a frame is observed, the attacker converts the frame to a QoS frame with priority 1, 2 or 3 and retransmits the frame twice. The attacker then observes the effect on the client that did not transmit the original broadcast frame. Since TKIP countermeasures are disabled in the AP, this experiment isolates the effect of MIC failures in the client. If hypotheses 1, 2 and 3 are true, then the client invokes countermeasures and will not be able to reconnect to the AP for at least 60 seconds.

The second experiment tests hypothesis 4. An AP is configured with support for both TKIP and CCMP. QoS and TKIP countermeasures are enabled in the AP. A client connects to the AP using TKIP as the pairwise cipher suite. The attacker listens for TKIP frames from the client to the AP. When such a frame is observed, the attacker converts the frame to a QoS frame if it is non-QoS, changes the frame priority to 1, 2 or 3 and retransmits the frame twice. The attacker then observes the effect on the AP. If the AP has invoked countermeasures, a client using CCMP as the pairwise cipher suite tries to connect to the AP. If hypothesis 4 is true, the client using CCMP should not be able to connect to the AP for at least 60 seconds after the countermeasures were invoked.

6   Results

The first experiment was performed with different clients to test whether the vulnerabilities were general or implementation specific. A Linksys WRT54GL wireless router with the OpenWrt [14] Kamikaze r19286 firmware was used as the AP. The hostapd [15] implementation in the firmware was modified so that it did not invoke TKIP countermeasures. The clients were configured to use CCMP as the pairwise cipher suite. The clients used for the experiment and the results are listed in Table 1. All of the clients that supported 802.11e QoS were vulnerable to the attack. Transmitting two modified broadcast or multicast frames from the attacker invoked countermeasures on all the clients, and caused the clients to send MIC failure notifications to the AP. The clients were unable to establish a new connection to the AP for 60 seconds.

Table 1. Equipment used in the first experiment. All 802.11e QoS supported clients were vulnerable to the DoS attack.

Hardware	Operating System	QoS Support	Vulnerable
Apple iMac 11.1	Mac OSX 10.6.2	Yes	Yes
Apple iPhone	iPhone OS	No	No
Apple iPhone 3G	iPhone OS	No	No
Asus EEE 901	Mandriva Linux 2010.0	Yes	Yes
Compaq 8510p	Windows Vista Ultimate	Yes	Yes
Compaq CQ60	Windows Vista Home Basic	Yes	Yes
Dell Latitude D620	Windows XP	Yes	Yes
Dell Latitude D630	Fedora Linux 11	Yes	Yes
Dell Latitude E4200	Windows 7	Yes	Yes
HTC Hero	Google Android	Yes	Yes
HTC S710	Windows Mobile 6	Yes	Yes
Nokia N810	Maemo Linux 4	Yes	Yes
Nokia N900	Maemo Linux 5	No	No

The second experiment was performed with different APs to test whether the vulnerabilities were general or implementation specific. The APs used for the experiment and the results are listed in Table 2. Once the AP invoked countermeasures, both clients using CCMP and clients using TKIP as the pairwise cipher suite were unable to establish a new connection for 60 seconds.

Table 2. APs used in the second experiment. All of the APs refused to establish new connections using CCMP as the pairwise cipher and TKIP as the group cipher while the attack countermeasures were active.

Hardware	Firmware	Vulnerable
Cisco 1242AG	4.1.192.35M	Yes
Linksys WRT54GL	Linksys 4.30.13	Yes
Linksys WRT54GL	OpenWrt 8.09.2 [14]	Yes
Linksys WRT54GL	Tomato 1.27 [16]	Yes
Netgear WNR1000	V1.0.1.5	Yes

7   Discussion

The test results confirmed all of the interpretations in Section 3. Networks that support both TKIP and CCMP are vulnerable to DoS attacks that invoke the TKIP countermeasures, and such attacks cause all clients to be disconnected for 60 seconds. Disabling QoS support on the AP does not prevent an attack against the clients. Furthermore, as long as at least one associated client supports QoS, the attack will cause it to send MIC failure notifications to the AP. The AP will invoke countermeasures and disconnect all clients, including those that do not support QoS.

Since this cryptographic DoS attack only affects networks using 802.11i, it could be used as a security rollback attack. When the attack is mounted, networks using weaker security mechanisms are functional, while networks using 802.11i are disrupted.

During the experiments, we observed that once clients reconnected after the 60 seconds of downtime, they transmitted several broadcast frames. The client operating systems used the Address Resolution Protocol (ARP) [17] and the Dynamic Host Configuration Protocol (DHCP) [18] to establish Internet Protocol (IP) connectivity. These protocols use broadcast messages that could be captured, modified and retransmitted as a new attack. Once the countermeasures were deactivated, a new attack immediately invoked the countermeasures again.

The vulnerability presented in this paper can be removed by modifying TKIP. If the QoS priority is used as input to the TKIP key mixing function, then a modified priority will result in a different RC4 key stream. A modified frame would then be rejected by the recipient due to a failed ICV check.

A network administrator could split a network that supports TKIP and CCMP into two logical networks to reduce the exposure to the vulnerability. One network would support TKIP only and the other would support CCMP only. This approach guarantees that attacks against TKIP do not affect clients using CCMP.

Another partial solution is to prevent broadcast and multicast traffic from the AP. By default, the Cisco Unified Wireless Network design [19] is configured so that the AP does not transmit any broadcast or multicast frames on the wireless network. As long as none of the clients use TKIP as the pairwise cipher suite, the attack does not work against such networks.

8   Conclusions

The DoS attack described in this paper is practical, easy to implement, and can be mounted using off the shelf hardware and readily available software. Transmitting two modified frames disrupts a TKIP/CCMP-based 802.11 network for 60 seconds. It is one of the most efficient known DoS attacks against 802.11. The attack works even if all clients use CCMP as the pairwise cipher and QoS support is disabled in the AP. There are several ways to mitigate or prevent the attack, and recommendations are given in Section 7.

Acknowledgments

Professor Stig F. Mjølsnes provided valuable discussions and feedback about the topics in this paper. Jing Xie provided valuable feedback for revising this paper. The following people and organization contributed their time and equipment for use in the experiments: Hans Almåsbakk, Steinar Andresen, Danilo Gligoroski, Linda Ariani Gunawan, Stig F. Mjølsnes, Pål S. Sæther, Benedikt Westermann, Jing Xie and the Wireless Trondheim project.

References

1. IEEE: IEEE Std 802.11-2007, New York, NY, USA (2007)
2. IEEE: IEEE Std 802.11i-2004, New York, NY, USA (2004)
3. Bellardo, J., Savage, S.: 802.11 denial-of-service attacks: Real vulnerabilities and practical solutions. In: Proceedings of the 12th USENIX Security Symposium. USENIX Association, Berkeley (2003)
4. Aime, M.D., Calandriello, G., Lioy, A.: Dependability in wireless networks: Can we rely on WiFi? IEEE Security and Privacy 5, 23–29 (2007)
5. Devine, C., d'Otreppe, T., Beck, M.: Aircrack-ng (2009),
   http://www.aircrack-ng.org
6. Smith, J.: Denial of service: Prevention, modelling and detection (2007)
7. Glass, S., Muthukkumarasamy, V.: A study of the TKIP cryptographic DoS attack. In: Proceedings of the 15th IEEE International Conference on Networks, ICON 2007, pp. 59–65. IEEE, New York (2007)
8. Tews, E., Beck, M.: Practical attacks against WEP and WPA. In: Proceedings of the Second ACM Conference on Wireless Network Security, WiSec 2009, pp. 79–86. ACM, New York (2009)
9. IEEE: IEEE Std 802.11e-2005, New York, NY, USA (2005)
10. Halvorsen, F.M., Haugen, O., Eian, M., Mjølsnes, S.F.: An improved attack on TKIP. In: Proceedings of the 14th Nordic Conference on Secure IT Systems, Nord-Sec 2009. LNCS, vol. 5838, pp. 120–132. Springer, Heidelberg (2009)
11. Könings, B., Schaub, F., Kargl, F., Dietzel, S.: Channel switch and quiet attack: New DoS attacks exploiting the 802.11 standard. In: Proceedings of the IEEE 34th Conference on Local Computer Networks, LCN 2009, pp. 14–21 (2009)
12. IEEE: IEEE Std 802.11h-2003, New York, NY, USA (2003)
13. Harkins, D.: Attacks against Michael and Their Countermeasures. In: IEEE 802.11 Working Group Document 03/211r0, New York, NY, USA (2003)

14. The OpenWrt Project: OpenWrt (2009), http://www.openwrt.org
15. Malinen, J.: hostapd: IEEE 802.11 AP, IEEE 802.1X / WPA / WPA2 / EAP / RADIUS Authenticator (2009), http://hostap.epitest.fi/hostapd
16. Zarate, J.: Tomato Firmware (2009), http://www.polarcloud.com/tomato
17. Plummer, D.C.: RFC 826: An Ethernet Address Resolution Protocol (1982), http://tools.ietf.org/html/rfc826
18. Droms, R.: RFC 2131: Dynamic Host Configuration Protocol (1997), http://tools.ietf.org/html/rfc2131
19. Cisco Systems Inc.: Enterprise Mobility 4.1 Design Guide, San Jose, CA, USA (2009)

A   Message Sequence Diagrams

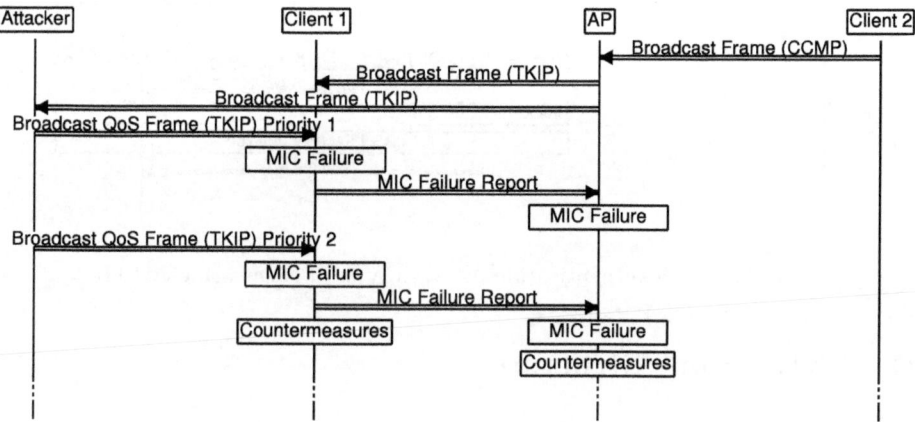

Fig. 3. The cryptographic DoS attack against clients using CCMP as the pairwise cipher suite. Two transmitted frames from the attacker invokes countermeasures in all clients except the originator of the broadcast frame. Countermeasures are also invoked in the AP due to the MIC failure reports from the clients.

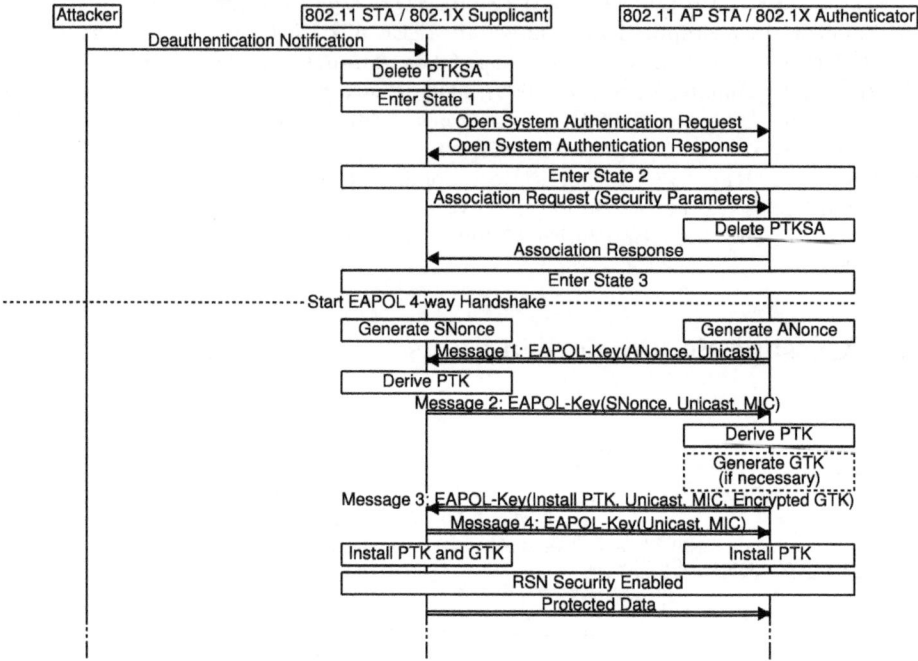

Fig. 4. The deauthentication attack against a client using 802.11i

B   TKIP Frame Structure

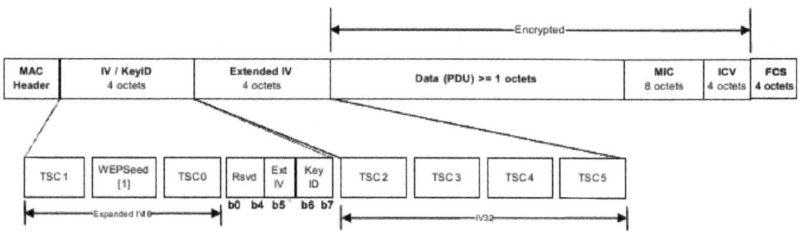

Figure 8-6—Construction of expanded TKIP MPDU

Fig. 5. An 802.11 TKIP frame [1]

C   Vulnerability Assessment Tool Source Code

```
qos = 0;
// z = MAC header length
z = ( ( h80211[1] & 3 ) != 3 ) ? 24 : 30;
if ( ( h80211[0] & 0x80 ) == 0x80 )
{
   qos = 1;
   z += 2;
}
if(qos == 0) // If the frame does not have a QoS Control
             // field then insert one
{
   // QoS data
   h80211[0] |= 0x80;
   // Move frame body 2 bytes to the right
   // to make room for QoS control field
   for(i=caplen+1; i>z+1; i--)
      h80211[i] = h80211[i-2];
   // Add 2 bytes QoS control field
   caplen += 2;
   h80211[24] = 0x00;
   h80211[25] = 0x00;
}

// QoS priority (TID)
tid = h80211[24] & 0x03;

if(tid >= opt.r_npkts)
   tid = opt.r_npkts;

for(i=0; i<=opt.r_npkts; i++)
{
   if(i != tid)
   {
      // Set QoS priority
      h80211[24] = i;
      // Send frame
      send_packet(h80211, caplen);
   }
}
```

D   Vulnerability Assessment Tool Command Line Parameters

```
usage: tkipdos-ng <options> <replay interface>
Filter options:
   -d dmac    : MAC address, Destination
   -s smac    : MAC address, Source
   -t tods    : frame control, To DS bit
   -f fromds  : frame control, From DS bit
   -D         : disable AP detection
Replay options:
   -a bssid   : set target AP MAC address
   -e essid   : set target AP SSID
   -n npkts   : number of replayed frames per frame captured [1-3]
   -m natks   : number of attacks (keep going forever if not set)

   --help     : Displays this usage screen
```

User Tracking Based on Behavioral Fingerprints

Günther Lackner[1], Peter Teufl[1], and Roman Weinberger[2]

[1] University of Technology Graz, Institute for Applied Information Processing
and Communications, Graz, Austria
`guenther.lackner@iaik.tugraz.at`, `peter.teufl@iaik.tugraz.at`
[2] Studio78.at, Graz, Austria
`roman.weinberger@studio78.at`

Abstract. The pervasiveness of wireless communications networks is advancing particularly in metropolitan areas. Broadband computer networks as IEEE 802.11 are seriously competing with cellular network technologies such as UMTS and HSDPA. Unfortunately, this increased mobility comes with privacy and security related issues. We are currently in the process of identifying possible attacks on the privacy of wireless network users, since the development of effective countermeasures is only possible with a thorough understanding of such attacks.

One serious threat we are discussing here, is the tracking of users in metropolitan networks by means of determining their physical location. Any individual user can be identified either by the devices she is using or by the behavior she is displaying. Suitable features range from single identifiers such as IP or MAC addresses to complex conglomerates of different values that provide valuable information due to their combination.

This article focuses on the extraction and analysis of features that are valuable for fingerprinting by employing *Activation Patterns*, a concept based on artificial intelligence and machine learning techniques. The concept is applied to email header data, since this allows for an effective illustration of the employed techniques. Furthermore, due to the human understandable data, we can easily evaluate the effectiveness of the concept before we start to analyze more complex data-sets.

1   Introduction

With the growing requirement of pervasive connectivity, new business models have emerged for network service providers. Cellular network based services are usually too expensive for exhaustive broadband usage. Therefore, wireless computer networks such as IEEE 802.11, or better known as *wireless fidelity* (Wi-Fi) offer higher bandwidth at a lower cost, especially if a high-speed wired backbone network is available. This is mostly the case in metropolitan areas.

More and more Internet service providers (ISP) are boarding this train and install large-scale wireless networks in metropolitan areas. But also cellular network operators are using their backbone infrastructure to install Wi-Fi access points in crowded areas such as city centers, shopping malls or airports. This

S.-H. Heng, R.N. Wright, and B.-M. Goi (Eds.): CANS 2010, LNCS 6467, pp. 76–95, 2010.
© Springer-Verlag Berlin Heidelberg 2010

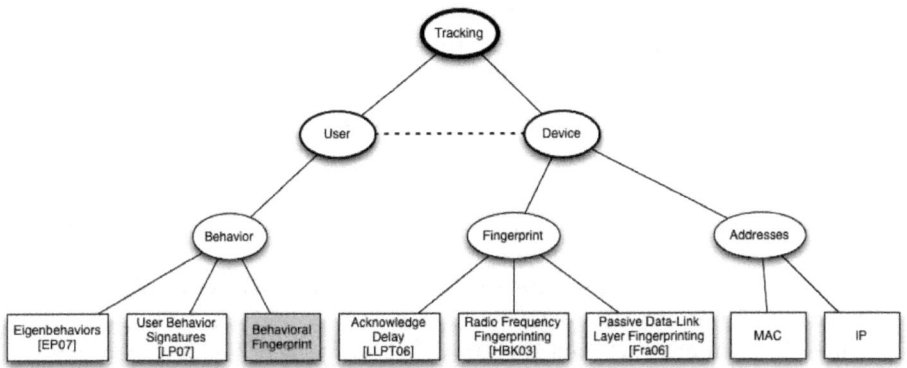

Fig. 1. Overview

trend is also promoted by the availability of low-cost mobile hardware like smart phones and netbooks, with excellent abilities to connect to Wi-Fi networks.

As many large metropolitan areas provide area-wide wireless network access, it is massively used by a growing crowd of people. However, a lot of these people are not aware of the security and privacy risks posed by wireless communications. Even if a user uses state-of-the-art encryption for her communication, a lot of personal information can be obtained from analyzing her traffic.

In this article we focus on privacy violations due to the threat of tracking the location of a user in a large-scale wireless computer network. Apart from determining the location of a user, the tracking process enables the collection of user-data on different times of days and locations. These data can then be used for further sophisticated analysis processes that disclose information about users or user groups.

The main problem of tracking in such a massive multi user setting is to identify the traffic of a particular user out of a myriad of network packets. It may seem possible to identify a user by the device she is using to connect to the network. In the best case, a device can be identified by its hardware address as shortly discussed in Section 3. But if we cannot bind a user to a single device or a device is forging its hardware address? In this case more sophisticated methods need to be applied.

Figure 1 provides an overview on how we define the term *tracking*. It also shows the connection between tracking a user by her behavior and by devices used by her. As the title of this paper indicates, we put our focus on identifying users by creating behavioral fingerprints.

2   Motivation

In order to understand and counter possible attacks on the privacy of users, we need to develop a thorough understanding in which way captured data of an user can be used to attack her privacy. Thereby, the identification/tracking of single users plays an important role, since the tracking of an user enables the collection of a large amount of data on different times of day and locations. This data can then be used for sophisticated analysis methods that violates user privacy:

- **Creation of fingerprints for user identification:** This allows the identification and thereby the tracking of users without having an unique ID. The process itself violates user privacy, since it can be used to determine the position of a user. Furthermore, it enables the capturing of more user-data than would be possible in one location.
- **Extracting knowledge about the relation between different features:** There is a wide range of features that provide data that can be captured (e.g. connection times, connections to servers, visited websites etc.). In order to understand how arbitrary features influence privacy, we need to understand how they are related to others and how the features themselves or their relations disclose information.
- **Finding groups of users with similar behavior:** Clustering users according to various combination of features gives a quick view of the analyzed data and allows to derive more general behavior models (e.g. users that go to similar restaurants, etc.).
- **Searching for users with similar behavior:** By utilizing semantic search queries we are able to execute queries that take the semantic relations between features into account.

In order to evaluate the proposed techniques, we apply theses analysis techniques to email headers captured from real traffic. Although, we are aware that plain email headers are not available in general, there are good reasons for choosing this data-set. These reasons will be discussed in Section 5.

Our work is organized as follows: Section 3 will shortly describe related work on creating user fingerprints and how to identify network devices by creating their fingerprints. Our approach, which allows the creation of behavioral fingerprints and subsequent analysis is based on *Activation Patterns*. The main idea behind this method is presented in Section 4 an a detailed explanation is given in the Appendix. Section 5 describes the analyzed data-set and the reason for choosing it. Section 6 explains in which way we can use these *Activation Patterns* to compromise user privacy. Finally, Section 7 concludes our work.

3 Related Work

The subject of privacy in wireless networks has become more and more of interest as their pervasiveness massively grows. Due to the immense scope of this topic we will focus on related work in the area of creating user fingerprints based on different approaches that we see related to our work. Further on, the concept of hardware fingerprinting and three major approaches are briefly described.

Eagle and Pentland [EP07] of MITs Media Lab developed a vector based scheme called *Eigenbehavior* that allows to quantify the behavior of a user in order to predict her next actions. This should introduce more interactivity in browsing webpages and allow networks and services to prepare contents in advance. As far as we know this approach has yet not been used for identifying users in order to attack their privacy.

Liu and Peng [LP07] from the University of North Carolina were addressing the problem of mutual trust in pervasive computer systems. They used unique identifiers as hardware addresses to identify devices. Further on they track their behavior by analyzing network event logs to find participants with hostile behavior and consequently level down trust to this nodes to prevent them from harming the rest of the network. Although the fingerprinting method is not very sophisticated, the approach of creating user profiles in order to assess trustworthiness is related to our ideas.

Pang et al. [PGG+07] followed a similar approach as we did as they tried to find characteristics in user behavior by analyzing their network traffic. The main difference is that most of their features are based on IEEE 802.11 MAC Layer properties while we focus on information extracted from higher layers. While Pang et al. need to analyze traffic captured in the wireless cell the target is actually stationed, we are able to collect all traffic at a central place in the backbone of the underlaying network.

A straight forward approach for device identification is to utilize the device addresses such as the MAC (Media Access Control) address (layer 2) or the assigned IP address (layer 3). This can easily be achieved by analyzing relevant ARP (Address Resolution Protocol) traffic [Plu82]. Unfortunately, this approach has two major drawbacks. On one hand most devices allow to modify their assigned MAC address with easy to use, free software tools and on the other hand devices may not be bound to a single user. This could be the case if hardware is used by a whole family or any other set of multiple users. The first problem might be tackled by creating fingerprints of network hardware. It allows the identification of any device by observing its external characteristics.

Radio Frequency Fingerprinting: This fingerprinting technique is based on the signal characteristics of turn-on transients of wireless transceivers. These transients are specific to each different transceiver and thus are perfectly suited as data source for fingerprint generation. Transient capturing and analysis requires a special infrastructure for signal capturing which is expensive and has to be operated by experts. Hall et al. evaluated the performance of the fingerprinting method with 30 transceivers. For each transceiver 120 signals were captured and used for the performance evaluation. The results indicate that the method is capable of achieving a very low false positive rate (0% during the evaluation) and a high detection accuracy (95% during the evaluation). However, the biggest disadvantage of this method is the special hardware needed for signal capturing which limits the broad deployment. [HBK06]

Passive Data Link Layer Fingerprinting: This paragraph describes the wireless NIC fingerprinting approach developed by Jason Franklin and his team, published in 2006 [Fra06]. Franklin identified an imprecision in the IEEE 802.11 Media Access Control specifications that has been differently interpreted by wireless NIC firmware developers. The time between sending two so called *beacon frames* used for network detection is not strictly defined. This method is able to classify different firmware versions instead of the underlying hardware. For creating a meaningful fingerprint a large number of probe-requests need to be captured. Due to the fact that a NIC willing to join a network, usually just needs a hand-full of these requests it could take a rather long time to obtain a suitable amount of data. Another significant drawback is that fingerprinting may easily be avoided by using passive-scanning or altering the device firmware. [Fra06] Some improvements to this approach have been developed by Loh et al. [DYPL08].

Acknowledge Delay Fingerprinting: Lackner et al. presented a passive fingerprinting technique in [LLPT06], which identifies WLAN chipset by analyzing the distribution of delay values between 802.11 packets and the corresponding acknowledgement (ACK) frames. Related work published by Guenther et al. [GH05] indicates that these delay values differ from chipset to chipset and thus could be used for chipset identification. The presented technique uses machine learning techniques to classify histograms which are created from delay time values extracted from passively observed WLAN traffic. The classification algorithm is based on Self Organizing Maps. This technique belongs to the area of neural networks and was developed by Kohonen [Koh01]. To increase accuracy, multiple SOMs arranged in a tree (SOM tree) are used for the classification of the different WLAN chipsets. [LLPT06] A similar approach also based on timing characteristics was developed by Sieka [Sie06].

4 Behavioral Fingerprints and Knowledge Mining

Based on the discussion in Section 3, we conclude that device fingerprints are rather unreliable or too costly to be applied in real world scenarios. Therefore, this paper concentrates on another approach – behavioral fingerprints. Taking such a fingerprint of any user means to find characteristic features that describe her behavior and thereby allows the identification and tracking of the user. Such features can be derived from a wide variety of available data, ranging from lower layer network packets to high level application related traffic. In addition the extracted features can be subject to further sophisticated analysis processes that extract information about users or user groups. Such analysis methods are also employed in other research areas that are focused on knowledge mining. In previous and ongoing work in the areas of e-participation [TPP09] , event correlation [TPF10] , malware analysis [TLP10] and semantic RDF analysis [TL10], we presented the concept of *Activation Patterns* (see Appendix B for a detailed description) that allows us to use a single model as a basis for a wide range of analysis methods. The basic idea behind this concept is to transform raw data into a new representation that models the relation between the analyzed features. Although, there are machine learning methods that could be used for particular analysis procedures, none of these methods has the flexibility of *Activation Patterns* and their wide range of applications.

5   Email Analysis

In this paper we utilize the *Activation Pattern* concept for the analysis of email data. In addition to generating behavioral fingerprints for user identification and tracking the technique enables us to extract further valuable information about the underlying dataset. The decision to utilize emails – to be precise the headers of emails – for this first evaluation is based on the following reasons:

- We need to get a better understanding of the capabilities of the employed *Activation Patterns* before we can apply them to other data. In case of emails the extracted features are easy to understand for humans and therefore the results gained by employing the *Activation Patterns* concept can easily be verified.
- The lessons learned by the application of *Activation Patterns* will be of benefit for future work that will concentrate on a wide range of features extracted from different abstraction layers.
- For emails, one could use the from address as unique ID for tracking, however this address can easily be forged. Therefore, we do not take it into consideration for the fingerprinting process.
- Although, VPNs, HTTPS and TLS POP/IMAP/SMTP connections are an effective countermeasure against extracting the analyzed features, there is still a large number of unencrypted POP3, IMAP and especially SMTP traffic that is vulnerable to this kind of analysis.
- In case of unencrypted connections, SMIME and other email encryption techniques are not an effective countermeasure, since the proposed method relies completely on information extracted from the email headers which are always transmitted in plaintext.

The remaining part of this section gives an introduction to the techniques employed by the *Activation Pattern* concept, the transformation process itself and how the whole framework is applied to the email data. For an in-detail description of the *Activation Pattern* technique, we refer to Appendix B.

5.1   Applying the Activation Patterns Concept to Email Data

The transformation of raw emails into *Activation Patterns* is based on five process layers depicted in Figure 2. After extracting and preprocessing the email features we apply the four layers L1-L4 to the raw feature data in order to determine the *Activation Patterns*. The techniques within these layers are based on various concepts related to machine learning and artificial intelligence: semantic networks for modeling relations within data [Qui68], [Fel98], [TVA07], spreading activation algorithms (SA) [Cre97] for extracting knowledge from semantic networks, and supervised/unsupervised learning algorithms to analyze data extracted from the semantic network [MS91], [QS04].

The basic idea is to create a semantic network that stores the feature values and the relations between these values (L2-L3). The *Activation Patterns* are

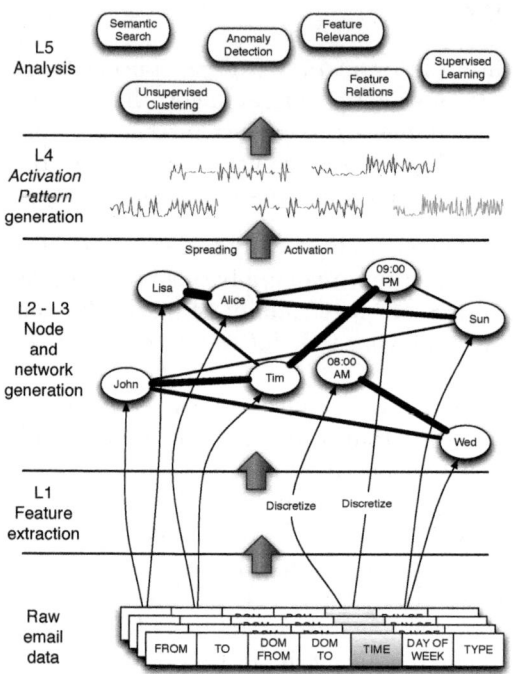

Fig. 2. Transformation of raw data into *Activation Patterns*

then generated by applying spreading activation algorithms (L4) to the semantic network. The *Activation Pattern* concept allows us to apply various analysis techniques in L5. All of these techniques can be used to gain knowledge about users, their behavior, the relations between data and to create fingerprints for user tracking/identification. All of these methods have serious implications on the privacy of users.

- *Supervised Learning:* The application of supervised learning algorithms allows the generation of behavioral fingerprints for each user. In reality, such fingerprints could be created at a Wi-Fi access point and then be used to identify/track users at other locations.
- *Semantic relations:* The analysis of the nodes and the links within the semantic network allows us to gain knowledge about the relations within the given data. For example, by specifying a given time-of-day we are able to retrieve the users that typically write emails at this time. Another example would be the retrieval of the most activate communication partners of a given user.
- *Semantic search:* Such search queries utilize the links (relations) within the semantic network to find concepts related to the search query. For example by specifying a user, we are able to find other users with a similar behavior or identify the same user without requiring a unique ID.

Table 1. Features

Feature	Type	Abbr.	Description
User From	Nominal	UF	Sender address without domain name
Domain From	Nominal	DF	Domain name of the sender address
User To	Nominal	UT	Receiver address
Domain To	Nominal	DT	Domain name of the receiver address
Time of Day	Distance Based	TD	Timestamp of the message without the date
Day of Week	Nominal	DW	Date of the message time stamp
Content Type	Nominal	CT	Content-type field of the email header

- *Unsupervised clustering:* By grouping (clustering) semantically related patterns, categories can be found that represent a certain behavior. An example would be the creation of categories that group users according to similar communication partners. The number of clusters (or model complexity) influences the grade of details covered by each category.
- *Feature relevance*: The relevance of a node representing a feature value within the network can be determined by the number of connections from this node. Nodes with high number of connections carry less information than those with fewer connections. This can be utilized to identify features that are perfectly suited for further analysis and therefore have an impact on privacy.

6 Fingerprinting and Further Analysis

In this section we apply various analysis methods to the transformed email *Activation Patterns*. For the analysis we have extracted 8 features (see Table 1 for a complete list) for each of the 1708 emails belonging to 13 users. For *Feature relations* and *Semantic Search* the UF (user from) feature was included in the analyzed patterns. For *Feature Relevance, Unsupervised Clustering* and *Supervised Learning* we excluded the UF feature in order to find out to what extend UF could be identified by the other features.

6.1 Supervised Learning for the Creation of Behavioral Fingerprints

The *Activation Patterns* of emails that belong to a given user represent a fingerprint that can be used to identify and thereby track users. Further knowledge, such as the feature relevance, extracted by the previous methods can be integrated into the *Activation Patterns* in order to create more robust fingerprints.

For the creation of fingerprints, we apply a neural network[1] to the *Activation Patterns* of 13 users. The training data is generated by randomly taking 50% of

---

[1] The Matlab™ Neural Network toolbox is utilized. Except for the validation check parameter, the standard parameters are used: Training: Scaled Conjugate Gradient, Performance: Mean Squared Error, Validation Checks: 20 instead of 6, Max Epochs: 1000, 20 hidden units, the training data is separated into 60% training data, 20% validation data and 20% independent test data.

Table 2. Performance per user: FN - false negatives, FP - false positives, TP - true positives. Since these are the mean values for the 10 iterations, the sum over the columns is not 1.

User	1	2	3	4	5	6	7	8	9	10	11	12	13
FN	0.0	0.31	0.23	0.16	0.12	0.40	0.07	0.23	0.01	0.14	0.1	0.1	0
FP	0.07	0.31	0.18	0.1	0.18	0.47	0.16	0.04	0.12	0.13	0.06	0.1	0.11
TP	1.0	0.69	0.77	0.84	0.88	0.60	0.93	0.77	0.99	0.86	0.9	0.9	1.0

the emails of each user. This results in 857 emails in the training data set and 851 in the test data set. The network is then trained and evaluated on these two data sets. In order to increase robustness, the whole process is repeated 10 times. Table 2 shows the results for each user. The mean value of correctly classified emails is calculated for the 10 iterations and yields 88.36%. We note, that user sessions, which typically contain more than one email, are not taken into account. By taking this information into account, the classification accuracy will be increased further.

6.2   Feature Relations

The links within the semantic network represent the relations between the different feature values. By activating one or more nodes within the network and applying spreading activation the related nodes receive activation energy. The strength of the received activation energy depends on the strength of the relations. In Table 3 we show several examples for analyzing the relations between different features. In all examples three nodes with the strongest activation values are extracted for each feature.

Relation 1: DT: The node representing the domain *wizzards.com* is activated and the links to other nodes are analyzed, the results reveal other users that are closely related to the given domain: *gandalf, merlin* and *saruman.*

Relation 2: TD: In this case the node representing the time 01:00 a.m. was selected. The results allow us to find users and domains that are typical for emails written at that time.

Relation 3: TD: In this example the time 10:00 a.m. was selected. When compared to the results of Relation 2, we can see that other users and domains are involved in the morning than during the night. This difference indicates that the time of day feature adds valuable information for discriminating users.

6.3   Semantic Search

Due to the relations stored in the semantic network, we are able to apply semantic search queries to the data in the following way: We activate one or more nodes within the network, spread their activation to neighboring nodes and extract the generated *Activation Pattern* from the network. For this extracted

Table 3. Relation between feature values

Relation 1	DT:wizzards.com
UF	gandalf (0.5), merlin (0.2), saruman (0.2)
DF	wizzards.com (1.0), dragons.com (0.1), hobbits.com (0.1)
DT	dwarfs.com (0.1), elfs.com (0.0), orks.com (0.0)
TD	08:46 (0.4), 16:08 (0.3), 13:12 (0.3)
DW	Mon (0.3), Wed (0.3), Tue (0.3)
CT	5 (1.0), 1 (0.4), 3 (0.3)
Relation 2	TD:60 (in minutes, meaning 01:00 a.m.)
UF	aragorn (0.2), ermurazor (0.1), fellowship (0.1)
DF	giants.com(0.3), nazgul.com (0.3), elfs.com (0.1)
UT	ermurazor (0.3), denetor (0.3), aragorn(0.3)
DT	nazgul.com (0.9), giants.com (0.6), elfs.com (0.2)
DW	Thu (0.4), Fri (0.4), Wed (0.3)
CT	5 (1.0), 1 (0.4), 3 (0.3)
Relation 3	TD:600 (in minutes, meaning 10:00 a.m.)
UF	tower (0.3), gandalf (0.2), gimli (0.2)
DF	wizzards.com (0.4), horadrim.com (0.2), dwarfs.com (0.2)
UT	tower (0.7), gandalf (0.3), mithrandir(0.1)
DT	dwarfs.com (0.6), wizzards.com (0.5), nazgul.com (0.4)
DW	Mon (0.6), Tue (0.6), Wed (0.5)
CT	1 (1.0), 5 (0.8), 3 (0.5)

Activation Pattern the distance to the other patterns can be calculated. As distance measure, the cosine similarity is used:

Query 1: UT: We activate the node for the user *lembas*, which only occurs once in the whole data set, spread the activation and compare the resulting *Activation Pattern* to the others. Obviously, the best matching result contains the user *lembas*. Since the *Activation Patterns* maintain the information about semantic relations, we are also able to retrieve other search results, that do not contain the given user, but are related to this user due to other features.

Query 2: DT: In this example we activate the node for the domain *wizzards.com* and search for emails that are semantically related. As the results show, we are also able to retrieve emails that contain other domains, but are still related to *wizzards.com* due to the involved users.

Query 3: DT and UT: In this case we activate the same nodes as in Relation 4 (UT: *gandalf* and DT: *wizzards.com*). Obviously, the first results contain the feature values specified in the search query. However, as we can see in results 648 and 650, we are also able to retrieve mails without those feature values, but that are still semantically related due to other features. For 648 this is the user *saruman* and for 650 it is the user *merlin*. We have already found out in Relation 3, that those users are strongly related to UT: *gandalf* and DT: *wizzards.com*.

Table 4. Semantic search queries

Query 1	Query for UT *lembas*						
Result	UF	DF	UT	DT	TD	DW	CT
1	gandalf	hobbits.com	lembas	hobbits.com	11:10 a.m.	Tue	3
2	gandalf	hobbits.com	frodo	hobbits.com	01:12 p.m.	Tue	3
9	frodo	hobbits.com	gandalf	hobbits.com	11:10 a.m.	Tue	6
50	frodo	hobbits.com	mithrandir	trolls.com	11:10 a.m.	Tue	1
1496	sauron	eagles.com	boromir	elfs.com	01:12 p.m.	Tue	2
Query 2	Query for DT *wizzards.com*						
Result	UF	DF	UT	DT	TD	DW	CT
1	gandalf	wizzards.com	merlin	wizzards.com	08:30 a.m.	Mon	3
35	saruman	wizzards.com	gandalf	wizzards.com	08:30 a.m.	Wed	6
210	gandalf	wizzards.com	tower	dwarfs.com	01:12 p.m.	Thu	5
321	saruman	wizzards.com	ankantoiel	gmail.com	06:42 p.m.	Tue	6
996	gollum	ents.com	tower	dwarfs.com	06:42 p.m.	Mon	3
Query 3	Query for DT *wizzards.com* and UT *gandalf*						
Result	UF	DF	UT	DT	TD	DW	CT
1	saruman	wizzards.com	gandalf	wizzards.com	08:30 a.m.	Mon	3
648	saruman	wizzards.com	durin	ringwraiths.com	01:12 p.m.	Mon	3
650	merlin	wizzards.com	faramir	urukhais.com	01:12 p.m.	Mon	3

Table 5. Examples for feature values with low relevance

CT	5 (unknown)
CT	1 (text-plain)
DT	nazgul.com
CT	3 (multipart-alternative)
DW	Tue
DW	Mon
DT	giants.com
DW	Fri

6.4   Feature Relevance

By analyzing the number and the strength of the links emanating from a given unit, we are able to filter out features that do not carry information. In this case the different content-types, certain days and domains are identified by the analysis (see Table 5). This is not surprising, since these features and values are shared by a large percentage of users and thus do not carry important information.

6.5   Unsupervised Clustering

Unsupervised clustering groups similar instances into clusters. These clusters enable the user to gain a quick overview of the whole data set. For this evaluation we apply the Neural Gas based RGNG algorithm [QS04] to the *Activation*

Table 6. Examples for clusters, the activation strength is normed and denoted within parentheses, 1.0 represents the strongest activation

Cluster 1	98 emails and 1 user
DF	nazgul.com (0.8), wizzards.com (0.1), giants.com (0.1)
UT	ermurazor (0.8), denetor (0.2), tower (0.2)
DT	nazgul.com (1.0), wizzards.com (0.2), dwarfs.com (0.2)
TD	03:33 p.m. (0.3), 10:57 a.m. (0.3), 01:03 p.m. (0.2)
DW	Wed (0.3), Thu (0.3), Tue (0.3)
CT	5 (0.9), 1 (0.4), 2 (0.2)
UF/emails	ermurazor/98
Cluster 2	101 emails and 4 users
DF	wizzards.com (0.9), dragons.com (0.1), hobbits.com (0.1)
UT	merlin (0.7), tower (0.2), gandalf (0.2)
DT	wizzards.com (1.0), dwarfs.com (0.2), nazgul.com (0.1)
TD	03:33 p.m. (0.4), 08:52 a.m. (0.3), 01:03 p.m.(0.3)
DW	Tue (0.5), Wed (0.3), Mon (0.3)
CT	6 (0.8), 5 (0.8), 1 (0.3)
UF/emails	gandalf/86, saruman/8, stormcraw/5, merlin/2

Patterns. However, it would also be possible to apply any other clustering algorithm to the patterns. In Table 6 we give two examples for the 22 clusters that were found by the RGNG algorithm. For each feature we extract the three *most activate feature values.* Cluster 1 covers the 98 emails of a single user, whereas Cluster 2 covers 4 users that have similar communication partners.

7 Conclusion

In this paper we apply the concept of *Activation Patterns* to email data in order to extract information related to the chosen set of features. The generated patterns and the information extracted by various analysis methods is then used for creating behavioral fingerprints. By employing such methods, a user's location privacy can be compromised. Although, the analyzed data is not always available, the lessons learned by applying *Activation Patterns* will be used to create more sophisticated fingerprints based on a wide range of features from various abstraction layers.

The final intention of our work is to develop privacy enhancing mechanisms for wirless networks by identifying and counteracting possible threads.

We especially want to thank P. N. Suganthan for providing the Matlab sources of RGNG [QS04].

References

[BDM04]   Berger, H., Dittenbach, M., Merkl, D.: An adaptive information retrieval system based on associative networks (2004)

88   G. Lackner, P. Teufl, and R. Weinberger

[BLC02]   Barbará, D., Li, Y., Couto, J.: Coolcat: an entropy-based algorithm for categorical clustering. In: Proceedings of the Eleventh International Conference on Information and Knowledge Management, CIKM 2002, pp. 582–589. ACM, New York (2002)

[Cou05]   Couto, J.: Kernel k-means for categorical data. In: Famili, A.F., Kok, J.N., Peña, J.M., Siebes, A., Feelders, A. (eds.) IDA 2005. LNCS, vol. 3646, pp. 46–56. Springer, Heidelberg (2005)

[Cre97]   Crestani, F.: Application of spreading activation techniques in information retrieval (1997)

[DYPL08]   Desmond, L.C.C., Yuan, C.C., Pheng, T.C., Lee, R.S.: Identifying unique devices through wireless fingerprinting. In: Proceedings of the First ACM Conference on Wireless Network Security, WiSec 2008, pp. 46–55. ACM, New York (2008)

[EP07]   Eagle, N., Pentland, A.S.: Eigenbehaviors: identifying structure in routine. Social Networks: New Perspectives (2007)

[Fel98]   Fellbaum, C.: Wordnet: An electronic lexical database (language, speech, and communication). Hardcover (May 1998)

[Fra06]   McCoy, D., Van Randwyk, J., Tabriz, P., Sicker, D., Neagoe, V., Franklin, J.: Passive data link layer 802.11 wireless device driver fingerprinting (2006)

[GH05]   Guenther, A., Hoene, C.: Measuring round trip times to determine the distance between wlan nodes. Technical report, Telecommunication Networks Group, TU-Berlin, Germans (2005)

[HBK06]   Hall, J., Barbeau, M., Kranakis, E.: Radio frequency fingerprinting for intrusion detection in wireless networks. In: IEEE Transactions On Dependable And Secure Computing (2006)

[KI96]   Kozima, H., Ito, A.: Context-sensitive measurement of word distance by adaptive scaling of a semantic space. volume cmp-lg/9601007 (1996)

[Koh95]   Kohonen, T.: Self-Organizing Maps. Springer Series in Information Sciences, vol. 30. Springer, Berlin (1995)

[Koh01]   Kohonen, T.: Self-organizing maps. Springer Series in Information Sciences, vol. 30. Springer, Heidelberg (2001)

[Koz93]   Kozima, H.: Similarity between words computed by spreading activation on an english dictionary. In: EACL, pp. 232–239 (1993)

[LLPT06]   Lackner, G., Lamberger, M., Payer, U., Teufl, P.: Wifi fingerprinting. In: DACH Mobility 2006 (September 2006)

[LP07]   Liu, Z., Peng, D.: User behavior identification for trust management in pervasive computing systems. In: Proceedings of the 11th IEEE International Workshop on Future Trends of Distributed Computing Systems, FTDCS 2007, Washington, DC, USA, pp. 65–72. IEEE Computer Society Press, Los Alamitos (2007)

[MS91]   Martinetz, T., Schulten, K.: A "neural gas" network learns topologies. In: Kohonen, T., Mäkisara, K., Simula, O., Kangas, J. (eds.) Artificial Neural Networks, pp. 397–402. Elsevier, Amsterdam (1991)

[PGG+07]   Pang, J., Greenstein, B., Gummadi, R., Seshan, S., Wetherall, D.: 802.11 user fingerprinting. In: Proceedings of the 13th Annual ACM International Conference on Mobile Computing and Networking, MobiCom 2007, pp. 99–110. ACM, New York (2007)

[Plu82]   Plummer, D.C.: Rfc 862 - an ethernet address resolution protocol or converting network protocol addresses to 48.bit ethernet address for transmission on ethernet hardware (1982)

[QS04]   Qin, A.K., Suganthan, P.N.: Robust growing neural gas algorithm with application in cluster analysis. Neural Netw. 17(8-9), 1135–1148 (2004)

[Qui68]   Quillian, M.R.: Semantic memory (1968)

[Ris89]   Rissanen, J.: Stochastic Complexity in Statistical Inquiry Theory. World Scientific Publishing Co., Inc., River Edge (1989)

[Sie06]   Sieka, B.: Active fingerprinting of 802.11 devices by timing analysis. In: 3rd IEEE Consumer Communications and Networking Conference, CCNC 2006. vol. 1, pp. 15–19 (January 2006)

[TL10]   Teufl, P., Lackner, G.: Rdf data analysis with activation patterns. In: Proceedings of the 10th International Conference on Knowledge Management and Knowledge Technologies (i-KNOW 2010), Graz, Austria (September 2010)

[TLP10]   Teufl, P., Lackner, G., Payer, U.: From NLP (natural language processing) to MLP (machine language processing). In: Kotenko, I., Skormin, V. (eds.) MMM-ACNS 2010. LNCS, vol. 6258, pp. 256–269. Springer, Heidelberg (2010)

[TPF10]   Teufl, P., Payer, U., Fellner, R.: Event correlation on the basis of activation patterns (2010)

[TPP09]   Teufl, P., Payer, U., Parycek, P.: Automated analysis of e-participation data by utilizing associative networks, spreading activation and unsupervised learning. pp. 139–150 (2009)

[TVA07]   Tsatsaronis, G., Vazirgiannis, M., Androutsopoulos, I.: Word sense disambiguation with spreading activation networks generated from thesauri (January 2007)

A   Machine Learning and Privacy

Using the Internet leaves tracks that might lead to the unwanted disclosure of information resulting in the violation of user privacy – ranging from user tracking over collecting web-usage data to modeling user behavior and putting it into relation to other users or user groups. When looking at the available information that can be logged (e.g. due to using a Wi-Fi access point) there are numerous features that might disclose information about the user's privacy. Although, there are obvious ones (e.g. the MAC address of a network interface card), there are other features or their combination that cannot be identified by simple methods[2]. However, in order to protect user privacy, we must understand how privacy information is disclosed. Only with this understanding it is feasible to deploy effective countermeasures. Here, machine learning can play an important role. Due to the fuzzy nature of machine learning algorithms, patterns within the data can be identified and analyzed without the need for a human understanding of the raw data. By combining various of these machine learning algorithms we have created the *Activation Patterns* technique that enables us to analyze data in an unsupervised way and to understand relations between single features.

This Appendix explains the *Activation Pattern* concept, gives details about the transformation process and provides a simple example that highlights the basic principles. The following definitions are used through this appendix.

---

[2] E.g. simple key word matching signatures, or searching for unique IDs within the data (e.g. MAC address) etc.

- *distance-based features:* These are features that are represented by continuous values for which it makes sense to define a distance measure (e.g temperature values, connection duration, etc.)
- *nominal features:* These are features that are represented by values that cannot be brought into relation via a distance measure (e.g. protocol identifiers such as UDP, ICMP, TCP or email addresses).
- *Activation Pattern:* This is as an n-dimensional vector that represents the activation values of the n nodes of an associative network.

A.1   Employed Machine Learning Methods

Activation Patterns are generated by utilizing three different techniques from the areas of machine learning and artificial intelligence. These building blocks include unsupervised learning algorithms, associative networks and SA algorithms. For the analysis and discretization of single features and feature groups we require unsupervised learning algorithms based on prototypes. Examples for such algorithms are Neural Gas (NG) [MS91] and its successors Growing Neural Gas, Robust Neural Gas and Robust Growing Neural Gas (RGNG) [QS04]).

Associative networks [Qui68] are directed or undirected graphs that store information in the network nodes and use edges (links) to present the relation between these nodes. Typically, these links are weighted according to a weighting scheme. Spreading activation (SA) algorithms [Cre97] can be used to extract information from associative networks. Associative networks and SA algorithms play an important role within Information Retrieval (IR) systems such as [Fel98], [Koz93], [KI96], [BDM04] and [TVA07]. By applying SA algorithms we are able to extract *Activation Patterns* from trained associative networks. These *Activation Patterns* can then be analyzed by arbitrary unsupervised learning algorithms such as Self Organizing Maps (SOM) [Koh95], Hierarchical Agglomerative Clustering (HAC), Expectation Maximization (EM), k-means, etc.

Unsupervised learning algorithms rely on some kind of distance measure to find clusters of similar data within a given dataset. It is easy to define such distance measures for datasets based on continuous features. However, as soon as categorical features need to be analyzed, these distance measures might not make sense. Typically, it is not possible to define a meaningful distance for the values of such features. Therefore, several unsupervised algorithms for the analysis of categorical data have been developed. Some examples are COOLCAT [BLC02], or Kernel K-Means [Cou05]. Typically such techniques analyze the co-occurences of attributes and use this information for unsupervised clustering. Couto [Cou05] introduces the Kernel K-Means algorithm for categorical data and gives a good overview of other unsupervised algorithms for categorical data. If continuous features need to be analyzed with such algorithms, the values need to discretized first. Such discretization methods range from very simple methods that put the categorical data into n bins to more complex methods based on entropy or fuzzy techniques.

The transformation of raw feature vectors into *Activation Patterns* involves the mapping of categorial and continuous features into an associative network.

Similar to the other methods the continuous features need to be discretized. Although there are several discretization methods available, we have decided to employ an NG based algorithm for the discretization. This comes with certain advantages that will be explained later.

As mentioned above, applying SA to the associative network generates the *Activation Patterns*. These patterns can then be analyzed by standard unsupervised learning algorithms with conventional distance measures. In addition to unsupervised analysis, the information stored in the associative network can be used directly to gain information about the relations between features. Furthermore, we are able to execute search queries that retrieve similar *Activation Patterns*. These additional benefits are not given by the other algorithms.

B   Activiation Patterns

B.1   From Feature Vectors to *Activation Patterns*

The process of generating and analyzing *Activation Patterns* is separated into five processing layers. The general idea is to extract the co-occurence information of different features (L1, L2), to store this information in an associative network (L3) and to generate *Activation Patterns* by applying SA strategies (L4). Various analysis techniques can then be applied to the generated patterns (L5).

L1 - Feature extraction: As mentioned before, features of any data set can be separated into the categories *distance-based features* and *nominal features*. These two types of features are handled differently by subsequent processing steps and need to be identified correctly at L1. For *distance-based features* groups that represent features with similar meanings and value ranges can be created. This grouping is not a requirement for further analysis, but reduces the computational complexity.

L2 - Node generation: This process layer creates the nodes of the associative network that will be generated in the next layer. The process of mapping feature values to nodes depends on the type of the particular feature. For *nominal features* the possible values are directly mapped to separate nodes. For *distance-based features* we need to apply some kind of discretization operation to map values onto nodes. Although there is a wide range of discretization algorithms available, we have chosen the RGNG algorithm. It is applied to the continuous values and the trained prototypes are used as nodes for the associative network.

Basically any prototype based unsupervised learning algorithm could be used for the discretization process. RGNG was selected, since it includes several advanced method and employs the Minimum Description Length (MDL) [Ris89] to automatically determine the model complexity. Since the performance of RGNG and similar algorithms has been evaluated by applying them to a wide range of datasets, we can assume that these algorithms will produce good results for the low dimensional data represented by single features or selected feature groups. Although the computational complexity of RGNG is high, the benefits justify its application and improve the employed analysis techniques. In other more specific scenarios, the RGNG algorithm can be replaced with a simple adequate discretization method.

The node generation process of L2 can be summarized as follows : Values of *nominal-features* are directly mapped to unique nodes within the associative network. For each of the feature groups or single features (defined in L1) of the category *distance-based features*, an RGNG-map is trained and prototypes are incorporated as new nodes into the associative network.

L3 - Network generation: In this layer, links are created between the nodes according to the relations between the nodes:

1. The features are analyzed according to the two categories determined in L1. *Nominal features* are directly mapped to nodes according to the mapping from the previous step. For *distance-based features* (single values or groups) the prototype of the corresponding RGNG-map with the smallest distance to the data vector, is located. This prototype is called the Best Matching Unit (BMU). Its corresponding node in the network is found according to the mapping generated in L2.
2. All these nodes are now linked within the associative network. Newly created links between two nodes are initialized with weight 1. The weight of existing links is increased by 1. This linking represents the co-occurence of different values of distinct features. The link weight represents the strength of this relation.

The weight of the links within the network represents the number of times two nodes co-occur. In order to apply the SA-algorithm in L4, we need to normalize the link weights within the associative network, so that the maximum weight is equal to 1. We can apply different strategies here that normalize the links locally or globally.

L4 - *Activation Pattern* Generation: The links of the associative network created in L3 represent the relations between features and values of the features are represented as the nodes of the associative network. The information about relations can be extracted by applying the SA-algorithm to the network. For each data vector, the nodes in the network that represent the values stored in the data vector, are determined. By activating these nodes for a given data vector, the activation can be spread over the network according to the links and their associated weights for a predefined number of iterations. After this spreading process, we can determine the activation value for each of the nodes in the network and present this information in a vector - the *Activation Pattern*. The areas of the associative network that are activated and the strength of the activation gives information about which feature values occurred and how they co-occur. Examples for different patterns are shown in Figure 3. By applying distance measures, such as the cosine similarity, the patterns can easily be compared.

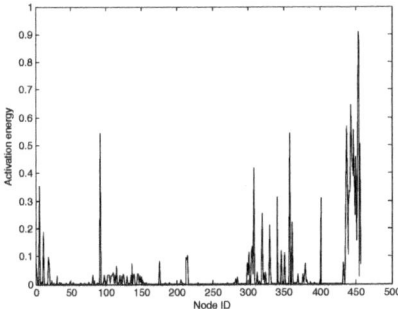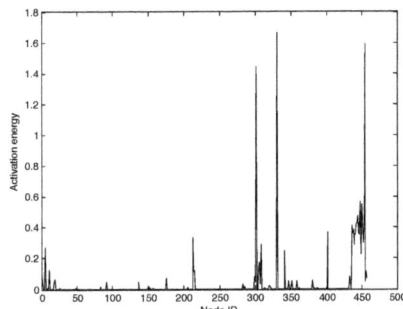

Fig. 3. Examples for two different *Activation Patterns*, the x-axis represents the nodes within the network, the y-axis represents the activation energy of these nodes after applying the spreading activation process to the activated nodes

L5 - Analysis

- *Unsupervised clustering and semantic search*: Due to the transformation of the raw data into *Activation Patterns* we can apply standard distance-based unsupervised clustering algorithms while keeping the information about semantic relations within the data. This allows us to find clusters of similar behavior patterns and to deduce common features within a cluster. By varying the model complexity, we are able to build a hierarchy from a very coarse grained categorization down to a very detailed representation of the analyzed data. The distance between the *Activation Patterns* can be used to implement semantic search algorithms that retrieve similar behavior patterns. These search queries can also be used to specify certain feature values and find closely related patterns (e.g. given a user: which other users use similar recipients within their emails).
- *Feature relevance*: The relations within the semantic network are created according to the co-occurrence of feature values within the analyzed data set. The strength of these relations are represented by the associated weights within the network. Given a feature value that is represented by a node and the number of emerging/incoming links and their weights, we are able to deduce the importance of the information carried by the node. Nodes that are connected to a large number of other nodes typically do not add information for subsequent analysis processes. This is highlighted by a simple example: Assuming a data set that describes features of various vehicles (bikes, cars, trains), a node that describes that the vehicle has wheels does not carry any information at all. The reason is that all the mentioned vehicles have wheels and thus the node is connected to all other possible feature values. In the analysis section we will show some examples of such values in the case of the analyzed email data.

    Nodes that do not carry information can be penalized by introducing so called fanout factors that attenuate the spread activation.

– *Feature relations*: The semantic network describes arbitrary relations between feature values. By activating one or more nodes (corresponding to
feature values) within the semantic network, and spreading their activation
via the links to the neighbors, we are able to extract details about the relations between various feature values (e.g. between a given time and the
typical users that write emails at this time).

B.2   A Simple Example - Artificial Clusters

This section presents a simple example based on four distinct clusters that does
not include *nominal-features* (see Figure **??**). Although this example data could
easily be analyzed with standard unsupervised analysis methods, we use it to
show the basic properties of the *Activation Patterns* and the influence of single parameters. For visualization of the clusters, Self Organizing Maps (SOMs)
[Koh95] are used.

– **L1:**   The data-set consists of 2D data-vectors representing two different
features. In this case there are only *distance-based features* meaning that the
values of these features can be related with a distance measure.
– **L2:**   In this layer we create the nodes for the associative network. Since
both features belong to the category of *distance-based features*, we cannot
map their values directly to nodes within the associative network. Instead,
we apply the RGNG algorithm to the values of the first feature (represented
by the x-axis) and to the values of the second feature (represented by the
y-axis). The RGNG employs the MDL in order to control model complexity
and finds one prototype per cluster.
– **L3:**   We now create the links of the associative network and determine
their strength by analyzing the co-occurence information of the two features
within the training data.
– **L4:**   We now make use of the SA algorithm to generate *Activation Patterns*
for each data vector within the training set.
– **L5:**   The generated *Activation Patterns* can now be analyzed with unsupervised learning techniques. In order to provide a meaningful visualization, we
train SOMs for both *Activation Pattern* sets.

Coping with different ranges of features: Due to the transformation of
the raw data into the *Activation Patterns*, we get information about the co-
occurence of different features. The values of the features are represented by
different nodes in the associative network. The information about the relations
between them is stored in the links between the nodes and their strength. Therefore, the framework does not require any kind of normalization operation applied
to the raw data. In order to show that we use the same data-set as in Figure **??**,
we multiply the value of the second feature with a constant (1000). Again, we
train SOMs on the unmodified and modified raw data-set. For the unmodified
data-set, the four different clusters are very easy to recognize within the trained
SOM (see Figure 4(a)). However, for the modified data-set the trained SOM

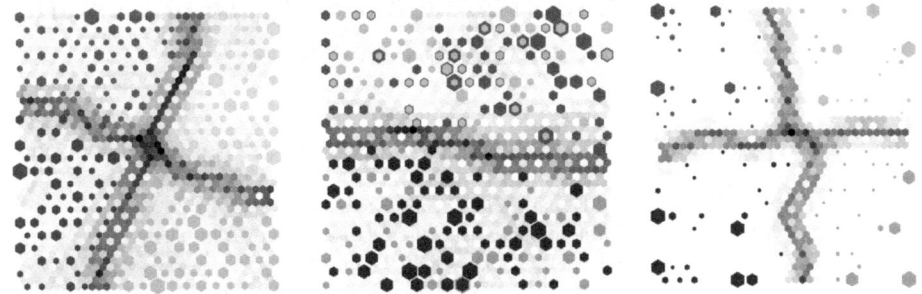

(a) SOM trained on 2D raw-data

(b) SOM trained on modi-fied 2D raw-data (Y data is *Activation Patterns* multiplied with 1000)

(c) SOM trained on 4D

Fig. 4. Due to different value ranges only 2 clusters can be identified for the raw data-set. Due to the inclusion of the co-occurrence information, the *Activation Patterns* can still be separated into 4 clusters.

only shows two distinct clusters (see Figure 4(b)). This behavior is due to the fact that the much larger values of the second feature have more influence on the Euclidean distance and hide the relative small distances between the values of the first feature. In contrast, by utilizing the *Activation Patterns*, the SOM is always able to find four clusters regardless of the range differences between the features (see Figure 4(c)).

On the Collision and Preimage Resistance of Certain Two-Call Hash Functions

Nasour Bagheri[1,3], Praveen Gauravaram[2], Majid Naderi[3], and Søren S. Thomsen[2]

[1] Electrical Engineering Department, Shahid Rajaee University, Tehran, Iran
[2] Department of Mathematics, Technical University of Denmark,
DK-2800 Kgs. Lyngby, Denmark
{P.Gauravaram,S.Thomsen}@mat.dtu.dk
[3] Electrical Engineering Department, Iran University of Science and Technology, Tehran, Iran
{N_Bagheri,M_Nderi}@iust.ac.ir

Abstract. In this paper we present concrete collision and preimage attacks on a large class of compression function constructions making two calls to the underlying ideal primitives. The complexity of the collision attack is above the theoretical lower bound for constructions of this type, but below the birthday complexity; the complexity of the preimage attack, however, is equal to the theoretical lower bound.

We also present undesirable properties of some of Stam's compression functions proposed at CRYPTO '08. We show that when one of the n-bit to n-bit components of the proposed $2n$-bit to n-bit compression function is replaced by a fixed-key cipher in the Davies-Meyer mode, the complexity of finding a preimage would be $2^{n/3}$. We also show that the complexity of finding a collision in a variant of the $3n$-bits to $2n$-bits scheme with its output truncated to $3n/2$ bits is $2^{n/2}$. The complexity of our preimage attack on this hash function is about 2^n. Finally, we present a collision attack on a variant of the proposed $m + s$-bit to s-bit scheme, truncated to $s - 1$ bits, with a complexity of $O(1)$. However, none of our results compromise Stam's security claims.

Keywords: cryptographic hash functions, information-theoretic security, permutation-based hash functions.

1 Introduction

A cryptographic hash function maps messages of arbitrary length to a fixed length digest. The design of a cryptographic hash function often involves two parts: designing a fixed input length compression function, and designing a mode of operation of this compression function. The most commonly used mode of operation is the Merkle-Damgård construction [2,5]. In this paper, we focus on the design of the compression function.

There are two general types of compression function designs; those that are based on an existing cryptographic primitive such as a block cipher, and those that are designed specifically for hashing. Of the first kind, many constructions are known, e.g., the so-called PGV schemes [9] and the MDC-2 construction [6]. Of the second kind, there are many examples, for instance MD4 [11], MD5 [12], SHA family [8] etc, although structures of many of these designs look like that of block cipher based.

S.-H. Heng, R.N. Wright, and B.-M. Goi (Eds.): CANS 2010, LNCS 6467, pp. 96–105, 2010.
© Springer-Verlag Berlin Heidelberg 2010

Some compression function designs are provably secure (e.g., collision resistant) assuming that the underlying cryptographic primitives are ideal. In the security proofs the adversary may be assumed to be computationally unbounded (or *information-theoretic*), which means that one only counts the required number of calls to the underlying primitives in the complexity estimate of the adversary. This type of security proof has received a good deal of attention in the last few years [1, 13, 15, 17, 16]. In this paper, we continue this line of research in the sense of providing cryptanalytic results on a number of compression function schemes.

1.1 Background

Black *et al.* showed [1] that in the ideal cipher model, compression functions making a single call to a permutation can never provide ideal collision resistance in the information-theoretic security model.

Rogaway and Steinberger [13] generalised the results of Black *et al.* by deriving upper bounds to collision and preimage attack complexities on compression functions using any number of underlying permutations, again assuming information-theoretic adversaries. Rogaway and Steinberger [14] continued this framework and proposed constructions that are provably secure assuming ideal underlying primitives and information-theoretic adversaries. Shrimpton and Stam [15] also proposed constructions that reach the security bounds presented by Rogaway and Steinberger [13].

Stam [16] further generalised (and commented upon) the results of Rogaway and Steinberger [13], and he proposed some constructions achieving provable collision resistance assuming ideal underlying primitives (in Stam's case, these primitives are usually modeled as random functions, not permutations).

1.2 Our Contributions

In this paper, we describe concrete collision and preimage attacks on a large class of compression function constructions making two calls to the underlying ideal primitives. The complexities of these attacks are less than that of for an ideal compression function, but in the case of the collision attack complexity of our attack is above the lower bound implied by the information-theoretic security proof.

In addition, we show that although the compression functions proposed in [16] are provably collision resistant in the information-theoretic security model, they have other undesirable properties. More precisely, we describe concrete preimage and near-collision attacks on some of these compression functions.

2   The Compression Function Model

We use Stam's model for a compression function based on a number of underlying ideal primitives [16] which is a generalization to the model presented by Rogaway and Steinberger [13]. This model, depicted in Figure 1, assumes that the compression function takes an m-bit message block and an s-bit chaining value as inputs, and produces an output of size s bits. It processes the input by making one call to each of r underlying

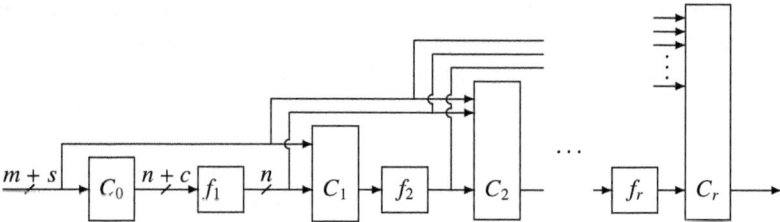

Fig. 1. The model of [16] for a compression function based on underlying primitives, which is a generalization to the model presented in [13]

primitives f_i, $1 \leq i \leq r$, that all compress $n + c$ bits down to n bits. The model also assumes the existence of $r + 1$ arbitrary functions C_i, $0 \leq i \leq r$, without any cryptographic property assumption. The first r of those functions take $(m + s + in)$ bits of input and produce $n + c$ output bits and the last function, C_r, takes $(m + s + rn)$ input bits and produces s output bits, that form the output of the compression function.

Stam [16] showed that assuming computationally unbounded (information-theoretic) adversaries, uniform compression functions following this model allow attacks of the expected complexities that can be seen in Table 1. The complexities are stated in terms of the number of queries that must be made to the underlying primitives f_i; calls to the functions C_i are considered "free of charge".

Table 1. Information-theoretic complexities of attacks on compression functions following the model of [16] (output size: s bits)

Collision	$2^{n+c-(m+s/2)/r}$
Preimage	$2^{n+c-m/r}$

For future reference, we recall that finding a collision in an ideal n-bit hash function requires about $2^{n/2}$ queries to the compression function, and finding a preimage requires about 2^n queries to the compression function. A near-collision is a collision in a pre-specified subset of the output bits. If the subset has size $k \neq 1$, then the expected complexity of the near-collision for an ideal hash function is $2^{k/2}$.

3 Attacks

Shrimpton and Stam [15], Rogaway and Steinberger [13], and Stam [16] describe attacks assuming information-theoretic adversaries. The attacks they described apparently cannot always be carried out with the same time complexity in practice. As an example, in the model of information-theoretic adversaries, the K-sum problem [18], with $K = 2^k$, has complexity $2^{n/K}$, but the best algorithm to solve this problem in practice (also [18]) has complexity $K2^{n/(1+k)}$. In many cases, there is a large gap between the theoretical complexity, and the best known "real" upper bound.

In this section we describe "real" attacks on some compression function constructions. First, we describe real collision and preimage attacks on a large class of constructions that make two calls to the underlying primitives. The collision attack does not quite close the gap to the information-theoretic attack complexity, but the preimage attack does. Then we describe preimage and near-collision attacks on some provably collision resistant constructions [16]. Recall that it is a common approach to truncate the output of the compression function at the output stage of hash functions following wide-pipe hash function design strategy [3], it is important to consider near-collision and near-preimage attacks in practice. Hence, the best attacks against a truncated compression function are expected to be generic attacks.

3.1 Attacks on a Class of Two-Call Constructions

In this section we describe attacks using a standard complexity model on a large class of two-call constructions. The constructions in question are $2n$-bit to n-bit constructions using two calls to the n-bit to n-bit underlying primitives. Hence, in the terminology introduced above, we have $m = s = n$, $c = 0$, and $r = 2$. We assume that the functions C_i, $0 \leq i \leq 2$, are linear over $\mathbb{F}_{2^n}$, but this is not strictly necessary; however, the attack does not work for arbitrary C_i (obviously, if C_i are random functions, then we really have a five-call construction). These constructions have collision resistance at most $2^{n/4}$ and preimage resistance at most $2^{n/2}$, in the model of information-theoretic adversaries (see Table1). In this section, we describe a collision attack of complexity $2^{n/3}$, and a preimage attack of complexity $2^{n/2}$, using standard complexity measures.

The constructions can be described as follows (the input is (M, V), and $|M| = |V| = n$).

$$W_1 \leftarrow C_0(M, V) = \mathbf{a}M + \mathbf{b}V$$
$$Y \leftarrow f_1(W_1)$$
$$W_2 \leftarrow C_1(M, V, Y) = \mathbf{c}M + \mathbf{d}V + \mathbf{e}Y$$
$$Z \leftarrow f_2(W_2)$$
$$W_3 \leftarrow C_2(M, V, Y, Z) = \mathbf{f}M + \mathbf{g}V + \mathbf{h}Y + \mathbf{i}Z.$$

The output is W_3. All arithmetic takes place in the field $\mathbb{F}_{2^n}$, and bold face symbols are constants in this field.

Collision attack. First, we describe a collision attack on the mentioned class of constructions. The attack is based on Wagner's generalized birthday attack [18]. This attack is able to find collisions in functions $F(x, y)$ that can be described as the sum of two subfunctions with separate inputs, i.e., $F(x, y) = F_1(x) + F_2(y)$. We show that the mentioned class of constructions can be described in this way.

Let F be the compression function based on f_1, f_2, and the linear functions C_i. We may expand the description of F somewhat to get

$$F(M, V) = \mathbf{f}M + \mathbf{g}V + \mathbf{h}f_1(W_1) + \mathbf{i}f_2(W_2), \tag{1}$$

where $W_1 = \mathbf{a}M + \mathbf{b}V$ and $W_2 = \mathbf{c}M + \mathbf{d}V + \mathbf{e}f_1(W_1)$.

Consider, on the other hand, the sum

$$\sigma = F_1(W_1) + F_2(W_2) = (\mathbf{x}W_1 + \mathbf{y}f_1(W_1)) + (\mathbf{z}W_2 + \mathbf{i}f_2(W_2)).$$

This sum expands into

$$\sigma = (\mathbf{ax}M + \mathbf{bx}V + \mathbf{y}f_1(W_1)) + (\mathbf{cz}M + \mathbf{dz}V + \mathbf{ez}f_1(W_1) + \mathbf{i}f_2(W_2)).$$

Collecting terms, one obtains

$$\sigma = (\mathbf{ax} + \mathbf{cz})M + (\mathbf{bx} + \mathbf{dz})V + (\mathbf{y} + \mathbf{ez})f_1(W_1) + \mathbf{i}f_2(W_2).$$

In order to have $\sigma = F(M, V)$, see (1), we need to solve (for $\mathbf{x}$, $\mathbf{y}$, and $\mathbf{z}$) the following three simultaneous equations:

$$\mathbf{ax} + \mathbf{cz} = \mathbf{f}$$
$$\mathbf{bx} + \mathbf{dz} = \mathbf{g}$$
$$\mathbf{y} + \mathbf{ez} = \mathbf{h},$$

which can be written in the matrix form as

$$\begin{pmatrix} \mathbf{a}\ 0\ \mathbf{c} \\ \mathbf{b}\ 0\ \mathbf{d} \\ 0\ 1\ \mathbf{e} \end{pmatrix} \cdot \begin{pmatrix} \mathbf{x} \\ \mathbf{y} \\ \mathbf{z} \end{pmatrix} = \begin{pmatrix} \mathbf{f} \\ \mathbf{g} \\ \mathbf{h} \end{pmatrix}.$$

This system of equations has a solution whenever $\mathbf{ad} \neq \mathbf{bc}$. Once $\mathbf{x}$, $\mathbf{y}$, and $\mathbf{z}$ have been found, the compression function F can be written as

$$F(M, V) = (\mathbf{x}W_1 + \mathbf{y}f_1(W_1)) + (\mathbf{z}W_2 + \mathbf{i}f_2(W_2)).$$

This means that Wagner's generalized birthday attack, with $2^{n/3}$ query/time-complexity, can be used to find the quadruple (W_1, W'_1, W_2, W'_2) such that:

$$(\mathbf{x}W_1 + \mathbf{y}f_1(W_1)) + (\mathbf{z}W_2 + \mathbf{i}f_2(W_2)) = (\mathbf{x}W'_1 + \mathbf{y}f_1(W'_1)) + (\mathbf{z}W'_2 + \mathbf{i}f_2(W'_2))$$

Now, From W_1 and W_2 one obtains M and V by solving the following equation:

$$\begin{pmatrix} \mathbf{a}\ \mathbf{b} \\ \mathbf{c}\ \mathbf{d} \end{pmatrix} \cdot \begin{pmatrix} M \\ V \end{pmatrix} = \begin{pmatrix} W_1 \\ W_2 - \mathbf{e}f_1(W_1) \end{pmatrix}.$$

Similarly, from W'_1 and W'_2 one obtains M' and V' which form a collision with M and V. As above, the system of equations has a solution if $\mathbf{ad} \neq \mathbf{bc}$.

If $\mathbf{ad} = \mathbf{bc}$, then preimages, and thereby collisions, can be found for F in constant time: The equation $\mathbf{ad} = \mathbf{bc}$ implies that the vector $(\mathbf{c}, \mathbf{d})$ is a multiple of $(\mathbf{a}, \mathbf{b})$, or $(\mathbf{a}, \mathbf{b}) = (0, 0)$. If $(\mathbf{a}, \mathbf{b}) = (0, 0)$, then the construction is in effect reduced to a single-call construction. Therefore, we assume that $(\mathbf{c}, \mathbf{d}) = \mathbf{k} \cdot (\mathbf{a}, \mathbf{b})$. Then, it can bee seen that $W_2 = \mathbf{k}W_1 + \mathbf{e}f_1(W_1)$ is a function of W_1 only. Given target image U, one finds a preimage simply by choosing arbitrary W_1 and solving the system of equations

$$\begin{pmatrix} \mathbf{a}\ \mathbf{b} \\ \mathbf{f}\ \mathbf{g} \end{pmatrix} \cdot \begin{pmatrix} M \\ V \end{pmatrix} = \begin{pmatrix} W_1 \\ U - \mathbf{h}f_1(W_1) - \mathbf{i}f_2(W_2) \end{pmatrix}.$$

This system of equations can be solved if $\mathbf{ag} \neq \mathbf{bf}$. If $\mathbf{ag} = \mathbf{bf}$, then the output of F is a function of W_1 only, and hence collisions in this variable extend to the compression function. Surprisingly, this case does not seem to allow preimages to be found faster than by brute force.

Preimage attack. The method of separating the function into the sum of two sub-functions, each having its own independent input, can also be used to find preimages. Assume we are given a target compression function output U. As above, write the compression function as

$$F(M, V) = (\mathbf{x}W_1 + \mathbf{y}f_1(W_1)) + (\mathbf{z}W_2 + \mathbf{i}f_2(W_2)).$$

Since we need $F(M, V) = U$, compute $U_1 = \mathbf{x}W_1+\mathbf{y}f_1(W_1)$ and $U_2 = -(\mathbf{z}W_2+\mathbf{i}f_2(W_2))+ U$ for $2^{n/2}$ different values of each of W_1 and W_2, and find (with good probability) a collision between U_1 and U_2. This gives $U_1 - U_2 = 0$, and hence $F(M, V) = U$. The time complexity is about $2^{n/2}$, and the memory requirements are the same; however, memoryless techniques [10, 7] can be used in the collision search.

3.2 Attacks on Some CRYPTO '08 Proposals

At CRYPTO '08, a number of compression function constructions [16], which are provably collision resistant in the information-theoretic complexity model, were presented. By provably collision resistant we mean that the information-theoretic complexity of finding collisions is $2^{n/2}$ when the output size of the hash function is n bits. Here, we show that these constructions have a number of undesirable properties.

Preimage attack on a two-call single-length construction. The first construction that we consider is a provably collision resistant single-length construction applying two underlying primitives, modelled as random n-bit to n-bit functions. The output size is $2n/3$ bits, and the size of a message block is n bits. This construction is called Construction 1.

Construction 1. *On input M and V ($|M| = n$, $|V| = 2n/3$), compute:*

$$X \leftarrow V\|0^{n/3}$$
$$Y \leftarrow f_2(f_1(M) \oplus X) \oplus X$$
$$Z \leftarrow \mathsf{msb}_{2n/3}(Y)$$

Output Z

Here, '$\|$' denotes concatenation, $|x|$ means the bitlength of x, $0^{n/3}$ is the string of $n/3$ '0'-bits, and $\mathsf{msb}_\ell(x)$ means the ℓ most significant bits of x. If f_1 is replaced by a permutation p applied in the Davies-Meyer mode [9, 4], i.e., $f_1(x) = p(x) \oplus x$, then the security proof still holds. Whether the same is true for f_2 is unclear [16].

Note that the attacks described above on a class of two-call constructions do not seem to apply to the Construction 1 due to the truncation. Without the truncation, however, they would.

Here, we show that replacing f_2 with a fixed-key blockcipher in Davies-Meyer mode would lead to a preimage attack in time $2^{n/3}$. Let $E_K(\cdot)$ be the fixed-key blockcipher. Then the construction can be described as follows:

On input M and V ($|M| = n$, $|V| = 2n/3$), compute:

$$X \leftarrow V\|0^{n/3}$$
$$Y \leftarrow E_K(f_1(M) \oplus X) \oplus f_1(M)$$
$$Z \leftarrow \mathsf{msb}_{2n/3}(Y).$$

Output Z.

Let us denote by $\langle x \rangle_\ell$ an ℓ-bit representation of the integer x. To find a preimage of a target image W, one may fix M to an arbitrary value, and compute $X^* = E_K^{-1}((W\|\langle i \rangle_{n/3}) \oplus f_1(M)) \oplus f_1(M)$ for increasing i (from 0 up to at most $2^{n/3} - 1$), until X^* has $n/3$ trailing '0' bits. Then, $V = \mathsf{msb}_{2n/3}(X^*)$ and M form a preimage of W. The success probability of receiving $n/3$ trailing '0' bits at the ith query, that have not been received for the previous queries, is $\frac{2^{2n/3}}{2^n - i}$. Hence, we can determine the success probability of finding a preimage for target image W, denoted by $Pr_{pre}(W)$, as follows:

$$Pr_{pre}(W) = 1 - \prod_{i=1}^{i=q}(1 - \frac{2^{2n/3}}{2^n - i}) \approx 1 - e^{-\sum_{i=1}^{i=q} \frac{2^{2n/3}}{2^n - i}} \geq$$

$$1 - e^{-\sum_{i=1}^{i=q} \frac{2^{2n/3}}{2^n}} = 1 - e^{\frac{-q \times 2^{2n/3}}{2^n}} = 1 - e^{-q \times 2^{-n/3}}.$$

Where we used the approximation $e^x \approx 1 - x$, for $x \ll 1$, through the calculations. If we replace q by $2^{n/3}$ we receive the success probability of our attack which is 0.63. Hence, the expected complexity of attack is $2^{n/3}$.

We note that when f_2 is replaced by a fixed-key block cipher, then the truncation of the chaining value from n bits to $2n/3$ bits is vital to the security of the construction – without it, collision and preimage attacks in constant time would be possible.

Preimage and near-collision attacks on a double-length construction. A provably collision resistant double-length construction based on an underlying random $3n$-bit to n-bit function is also proposed in [16]. This construction has input size $3n$ bits and output size $2n$. We call it Construction 2.

Construction 2. *On input M, V_1, and V_2 ($|M| = |V_1| = |V_2| = n$), compute:*

$$Y \leftarrow f(M, V_1, V_2)$$
$$Z \leftarrow Y\|(V_2 Y^2 + V_1 Y + M).$$

Output Z.

All arithmetic takes place in the finite field $\mathbb{F}_{2^n}$.

A variant of Construction 2 applies two $2n$-bit to n-bit functions f_1 and f_2, and defines $f(M, V_1, V_2) = f_2(f_1(M, V_1), V_2)$. The properties that we describe below for Construction 2 also apply to this variant.

Construction 2 allows near-collisions in $3n/2$ out of $2n$ bits to be found in time about $2^{n/2}$, significantly below the ideal complexity of $2^{3n/4}$.

The idea of the attack is to fix V_1 and V_2 to zero, yielding $Y\|M$ as the output. Now, one may vary half the bits in M until a collision occurs in Y, which would yield a near-collision in $3n/2$ bits of the output. More precisely, the attack proceeds as follows (where C is an arbitrary $n/2$-bit string).

1. Assign $V_1 \leftarrow 0$ and $V_2 \leftarrow 0$.
2. For i from 0 to $2^{n/2}$ do
   (a) Assign $M_i \leftarrow C\|\langle i\rangle_{n/2}$
   (b) Compute $Y_i = f(M_i, V_1, V_2)$ and assign $Z_i \leftarrow Y_i\|M_i$
3. Find (with good probability) a collision in the first (most significant) $3n/2$ bits, i.e., $i \neq j$ such that $\mathsf{msb}_{3n/2}(Z_i) = \mathsf{msb}_{3n/2}(Z_j)$.

We note that any set of $n/2$ bits of M can be varied, and the collision will occur in the remaining $3n/2$ bits.

Furthermore, a preimage attack in time about 2^n can be launched on Construction 2. Assume that we are given the target image $Z = Y\|W$. As an example, we may fix $V_2 = 0$, and search for a preimage of Y under f, using V_1 and $M = W + V_1 Y$. When the preimage of Y has been found, we have also found a preimage of Z under the compression function, since $V_2 Y^2 + V_1 Y + M = W$. Since this technique requires finding a preimage in an n-bit value, the expected complexity is 2^n.

We note that $2n$ bits of the input to the compression function can be chosen freely in every iteration, and the remaining n bits are fixed by the equation $V_2 Y^2 + V_1 Y + M = W$. We also note that this attack is non-adaptive, meaning that all queries can be fixed before any response is given.

Near-collision attack on a variable-length construction. A third construction proposed in [16] is a generic single-call construction having collision resistance up to $2^{(n-c+m)/2}$ queries, where $m \leq n + c$, $n \leq s$, and $c \leq m$. This construction, which we call Construction 3, provides a range of efficiency/security trade-offs.

Construction 3. *On input $M = M_0\|M_1$ and $V = V_0\|V_1$, $|M_0| = n + c - s$, $|M_1| = |V_1| = m - n - c + s$, and $|V_0| = n + c - m$, compute*

$$X \leftarrow M_0\|(M_1 \oplus V_1)\|V_0$$
$$Y \leftarrow f(X)$$
$$Z \leftarrow Y_0\|((Y_1\|0^{s-n}) \oplus V_1).$$

Here, $|Y_0| = n + c - m$, and $|Y_1| = m - c$. Output Z.

The parameters m, s, n, and c can be varied; e.g., $c = 0$ and $m = n = s$ gives a $2n$-bit to n-bit compression function using a single call to an n-bit to n-bit function – however, this construction allows collisions to be found using a single call to f. At the other end of the spectrum, by using $c = m = s = n$ one obtains an optimally collision resistant $2n$-bit to n-bit compression function; however, this requires a $2n$-bit to n-bit random function. Hence, this approach does not make the task of the designer to construct a $2n$-bit to n-bit collision resistant compression function any easy.

We observe that a fixed XOR difference in the two inputs to Construction 3 may propagate unchanged to the output. To be more precise, choose M and V arbitrarily, and let d be an arbitrary bitstring of length $m - n - c + s$. Assign $M^* = M \oplus d$ and $V^* = V \oplus d$, where d is prepended with an appropriate number of '0' bits. Now the two inputs yield the same input to the function f, and hence the two outputs of the compression function have the difference d in the last (least significant) $m - n - c + s$ bits. Therefore, if $m - n - c + s > 0$, a near-collision in all but a single output bit can be found by selecting (e.g.) $d = \langle 1\rangle_{m-n-c+s}$.

4   Conclusion

In this paper we presented collision and preimage attacks on a large class of compression function constructions making two calls to the underlying ideal primitives.

We also described some undesirable properties in a number of provably collision resistant compression function constructions proposed at CRYPTO '08. These attacks underline that designing an efficient compression function from small ideal components is not an easy task.

References

1. Black, J., Cochran, M., Shrimpton, T.: On the Impossibility of Highly-Efficient Blockcipher-Based Hash Functions. In: Cramer, R. (ed.) EUROCRYPT 2005. LNCS, vol. 3494, pp. 526–541. Springer, Heidelberg (2005)
2. Damgård, I.: A Design Principle for Hash Functions. In: Brassard, G. (ed.) CRYPTO 1989. LNCS, vol. 435, pp. 416–427. Springer, Heidelberg (1990)
3. Lucks, S.: A failure-friendly design principle for hash functions. In: Roy, B. (ed.) ASIACRYPT 2005. LNCS, vol. 3788, pp. 474–494. Springer, Heidelberg (2005)
4. Menezes, A.J., van Oorschot, P.C., Vanstone, S.A.: Handbook of Applied Cryptography. CRC Press, Boca Raton (1997)
5. Merkle, R.C.: One Way Hash Functions and DES. In: Brassard, G. (ed.) CRYPTO 1989. LNCS, vol. 435, pp. 428–446. Springer, Heidelberg (1990)
6. Meyer, C.H., Schilling, M.: Secure program load with manipulation detection code. In: Proceedings SECURICOM 1988, pp. 111–130 (1988)
7. Morita, H., Ohta, K., Miyaguchi, S.: A Switching Closure Test to Analyze Cryptosystems. In: Feigenbaum, J. (ed.) CRYPTO 1991. LNCS, vol. 576, pp. 183–193. Springer, Heidelberg (1992)
8. National Institute of Standards and Technology. FIPS PUB 180-2, Secure Hash Standard. Federal Information Processing Standards Publication 180-2, U.S. Department of Commerce (August 2002)
9. Preneel, B., Govaerts, R., Vandewalle, J.: Hash Functions Based on Block Ciphers: A Synthetic Approach. In: Stinson, D.R. (ed.) CRYPTO 1993. LNCS, vol. 773, pp. 368–378. Springer, Heidelberg (1994)
10. Quisquater, J.-J., Delescaille, J.-P.: How Easy is Collision Search. New Results and Applications to DES. In: Brassard, G. (ed.) CRYPTO 1989. LNCS, vol. 435, pp. 408–413. Springer, Heidelberg (1990)
11. Rivest, R.L.: The MD4 Message Digest Algorithm. In: Menezes, A., Vanstone, S.A. (eds.) CRYPTO 1990. LNCS, vol. 537, pp. 303–311. Springer, Heidelberg (1991)
12. Rivest, R.L.: The MD5 Message-Digest Algorithm, Network Working Group, Request For Comments: 1321 (April 1992)
13. Rogaway, P., Steinberger, J.: Security/Efficiency Tradeoffs for Permutation-Based Hashing. In: Smart, N. (ed.) EUROCRYPT 2008. LNCS, vol. 4965, pp. 220–236. Springer, Heidelberg (2008)
14. Rogaway, P., Steinberger, J.P.: Constructing Cryptographic Hash Functions from Fixed-Key Blockciphers. In: Wagner, D. (ed.) CRYPTO 2008. LNCS, vol. 5157, pp. 433–450. Springer, Heidelberg (2008)

15. Shrimpton, T., Stam, M.: Building a Collision-Resistant Compression Function from Non-compressing Primitives. In: Aceto, L., Damgård, I., Goldberg, L.A., Halldórsson, M.M., Ingólfsdóttir, A., Walukiewicz, I. (eds.) ICALP 2008, Part II. LNCS, vol. 5126, pp. 643–654. Springer, Heidelberg (2008)

16. Stam, M.: Beyond Uniformity: Better Security/Efficiency Tradeoffs for Compression Functions. In: Wagner, D. (ed.) CRYPTO 2008. LNCS, vol. 5157, pp. 397–412. Springer, Heidelberg (2008)

17. Stam, M.: Blockcipher-Based Hashing Revisited. In: Dunkelman, O. (ed.) FSE 2009. LNCS, vol. 5665, pp. 67–83. Springer, Heidelberg (2009)

18. Wagner, D.: A Generalized Birthday Problem. In: Yung, M. (ed.) CRYPTO 2002. LNCS, vol. 2442, pp. 288–303. Springer, Heidelberg (2002)

Integral Distinguishers of Some SHA-3 Candidates

Marine Minier[1], Raphael C.-W. Phan[2], and Benjamin Pousse[3]

[1] Universit de Lyon, INRIA
INSA-Lyon, CITI, F-69621, Villeurbanne, France
marine.minier@insa-lyon.fr
[2] Electronic and Electrical Engineering, Loughborough University
LE11 3TU Leicestershire - UK
R.Phan@lboro.ac.uk
[3] XLIM (UMR CNRS 6172), Université de Limoges
23 avenue Albert Thomas, F-87060 Limoges Cedex - France
benjamin.pousse@unilim.fr

Abstract. In this paper, we study structural Integral properties on reduced versions of the compression functions of some SHA-3 candidates:
Hamsi-256, LANE-256 and Grøstl-512. More precisely, we improve on
the Integral distinguishers of Hamsi-256 (less time complexity or deterministic instead of probabilistic) and present the first known Integral
distinguishers for LANE-256 and improved Integral distinguisher for
Groestl-512. Whereas the SHA-3 competition focuses the cryptographic
world attention on the design and the attacks of hash functions, results
in this paper analyze the resistance of some SHA-3 candidates against
structural properties built on Integral distinguishers.

Keywords: hash functions, cryptanalysis, integral distinguishers,
SHA-3 candidates.

1 Introduction

Today, with the cryptographic world focus on the SHA-3 competition[1], recent
cryptanalytic attacks are mounted across both hash function primitives and
underlying ciphers/permutations. Notably, the joint analysis of hash functions
and underlying block ciphers or permutations has led to considerations of new
attack models for block ciphers, e.g. known key [20] or chosen key [5]. In fact,
when block cipher inspired permutation structures are used as building blocks
within hash functions, there is no secret (key) input, thus the building block is not
only a known transformation, it is also a computable one. Interestingly enough,
doing so has led to the discovery of more powerful new techniques to construct
distinguishers and/or mount key recovery attacks for block ciphers back in the
conventional unknown key model where ciphers are standalone constructs instead

---

[1] http://csrc.nist.gov/groups/ST/hash/sha-3/index.html

S.-H. Heng, R.N. Wright, and B.-M. Goi (Eds.): CANS 2010, LNCS 6467, pp. 106–123, 2010.
© Springer-Verlag Berlin Heidelberg 2010

of underlying hash functions, including the first known related-key attacks on the full version of AES-256 and AES-192 [5,4].

In this paper, we mainly focus on Integral properties and their applications in the known transformation model to find structural properties of some of the SHA-3 candidates: Hamsi-256, LANE-256 and Grøstl-512. As shown in [25,2], particular Integral properties essentially depending on the structure of the linear part of the cipher could be exhibited and are powerful tool for cryptanalytic studies. Those integral properties starting from the middle of the cipher lead to distinguishers in the known key settings defined in [20] and for the underlying compression functions of hash functions. In more details, our contributions are as follows. For Hamsi-256, we present improved Integral distinguishers for its full compression function: for its P_f permutation, our distinguisher requires less time complexity, for its P permutation our distinguisher is deterministic versus the previous known distinguisher that is probabilistic. For LANE-256, we present the first known Integral distinguishers. For Groestl-512, we present an Integral distinguisher that is improved over the best previously known Integral distinguisher.

This paper is organized as follows: Section 2 introduces the related work in the field of integral cryptanalysis and the notations that will be used in the rest of this paper. Sections 3, 4 and 5 respectively present integral properties and distinguishers on the Hamsi-256, LANE-256 and Grøstl-512 compression functions. Finally, Section 6 concludes this paper.

2   Related Work and Notations

Integral cryptanalysis was first introduced by Knudsen against the block cipher Square in the original paper [10] in the unknown key model to retrieve information on some key bytes. Then, it was applied to the AES in the original submission paper [11,12] and later the distinguisher was extended by one round by Ferguson et al. in [14].

After those first attacks, many ciphers especially the ones that use an SPN structure have been studied in regard of this kind of distinguishers. Among the integral cryptanalyses proposed in the literature, we could cite the attacks against SAFER [3], CRYPTON [13] and more recently on PRESENT [9]. The different Rijndael versions (Rijndael-192 and Rijndael-256) have also been attacked using integral properties [19,15]. Other contributions also analyze the general framework of Integral cryptanalysis and especially focus on the required conditions for a block cipher to be attacked using this method [21,6]. In [21], Knudsen and Wagner analyze integral cryptanalysis as a dual to differential attacks notably applicable to block ciphers with bijective components. A first-order integral cryptanalysis considers a particular collection of m words in the plaintexts and ciphertexts that differ on a particular word. The aim of this attack is thus to predict the values in the sums (i.e. the integral) of the chosen words after a certain number of rounds of encryption. The same authors also generalize this approach to higher-order integrals: the original set to consider becomes a set of m^d vectors which differ in d words and where the sum of this set is predictable after a certain number of rounds. The sum of this set is called a dth-order integral.

More recently, in [20] Integral cryptanalysis has been proposed in the new cryptanalysis model called known key setting where the key is known to the attacker. In the same setting, compression functions of hash functions could also be analyzed and some distinguishers have been proposed against SHA-3 candidates using Integral properties. As example we could cite integral distinguishers on the compression functions of Hamsi-256 [2] and Keccak [7].

In the rest of this paper, we use the consistent notations introduced in [21] and extend them for expressing word-oriented integral attacks. For a dth order integral, we have:

- The symbol '$\mathcal{C}$' (for "Constant") in the ith entry, means that the values of all the ith words in the collection of texts are equal.
- The symbol '$\mathcal{P}$' (for "Permutation") means that all words in the collection of texts are different.
- The symbol '?' means that the sum of words cannot be predicted.
- The symbol '$\mathcal{P}^d$' corresponds with the components that participate in a dth-order integral, i.e. if a word can take m different values then $\mathcal{P}^d$ means that in the integral, the particular word takes all values exactly m^{d-1} times.
- The symbol '$\mathcal{B}$' (for "Balanced") means that the sum of all values is zero.
- The symbol 'Eq_i' (for "Equality') found for two different words means that the sums of all values taken on those particular words matched (i.e. are equals).

3   Hamsi-256

In this section, we will introduce new Integral properties first on the compression function P of Hamsi-256 that happens with probability 1 contrary to the one proposed in [2] and then on the final transformation P_f. Many recent results concerning Hamsi cryptanalysis have been presented in [8] and in [2]. As previously mentioned, some probabilistic integral properties have been presented in [2] and also an integral distinguisher on P_f which has a complexity of 2^{28} computations. The rest of the cryptanalytic results on Hamsi-256 essentially concerns differential paths that happen with low probabilities to build message-recovery attacks and pseudo-second-preimage attacks.

3.1   Description of the Hamsi-256 Hash Function

This section describes the hash function Hamsi-256. We refer to [22] for a complete specification.

General View. Hamsi is based on the Concatenate-Permute-Truncate design strategy. It uses in addition a message expansion and a feed forward of the chaining value in each iteration. Thus the compression function of Hamsi can be divided into four mappings:

- **Message Expansion:** $E : \{0,1\}^{32} \to \{0,1\}^{256}$
- **Concatenation:** $C : \{0,1\}^{256} \times \{0,1\}^{256} \to \{0,1\}^{512}$
- **Non linear permutations** P **and** P_f: $\{0,1\}^{512} \to \{0,1\}^{512}$
- **Truncation:** $T : \{0,1\}^{512} \to \{0,1\}^{256}$

The message M to hash is properly padded and cut into l 32-bit blocks $M_1, \cdots M_l$. Each block is transformed using the message expansion into a 256-bit block seen as a 4×4 matrix of 32-bit words. Once expanded, each block is processed by the compression function. Starting from a predefined initial value h_0, the compression function H (or H_f for the final transformation) could be written as follows to compute the digest h of M:

$$h_i = H(h_{i-1}, M_i) = (T \circ P \circ C(E(M_i), h_{i-1})) \oplus h_{i-1} \text{ for } 0 < i < l$$
$$h = H(h_{l-1}, M_l) = (T \circ P \circ C(E(M_l), h_{l-1})) \oplus h_{l-1}$$

Internal mappings

Message expansion. The message expansion of Hamsi-256 is based on a linear code given by its generator matrix G (see [22] for more details). Thus:

$$E(M_i) = (m_0, \cdots, m_7) = (M_i \times G)$$

where $(m_0, \cdots, m_7)$ are eight 32-bit words.

Concatenation. The concatenation function C builds a 512-bit word from the 256-bit expanded message $(m_0, \cdots, m_7)$ and the 256-bit chaining value $h_i = (c_0, \cdots, c_7)$ at 32-bit word level:

$$C(m_0, \cdots, m_7, c_0, \cdots, c_7) = (m_0, m_1, c_0, c_1, m_2, m_3, c_2, c_3, m_4, m_5, c_4, c_5, m_6, m_7, c_6, c_7)$$

corresponding, when looking at the matrix representation, with rows that are composed of two message words and of two chaining value words.

Truncation. The truncation function T selects eight 32-bit words among the 16 from the internal state to form the new chaining value after feed forward:

$$T(s_0, s_1, s_2, \cdots, s_{14}, s_{15}) = (s_0, s_1, s_2, s_3, s_8, s_9, s_{10}, s_{11})$$

corresponding, when looking at the matrix representation, with choosing the second and the fourth rows.

Non-linear permutations. P and P_f use the same size parameters (the input and output blocks are 512-bit long) and the same round function. They differ by the round constants and by the number of rounds: three rounds for P and six rounds for P_f (for this latter in fact 8 rounds are the recommended parameters of the designers even if the specifications are always provided with 6 rounds). The round function is composed of three layers:

- Addition of constants and counter: predefined constants are first XORed to the whole internal state. Then, a counter is XORed to s_1, the second 32-bit word of the internal state.
- Substitution layer: it uses the 4-bit Sbox of the block cipher Serpent [1] in a bitslice mode. From each of the four 32-bit words of a same column, the four bits at the same position are extracted and replaced by the corresponding value of the Sbox.
- Linear diffusion layer: it applies the Serpent linear transform L from $\{0,1\}^{128}$ into itself to each diagonal of the state matrix:

$$(s_0, s_5, s_{10}, s_{15}) = L(s_0, s_5, s_{10}, s_{15})$$
$$(s_1, s_6, s_{11}, s_{12}) = L(s_1, s_6, s_{11}, s_{12})$$
$$(s_2, s_7, s_8, s_{13}) = L(s_2, s_7, s_8, s_{13})$$
$$(s_3, s_4, s_9, s_{14}) = L(s_3, s_4, s_9, s_{14})$$

The linear transformation L could be written in pseudo-code as follows for a four 32-bit words input (a, b, c, d):

$a := a <<< 13; \quad c := c <<< 3; \quad b := b \oplus a \oplus c; \quad d := d \oplus c \oplus (a << 3); \quad b := b <<< 1;$
$d := d <<< 7; \quad a := a \oplus b \oplus d; \quad c := c \oplus d \oplus (b << 7); \quad a := a <<< 5; \quad c := c <<< 22;$

where $<<<$ denotes left rotation and $<<$ denotes left shifting.

3.2   Integral Properties of Hamsi-256

We first give an Integral property on the permutation P that only depends on the chaining value h_i then we analyze an integral distinguisher on the permutation P_f. When looking in detail at the diffusion function of P, we could see that a complete diffusion could not be reached after three rounds. So, we have exhibited the integral property described in Fig. 1, which was tested and verified on 2×10^6 random values. This particular integral property leads to a distinguisher that uses $(2^2)^4 = 2^8$ values and that leads to values that sum to zero on 444 bits. In the same way when trying to limit the number of inputs, if the same four particular words take all possible values only on the least significant bit, this integral property leads to sums equal to 0 on about 30 bits after the P application.

When looking at an integral distinguisher for the final transformation P_f, we start from the middle as done when known key settings are considered. Thus, we are looking for integral properties in the forward and in the backward senses on three rounds. Fig. 2 presents the integral property found using intensive computations in the forward sense for three P_f rounds. Thus, we obtain a three forward rounds integral distinguisher that uses 2^{16} values that sum to zero everywhere.

In the same way, we have exhibited the three backward rounds integral property shown in Fig. 3. Thus, we obtain a three backward rounds integral distinguisher that uses 2^{16} values that sum to zero everywhere.

Thus, we could combine those two properties in the forward and in the backward senses starting from the middle using 2^{16} middletexts that are constant

C	C	C	C		4124000	a0000	904	82820000
P^8	P^8	C	C	$\xrightarrow{P}$	4001000	8200	5000	20000040
C	C	C	C		2040008	702	1854	a00
P^8	P^8	C	C		0	8	140000	201000

Fig. 1. The Integral property for P where P^8 means that the four considered 32-bit words are constant everywhere except on the 2 least significant bits that take all possible values (between 0 and 3) and where the values given in hexadecimal notation for the output corresponds with the output mask (i.e. a 0 value means that the sum on this byte is always equal to 0)

P^{16}	C	C	C			B	B	B	B
P^{16}	C	C	C	3-round		B	B	B	B
P^{16}	C	C	C	$\rightarrow$		B	B	B	B
P^{16}	C	C	C			B	B	B	B

Fig. 2. The Integral property for 3 P_f rounds in the forward sense where P^{16} means that the four considered 32-bit words are constant everywhere except on the 4 least significant bits that take all possible values (between 0 and 15). All the outputs have the property $\mathcal{B}$ and sum to zero.

B	B	B	B			P^{16}	C	C	C
B	B	B	B	3-round		P^{16}	C	C	C
B	B	B	B	$\leftarrow$		P^{16}	C	C	C
B	B	B	B			P^{16}	C	C	C

Fig. 3. The Integral property for 3 P_f rounds in the backward sense where P^{16} means that the four considered 32-bit words are constant everywhere except on the 4 least significant bits that take all possible values (between 0 and 15). All the outputs have the property $\mathcal{B}$ and sum to zero.

everywhere except on the 4 least significant bits of four particular words. From this set of texts, if we go forward, we obtain a set of values that sum to zero everywhere and if we go backward, we also obtain a set of values that sum to zero everywhere. Thus, we have exhibited a particular integral distinguisher that is such that starting from 2^{16} particular middletexts leads to input and output values that sum everywhere to 0. The computational cost for this distinguisher is 2^{16} calls to the transformation P_f.

4 LANE-256

4.1 Description of the LANE-256 Hash Function

The cryptographic hash function LANE [18] has been submitted to the NIST SHA-3 competition and has been discarded after round 1 due to the attack presented in [23]. This attack against the whole version of LANE-256 presents semi-free-start collisions that could be constructed in 2^{96} computations using 2^{88} memory.

LANE-256 is an iterated hash function that supports four digest sizes (224, 256, 384 and 512 bits) and the use of a salt. We focus here on LANE-256 where the initial chaining value H_{-1} has a 256 bits size. The message is padded and split into message blocks M_i of length 512 bits for LANE-256. Then the compression function f of Lane-256 transforms iteratively 256 bits of the chaining value and 512 bits of the message block into a new chaining value of 256 bits $H_i = f(H_{i-1}, M_i, C_i)$ where C_i is a counter that indicates the number of message bits processed so far. Finally, after all the message blocks are processed, the final digest is derived from the last chaining value, the message length and the salt by an additional call to the compression function. For the detailed structure of the compression function we refer to the specification of LANE [18]. First, the chaining value and the message block are processed by a message expansion that produces an expanded state with doubled size. Then, this expanded state is processed in two layers. The first layer is composed of six permutations $P_0, \cdots, P_5$ applied in parallel, and the second layer of two parallel layers Q_0, Q_1.

The message expansion of LANE takes a message block M_i and a chaining value H_{i-1} and produces the input to six permutations $P_0, \cdots, P_5$. In LANE-256, the 512-bit message block M_i is split into four 128-bit blocks $m_0 || m_1 || m_2 || m_3$ and the 256-bit chaining value H_{i-1} is split into two 128-bit words $h_0 || h_1$. Then, six more 128-bit words $a_0, a_1, b_0, b_1, c_0, c_1$ are computed:

$$a_0 = h_0 \oplus m_0 \oplus m_1 \oplus m_2 \oplus m_3, \ a_1 = h_1 \oplus m_0 \oplus m_2,$$
$$b_0 = h_0 \oplus h_1 \oplus m_0 \oplus m_2 \oplus m_3, \ b_1 = h_0 \oplus m_1 \oplus m_2,$$
$$c_0 = h_0 \oplus h_1 \oplus m_0 \oplus m_1 \oplus m_2, \ c_1 = h_0 \oplus m_0 \oplus m_3.$$

Each of these 128-bit values, as in AES, can be seen as 4×4 matrix of bytes. In the following, we will use the notation $x[i, j]$ when we refer to the byte of the matrix x with row index i and column index j, starting from 0. The values $a_0 || a_1, b_0 || b_1, c_0 || c_1, h_0 || h_1, m_0 || m_1, m_2 || m_3$ become the inputs of the six permutations $P_0, \cdots, P_5$ described below.

Each permutation P_i operates on a state that can be seen as a double AES state ($2\times$ 128-bits). The permutation reuses the transformations SubBytes (SB), ShiftRows (SR) and MixColumns (MC) of the AES with the only exception, that due to the larger state size, they are applied twice in parallel. Additionally, there are three new round transformations introduced in LANE. AddConstant (AC) adds a different value to each column of the state and AddCounter (ACO) adds part of the counter C_i to the state. The third transformation is SwapColumns (SC) used for mixing parallel AES states. It swaps the two right columns of the left half-state with the two left columns of the right half-state.

The complete round transformation consists of the sequential application of all these transformations in the given order. The last round omits AddConstant and AddCounter. Each of the permutations P_i consists of six rounds for LANE-256.

The permutations Q_0 and Q_1 are composed of three previously defined rounds where the last round does not contain the AddConstant and AddCounter operations.

4.2   Integral Properties of LANE-256

As the diffusion layer of LANE-256 has a really slow diffusion, we could expect that the same kind of properties as the ones presented in [25] exist on 5 rounds or more. When studying in detail the integral properties, we obtain a 2nd order integral property on 4-round of each P_i as shown in Fig. 9 in Appendix A. This 4 forward rounds integral property could be extended by two rounds at the beginning using the two extensions (an 8th order and a 16th order integral respectively) shown in Fig. 11 in Appendix A. Thus, using 2^{128} chosen plaintexts we are able to distinguish the output after 6 forward rounds of all the P_i functions from a random permutation because the sums taken at byte level over all the inputs are equal to 0 for all the output bytes. The complexity of the distinguisher is equal to 2^{128} P_i operations.

Similarly in the backward sense, we found a 2nd order integral property on 3 backward rounds presented in Fig. 10 in Appendix A. This property leads to a distinguisher on 3 backward rounds where the sums taken at byte level over all the inputs are equal to 0. It requires 2^{16} chosen texts to work and has a complexity equal to 2^{16} operations. This property could be extended by one backward rounds at the beginning using an 8th order integral property as shown on Fig. 12 in Appendix A. This leads to an integral distinguisher that uses 2^{64} chosen plaintexts with a complexity equal to 2^{64} operations to test if the sums taken at byte level over the 256 bits are equal to 0 or not.

We could combine those two properties (in the backward and in the forward senses) starting from the middle of one particular P_i (say P_4) and the corresponding Q_i (say Q_1) to build a structural property on the right part of the LANE compression function (see Fig. 13 in Appendix A). For the permutation P_4, start in the middle (after 4 rounds) with 2^{112} middletexts with 14 active bytes (the other are taken equal to a constant) then, go backward on four rounds to obtain inputs that sum to 0 everywhere and go forward on 5 rounds to obtain outputs that sum to 0 everywhere.

In the same way, using the previous property on the right part of the compression function, we could obtain a complete property on the LANE-256 compression function when P_0, P_1 and P_2 are limited to 3 rounds (instead of 6 in the original version) using our 16th order integral property on the left part composed of 6 rounds when concatenating P_0 and Q_0. To do so, repeat the property on the right part considering that h_0 and h_1 are constants, for 2^{128} $h_0 || h_1$ values that correspond to a 16th order integral. Thus, and as shown in Fig. 14 in Appendix A, we have exhibited an intrinsic property of the compression function of LANE when P_0 has only 3 rounds (instead of 6) using 2^{240} values seen as in the one hand 2^{128} copies of the 2^{112} middletexts that lead to integral properties on the right part of LANE-256 and on the other hand 2^{112} copies of a 16th order integral 6-round property. The complexity required to exhibit this property is equal to 2^{240} P_i operations whereas the memory requirements are small.

Note also that the same kind of properties could be directly deduced for LANE-512.

5   Grøstl-512

5.1   Description of the Grøstl-512 Hash Function

Grøstl [16] is a SHA-3 candidate designed by Guaravaram *et al.*, notably Grøstl-256 outputs hash value of length 224 and 256 bits whereas Grøstl-512 outputs hash value of length 384 or 512 bits. We mainly focus here on Grøstl-512. It is an itcrated hash function with a compression function built from two distinct permutations P and Q. A t-block message M (after padding) is hashed using the compression function $f(H_{i-1}, M_i)$ and output tranformation $g(H_t)$ as follows:

$$H_0 = IV$$
$$H_i = f(H_{i-1}, M_i) = H_{i-1} \oplus P(H_{i-1} \oplus M_i) \oplus Q(M_i) \text{ for } 1 \le i \le t$$
$$h = g(H_t) = trunc(H_t \oplus P(H_t))$$

The two permutations P and Q are constructed using the wide trail strategy, their design is very similar to the AES with a fixed key input. Both permutations of Grøstl-512 act on a 1024-bit state represented as a 8×16 matrix of bytes and have 14 rounds. The round transformations of Grøstl-512 are the following ones:

- AddRoundConstant (AC) adds different one-byte round constants to the 8×16 states of P and Q.
- SubBytes (SB) is the non-linear layer that applies the AES Sbox to each byte of the state.
- ShiftBytes (ShB) rotates the bytes of row j in the following way: 0 for $j = 1$, 1 for $j = 2, \cdots$ 6 for $j = 7$ and 11 for $j = 8$.
- MixBytes (MB) is the linear diffusion layer where each column of the state is multiplied by a constant matrix B.

5.2   Grøstl Analysis So Far

Grøstl has attracted significant amount of cryptanalysis. Among the best cryptanalytic results on the two Grøstl version (Grøstl-256 and Grøstl-512), we could mention semi-free-start collisions on the compression function of Grøstl-256 reduced to seven rounds presented in [17] that have a complexity of 2^{120} computations and 2^{64} in memory; and semi-free-start collisions on the compression function of Grøstl-512 reduced to eight rounds presented in [24] that have a complexity of 2^{152} computations and 2^{64} in memory.

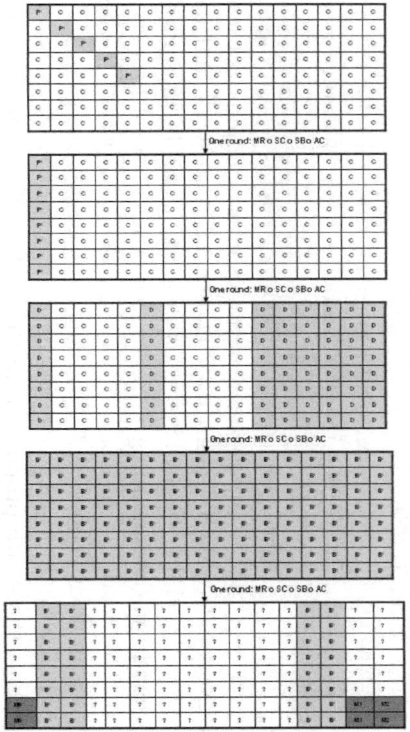

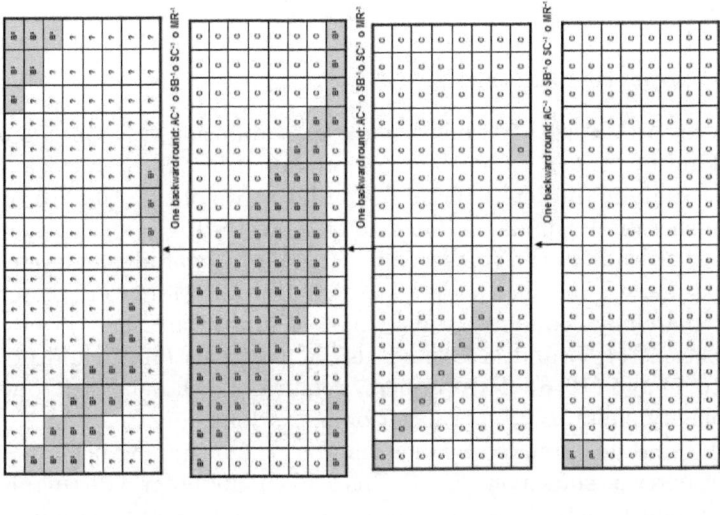

Fig. 5. The 2nd order Integral property on 3 backward rounds of P and Q

Fig. 4. The 5th order Integral property on 4 rounds of P and Q

Concerning particular distinguishers, structural non-random properties have been observed e.g. fixed points [16] and q-multicollisions in the original submission; and distinguishers have been presented on Grøstl-256 reduced to 8 out of 10 rounds. In a more recent article [26], Thomas Peyrin presents distinguishers using truncated differential properties on the 10 rounds of the compression function of Grøstl-256 requiring 2^{192} computations and 2^{64} memory; and on 11 rounds of the compression function of Grøstl-512 requiring 2^{640} computations and 2^{64} memory.

5.3 Integral Properties of Grøstl-512

As the diffusion layer of Grøstl-512 has a really slow diffusion, we could expect that the same kind of properties as the ones presented in [25] may exist on 5 rounds or more. Studying in detail the diffusion layer, we obtain the 5th order integral distinguisher on 4 rounds of P and of Q shown in Fig. 4.

This distinguisher leads to the sums taken at byte level over all the inputs on the four columns marked in blue in Fig. 4 are equal to 0. Moreover, there are some particular matching (equality) sums (marked in gray in Fig. 4) between the bytes at positions $(1, 7)$ and $(1, 8)$ and also at positions $(15, 7)$ and $(15, 8)$ and positions $(16, 7)$ and $(16, 8)$. This property requires 2^{40} chosen texts to work and has a complexity equal to 2^{40} permutation operations.

This property could be extended by one round at the beginning using a 40th order integral property as shown on Fig. 6. This leads to an integral distinguisher that uses 2^{320} chosen plaintexts with a complexity equal to 2^{320} permutation operations to test if the sums taken at byte level over $8 \times 8 \times 4 = 256$ bits are equal to 0 or not. The sums taken on the equalities could also be considered leading to 0-sums on 230 bits.

Let us now analyze which are the integral properties that exist for the backward sense. We found the 2nd order integral property on 3 backward rounds presented in Fig. 5.

This property leads to a distinguisher on 3 backward rounds where the sums taken at byte level over all the inputs on the three shifted columns marked in blue in Fig. 5 are equal to 0. It requires 2^{16} chosen texts to work and has a complexity equal to 2^{16} cipher operations. This property could be extended by two backward rounds at the beginning using a 64th order integral property as shown on Fig. 5. This leads to an integral distinguisher that uses 2^{512} chosen plaintexts with a complexity equal to 2^{512} permutation operations to test if the sums taken at byte level over $3 \times 8 \times 4 = 192$ bits are equal to 0 or not.

We could combine those two properties (in the backward and in the forward senses) starting from both the middle of P and the middle of Q to build a structural property on the compression function of Grøstl-512 when 10 rounds are considered (see Fig. 8). For the permutation P, start from the middle with 2^{512} middletexts with 64 active bytes (the other are taken equal to a constant) then, go backward on five rounds to obtain inputs that sum to 0 on 3 shifted columns

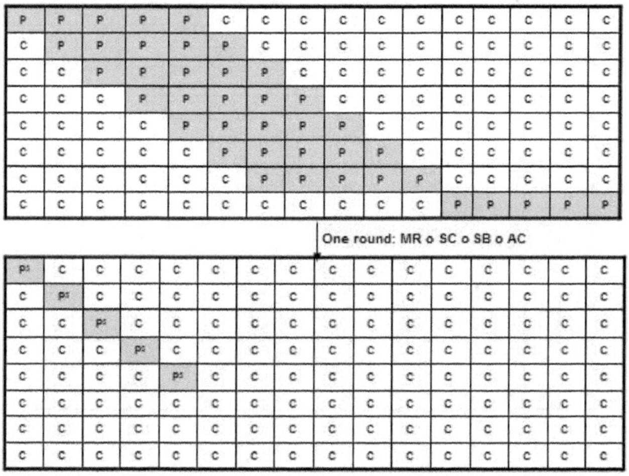

One round: MR o SC o SB o AC

Fig. 6. Extension by one round of the previous distinguisher using a 40th order Integral property

One backward round: AC⁻¹ o SB⁻¹o SC⁻¹ o MR⁻¹

One backward round: AC⁻¹ o SB⁻¹o SC⁻¹ o MR⁻¹

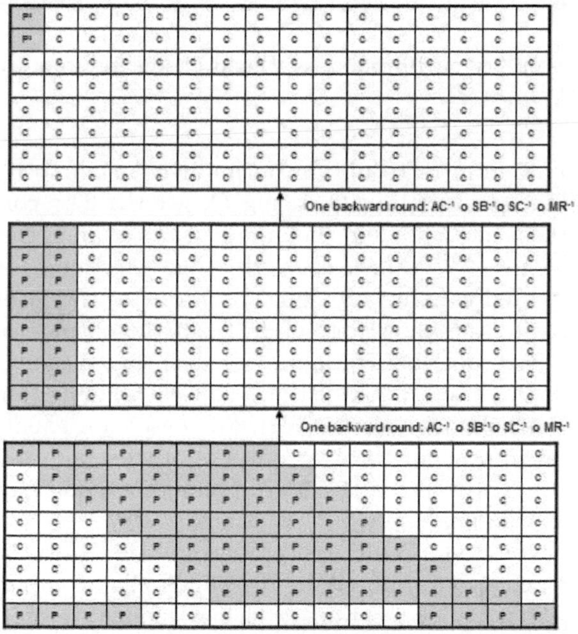

Fig. 7. Extension by two backward rounds of the 3 backward rounds distinguisher using a 64th order Integral property for P and of Q

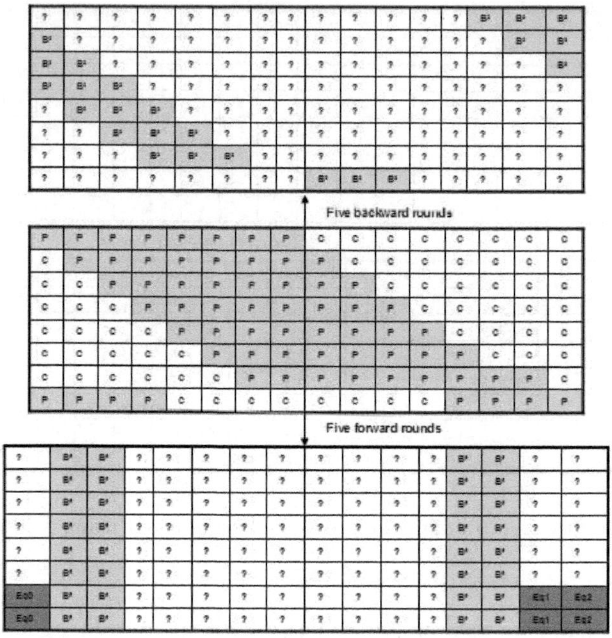

Fig. 8. Complete property on 10 rounds of P and Q starting from the middle with a 64th order integral property

and go forward on 5 rounds to obtain outputs that sum to 0 on 4 columns. Do the same for the permutation Q. Using Q, get the 2^{512} corresponding M_t messages. Using those messages and the inputs of P, compute the corresponding 2^{512} H_{t-1} values. Those 2^{512} values also verify that their sums taken over all the 2^{512} values on 3 shifted columns are equal to 0 (due to the linearity of the xor operation). Then with the knowledge of H_{t-1}, of the outputs of P and of the outputs of Q, the corresponding H_t values are such that the sums taken over all the 2^{512} values on the intersection of the 3 shifted columns (for the backward sense) and of the 4 columns (for the forward sense) are equal to 0. In other words, the sum taken over all the 2^{512} outputs of the compression function is zero at 7 byte positions whereas the corresponding inputs H_{t-1} and M_t have 0-sum on 3 shifted columns.

Thus, we have exhibited a structural property of the Grøstl-512 compression function when P and Q are limited to 10 rounds. The computational cost of this property is about 2^{513} operations with few memory requirements to find some 0-sums at particular positions (7 bytes at the output of the compression function and 24 bytes at the input). Note that this new structural property really improves the one described in the original proposal of Grøstl [16] that reaches 9 rounds with a complexity equal to 2^{704} cipher operations.

Table 1. Summary of distinguishers and of attacks on the three studied candidates

Hash functions	Nb rounds	Type of Attack	Time	Memory	Source
Hamsi-256	P_f	Integral Dist.	2^{28}	small	[2]
Hamsi-256	P	Prob. Int. Dist.	2^2	small	[2]
Hamsi-256	P_f	Integral Dist.	2^{16}	small	this paper
Hamsi-256	P	Integral Dist.	2^8	small	this paper
LANE-256	complete	Semi-free-start Coll.	2^{96}	2^{88}	[23]
LANE-256	(3,3,6,3)	Integral Dist.	2^{240}	small	this paper
Grøstl-512	8	Semi-free-start Coll.	2^{152}	2^{64}	[24]
Grøstl-512	9	Integral Dist.	2^{704}	small	[16]
Grøstl-512	11	Trunc. Diff. Dist.	2^{640}	2^{64}	[26]
Grøstl-512	10	Integral Dist.	2^{513}	small	this paper

6 Conclusion

In this paper, we analyzed some SHA-3 candidates (Hamsi-256, LANE-256, Groestl-512) in regard of Integral properties. Due to a slow diffusion of the linear part of the proposed candidates, integral properties could be exhibited for a number of rounds greater than expected. We sum up our results and the related works concerning distinguishers on the compression functions of the SHA3 candidates in Table 1 where Prob. Int. Dist. means Probabilistic Integral Distinguisher.

References

1. Anderson, R.J., Biham, E., Knudsen, L.R.: Serpent : A proposal for the advanced encryption standard. In: The First Advanced Encryption Standard Candidate Conference. N.I.S.T. (1998), http://www.cl.cam.ac.uk/~rja14/serpent.html
2. Aumasson, J.-P., Käsper, E., Knudsen, L.R., Matusiewicz, K., Ødegård, R., Peyrin, T., Schläffer, M.: Distinguishers for the compression function and output transformation of hamsi-256. Cryptology ePrint Archive, Report 2010/091, to appear in ACISP 2010 (2010), http://eprint.iacr.org/
3. Biryukov, A., De Cannière, C., Dellkrantz, G.: Cryptanalysis of safer++. In: Boneh, D. (ed.) CRYPTO 2003. LNCS, vol. 2729, pp. 195–211. Springer, Heidelberg (2003)
4. Biryukov, A., Khovratovich, D.: Related-key cryptanalysis of the full AES-192 and AES-256. In: Matsui, M. (ed.) ASIACRYPT 2009. LNCS, vol. 5912, pp. 1–18. Springer, Heidelberg (2009)
5. Biryukov, A., Khovratovich, D., Nikolic, I.: Distinguisher and related-key attack on the full AES-256. In: Halevi, S. (ed.) CRYPTO 2009. LNCS, vol. 5677, pp. 231–249. Springer, Heidelberg (2009)
6. Biryukov, A., Shamir, A.: Structural cryptanalysis of SASAS. In: Pfitzmann, B. (ed.) EUROCRYPT 2001. LNCS, vol. 2045, pp. 394–405. Springer, Heidelberg (2001)
7. Boura, C., Canteaut, A.: A zero-sum property for the keccak-f permutation with 18 rounds. NIST mailing list (2010)

8. Calik, C., Turan, M.S.: Message recovery and pseudo-preimage attacks on the compression function of hamsi-256. Cryptology ePrint Archive, Report 2010/057
9. Collard, B., Standaert, F.-X.: A statistical saturation attack against the block cipher present. In: Fischlin, M. (ed.) CT-RSA 2009. LNCS, vol. 5473, pp. 195–210. Springer, Heidelberg (2009)
10. Daemen, J., Knudsen, L.R., Rijmen, V.: The block cipher square. In: Biham, E. (ed.) FSE 1997. LNCS, vol. 1267, pp. 149–165. Springer, Heidelberg (1997)
11. Daemen, J., Rijmen, V.: Aes proposal: Rijndael. In: The First Advanced Encryption Standard Candidate Conference. N.I.S.T. (1998)
12. Daemen, J., Rijmen, V.: The Design of Rijndael. Springer, Heidelberg (2002)
13. D'Halluin, C., Bijnens, G., Rijmen, V., Preneel, B.: Attack on six rounds of crypton. In: Knudsen, L.R. (ed.) FSE 1999. LNCS, vol. 1636, pp. 46–59. Springer, Heidelberg (1999)
14. Ferguson, N., Kelsey, J., Lucks, S., Schneier, B., Stay, M., Wagner, D., Whiting, D.: Improved cryptanalysis of rijndael. In: Schneier, B. (ed.) FSE 2000. LNCS, vol. 1978, pp. 213–230. Springer, Heidelberg (2001)
15. Galice, S., Minier, M.: Improving integral attacks against rijndael-256 up to 9 rounds. In: Vaudenay, S. (ed.) AFRICACRYPT 2008. LNCS, vol. 5023, pp. 1–15. Springer, Heidelberg (2008)
16. Gauravaram, P., Knudsen, L.R., Matusiewicz, K., Mendel, F., Rechberger, C., Schläffer, M., Thomsen, S.S.: Grøstl – a sha-3 candidate. Submission to NIST (2008)
17. Gilbert, H., Peyrin, T.: Super-sbox cryptanalysis: Improved attacks for aes-like permutations. Cryptology ePrint Archive, Report 2009/531, to appear at FSE 2010 (2009), http://eprint.iacr.org/
18. Indesteege, S.: The lane hash function. Submission to NIST (2008)
19. Nakahara Jr., J., de Freitas, D.S., Phan, R.C.-W.: New multiset attacks on rijndael with large blocks. In: Dawson, E., Vaudenay, S. (eds.) MYCRYPT 2005. LNCS, vol. 3715, pp. 277–295. Springer, Heidelberg (2005)
20. Knudsen, L.R., Rijmen, V.: Known-key distinguishers for some block ciphers. In: Kurosawa, K. (ed.) ASIACRYPT 2007. LNCS, vol. 4833, pp. 315–324. Springer, Heidelberg (2007)
21. Knudsen, L.R., Wagner, D.: Integral cryptanalysis. In: Daemen, J., Rijmen, V. (eds.) FSE 2002. LNCS, vol. 2365, pp. 112–127. Springer, Heidelberg (2002)
22. Küçük, Ö.: The hash function hamsi. Submission to NIST (updated) (2009)
23. Matusiewicz, K., Naya-Plasencia, M., Nikolic, I., Sasaki, Y., Schläffer, M.: Rebound attack on the full lane compression function. In: Matsui, M. (ed.) ASIACRYPT 2009. LNCS, vol. 5912, pp. 106–125. Springer, Heidelberg (2009)
24. Mendel, F., Rechberger, C., Schläffer, M., Thomsen, S.S.: Rebound attacks on the reduced grøstl hash function. In: Pieprzyk, J. (ed.) CT-RSA 2010. LNCS, vol. 5985, pp. 350–365. Springer, Heidelberg (2010)
25. Minier, M., Phan, R.C.-W., Pousse, B.: Distinguishers for ciphers and known key attack against rijndael with large blocks. In: Preneel, B. (ed.) AFRICACRYPT 2009. LNCS, vol. 5580, pp. 60–76. Springer, Heidelberg (2009)
26. Peyrin, T.: Improved differential attacks for echo and grøstl. Cryptology ePrint Archive, Report 2010/223, to appear in Crypto 2010 (2010), http://eprint.iacr.org/

A   The Integral LANE Properties

Here are all the figures for a better illustration of Section 4.

P²	C	C	C		C	C	C	C
C	P²	C	C		C	C	C	C
C	C	C	C		C	C	C	C
C	C	C	C		C	C	C	C

One round: SC o ACO o AC o MC o SR o SB

P²	C	C	C		C	C	C	C
P²	C	C	C		C	C	C	C
P²	C	C	C		C	C	C	C
P²	C	C	C		C	C	C	C

One round: SC o ACO o AC o MC o SR o SB

D	D	C	C		D	D	C	C
D	D	C	C		D	D	C	C
D	D	C	C		D	D	C	C
D	D	C	C		D	D	C	C

One round: SC o ACO o AC o MC o SR o SB

B²	B²	B²	B²		B²	B²	B²	B²
B²	B²	B²	B²		B²	B²	B²	B²
B²	B²	B²	B²		B²	B²	B²	B²
B²	B²	B²	B²		B²	B²	B²	B²

One round: SC o ACO o AC o MC o SR o SB

B²	B²	B²	B²		B²	B²	B²	B²
B²	B²	B²	B²		B²	B²	B²	B²
B²	B²	B²	B²		B²	B²	B²	B²
B²	B²	B²	B²		B²	B²	B²	B²

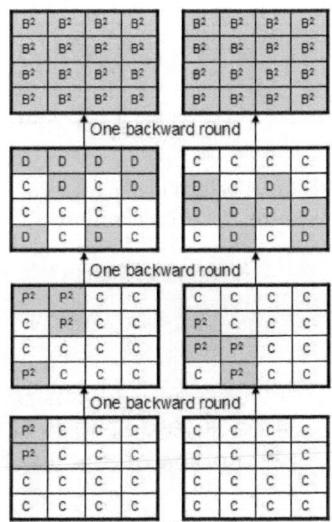

Fig. 9. The 2nd order Integral property on 4 forward rounds of P_i

Fig. 10. The 2nd order Integral property on 4 backward rounds of P_i or Q_i

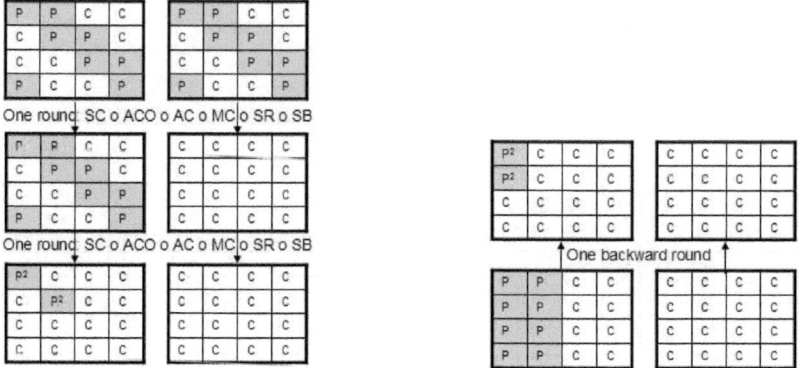

Fig. 11. The 8th order Integral extension on 4 forward rounds of P_i

Fig. 12. The 8th order Integral extension on 3 backward rounds of P_i

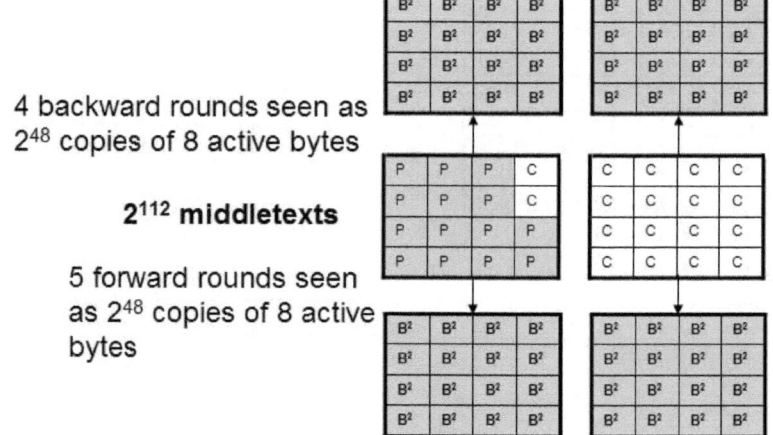

Fig. 13. The 9 rounds integral property using 2^{112} middletexts

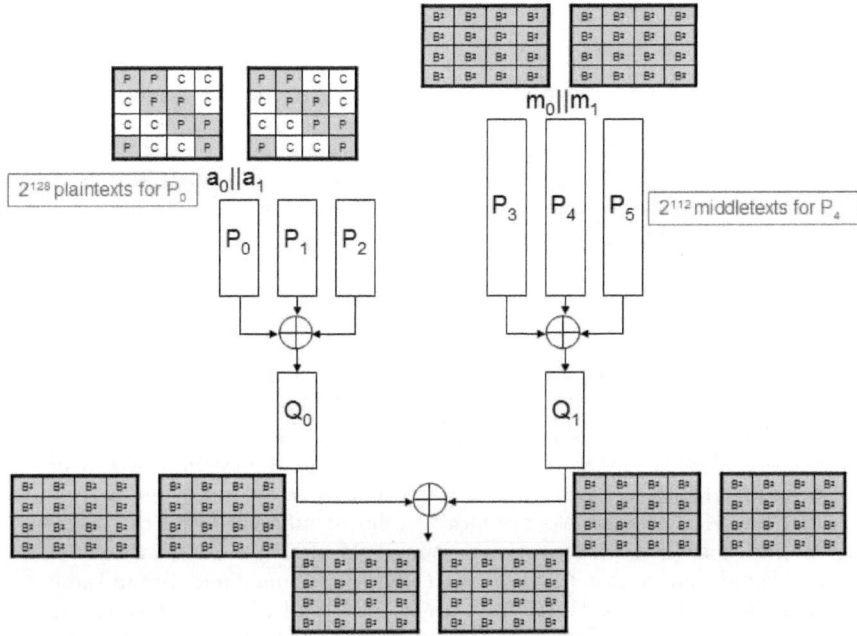

Fig. 14. The complete attack on the compression function where P_0, P_1 and P_2 are limited to 3 rounds using 2^{240} texts

Near-Collisions on the Reduced-Round Compression Functions of Skein and BLAKE

Bozhan Su, Wenling Wu, Shuang Wu, and Le Dong

State Key Laboratory of Information Security,
Institute of Software, Chinese Academy of Sciences, Beijing 100190, P.R. China
Graduate University of Chinese Academy of Sciences, Beijing 100049, P.R. China
{subozhan,wwl,wushuang,dongle}@is.iscas.ac.cn

Abstract. The SHA-3 competition organized by NIST [1] aims to find a new hash standard as a replacement of SHA-2. Till now, 14 submissions have been selected as the second round candidates, including Skein and BLAKE, both of which have components based on modular addition, rotation and bitwise XOR (ARX). In this paper, we propose improved near-collision attacks on the reduced-round compression functions of Skein and BLAKE. The attacks are based on linear differentials of the modular additions. The computational complexity of near-collision attacks on a 4-round compression function of BLAKE-32, 4-round and 5-round compression functions of BLAKE-64 are 2^{21}, 2^{16} and 2^{216} respectively, and the attacks on 20-round compression functions of Skein-256, Skein-512 and a 24-round compression function of Skein-1024 have a complexity of 2^{97}, 2^{52} and 2^{452} respectively.

Keywords: Hash function, Near-collision, SHA-3 candidates, Skein, BLAKE.

1 Introduction

Hash function, a very important component in cryptology, is a function of creating a short digest for a message of arbitrary length. The classical security requirements for such a function are preimage resistance, second-preimage resistance and collision resistance. In other words, it should be impossible to find a collision in less hash computations than birthday attack, or a (second)-preimage in less hash computations than brute force attack.

In recent years, the popular hash functions (MD4, MD5, RIPEMD, SHA-0 and SHA-1) have been seriously attacked [2–5]. As a response to advances in the cryptanalysis of hash functions, NIST launched a public competition to develop a new hash function called SHA-3. Till now, 14 submissions have been selected as the second round candidates.

Skein and BLAKE are two of the second round candidates of SHA-3. Skein uses the UBI chaining mode, while BLAKE uses HAIFA approach. Both of them are of the ARX (Addition-Rotate-XOR) type. More specifically, their design primitives use only addition, rotation and XOR.

S.-H. Heng, R.N. Wright, and B.-M. Goi (Eds.): CANS 2010, LNCS 6467, pp. 124–139, 2010.
© Springer-Verlag Berlin Heidelberg 2010

Previous works studied the linear differential trails of non-linear operations such as boolean functions and modular additions. Linear differential trails can be constructed to find near-collisions of these hash functions [7, 9, 10, 13]. Recently, linear differential attacks have been applied to many SHA-3 candidates, such as EnRUPT, CubeHash, MD6, and BLAKE [8–10].

In this paper, we further study the linear differential techniques and propose near-collision attacks on the reduced-round compression functions of Skein and BLAKE. Our strategy to find optimal linear differential trails can be described in three steps. First, linear approximations of reduced-round compression functions of Skein and BLAKE is constructed. In this step, all the addition modulo 2^{64} components of Skein and BLAKE are approximated by bitwise XOR of the inputs. Second, we select some intermediate state as a starting point and place a low Hamming weight difference in it. Third, the difference above propagates in both forward and backward directions until the probability becomes too small to obtain near collisions. Table 1 summarizes our attack along with the previously known ones on the reduced-round compression functions of Skein and BLAKE.

Table 1. Comparison of results on the reduced-round compression functions of Skein and BLAKE

Target	Rounds	Time	Memory	Type	Authors
Skein-512	17	2^{24}	-	434-bit near-collision	[12]
Skein-256	20	2^{97}	-	130-bit near-collision	✓
Skein-512	20	2^{52}	-	266-bit near-collision	✓
Skein-1024	24	2^{452}	-	512-bit near-collision	✓
BLAKE-32	4	2^{56}	-	232-bit near-collision	[13]
BLAKE-32	4	2^{21}	-	152-bit near-collision	✓
BLAKE-64	4	2^{16}	-	396-bit near-collision	✓
BLAKE-64	5	2^{216}	-	306-bit near-collision	✓

The paper is organized as follows. In Section 2, we describe Skein and BLAKE hash functions. In Section 3, the linear differential technique is applied to Skein and present near-collisions for Skein's compression function with reduced-round Threefish-256, Threefish-512 and Threefish-1024. In Section 4, we apply the linear differential technique to BLAKE and obtain near-collisions for reduced-round compression functions of BLAKE. Finally, Section 5 summarizes this paper.

2 Description of Skein and BLAKE

2.1 Skein

Skein is a family of hash functions based on the tweakable block cipher Threefish, which has equal block and key size of either 256, 512, or 1,024 bits. The MMO (Matyas-Meyer-Oseas) mode is used to construct the Skein compression function from Threefish. The format specification of the tweak and a padding scheme defines the so-called Unique Block Iteration (UBI) chaining mode. UBI is used for IV generation, message compression, and as output transformation.

Threefish consists of a number of similar rounds, which is based on three simple operations: Addition modulo 2^{64}, Rotation and XOR. The intermediate state of Threefish is organized as a number of 64-bit words. The letter Δ stands for a difference in the most significant bit (MSB), i.e., $\Delta = 0x8000000000000000$. Subkeys are derived from the cipher key K and tweak $T = (t_0, t_1)$ through a simple key schedule.

Let N_w denote the number of words in the key and the plaintext block, N_r be the number of rounds. For Threefish-256, $N_w = 4$ and $N_r = 72$. Let $v_{d,i}$ be the value of the ith word of the encryption state after d rounds. The procedure of Threefish-256 encryption is:

1. $(v_{0,0}, v_{0,1}, \cdots, v_{0,N_w-1}) := (p_0, p_1, \cdots, p_{N_w-1})$, where (p_0, p_1, p_2, p_3) is the 256-bit plaintext.

2. For each round, we have

$$e_{d,i} := \begin{cases} (v_{d,i} + k_{d/4,i}) \bmod 2^{64} & \text{if } d \bmod 4=0, \\ v_{d,i} & \text{otherwise.} \end{cases}$$

Where $k_{d/4,i}$ is the i-th word of the subkey added to the d-th round. For $i = 0, 1, \cdots, N_w - 1$, $d = 0, 1, \cdots, N_r - 1$.

3. Mixing and word permutations followed:

$$(f_{d,2j}, f_{d,2j+1}) := \text{MIX}_{d,j}(e_{d,2j}, e_{d,2j+1}), \qquad j = 0, \cdots, N_w/2 - 1,$$
$$v_{d+1,i} := f_{d,\pi(i)}, \qquad\qquad\qquad i = 0, \cdots, N_w - 1,$$

where the MIX operation depicted in Figure 1 transforms two of these 64-bit words and is common to all Threefish variants, with $R_{d,i}$ rotation constant depending on the Threefish block size, the round index d and the position of the two 64-bit words i in the Threefish state. The permutation $\pi(.)$ and the rotation constant $R_{d,i}$ can be referred to [14].

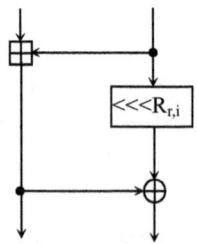

Fig. 1. The MIX function

After N_r rounds, the ciphertext $C = (c_0, c_1, \cdots, c_{N_w-1})$ is given as follows:

$$c_i := (v_{N_r,i} + k_{N_r/4,i}) \bmod 2^{64} \qquad\qquad \text{for } i = 0, 1, \cdots, N_w - 1.$$

The s-th keying ($d = 4s$) uses subkeys $k_{s,0}, \cdots, k_{s,N_w-1}$. These are derived from the key $k_0, \cdots, k_{N_w-1}$ and from the tweak t_0, t_1 as follows:

$$
\begin{aligned}
k_{s,i} &:= k_{(s+i) \bmod (N_w+1)} & \text{for } i = 0, \cdots, N_w - 4 \\
k_{s,i} &:= k_{(s+i) \bmod (N_w+1)} + t_{s \bmod 3} & \text{for } i = N_w - 3 \\
k_{s,i} &:= k_{(s+i) \bmod (N_w+1)} + t_{(s+1) \bmod 3} & \text{for } i = N_w - 2 \\
k_{s,i} &:= k_{(s+i) \bmod (N_w+1)} + s & \text{for } i = N_w - 1
\end{aligned}
$$

where $k_{N_w} := \lfloor 2^{64}/3 \rfloor \oplus \bigoplus_{i=0}^{N_w-1} k_i$ and $t_2 := t_0 \oplus t_1$.

2.2 BLAKE

The BLAKE family of hash functions is designed by Aumasson et al. [11] and follows HAIFA structure [6] with internal wide-pipe design strategy. Two versions of BLAKE are available: a 32-bit version (BLAKE-32) for message digests of 224 bits and 256 bits operates on 32-bit words, and a 64-bit version (BLAKE-64) for message digests of 384 bits and 512 bits operates on 64-bit words.

BLAKE operates on a large inner state v which is represented as a 4×4 matrix of words. The compression function consists of three steps: Initialization, 14 iterations of Rounds and Finalization as illustrated in Figure 2.

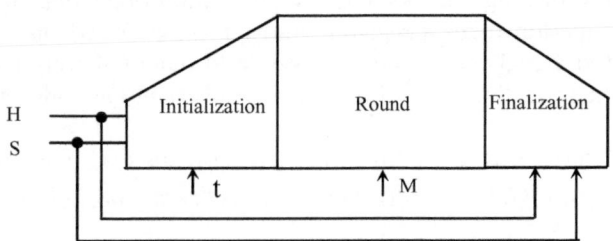

Fig. 2. Overall Structure of Compression Function of BLAKE

During the First step, the inner state v is initialized from 8 words of the chaining value $h = h_0, \cdots, h_7$, 4 words of the salt S and 2 words of block index (t_0, t_1) as follows:

$$
\begin{pmatrix}
v_0 & v_1 & v_2 & v_3 \\
v_4 & v_5 & v_6 & v_7 \\
v_8 & v_9 & v_{10} & v_{11} \\
v_{12} & v_{13} & v_{14} & v_{15}
\end{pmatrix}
\longleftarrow
\begin{pmatrix}
h_0 & h_1 & h_2 & h_3 \\
h_4 & h_5 & h_6 & h_7 \\
s_0 \oplus c_0 & s_1 \oplus c_1 & s_2 \oplus c_2 & s_3 \oplus c_3 \\
t_0 \oplus c_4 & t_0 \oplus c_5 & t_1 \oplus c_6 & t_1 \oplus c_7
\end{pmatrix}
$$

Then, a series of 14 rounds is performed. Each round is based on the stream cipher ChaCha [15] and consists of the eight round-dependent transformations $G_0, \cdots, G_7$. Figure 3 and Figure 4 show the G function of BLAKE-32 and

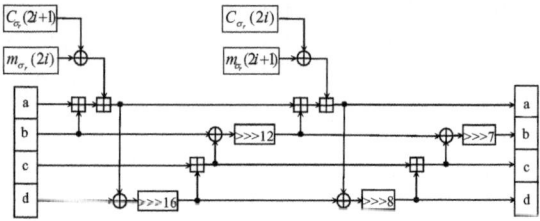

Fig. 3. The G function of BLAKE-32 for index i

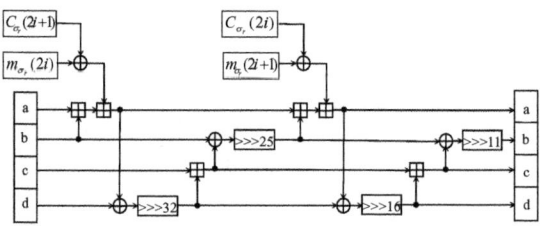

Fig. 4. The G function of BLAKE-64 for index i

BLAKE-64 for index i respectively, where σ_r is a fixed permutation used in round r, M_{σ_r} are message blocks and C_{σ_r} are round-dependent constants. The $G_i(0 \le i \le 7)$ function takes 4 registers and 2 message words as input and outputs the updated 4 registers. A column step and diagonal step update the four columns and the four diagonals of matrix v respectively as follows:

$$G_0(v_0, v_4, v_8, v_{12}) \quad G_1(v_1, v_5, v_9, v_{13}) \quad G_2(v_2, v_6, v_{10}, v_{14}) \quad G_3(v_3, v_7, v_{11}, v_{15})$$
$$G_4(v_0, v_5, v_{10}, v_{15}) \quad G_5(v_1, v_6, v_{11}, v_{12}) \quad G_6(v_2, v_7, v_8, v_{13}) \quad G_7(v_3, v_4, v_9, v_{14})$$

In the last step, the new chaining value $h' = h'_0, \cdots, h'_7$ is computed from the internal state v and the previous chain value h (Finalization step):

$$
\begin{array}{l|l}
h'_0 \leftarrow h_0 \oplus s_0 \oplus v_0 \oplus v_8 & h'_4 \leftarrow h_4 \oplus s_4 \oplus v_4 \oplus v_{12} \\
h'_1 \leftarrow h_1 \oplus s_1 \oplus v_1 \oplus v_9 & h'_5 \leftarrow h_5 \oplus s_5 \oplus v_5 \oplus v_{13} \\
h'_2 \leftarrow h_2 \oplus s_2 \oplus v_2 \oplus v_{10} & h'_6 \leftarrow h_6 \oplus s_6 \oplus v_6 \oplus v_{14} \\
h'_3 \leftarrow h_3 \oplus s_3 \oplus v_3 \oplus v_{11} & h'_7 \leftarrow h_7 \oplus s_7 \oplus v_7 \oplus v_{15}
\end{array}
$$

3   Near-Collisions for the Reduced-Round Compression Function of Skein

Skein is based on the UBI (Unique Block Iteration) chaining mode that uses Threefish block cipher to build a compression function. The compression function outputs $E_k(t, m) \oplus m$, where E is Threefish.

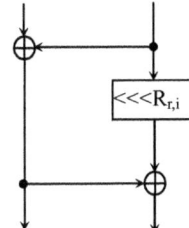

Fig. 5. linearized MIX function in Threefish

Since the MIX function is the only non-linear component in the Threefish block cipher, the first step is to linearize the MIX function to obtain linear approximations of the Compression Function of Skein. To Linearize the MIX function, We replace the modular addition with XOR. The linearized MIX function is illustrated in Figure 5.

3.1 Near Collisions for the 20-Round Compression Function of Skein-256

After linearizing the Compression Function of Skein-256, we need to choose the starting point. Since Skein-256 has 72 rounds, there are $72 \approx 2^6$ possible choices. Then we place one or two bits of differences in the message blocks and certain round of the intermediate state at the starting point. Since compression function of Skein-256 uses 256-bit message and 256-bit state, there are $\binom{512}{1} + \binom{512}{2} \approx 2^{17}$ choices of positions for the one or two bits above. Therefore, the search space is less than 2^{23}, which can be searched exhaustively.

Our aim is to find one path with the highest probability in the search space. As introduced in [9], we can calculate probability of one differential trail by counting Hamming weight of the differences. We search for 24-round differential trail and the results are introduced as follows.

The difference Δ in k_2 and t_0 gives a difference $(\Delta, \Delta, 0, 0)$ at the third subkey, and $(0, 0, 0, 0)$ after the fourth. The difference in the state of round 8 is canceled out at the third subkey which is then turned into an eight-round local collision from round 9 to round 16. After 20 rounds, the Hamming weight of the difference becomes too large to obtain near collisions. In the 20-th round, after adding the final subkey and feedforward value, one obtains a collision on $256 - 126 = 130$ bits. Table 2 shows the corresponding differential trail of the key and the tweak from the 0-th round to the 19-th round. Table 3 presents the corresponding trail from the 0-th round to the 19-th round. In the table, the probability for all rounds are given, except for the first round, which are indicated with M as we will use message modification techniques to make sure the first round of the trail fulfills.

Table 2. Details of the subkeys and of their differences of Skein-256, given a difference in k_2 and t_0

Rd	d	$k_{s,0}$	$k_{s,1}$	$k_{s,2}$	$k_{s,3}$
0	0	k_0	$k_1 + t_0$	$k_2 + t_1$	k_3
		0	Δ	Δ	0
1	4	k_1	$k_2 + t_1$	$k_3 + t_2$	k_4
		0	Δ	Δ	Δ
2	8	k_2	$k_3 + t_2$	$k_4 + t_0$	k_0
		Δ	Δ	0	0
3	12	k_3	$k_4 + t_0$	$k_0 + t_1$	k_1
		0	0	0	0
4	16	k_4	$k_0 + t_1$	$k_1 + t_2$	k_2
		Δ	0	Δ	Δ
5	20	k_0	$k_1 + t_2$	$k_2 + t_0$	k_3
		0	Δ	0	0

Table 3. Differential trail used for near collision of a 20-round compression function of Skein-256, with probability of 2^{-97}

Rd	Difference				Pr
0	b0dff57c25c19314	a5b2b6692bd196c8	861349393b7673c0	3c708bb2d1caf2d2	-
1	e82d8c56764c8096	956d43150e1005dc	601166d49d04b503	3a63c28beabc8112	M
2	0a44a5491af1e45a	7d40cf43785c854a	5090945bd4b01c4b	5a72a45f77b83411	M
3	2708680a86a06010	77046a0a62ad6110	86e030002608280a	0ae23004a308285a	M
4	5004000044050100	500c0200e40d0100	8400000405000050	8c02000485000050	M
5	0008000020080000	80080200a0080000	0802000000000000	0802000080000000	2^{-58}
6	0000020000000000	8000020080000000	0000000000000000	0000000080000000	2^{-8}
7	0000000080000000	8000000080000000	0000000080000000	0000000080000000	2^{-3}
8	8000000000000000	8000000000000000	0000000000000000	0000000000000000	2^{-2}
	no differences in round 9 - 16				1
17	0000000000002000	8000000000000000	8000000000008000	0000000000000000	1
18	8008000000008008	8000000000002000	8000000000002040	8000000000008000	2^{-2}
19	000000102040a040	000800000000a008	008808800800a008	000000000000a040	2^{-7}
20	a156edfd2dd5925c	25bab6790b919680	8e0f41291b36718c	3cf88332d9caf29a	2^{-17}

The message modification are applied to the most expensive part in our trail, namely the first round. Freedom degrees in chaining value and the message can be used to fulfill the first round of the trail. We use techniques introduced in [9] to derive sufficient conditions for each modular addition of the first round of the trail. Then the message block and the chaining value are chosen according to the conditions.

Table 4. Details of the subkeys and of their differences of Skein-512, given a difference in k_4, k_5 and t_0 (leading to a differences in t_2)

Rd	d	$k_{s,0}$	$k_{s,1}$	$k_{s,2}$	$k_{s,3}$	$k_{s,4}$	$k_{s,5}$	$k_{s,6}$	$k_{s,7}$
5	20	k_5	k_6	k_7	k_8	k_0	k_1+t_2	k_2+t_0	k_3
		0	0	0	0	0	Δ	0	Δ
6	24	k_6	k_7	k_8	k_0	k_1	k_2+t_0	k_3+t_1	k_4
		0	0	0	0	0	0	0	Δ
7	28	k_7	k_8	k_0	k_1	k_2	k_3+t_1	k_4+t_2	k_5
		0	0	0	0	0	0	0	0
8	32	k_8	k_0	k_1	k_2	k_3	k_4+t_2	k_5+t_0	k_6
		0	0	0	0	Δ	0	0	0
9	36	k_0	k_1	k_2	k_3	k_4	k_5+t_0	k_6+t_1	k_7
		0	0	0	Δ	Δ	0	Δ	0
10	40	k_1	k_2	k_3	k_4	k_5	k_6+t_1	k_7+t_2	k_8
		0	0	Δ	Δ	0	Δ	Δ	0

Table 5. Details of the subkeys and of their differences of Skein-1024, given a difference in k_0, k_2 and t_1 (leading to a differences in t_2)

Rd	d	$k_{s,0}$	$k_{s,1}$	$k_{s,2}$	$k_{s,3}$	$k_{s,4}$	$k_{s,5}$	$k_{s,6}$	$k_{s,7}$	$k_{s,8}$	$k_{s,9}$	$k_{s,10}$	$k_{s,11}$	$k_{s,12}$	$k_{s,13}$	$k_{s,14}$	$k_{s,15}$
0	0	k_0	k_1	k_2	k_3	k_4	k_5	k_6	k_7	k_8	k_9	k_{10}	k_{11}	k_{12}	$k_{13}+t_0$	$k_{14}+t_1$	k_{15}
		0	Δ	0	0	0	0	0	0	0	0	0	0	0	Δ	Δ	0
1	4	k_1	k_2	k_3	k_4	k_5	k_6	k_7	k_8	k_9	k_{10}	k_{11}	k_{12}	k_{13}	$k_{14}+t_1$	$k_{15}+t_2$	k_0
		Δ	0	0	0	0	0	0	0	0	0	0	0	0	Δ	0	Δ
2	8	k_2	k_3	k_4	k_5	k_6	k_7	k_8	k_9	k_{10}	k_{11}	k_{12}	k_{13}	k_{14}	$k_{15}+t_2$	k_0+t_0	k_1
		0	0	0	0	0	0	0	0	0	0	0	0	0	0	0	0
3	12	k_3	k_4	k_5	k_6	k_7	k_8	k_9	k_{10}	k_{11}	k_{12}	k_{13}	k_{14}	k_{15}	k_0+t_0	k_1+t_1	k_2
		0	0	0	0	0	0	0	0	0	0	0	0	0	0	Δ	Δ
4	16	k_4	k_5	k_6	k_7	k_8	k_9	k_{10}	k_{11}	k_{12}	k_{13}	k_{14}	k_{15}	k_0	k_1+t_1	k_2+t_2	k_3
		0	0	0	0	0	0	0	0	0	0	0	0	Δ	Δ	Δ	0
5	20	k_5	k_6	k_7	k_8	k_9	k_{10}	k_{11}	k_{12}	k_{13}	k_{14}	k_{15}	k_0	k_1	k_2+t_2	k_3+t_0	k_4
		0	0	0	0	0	0	0	0	0	0	0	Δ	0	Δ	Δ	0
6	24	k_6	k_7	k_8	k_9	k_{10}	k_{11}	k_{12}	k_{13}	k_{14}	k_{15}	k_0	k_1	k_2	k_3+t_0	k_4+t_1	k_5
		0	0	0	0	0	0	0	0	0	0	Δ	0	Δ	Δ	Δ	0

3.2 Near Collisions for the 20-Round Compression Functions of Skein-512 and Skein-1024

Ideas for near collision attacks on the reduced-round compression functions of Skein-512 and Skein-1024 are similar to the one of Skein-256. So we skip explanations here. In Table 4 and Table 5, we propose difference in the key schedule of Skein-512 and Skein-1024. The differential trails for them are illustrated in Table 6 and Table 7 in the appendix.

4 Near Collisions for the Reduced-Round Compression Function of BLAKE

4.1 Linearizing G Function of BLAKE-32 and BLAKE-64

In order to linearize the G function, modular additions are replaced with XORs. Near collision attack for a 4-round compression function of BLAKE-32 in [13] also uses the linearization technique. The cyclic rotation constants in BLAKE-32 are 16,12,8,7. Notice that three of the constants 16,12 and 8 have a greatest common divisor 4, so difference $0xAAAAAAAA$ is cyclic invariant with these rotation constants, where A is a 4-bit value. In the linearized BLAKE-32, if all differences in registers are restricted to this pattern, cyclic rotations difference $>>> 16, >>> 12$ and $>>> 8$ can be removed. If zero differences pass through $>>> 7$, the only possible difference pattern in registers is either $0xAAAAAAAA$ or zero which can be indicated as 1-bit value. So the linear differential trails with this difference pattern form a small space of size 2^{32}, which can be searched by brute force. The linear differential trail in [13] is the best one in this space. But this attack doesn't work on BLAKE-64, because the cyclic rotation constants are different. BLAKE-64 uses the number of rotations 32, 25, 16 and 11. Two of them are not multiples of 4, which implies more restrictions of the differential trail.

To obtain near collisions for a reduced-round compression function of BLAKE-64 and improve the previous near-collision attack on a reduced-round compression function of BLAKE-32 in [13], we have to release the restrictions. This can be done in two ways: using non-linear differential trail instead of linear one, or still using linear differential trail but releasing restrictions on the differential pattern. In this paper, we use linear differential trail and try to release restrictions on the differential pattern. Instead of using cyclic invariant differences, we use a random difference of Hamming weight less than or equal to two in the intermediate states.

Since we intend to release restrictions on the differential pattern, the cyclic invariant differential pattern in previous works is not used. So the cyclic rotations can not be removed.

Figure 6 and Figure 7 show the linearized G function of BLAKE-32 and BLAKE-64 respectively.

4.2 Searching for Differential Trails with High Probability

We need to choose the starting point after linearizing G function. Since BLAKE-32 has 10 rounds and BLAKE-64 has 14 rounds, there are less than 2^4 possible choices. Then we place one or two bits of differences in the message blocks and

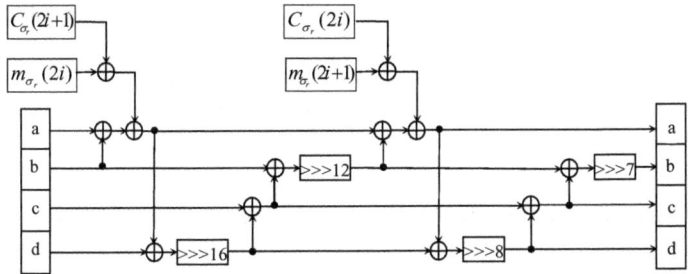

Fig. 6. linearized G function in BLAKE-32

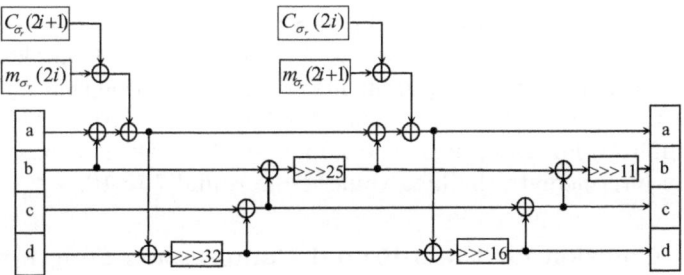

Fig. 7. linearized G function in BLAKE-64

certain round of the intermediate state at the starting point. Because compression function of BLAKE-32 uses 512-bit message and 512-bit state and compression function of BLAKE-64 uses 1024-bit message and 1024-bit state, there are $\binom{1024}{1} + \binom{1024}{2} \approx 2^{19}$ and $\binom{2048}{1} + \binom{2048}{2} \approx 2^{21}$ choices of positions for the pair of bits on BLAKE-32 and BLAKE-64 respectively. Therefore, the search spaces for BLAKE-32 and BLAKE-64 are less than 2^{23} and 2^{25} respectively, which can be explored exhaustively.

Our aim is to find one path with the highest probability in the search space. Furthermore, following Section 3.1, we calculate probability of one differential trail by counting Hamming weight in the differences. We search for differential trails of 4-round compression function of BLAKE-32, 4-round and 5-round compression functions of BLAKE-64. And the results are introduced in the following sections.

4.3   Near Collision for 4-Round Compression Function of BLAKE-32

We search with the configuration where differences are in $m[0] = 0x80008000$ and $v[0, 2, 4, 8, 10]$ and find that a starting point at round 4 leads to a linear differential trail whose total Hamming weight is 21. We don't need to count for the last round, since it can be fulfilled by message modifications with similar techniques used in attacks on Skein.

So, This trail can be fulfilled with probability of 2^{-21}. Complexity of this attack is 2^{21} with no memory requirements. With assumption that no differences in the salt value, this configuration has a final collision on $256 - 104 = 152$ bits after the finalization. Table 8 in the appendix demonstrates how differences propagate in intermediate chaining values from round 4 to 7.

4.4   Near Collision for the 4-Round Compression Function of BLAKE-64

We search with the configuration where differences are in $m[11] = 0x80000000$ 80000000 and $v[0, 2, 4, 8, 10]$ and find that a starting point at round 7 leads to a linear differential trail whose total Hamming weight is equal to 16. We don't need to count for the last round, since it can be fulfilled by message modifications with similar techniques used in attacks on Skein.

So, This trail can be fulfilled with probability of 2^{-16}. Complexity of this attack is 2^{16} with no memory requirements. With assumption that no differences in the salt value, this configuration has a final collision on $512 - 116 = 396$ bits after the finalization. Table 9 in the appendix demonstrates how differences propagate in intermediate chaining values from round 7 to 10.

4.5   Near Collision for the 5-Round Compression Function of BLAKE-64

Then we search for 5-round differential trails, with the configuration where differences are placed in $m[11] = 0x8000000080000000$ and $v[0, 2, 4, 8, 10]$. We find that a starting point at round 7 leads to a linear differential trail whose total Hamming weight is 216. This trail with probability of 2^{-216} is illustrated in Table 10 of the appendix, which leads to a $512 - 206 = 306$-bit collision after feedforward. The message modifications are also applied to the last round.

5   Conclusion

In this paper, we revisited the linear differential techniques and applied it to two ARX-based hash functions: Skein and BLAKE. Our attacks include near-collision attacks on the 20-round compression functions of Skein-256, Skein-512 and the 24-round compression function of Skein-1024, the 4-round compression function of BLAKE-32, and the 4-round and 5-round compression functions of BLAKE-64. Future works might apply some non-linear differentials for integer addition besides XOR differences to improve our results.

Acknowledgment

The authors would like to thank the anonymous referees for their valuable comments. Furthermore, this work is supported by the National Natural Science Foundation of China (No. 60873259, and No. 60903212) and the Knowledge Innovation Project of The Chinese Academy of Sciences.

References

1. National Institute of Standards and Technology: Announcing Request for Candidate Algorithm Nominations for a New Cryptographic Hash Algorithm (SHA-3) Family. Federal Register 27(212), 62212–62220 (2007), http://csrc.nist.gov/groups/ST/hash/documents/FR_Notice_Nov07.pdf (17/10/2008)
2. Wang, X., Lai, X., Feng, D., Chen, H., Yu, X.: Cryptanalysis of the Hash Functions MD4 and RIPEMD. In: Cramer, R. (ed.) EUROCRYPT 2005. LNCS, vol. 3494, pp. 1–18. Springer, Heidelberg (2005)
3. Wang, X., Yu, H.: How to Break MD5 and Other Hash Functions. In: Cramer, R. (ed.) EUROCRYPT 2005. LNCS, vol. 3494, pp. 19–35. Springer, Heidelberg (2005)
4. Wang, X., Yu, H., Yin, Y.L.: Efficient Collision Search Attacks on SHA-0. In: Shoup, V. (ed.) CRYPTO 2005. LNCS, vol. 3621, pp. 1–16. Springer, Heidelberg (2005)
5. Wang, X., Yin, Y.L., Yu, H.: Finding Collisions in the Full SHA-1. In: Shoup, V. (ed.) CRYPTO 2005. LNCS, vol. 3621, pp. 17–36. Springer, Heidelberg (2005)
6. Biham, E., Dunkelman, O.: A Framework for Iterative Hash Functions - HAIFA. In: Second NIST Cryptographic Hash Workshop, Santa Barbara, California, USA, August 24-25 (2006)
7. Chabaud, F., Joux, A.: Differential Collisions in SHA-0. In: Krawczyk, H. (ed.) CRYPTO 1998. LNCS, vol. 1462, pp. 56–71. Springer, Heidelberg (1998)
8. Indesteege, S., Preneel, B.: Practical Collisions for EnRUPT. In: Dunkelman, O. (ed.) FSE 2009. LNCS, vol. 5665, pp. 122–138. Springer, Heidelberg (2009)
9. Brier, E., Khazaei, S., Meier, W., Peyrin, T.: Linearization Framework for Collision Attacks: Application to Cubehash and MD6. In: Matsui, M. (ed.) ASIACRYPT 2009. LNCS, vol. 5912, pp. 560–577. Springer, Heidelberg (2009)
10. Rijmen, V., Oswald, E.: Update on SHA-1. In: Menezes, A. (ed.) CT-RSA 2005. LNCS, vol. 3376, pp. 58–71. Springer, Heidelberg (2005)
11. Aumasson, J.-P., Henzen, L., Meier, W., Phan, R.C.-W.: SHA-3 proposal BLAKE, version 1.3 (2008), http://131002.net/blake/blake.pdf
12. Aumasson, J.-P., Çalik, Ç., Meier, W., Özen, O., Phan, R.C.-W., Varici, K.: Improved Cryptanalysis of Skein. In: Matsui, M. (ed.) ASIACRYPT 2009. LNCS, vol. 5912, pp. 542–559. Springer, Heidelberg (2009)
13. Aumasson, J.-P., Guo, J., Knellwolf, S., Matusiewicz, K., Meier, W.: Differential and Invertibility Properties of BLAKE. In: Beyer, I. (ed.) FSE 2010. LNCS, vol. 6147, pp. 318–332. Springer, Heidelberg (2010)
14. Ferguson, N., Lucks, S., Schneier, B., Whiting, D., Bellare, M., Kohno, T., Callas, J., Walker, J.: The Skein Hash Function Family. Submission to NIST (2008)
15. Bernstein, D.J.: ChaCha, a variant of Salsa20 (January 2008), http://cr.yp.to/chacha/chacha-20080128.pdf

A   Differential Trails of Reduced-Round Skein and BLAKE

Table 6. Differential trail used for near collision of 20-round Skein-512, with probability of 2^{-52}

Rd	Difference				Pr
20	0000000010004800	0020001000004000	0002201000080000	0000200000080000	-
	8000000020000200	8000000020000200	0000088000080000	8000008000080000	
21	0002001000000000	0000001000000000	8000000000000000	8000000000000000	2^{-35}
	0000080000000000	0000080000000000	0020001010000800	0000001000000800	
22	0000000000000000	0000000000000000	0000000000000000	0000000000000000	2^{-7}
	0020000010000000	0020000000000000	0002000000000000	0002000000000000	
23	0000000000000000	0000000000000000	0000000010000000	0000000010000000	2^{-3}
	0000000000000000	0000000000000000	0000000000000000	0000000000000000	
24	0000000000000000	0000000000000000	0000000000000000	0000000000000000	2^{-1}
	0000000000000000	0000000000000000	0000000000000000	8000000000000000	
	no differences in round 25 - 32				1
33	0000000000000000	0000000000000000	8000000000000000	0000000000000000	1
	0000000000000000	8000000000000000	0000000000000000	0000000000000000	
34	8000000000000000	0000000000000000	8000000000000000	0000000000000000	1
	0000000000000000	8000000000002000	0000000000000000	8000000000000000	
35	8000000000000000	8000000000000000	8000000000002000	8000004000000000	2^{-1}
	8000000000000000	8002000800002000	8000000000000000	8000000000000000	
36	0000004000002000	0000080000000000	0002000800002000	0080000000000000	2^{-5}
	0000000000000000	0022008802002008	0000000000000000	0000804000002100	
37	8082000800002000	0000084000042000	8022008802002008	c000806100002180	M
	8000804000002100	882280a802882228	0000084000002000	8082000820202000	
38	402280e902000188	818a084884040000	082200e802880328	8092480860210104	M
	8082084820200000	8220a0e22200a108	8082084800040000	c62180eb03840188	
39	88b048e062a9022c	50a080a187071598	02a2a8aa0220a108	66afce920f875994	M
	46a388a303800188	02f22ceb1270d019	c1a888a186040188	84b468c0f2bb4b2d	
40	640d66381da7b09c	78b069d6e2bbcfe4	c453845811f8d191	f5206eb3bfd667bf	M
	c51ce06154bf48a5	5d535664dae2a341	5810c0c1e5a617b4	9837aa1b38d18c0c	

Table 7. Differential trail used for near collision of Skein-1024, of probability 2^{-452}

Rd	Difference				Pr
0	8140008142000042	8040008100000042	0000000000080040	0000000000000040	
	0000000000000080	0000000000000080	4100000100488224	4000000100480200	
	0001000000024040	0001000000020040	0010208010000000	0010008010000000	-
	2000000000000000	a000000000000000	8000044000008002	0000040000000002	
1	8100000042000000	0100000002000000	0000000000080000	0000000000080000	
	0100000000008024	0100000000000020	0000000000000000	0000000000000000	2^{-87}
	0000200000000000	0000200000000000	0000000000000000	0000000000000000	
	0000040000008000	0000000000008000	0000000000004000	0000000000004000	
2	8000000040000000	0000000040000000	0000000000000000	0000000000000000	
	0000000000000000	0000000000000000	0000000000008004	0000000000000004	2^{-12}
	0000000000000000	0000000000000000	0000040000000000	0000040000000000	
	0000000000000000	0000000000000000	0000000000000000	0000000000000000	
3	8000000000000000	0000000000000000	0000000000000000	0000000000000000	
	0000000000008000	0000000000008000	0000000000000000	0000000000000000	2^{-4}
	0000000000000000	0000000000000000	0001008000000000	0000000000000000	
	0000000000000000	0000000000000000	0000000000000000	0000000000000000	
4	8000000000000000	0000000000000000	0000000000a00000	0000000000000000	
	0000000000000000	0000000000000000	0000000000000000	0000000000000000	2^{-1}
	0000000000000000	0000000000000000	0000000000000000	0000000000000000	
	0000000000000000	8000000000000000	0000000000000000	8000000000000000	
	no differences in round 5 - 12				1
13	0000000000000000	0000000000000000	0000000000000000	0000000000000000	
	0000000000000000	0000000000000000	0000000000000000	0000000020000000	1
	0000000000000000	0000000000000000	0000000000000000	0000000000000000	
	0000000000000000	0000000000000000	0000000000000000	0000000000000000	
14	0000000000000000	0000000000000000	0000000000000000	0000000000000000	
	0000000020000000	0000000000000000	0000000000000000	0000000000000000	2^{-1}
	0000000000000000	0000000020010000	0000000000000000	0000000000000000	
	0000000000000000	0000000000000000	0000000000000000	0000000000000000	
15	0000000000000000	0001000820010000	0000000000000000	0000000000000000	
	0000000000000000	0000000000000000	0000000020000000	0000000000000000	2^{-3}
	0000000000000000	0000000000000000	0000000000000000	0000000000000000	
	0000000000000000	0000000020000000	0000000020010000	0000000000000000	
16	0001000820010000	0000000000000000	0000000000000000	0000000020000004	
	0000000020000000	0000000000000000	0000000000000000	0000000020010000	2^{-8}
	0000000000000000	0000000020000000	0000000020000000	0000000000000000	
	0000000020010000	0000000000000000	0000000000000000	0201104822010000	
17	0001000820010000	0000002020000000	0000000020000004	0000000020210000	
	0000000020010000	0000000020000000	0000000020000000	c221104862230904	2^{-42}
	0000000020000000	8000000020011000	0000000020010000	0000040020008004	
	8201104822010000	0000000020000000	0000000020000000	0001000820010000	
18	0001002800010000	a001000000015002	0000000000210004	8211104802010000	
	c221104842230904	1000840200118004	0000000000010000	0001001800830010	2^{-47}
	0000040000018004	404000c066121880	8201104802010000	0001010800210004	
	0001000800010000	0000008000010000	8000000000011000	0001002800010808	
19	a000002800005002	d80866c167139b85	8211104802200004	0001008c00000800	
	0001001800820010	c000194000221004	d221944a42328900	8051002a1010180a	2^{-84}
	8200114002200004	2002000860800010	0001008800000000	a30014c82230000c	
	8001002800001808	d201144e6222898c	404004c066139884	a002a02d40025002	
20	780866e96713cb87	a20a014962a43054	821010c402200804	464e7644ebae4385	
	527094605222910a	21221440e8000140	c001195800a01014	bac2a04d2351cdc6	2^{-163}
	a30114402230000c	72408160f022c52a	5200146662229184	8ad010c482200814	
	e042a4ed2611c886	d005d95819e01036	a202114862a00014	7904bec58560bb3c	
21	da0267a005b7fbd3	d9579a22fe406202	c45e6680e98e4b81	b14418b56264592e	
	7ac3b91523f1ddd2	5cda01cae860d880	73528020ba22904a	3a5e8142f9819499	M
	58d004a2e0029990	eb6b5ff67e908df4	b0477db53ff1d8b0	58b2ef57b509410d	
	5b06af8de7c0bb28	7352a8249e601857	d1419520d212c526	9262cee411b56916	
22	0355fd82fbf799d1	050ea433779acb2a	751a7e358bea12af	a8355a6433003106	
	490c016243a304d3	c95e84bf600e3895	2619b8dfcb910552	29f176e0063a6413	M
	e8f592e28af899bd	edc5d39649b4c8df	285407a979a0a37f	fc8b1a8f4efa707a	
	43235bc4c3a7ac30	a64562de0179658a	b3bb5b549e921464	997703c299f54086	
23	065b59b18c6d52fb	99820cd285b33f4c	dd2f2451b8ea23a9	733e937e94f329ad	
	0fe8ce3fcdab6141	3d1ef6d41b30ee3e	805285dd23ad3c46	afffb2170a55bae5	M
	d4df1d26375ad305	96ec0443901360cf	e566391ac2dec9ba	de3f4ed2ed4c6099	
	2acc5896076754e2	9c18617261f28c41	05304174c34c5162	bd963c248eea00e2	
24	9fee540b09e9742f	63fb10d5c082c5c8	ae11bf272c2e139c	88b2be9fe5aeef4f	
	2fad3fc229cf87db	4dc84784c08d0ee2	32f638ebd6897253	067c7ad0439f7753	M
	c4a688375301a8c3	81b79521741b2223	36d439ed66a2d8a3	85f11291bf6796f7	
	38b482904da65194	6b71411a3e2c0f92	bea1c00ba749b3ce	9b8060686fe0cc74	

Table 8. Differential trail used for near collision of 4-round BLAKE-32, with probability of 2^{-21}

Rd	Difference	Pr
4	88008800 00000000 80008000 00000000	
	88008800 00000000 00000000 00000000	-
	80008000 00000000 80008000 00000000	
	00000000 00000000 00000000 00000000	
5	00000000 00000000 80008000 00000000	
	00000000 00000000 00000000 00000000	2^{-12}
	00000000 00000000 00000000 00000000	
	00000000 00000000 00000000 00000000	
6	00000000 00000000 00000000 00000000	
	00000000 00000000 00000000 00000000	2^{-1}
	00000000 00000000 00000000 00000000	
	00000000 00000000 00000000 00000000	
7	80088008 00000000 00000000 00000000	
	00000000 11101110 00000000 00000000	2^{-8}
	00000000 00000000 88008800 00000000	
	00000000 00000000 00000000 08000800	
8	28222822 18981898 11111111 19181918	
	33123312 44414441 02230223 32233223	M
	91919191 10101010 28222822 08080808	
	89918991 08800880 89918991 08880888	

Table 9. Differential trail used for near collision of 4-round BLAKE-64, with probability of 2^{-16}

Rd	Difference	Pr
7	8100000081000000 0000000000000000 8000000080000000 0000000000000000	
	8100000081000000 0000000000000000 0000000000000000 0000000000000000	-
	8000000080000000 0000000000000000 8000000080000000 0000000000000000	
	0000000000000000 0000000000000000 0000000000000000 0000000000000000	
8	0000000000000000 0000000000000000 8000000080000000 0000000000000000	
	0000000000000000 0000000000000000 0000000000000000 0000000000000000	2^{-12}
	0000000000000000 0000000000000000 0000000000000000 0000000000000000	
	0000000000000000 0000000000000000 0000000000000000 0000000000000000	
9	0000000000000000 0000000000000000 0000000000000000 0000000000000000	
	0000000000000000 0000000000000000 0000000000000000 0000000000000000	2^{-1}
	0000000000000000 0000000000000000 0000000000000000 0000000000000000	
	0000000000000000 0000000000000000 0000000000000000 0000000000000000	
10	8000000080000000 0000000000000000 0000000000000000 0000000000000000	
	0000000000000000 0000001000000010 0000000000000000 0000000000000000	2^{-3}
	0000000000000000 0000000000000000 0000800000008000 0000000000000000	
	0000000000000000 0000000000000000 0000000000000000 0000800000008000	
11	8240204082402040 a8402040a8402040 0850085008500850 2850200028502000	
	0a0002000a000200 0004400400044004 0010080000100800 0a110a010a110a01	M
	8850081088500810 2010285020102850 2240000022400000 a0002840a0002840	
	2840a0002840a000 0040000000400000 2840200028402000 2040804020408040	

Table 10. Differential trail used for near collision of 5-round BLAKE-64, with probability of 2^{-216}

Rd	Difference				Pr
7	8100000081000000	0000000000000000	8000000080000000	0000000000000000	
	8100000081000000	0000000000000000	0000000000000000	0000000000000000	
	8000000080000000	0000000000000000	8000000080000000	0000000000000000	-
	0000000000000000	0000000000000000	0000000000000000	0000000000000000	
8	0000000000000000	0000000000000000	8000000080000000	0000000000000000	
	0000000000000000	0000000000000000	0000000000000000	0000000000000000	
	0000000000000000	0000000000000000	0000000000000000	0000000000000000	2^{-12}
	0000000000000000	0000000000000000	0000000000000000	0000000000000000	
9	0000000000000000	0000000000000000	0000000000000000	0000000000000000	
	0000000000000000	0000000000000000	0000000000000000	0000000000000000	
	0000000000000000	0000000000000000	0000000000000000	0000000000000000	2^{-1}
	0000000000000000	0000000000000000	0000000000000000	0000000000000000	
10	8000000080000000	0000000000000000	0000000000000000	0000000000000000	
	0000000000000000	0000001000000010	0000000000000000	0000000000000000	
	0000000000000000	0000000000000000	0000800000008000	0000000000000000	2^{-3}
	0000000000000000	0000000000000000	0000000000000000	0000800000008000	
11	8240204082402040	a8402040a8402040	0850085008500850	2850200028502000	
	0a0002000a000200	0004400400044004	0010080000100800	0a110a010a110a01	2^{-200}
	8850081088500810	2010285020102850	2240000022400000	a0002840a0002840	
	2840a0002840a000	0040000000400000	2840200028402000	2040804020408040	
12	8a14284d8a14284d	8285222482852224	c2a442e0c2a442e0	4881023048810230	
	001d0aac001d0aac	1b001a111b001a11	4aa500044aa50004	0c284c3c0c284c3c	M
	6ab4c0e56ab4c0e5	c26048d1c26048d1	2851a04d2851a04d	0a6122d00a6122d0	
	0081aa700081aa70	28c0209128c02091	2885223428852234	0091a8950091a895	

Practical Algebraic Cryptanalysis for Dragon-Based Cryptosystems

Johannes Buchmann[1], Stanislav Bulygin[2], Jintai Ding[3],
Wael Said Abd Elmageed Mohamed[1], and Fabian Werner[4]

[1] TU Darmstadt, FB Informatik Hochschulstrasse 10, 64289 Darmstadt, Germany
{buchmann,wael}@cdc.informatik.tu-darmstadt.de
[2] Center for Advanced Security Research Darmstadt (CASED)
Stanislav.Bulygin@cased.de
[3] Department of Mathematical Sciences, University of Cincinnati,
Cincinnati OH 45220, USA
jintai.ding@uc.edu
[4] TU Darmstadt
fw@cccmz.de

Abstract. Recently, the Little Dragon Two and Poly-Dragon multivariate based public-key cryptosystems were proposed as efficient and secure schemes. In particular, the inventors of the two schemes claim that Little Dragon Two and Poly-Dragon resist algebraic cryptanalysis. In this paper, we show that MXL2, an algebraic attack method based on the XL algorithm and Ding's concept of Mutants, is able to break Little Dragon Two with keys of length up to 229 bits and Poly-Dragon with keys of length up to 299. This contradicts the security claim for the proposed schemes and demonstrates the strength of MXL2 and the Mutant concept. This strength is further supported by experiments that show that in attacks on both schemes the MXL2 algorithm outperforms the Magma's implementation of F4.

1   Introduction

The multivariate-based public-key cryptosystems (MPKCs) are public-key cryptosystems that are based on the problem of solving multivariate quadratic equations over finite fields. This problem is called "MQ-problem" and it is NP-complete [1]. Several MPKCs based on the MQ-problem have been proposed in the last two decades. An overview of MPKCs can be found in [2,3].

Recently, in the International Journal of Network Security & Its Applications (IJNSA), Singh et al. presented a new multivariate-based public-key encryption scheme which is called Little Dragon Two (LD2 for short) that is constructed using permutation polynomials over finite fields [4]. According to the authors, LD2 is as efficient as Patarin's Little Dragon [5], but secure against all the known attacks. Shortly after the publication [4] appeared, linearization equations were found by Lei Hu, which became known in the private communication with the first author.

S.-H. Heng, R.N. Wright, and B.-M. Goi (Eds.): CANS 2010, LNCS 6467, pp. 140–155, 2010.

Due to these linearization equations, the authors of the LD2 scheme presented another scheme called Poly-Dragon [6]. Poly-Dragon as well as LD2 is constructed using permutation polynomials over finite fields. It is considered as an improved version of Patarin's Big Dragon cryptosystem [5]. The Poly-Dragon scheme was also proposed as an efficient and secure scheme. In particular, the inventors of the Poly-Dragon scheme claim that Poly-Dragon resist algebraic cryptanalysis.

In this paper, we present an algebraic attack for the LD2 and Poly-Dragon schemes. We present experiments that show the weakness of these schemes by solving corresponding multivariate quadratic equation systems over $\mathbb{F}_2$ up to number of variables equal to 229 for LD2 and 299 for Poly-Dragon. In this attack, we use an improved implementation of the MXL2 algorithm. This algorithm is an improvement of the MutantXL [7] algorithm which was used to break the MQQ scheme in [8].

We analyzed the reason why MXL2 is able to break the two schemes efficiently. We use also two different versions of Magma's implementation of F4 to compare our results. For all the instances that we have, MXL2 outperforms Magma's F4 in terms of memory and time. We discuss linearization equations for both schemes and present a way of obtaining these linearization equations.

This paper is organized as follows. In Section 2 we give an overview of the LD2 scheme. Section 3 is a description for Poly-Dragon. We then briefly present the MXL2 algorithm in Section 4. Section 5 presents the experimental results. We discuss linearization equations in Section 6. Finally, we conclude the paper in Section 7.

2   LD2: Little Dragon Two Multivariate Public-Key Cryptosystem

The Little Dragon Two, LD2, multivariate public-key cryptosystem is a mixed type scheme that has a public key in which plaintext and ciphertext variables are "mixed" together. LD2 is a modified version of Patarin's Little Dragon cryptosystem and it is constructed using permutation polynomials over finite fields. In this section, we present an overview of the LD2 scheme. In multivariate public-key cryptosystems, the main security parameters are the number of equations and the number of variables. The authors of LD2 did not propose any such security parameters. For a more detailed explanation see [4].

Definition 1. *Let $\mathbb{F}_q$ be the finite field of $q = p^n$ elements where p is prime and n is a positive integer. A polynomial $f \in \mathbb{F}_q[x_1, \ldots, x_n]$ is called a permutation polynomial in n variables over $\mathbb{F}_q$ if and only if one of the following equivalent conditions holds:*

1. *the function f is onto.*
2. *the function f is one-to-one.*
3. *$f(x) = a$ has a solution in $\mathbb{F}_q$ for each $a \in \mathbb{F}_q$.*
4. *$f(x) = a$ has a unique solution in $\mathbb{F}_q$ for each $a \in \mathbb{F}_q$.*

Simply speaking, this means that a polynomial $f \in \mathbb{F}_q[x_1, \ldots, x_n]$ is a permutation polynomial over $\mathbb{F}_q$ if it induces a bijective map from $\mathbb{F}_q$ to itself.

Lemma 1. *Let $Tr(x)$ denotes the trace function on the field $\mathbb{F}_{2^n}$ i.e. $Tr : \mathbb{F}_{2^n} \to \mathbb{F}_2$ is defined by $Tr(x) = x + x^2 + x^{2^2} + \cdots + x^{2^{n-1}}$. The polynomial $g(x) = (x^{2^r k} + x^{2^r} + \alpha)^\ell + x$ is a permutation polynomial of $\mathbb{F}_{2^n}$, when $Tr(\alpha) = 1$ and $\ell \cdot (2^{2^r k} + 2^r) = 1 \mod 2^n - 1$.*

A proof of Lemma 1 is presented by the authors of [4]. A more detailed explanation for permutation polynomials and trace representation on finite fields can be found in [9].

Suppose that $X = (x_1, x_2, \ldots, x_n)$ denotes the plaintext variables and $Y = (y_1, y_2, \ldots, y_n)$ denotes the ciphertext variables. In the LD2 scheme, the public key equations are multivariate polynomials over $\mathbb{F}_2$ of the form:

$$
\begin{cases}
P_1(x_1, x_2, \ldots, x_n, y_1, y_2, \ldots, y_n) = 0, \\
P_2(x_1, x_2, \ldots, x_n, y_1, y_2, \ldots, y_n) = 0, \\
\vdots \qquad\qquad\qquad\qquad\qquad \vdots \\
P_\lambda(x_1, x_2, \ldots, x_n, y_1, y_2, \ldots, y_n) = 0,
\end{cases}
$$

where $P_1, P_2, \ldots, P_\lambda$ are polynomials of $\mathbb{F}_2^n \times \mathbb{F}_2^n \to \mathbb{F}_2$ of total degree two.

Due to the restrictions on r, k and ℓ placed by Lemma 1, there are a few choices for r, k and ℓ to produce a permutation polynomial $g(x) = (x^{2^r k} + x^{2^r} + \alpha)^\ell + x$ and to use $g(x)$ to design a public-key scheme with a quadratic public key. We can choose $r = 0, n = 2m - 1, k = m$ and $\ell = 2^m - 1$ for example, then $G(x) = (x^{2^m} + x + \alpha)^{2^m - 1} + x$ is a permutation polynomial. By choosing $r = 0, n = 2m - 1, k = m$ and $\ell = 2^m + 1$, the authors of [4] stated that it is not clear whether $G'(x) = (x^{2^m} + x + \alpha)^{2^m + 1} + x$ is a permutation polynomial or not while it can produce a quadratic public key.

Let S and T be two invertible affine transformations. Then the plaintext can be permuted to a ciphertext using the relation $G(S(x_1, \ldots, x_n)) = T(y_1, \ldots, y_n)$. Suppose $S(x_1, \ldots, x_n) = u$ and $T(y_1, \ldots, y_n) = v$. Therefore the relation between plaintext and ciphertext can be written as follows:

$$
(u^{2^m} + u + \alpha)^{2^m - 1} + u = v
$$
$$
(u^{2^m} + u + \alpha)^{2^m - 1} + (u + v) = 0
$$
$$
(u^{2^m} + u + \alpha)^{2^m} + (u + v)(u^{2^m} + u + \alpha) = 0
$$
$$
((u^{2^m} + u) + \alpha)^{2^m} + u^{2^m + 1} + u^2 + u\alpha + vu^{2^m} + vu + v\alpha = 0
$$
$$
(u^{2^m} + u)^{2^m} + \alpha^{2^m} + u^{2^m + 1} + u^2 + u\alpha + vu^{2^m} + vu + v\alpha = 0
$$

$$
\vdots
$$

$$
u^{2^m} + \alpha^{2^m} + u^{2^m + 1} + u\alpha + vu^{2^m} + vu + v\alpha = 0
$$
$$
u^{2^m + 1} + u^{2^m} v + uv + u\alpha + u^{2^m} + v\alpha + \alpha^{2^m} = 0 \qquad (1)
$$

It is known that the extension field $\mathbb{F}_{2^n}$ can be viewed as a vector space over $\mathbb{F}_2$. Let $\beta = \{\beta_1, \beta_2, \ldots, \beta_n, \}$ be a normal basis of $\mathbb{F}_{2^n}$ over $\mathbb{F}_2$ for some $\beta \in \mathbb{F}_{2^n}$.

Therefore any $z \in \mathbb{F}_{2^n}$ can be expressed as $z = \sum_{i=1}^{n} z_i \beta_i$, where $z \in \mathbb{F}_2$. By substituting $u = S(x_1, \ldots, x_n)$ and $v = T(y_1, \ldots, y_n)$, Equation (1) can be represented as n quadratic polynomial equations of the form:

$$\sum a_{ij} x_i x_j + \sum b_{ij} x_i y_j + \sum c_k y_k + \sum d_k x_k + e_l = 0 \tag{2}$$

where the coefficients $a_{ij}, b_{ij}, c_k, d_k, e_l \in \mathbb{F}_2$.

In Equation (2), the terms of the form $\sum x_i x_j + \sum x_k + c_1$ are obtained from u^{2^m+1}, the terms of the form $\sum x_i y_j + \sum x_k + \sum y_k + c_2$ are obtained from $u^{2^m} v + uv$, the terms of the form $\sum x_i + c_3$ is obtained from $u\alpha + u^{2^m}$ and $v\alpha$ gives the terms of the form $\sum y_i + c_4$, , where c_1, c_2, c_3 and c_4 are constants.

The secret parameters are the finite field element α and the two invertible affine transformations S and T. A plaintext $X = (x_1, x_2, \ldots, x_n) \in \mathbb{F}_2^n$ is encrypted by applying Algorithm 1. Decryption is accomplished by using Algorithm 2.

A discussion for the security and the efficiency of the proposed scheme is presented in [4]. As a conclusion, the authors claimed that they present an efficient and secure multivariate public key cryptosystem that can be used for both encryption and signatures.

Algorithm 1. Encryption

1: **Inputs**
2:   A plaintext message $X = (x_1, x_2, \ldots, x_n)$ of length n.
3: **Output**
4:   A ciphertext message $Y = (y_1, y_2, \ldots, y_n)$ of length n.
5: **Begin**
6: Substitute the plaintext $(x_1, x_2, \ldots, x_n)$ in the public key.
7: Get n linear equations in the ciphertext variables $(y_1, y_2, \ldots, y_n)$.
8: Solve these linear equations by Gaussian elimination method to obtain the correct ciphertext $Y = (y_1, y_2, \ldots, y_n)$.
9: **End**

Algorithm 2. Decryption

1: **Inputs**
2:   A ciphertext message $Y = (y_1, y_2, \ldots, y_n)$ of length n and the secret parameters (S, T, α).
3: **Output**
4:   A plaintext message $X = (x_1, x_2, \ldots, x_n)$ of length n.
5: **Begin**
6: Let $v = T(y_1, y_2, \ldots, y_n)$.
7: Let $z_1 = \alpha + 1 + v + v^{2^m}$.
8: Let $z_2 = z_1^{2^m-1}$.
9: Let $z_3 = v + 1 + z_2$.
10: Let $X_1 = S^{-1}(v + 1)$.
11: Let $X_2 = S^{-1}(z_3)$.
12: **Return** (X_1, X_2), Either X_1 or X_2 is the required secret message.
13: **End**

3   Poly-Dragon Multivariate Public-Key Cryptosystem

The Poly-Dragon multivariate public-key cryptosystem is a mixed type scheme
that has a public key of total degree three, two in plaintext and one in ciphertext.
Poly-Dragon is based on permutation polynomials and is supposed to be as
efficient as Patarin's Big Dragon [5]. As well as LD2, Poly-Dragon did not have
a proposed security parameters. In this section, we introduce an overview of the
Poly-Dragon scheme. See [6] for more details.

Definition 2. *Let* $\mathbb{F}_q$ *be the finite field of of characteristic* p. *A polynomial of
the form*

$$L(x) = \sum_i \alpha_i x^{p^i}$$

with coefficients in an extension field $\mathbb{F}_q$ *of* $\mathbb{F}_p$ *is called a* p-*polynomial over* $\mathbb{F}_q$.

Simply speaking, a polynomial over $\mathbb{F}_q$ is said to be a p-polynomial over $\mathbb{F}_q$ if
each of its terms has a degree equal to a power of p. A p-polynomial is also
called linearized polynomial because for all $\beta, \gamma \in \mathbb{F}_q$ and $a \in \mathbb{F}_p$ it satisfies the
following properties: $L(\beta + \gamma) = L(\beta) + L(\gamma)$ and $L(a\beta) = aL(\beta)$. In [9], it is
proved that $L(x)$ is a permutation polynomial of $\mathbb{F}_q$ if and only if the only root
of $L(x)$ in $\mathbb{F}_q$ is 0.

Proposition 1. *Let* $L_\beta(x) = \sum_{i=0}^{n-1} \beta_i x^{2^i} \in \mathbb{F}_{2^n}$ *be a* p-*polynomial defined with*
n *an odd positive integer and* $\beta = (\beta_1, \beta_2, \ldots, \beta_n) \in \mathbb{F}_{2^n}$ *such that the weight of*
β *is even and that 0 and 1 are the only roots of* $L_\beta(x)$. *Then*

$$f(x) = (L_\beta(x) + \gamma)^\ell + Tr(x)$$

is a permutation polynomial of $\mathbb{F}_{2^n}$, *where* l *is any positive integer with* $(2^{k_1} +$
$2^{k_2}) \cdot \ell \equiv 1 \pmod{2^n - 1}, \gamma \in \mathbb{F}_{2^n}$ *with* $Tr(\gamma) = 1$ *and* k_1, k_2 *are non negative
integers such that* $\gcd(2^{k_1} + 2^{k_2}, 2^n - 1) = 1$.

Proposition 2. *The polynomial* $g(x) = (x^{2^{k2^r}} + x^{2^r} + \alpha)^\ell + x$ *is a permutation
polynomial of* $\mathbb{F}_{2^n}$ *if* $Tr(\alpha) = 1$ *and* $(2^{k2^r} + 2^r) \cdot \ell \equiv 1 \pmod{2^n - 1}$.

The two permutation polynomials $g(x) = (x^{2^{k2^r}} + x^{2^r} + \alpha)^\ell + x$ and $f(x) =$
$(L_\beta(x) + \gamma)^\ell + Tr(x)$ from Proposition 1 and Proposition 2 are used in Poly-
Dragon public-key cryptosystem. The permutation polynomials in which ℓ is of
the form $2^m - 1$ and $r = 0, n = 2m - 1, k = m, k_2 = m$ and $k_1 = 0$ are used
to generate the public key. Therefore, for key generation $G(x) = (x^{2^m} + x +$
$\alpha)^{2^m-1} + x$ and $F(x) = (L_\beta(x) + \gamma)^{2^m-1} + Tr(x)$ are used where α, β, γ are
secret.

The relation between plaintext $X = (x_1, x_2, \ldots, x_n)$ and ciphertext $Y =$
$(y_1, y_2, \ldots, y_n)$ can be written as $G(S(x_1, x_2, \ldots, x_n)) = F(T(y_1, y_2, \ldots, y_n))$,

where S and T are two invertible affine transformations. This relation can be written as $(u^{2^m} + u + \alpha)^{2^m-1} + u = (L_\beta(v) + \gamma)^{2^m-1} + Tr(v)$ such that $S(x_1, x_2, \ldots, x_n) = u$ and $T(y_1, y_2, \ldots, y_n) = v$. Multiplying by $(u^{2^m} + u + \alpha) \times (L_\beta(v) + \gamma)$, which is a nonzero in the field $\mathbb{F}_{2^n}$, we obtain:

$$(u^{2^m} + u + \alpha)^{2^m}(L_\beta(v) + \gamma) + u(u^{2^m+u+\alpha})(L_\beta(v) + \gamma)$$
$$+(u^{2^m} + u + \alpha)(L_\beta(v) + \gamma)^{2^m} + Tr(v)(u^{2^m+u+\alpha})(L_\beta(v) + \gamma) = 0 \qquad (3)$$

The extension field $\mathbb{F}_{2^n}$ can be viewed as a vector space over $\mathbb{F}_2$. Then we can identify $\mathbb{F}_{2^n}$ with $\mathbb{F}_2^n$. Let $Tr(v) = \zeta_y \in \{0,1\}$ and by substituting $u = S(x_1, x_2, \ldots, x_n)$ and $v = T(y_1, y_2, \ldots, y_n)$, in Equation (3), we obtain n nonlinear polynomials equations of degree three of the form:

$$\sum a_{ijk}x_ix_jy_k + \sum b_{ij}x_ix_j + \sum (c_{ij}+\zeta_y)x_iy_j + \sum (d_k+\zeta_y)y_k + \sum (e_k+\zeta_y)x_k + f_l, \qquad (4)$$

where $a_{ijk}, b_{ij}, c_{ij}, d_k, e_k, f_l \in \mathbb{F}_2$.

The secrete parameters are the finite field elements α, β, γ and the two invertible affine transformations S and T. A plaintext $X = (x_1, x_2, \ldots, x_n) \in \mathbb{F}_2^n$ is encrypted by applying Algorithm 3. Decryption is accomplished by using Algorithm 4.

In [6], the authors stated a proof for the validity of the generated plaintext by the decryption algorithm. A discussion for the security and the efficiency of the proposed scheme is also presented. As a conclusion, the authors claimed that they presented an efficient and secure multivariate public key cryptosystem that can be used for encryption as well as for signature.

Algorithm 3. Encryption

1: **Inputs**
2:    A plaintext message $X = (x_1, x_2, \ldots, x_n)$ of length n.
3: **Output**
4:    A ciphertext message pair (Y', Y'').
5: **Begin**
6: Substitute the plaintext variables $(x_1, x_2, \ldots, x_n)$ and $\zeta_y = 0$ in the public key.
7: Get n linear equations in the ciphertext variables $(y_1, y_2, \ldots, y_n)$.
8: Solve these linear equations by Gaussian elimination method to obtain the ciphertext variables $Y' = (y_1, y_2, \ldots, y_n)$.
9: Substitute the plaintext $(x_1, x_2, \ldots, x_n)$ and $\zeta_y = 1$ in the public key.
10: Get n linear equations in the ciphertext $(y_1, y_2, \ldots, y_n)$.
11: Solve these linear equations by Gaussian elimination method to obtain the ciphertext variables $Y'' = (y_1, y_2, \ldots, y_n)$.
12: Return the ordered pair (Y', Y'') as the required ciphertext.
13: **End**

Algorithm 4. Decryption

1: **Inputs**
2:    A ciphertext message (Y', Y'') and the secrete parameters $(S, T, \alpha, \beta, \gamma)$.
3: **Output**
4:    A plaintext message $X = (x_1, x_2, \ldots, x_n)$ of length n.
5: **Begin**
6: Let $v_1 = T(Y')$ and $v_2 = T(Y'')$.
7: Let $z_1 = L_\beta(v_1) + \gamma$ and $z_2 = L_\beta(v_2) + \gamma$.
8: Let $\bar{z}_3 = z_1^{2^m - 1}$ and $\bar{z}_4 = z_2^{2^m - 1}$.
9: Let $z_3 = \bar{z}_3 + Tr(v_1)$ and $z_4 = \bar{z}_4 + Tr(v_2)$.
10: Let $z_5 = z_3^{2^m} + z_3 + \alpha + 1$ and $z_6 = z_4^{2^m} + z_4 + \alpha + 1$.
11: Let $z_7 = z_5^{2^m - 1}$ and $z_8 = z_6^{2^m - 1}$.
12: Let $X_1 = S^{-1}(z_3 + 1)$.
13: Let $X_2 = S^{-1}(z_4 + 1)$.
14: Let $X_3 = S^{-1}(z_3 + z_7 + 1)$.
15: Let $X_4 = S^{-1}(z_4 + z_8 + 1)$.
16: **Return** (X_1, X_2, X_3, X_4), Either X_1, X_2, X_3 or X_4 is the required secret message.
17: **End**

---

4   MXL2: The MutantXL2 Algorithm

MXL2 [10] is an algorithm for solving systems of quadratic multivariate equations over $\mathbb{F}_2$ that was proposed at *PQC*2008. It is a variant of MutantXL [7] which improves on the XL algorithm [11]. The MXL2 and MutantXL algorithms are similar in using the concept of Mutants that is introduced by Ding [12], while MXL2 uses two substantial improvements over $\mathbb{F}_2$. In this section, we present a brief overview of the Mutant strategy, the MXL2 algorithm, MXL2 improvements and MXL2 implementation.

The main idea for the Mutant strategy is to maximize the effect of lower-degree polynomials occurring during the linear algebra step for the linearized representation of the polynomial system. Throughout this section we will use $x := \{x_1, \ldots, x_n\}$ to be a set of n variables. We consider

$$R := \mathbb{F}_2[x_1, \ldots, x_n]/\langle x_1^2 - x_1, \ldots, x_n^2 - x_n \rangle$$

the Boolean polynomial ring in x with graded lexicographical ordering $<_{gradlex}$ on the monomials of R. We consider elements of R as polynomials over $\mathbb{F}_2$ where the degree of each term w.r.t any variable is 0 or 1.

Let $P := (p_1, \ldots, p_m)$ be a system of m quadratic polynomial equations in R. We are interested in finding a solution for $P(x) = 0$ that is based on creating further elements of the ideal generated by the polynomials of P. In this context, Mutants are defined as follows.

Definition 3. *Let I be the ideal generated by the finite set of polynomials P, $f \in I$. For any representation of f, $f := \sum_{p \in P} f_p p$, we define the level of this representation to be $max\{deg(f_p p) : p \in P, f_p \neq 0\}$. Let $Rep(f)$ be the set of all representations of f. Then the level of f with respect to P is defined to be the* *minimum of levels of all representations in Rep(f). The polynomial f is called a Mutant with respect to P if its degree is less than its level.*

From a practical point of view, the concept of Mutants can be applied to the linear algebra step in the matrix-based algorithms for solving systems of multivariate polynomial equations, for example F4 and XL. In [10,7,13,14], during the linear algebra step, the new polynomials that appear having a lower degree are mutants. By using these mutants, MutantXL as well as MXL2 can solve multivariate quadratic polynomial equations at a lower degree than the usual XL.

The MXL2 algorithm performs the following steps:

- *Initialization:* Set $P = \{p_1, \ldots, p_m\}$, $D = Max\{deg(p) : p \in P\}$, the elimination degree $ED = min\{deg(p) : p \in P\}$, the set of Mutants $M = \emptyset$.
- *Echelonize:* Consider each term in P as a new variable. Set $P = \tilde{P}$ where $\tilde{P}$ is the row echelon form of P.
- *Solve:* If there are univariate polynomials in P, then determine the values of the corresponding variables and substitute in P. If the system is solved then return solution and terminate. Otherwise, set $D = Max\{deg(p) : p \in P\}$ and $ED = min\{deg(p) : p \in P\}$.
- *ExtractMutants:* Add all new polynomials of degree less than D in P to M.
- *MultiplyMutants:* If $M \neq \emptyset$, then select the necessary number of Mutants that have degree $k = min\{deg(p) : p \in M\}$, multiply lower degree Mutants by all terms up to degree D, remove the multiplied Mutants from M, add the new polynomials obtained from the multiplication to P, set $ED = k + 1$ then go back to *Echelonize.*
- *Extend:* Add all the polynomials that are obtained by multiplying a subset of the degree D polynomials in P by all variables that are smaller than the leading variable of the partition leading variable, set $D = D + 1$, $ED = D$. Go to back to *Echelonize.*

As stated in [10], MXL2 has two important advantages over MutantXL. The first is the use of the necessary number of mutants and the second is extending the system only partially to higher degrees. The main idea for the first improvement is to add only a necessary number of mutants to the given system. This number is numerically computed. By using not all the emerged mutants, the efficiency is increased, for space efficiency only a few mutants are used and for the time efficiency the multiplications to generate higher degree polynomials do not have to be performed to all mutants.

The second improvement is the usage of the partial enlargement technique in which the polynomials at degree D are divided according to their so-called "leading variable". The leading variable of a polynomial $p \in R$ is the smallest variable in the leading term with respect to the order defined on the variables. Instead of enlarging all the partitions, only the non-empty partitions are multiplied by all the variables that are smaller than the leading variable. This is accomplished partition by partition. This partial enlargement technique gives also an improvement in time and space since the system can be solved using a lower number of polynomials. Moreover, in some systems mutants may appear in the last step together with the solution of the system. These mutants are not fully utilized. Using partial enlargement technique enforces these mutants to appear before the system is solved.

As a result of using the two MXL2 improvements, MXL2 generates the same solution as if all Mutants would have been used and all partitioned would have been multiplied. MXL2 solves multivariate quadratic polynomial equations using a smaller number of polynomials than MutantXL.

The MXL2 algorithm has been implemented in C/C++ based on the latest version of M4RI package [15]. In this package, there exist three different algorithms for computing row echelon form. In this paper, we use the Method of Four Russians Inversion (M4RI) algorithm [16].

5   Experimental Results

In this section we present experimental results of the attack on LD2 and Poly-Dragon by using MXL2 and compare the performance of MXL2 with two versions of Magma's implementation of F4 namely V2.13-10 and V2.16-1. The reason for using these two versions is that when we used Magma's version (V2.16-1), we found that this version solves the LD2 and Poly-Dragon systems at degree 4 while MXL2 as well as Magma V2.13-10 solves at degree 3. In this context, it is not fair to use only this version (V2.16-1) in the comparison.

The main task for a cryptanalyst is to find a solution of the systems of equations that represent the LD2 and Poly-Dragon schemes. These systems were essentially implemented as described in [4] and [6] respectively. Magma version (2.16-1) has been used for the implementation. Due to the high number of variables, this direct approach is not very efficient but it is sufficient for modeling purposes. For real-life applications, there are more elegant ways to create the public key using specialized software and techniques like polynomial interpolation (see [17] for example).

All the experiments are done on a Sun X4440 server, with four "Quad-Core AMD Opteron^{TM} Processor 8356" CPUs and 128GB of main memory. Each CPU is running at 2.3 GHz. In these experiments we used only one out of the 16 cores.

We tried to solve different systems with the same number of variables. As a result of our experiments, we noticed that the complexity for different systems of LD2 and Poly-Dragon schemes with the same number of variable will be, essentially, the same. In this context, the results given in this section are for one particular instance for each system.

Table 1 presents the required steps of solving an LD2 instance of $n = 229$ using MXL2. In this table, for each step (Round) we present the elimination degree (ED), the matrix dimensions (Matrix), the rank of the matrix (Rank), the total number of mutants found (#Mutants), the number of linear mutants found (#LM) and the number of univariate polynomials found (#Uni).

Table 2 and Table 3 show results of the LD2 systems for the range 79-229 equations in 79-229 variables and results of the Poly-Dragon systems for the

Table 1. MXL2: Results for LD2-229Var

Round	ED	Matrix	Rank	#Mutants	#LM	#Uni
1	2	229 × 26336	229	0	0	0
2	3	457 × 1975812	457	0	0	0
3	3	52670 × 2001690	52669	686	228	2
4	2	915 × 26336	913	226	226	108
5	2	913 × 26336	805	118	118	0
6	2	14847 × 26336	7140	1	1	1
7	2	7140 × 26336	7021	118	118	118

Table 2. Performance of MXL2 versus F4 for Little-Dragon-Two

Sys	F4$_{v2.13}$			F4$_{v2.16}$			MXL2		
	D	Mem	Time	D	Mem	Time	D	Mem	Time
79	3	490	29	4	321	26	3	211	22
89	3	841	116	4	1600	203	3	346	40
99	3	1357	238	4	2769	411	3	545	73
109	3	2092	500	4	2046	331	3	844	122
119	3	3102	998	4	6842	1142	3	1251	217
129	3	4479	1827	4	11541	2529	3	2380	458
139	3	6280	3134	4	8750	1723	3	3387	742
149	3	8602	4586	4	23325	5795	3	3490	692
159	3	11547	7466	4	30178	7845	3	4545	1146
169	3	15191	11478	4	46381	18551	3	6315	1613
179	3	19738	17134	4	46060	17502	3	8298	2025
189	3	25234	28263	4	91793	54655	3	10697	2635
199	3	31848	11.24H	4	134159	1.47D	3	13772	1H
209	3	39800	16.36H	4	97834	13.19H	3	17431	1.32H
219	3	49,134	1.11D	4	184,516	2.92D	3	29,856	2.81H
229	3	60,261	1.56D	Ran out of memory			3	25,847	2.60H

range 79-299 equations in 79-299 variables respectively. The first column "Sys" denotes the number of variables and the number of equations for each system. The highest degree of the elements of the system that occurred during the computation is denoted by "D". The used memory in Megabytes and the execution time in seconds can be found in the columns represented by "Mem" and "Time" respectively except for bigger systems it is in hours (H) or days (D). In both tables we can see that MXL2 always outperforms both versions of Magma's F4 in terms of memory and time.

Table 4 shows the required rounds of solving an Poly-Dragon instance of $n = 259$ using MXL2. The columns are represented as the same as in Table 1. From Table 4 we can see that MXL2 can solve the Poly-Dragon instance with 259 variables in 7 rounds. In the first round of the algorithm, there was no dependency in the original 259 polynomials and no mutants were found. Therefore, MXL2 extended the system partially to generate new 258 cubic equations. In

Table 3. Performance of MXL2 versus F4 for Poly-Dragon

Sys	$F4_{v2.13}$			$F4_{v2.16}$			MXL2		
	D	Mem	Time	D	Mem	Time	D	Mem	Time
79	3	488	34	4	301	26	3	224	31
89	3	841	82	4	1519	153	3	285	39
99	3	1360	164	4	2883	320	3	454	71
109	3	2093	328	4	2039	241	3	674	117
119	3	3103	622	4	6873	972	3	1473	347
129	3	4475	1194	4	11542	1899	3	1573	312
139	3	6277	2113	4	8701	1238	3	2253	451
149	3	8606	3686	4	24105	5397	3	3151	659
159	3	11546	6645	4	30177	6734	3	4318	960
169	3	15195	10451	4	47787	14800	3	5812	1449
179	3	19741	15801	4	46064	14122	3	7698	1907
189	3	25262	25386	4	91720	41805	3	14154	3782
199	3	31852	40618	4	134144	124278	3	12944	3633
209	3	39813	64753	4	97898	69144	3	16472	6730
219	3	49129	85635	Ran out of memory			3	20736	8165
229	3	60231	1.83D	Ran out of memory			3	36617	4.13H
239	3	73006	2.36D	Ran out of memory			3	31922	3.62H
249	3	87,908	3.42D	Ran out of memory			3	39,098	4.34H
259	3	105,012	4.15D	Ran out of memory			3	47,512	6.51H
...	.	....	.....	.	.....	.....	.	......	......
299	Ran out of memory			Ran out of memory			3	95,317	11.28H

Table 4. MXL2: Results for Poly-Dragon-259Var

Round	ED	Matrix	Rank	#Mutants	#LM	#Uni
1	2	259 × 33671	259	0	0	0
2	3	517 × 2862727	517	0	0	0
3	3	67340 × 2895880	67339	776	258	2
4	2	1035 × 33671	792	256	256	138
5	2	1033 × 33671	895	118	118	0
6	2	14937 × 33671	7140	1	1	1
7	2	7140 × 33671	7021	118	118	118

the second round, after applying the *Echelonize* step to the extended system (517 equations), all the equations were independent and there were no mutants found. The MXL2 extended the system again by applying *Extend* step to generate 66823 new cubic equations. By echelonizing the resulting extended system (67340 equations), we obtained a system of rank 67339, 518 quadratic mutants, 258 linear mutants in which 2 equations are univariate. After simplifying with the two univariate polynomials and modifying the elimination degree to two, we obtain a quadratic system of (1035 equations). Then third round is finished. In the fourth round, echelonizing the system of 1035 equations at degree two,

Table 5. F4: Results for Poly-Dragon-259Var

Step	SD	Matrix	#Pairs
1	2	259 × 33671	251
2	3	67349 × 2895880	4694
3	2	34446 × 33671	777
4	3	2835604 × 2832190	20423

yielded a system of rank 1033 and 256 linear mutants, 138 out of them are univariate. Substituting with the 138 univariate equations and eliminating, we obtained a system of rank 895 and 118 linear mutants in round 5. The necessary number of mutants that are required at this round is 250. Therefore, all the 118 linear mutants are multiplied by variables using *MultiplyMutants*.In round 6, we obtained 1 linear mutant which is also univariate polynomial from eliminating the extended system of total 14937 equations and rank of 7140. In round 7, after substituting with the univariate equation, we started with 7140 equations and the elimination degree is the same, 2. We obtained a rank of 7021 and the rest 118 univariate equations after applying the *Echelonize* step.

Table 5 shows the required steps of solving the same Poly-Dragon instance as in Table 4 using F4 version (V2.13.10). In each step, we show the step degree (SD), the matrix dimensions (Matrix) and the number of pairs (#Pairs).

Experimentally, as far as we noticed, the main new feature of Magma's F4 version (V2.16-1) is that if there exist some predetermined number of linear polynomials, at certain degree, the program is interrupted and then a new phase is started with extended basis by adding the linear polynomials to the original

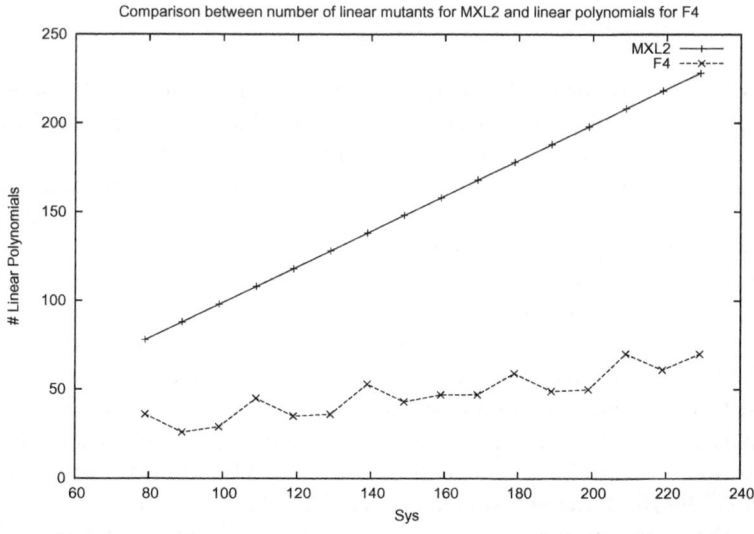

Fig. 1. LD2:Number of linear Mutants for MXL2 and linear polynomials for $F4_{v2.16-1}$

ones then compute a Gröbner basis to the extended new system. Figure 1 shows a comparison between the number of linear mutants that are generated at degree 3 and the number of linear polynomials generated by F4 (V2.16-1) at degree 3 for each LD2 system. These mutants as well as linear polynomials generated by F4 show that there is an algebraic hidden structure in the LD2 scheme that distinguishes it from a random system.

These predetermined number of linear polynomials at which Magma's F4 version (2.16-1) interrupts the first phase of computation is not enough to finish computing a Gröbner basis at the same degree at which these linear polynomials appear. The usage of the necessary number of mutants improvement for MXL2 could help new Magma's F4 to recover its defect.

6   Linearization Equations for the Dragons

The authors of [4] and [6] point out in [6] that both LD2 and Poly-Dragon possess high order linearization equations (second order in the case of LD2). Recall that a (high-order) linearization equation in plaintext variables X and ciphertext variables Y is a polynomial $F(X,Y)$, linear in the X-variables. If one is able to obtain such polynomials relating inputs and outputs of an MPKC, an attack can be undertaken by plugging in known ciphertexts and solving linear systems in the plaintext variables. Using notation of Sections 2 and 3, Poly-Dragon possesses linearization equations of the form

$$(z + u + 1)(z^{2^m} + z + \alpha + 1) + (z^{2^m} + z + \alpha + 1)^{2^m} = 0, \tag{5}$$

where $z = (L_\beta(v) + \gamma)^{2^m-1} + Tr(v)$. For the case of LD2 the equations are simply

$$(v + u + 1)(v^{2^m} + v + \alpha + 1) + (v^{2^m} + v + \alpha + 1)^{2^m} = 0. \tag{6}$$

Since equations (6) are only of degree 2 in ciphertext for LD2, a linearization attack is very feasible. The authors of [6] claim that the degree of (5) in ciphertext variables is too high to be used in a practical linearization attack.

It is interesting to note that even existence of equations (6) and (5) for LD2 and Poly-Dragon respectively does not explain the nice behavior of the direct attack using MXL2. Namely, the fact that we are able to solve at degree 3. This is unlike the classical attack on Matsumoto-Imai scheme. There it is possible to obtain linearization equations that are also linear in the ciphertext variables. Considering that ciphertext variables depend explicitly on the plaintext variables via quadratic equations, it is then clear why one is able to solve at degree 3 with XL and Gröbner basis techniques: one simply plugs in public key equations in the linearization equations and obtains that such degree-3 equations in the plaintext variables lie in the linear span of degree 3 XL-extension of the public key with the ciphertext variables fixed, see [18]. In the case of LD2 we have a mixed system, therefore there are no explicit quadratic relations between a ciphertext and a plaintext. Even if they existed, they would provide equations of degree 4 in the plaintext, which by no means explains solving at degree 3. Even "worse" is the situation for the Poly-Dragon. This cryptosystem seems to be immune to linearization attacks, but is very susceptible to algebraic attacks.

There is a simple known way of obtaining linearization equations. One assumes that equations of the form $F(x_1, \ldots, x_n, y_1, \ldots, y_n) = \sum_i a_i x_i f_i(Y) + \sum_i b_i g_i(Y) + c$ with $Y = (y_1, \ldots, y_n)$ exist with certain requirements on degrees of f_i, g_i. Then by plugging in known plaintext/ciphertext pairs for $X = (x_1, \ldots, x_n)$ and Y one tries to find coefficients for the polynomials f_i, g_i as well as coefficients a_i, b_i, c. One should be lucky to get enough linearly independent equations relating the above coefficients. In practice this method works pretty well and has practical complexity polynomial in n. We present next a method of finding linearization equations based on Gröbner bases.

Proposition 3. *Let I be an ideal in the ring $\mathbb{F}_2[X, Y], X = (x_1, \ldots, x_n), Y = (y_1, \ldots, y_n)$. Let $D > 0$ be an integer. Let G be a Gröbner basis of I w.r.t weighted degree lexicographic ordering with $x_1 > \cdots > x_n > y_1 > \cdots > y_n$ with weights D for each X-variable and weight 1 for each Y-variable. If in I there exists a linearization equation $f(x_1, \ldots, x_n, y_1, \ldots, y_n) = \sum_i a_i x_i f_i(Y) + \sum_i b_i g_i(Y) + c$ with $\deg(f_i) \leq D, \deg(g_i) \leq D \; \forall \; i$, then there is an element $g \in G$ which is also a linearization equation with leading monomial dividing the leading monomial of f: $lm(g)|lm(f)$.*

Proof. First of all, since G is a Gröbner basis of I and $f \in I$ there exists $g \in G$ such that $lm(g)|lm(f)$. From the form of f and the monomial ordering on $\mathbb{F}_2[X, Y]$ that we have chosen, we see that $\deg(g) = \deg(lm(g)) \leq \deg(lm(f)) = \deg(f) \leq 2D$, where by degree we mean weighted degree. If g has a monomial of the form $x_i x_j$ for some i and j, then weighted degree of this monomial is $2D$ and such monomial is larger than $lm(f)$ w.r.t weighted degree lexicographic ordering and so is larger than $lm(g)$, which yields a contradiction with the fact that $lm(f)$ is linear in the X-variables. Similarly one comes to a contradiction with other monomials that are non-linear in X-variables. This means that g is linear in the X-variables and so is a linearization equation in X.

So with the use of the above proposition we are able to get "basis" elements for the linearization equations in I. The method requires finding a Gröbner basis w.r.t a certain monomial ordering, therefore has higher complexity than the method described earlier. Still the latter method does not involve any probabilistic arguments and does not depend on a choice of plaintext/ciphertext pairs. It is a matter of future work to realize if such a method has practical implications on MPKCs.

7   Conclusion

We present an efficient algebraic cryptanalysis for the Little Dragon Two, LD2, and Poly-Dragon public-key cryptosystems. Both cryptosystems were proposed as efficient and secure schemes. In our attack we are able to break LD2 with key length up to 229 bits and Poly-Dragon with key length up to 299 bits using both Magma's F4 and MXL2. In all experiments, MXL2 outperforms the used versions of Magma's F4. We realized that the last version of Magma's F4 is not so well suitable for solving LD2 and Poly-Dragon systems.

In MXL2 algebraic attack, the LD2 and Poly-Dragon schemes are solved at degree three which reflexes the weakness and contradicts the security claims for these two schemes. We expect that MXL2 can attack a system that represent Little Dragon Two up to 389 variables in less than one day using the same memory resources that we have, 128GB memory. We claim also that MXL2 can solve a systems that represent Poly-Dragon up to 339 variables in less than 20 hours.

References

1. Garey, M.R., Johnson, D.S.: Computers and Intractability: A Guide to the Theory of NP-Completeness. W. H. Freeman & Co., New York (1979)
2. Ding, J., Gower, J.E., Schmidt, D.: Multivariate Public Key Cryptosystems (Advances in Information Security). Springer, New York (2006)
3. Ding, J., Yang, B.Y.: Multivariate Public Key Cryptography. In: Bernstein, D.J., et al. (eds.) Post Quantum Cryptography, pp. 193–234. Springer, Heidelberg (2008)
4. Singh, R.P., Saikia, A., Sarma, B.K.: Little Dragon Two: An Efficient Multivariate Public Key Cryptosystem. International Journal of Network Security and Its Applications (IJNSA) 2, 1–10 (2010)
5. Jacques, P.: Asymmetric Cryptography with a Hidden Monomial. In: Koblitz, N. (ed.) CRYPTO 1996. LNCS, vol. 1109, pp. 45–60. Springer, Heidelberg (1996)
6. Singh, R.P., Saikia, A., Sarma, B.: Poly-Dragon: An efficient Multivariate Public Key Cryptosystem. Cryptology ePrint Archive, Report 2009/587 (2009), http://eprint.iacr.org/
7. Ding, J., Buchmann, J., Mohamed, M.S.E., Moahmed, W.S.A., Weinmann, R.P.: MutantXL. In: Proceedings of the 1st International Conference on Symbolic Computation and Cryptography (SCC 2008), Beijing, China, pp. 16–22. LMIB (2008), http://www.cdc.informatik.tu-darmstadt.de/reports/reports/MutantXL_Algorithm.pdf
8. Mohamed, M.S., Ding, J., Buchmann, J., Werner, F.: Algebraic Attack on the MQQ Public Key Cryptosystem. In: Garay, J.A., Miyaji, A., Otsuka, A. (eds.) CANS 2009. LNCS, vol. 5888, pp. 392–401. Springer, Heidelberg (2009)
9. Lidl, R., Niederreiter, H.: Finite Fields, 2nd edn. Encyclopedia of Mathematics and its Applications, vol. 20. Cambridge University Press, Cambridge (1997)
10. Mohamed, M.S.E., Mohamed, W.S.A.E., Ding, J., Buchmann, J.: MXL2: Solving Polynomial Equations over GF(2) Using an Improved Mutant Strategy. In: Buchmann, J., Ding, J. (eds.) PQCrypto 2008. LNCS, vol. 5299, pp. 203–215. Springer, Heidelberg (2008)
11. Courtois, N.T., Klimov, A., Patarin, J., Shamir, A.: Efficient Algorithms for Solving Overdefined Systems of Multivariate Polynomial Equations. In: Preneel, B. (ed.) EUROCRYPT 2000. LNCS, vol. 1807, pp. 392–407. Springer, Heidelberg (2000)
12. Ding, J.: Mutants and its Impact on Polynomial Solving Strategies and Algorithms. Privately distributed research note, University of Cincinnati and Technical University of Darmstadt (2006)

13. Ding, J., Cabarcas, D., Schmidt, D., Buchmann, J., Tohaneanu, S.: Mutant Gröbner Basis Algorithm. In: Proceedings of the 1st International Conference on Symbolic Computation and Cryptography (SCC 2008), Beijing, China, pp. 23–32. LMIB (2008)
14. Mohamed, M.S.E., Cabarcas, D., Ding, J., Buchmann, J., Bulygin, S.: MXL3: An Efficient Algorithm for Computing Gröbner Bases of Zero-dimensional Ideals. In: Lee, D., Hong, S. (eds.) ICISC 2009. LNCS, vol. 5984, pp. 87–100. Springer, Heidelberg (2010)
15. Albrecht, M., Bard, G.: The M4RI Library– Linear Algebra over GF(2) (2008), http://m4ri.sagemath.org
16. Bard, G.V.: Algebraic Cryptanalysis. Springer Publishing Company, Incorporated, Heidelberg (2009)
17. Wolf, C.: Efficient Public Key Generation for HFE and Variations. In: Dawson, E., Klemm, W. (eds.) Cryptographic Algorithms and their Uses, Queensland University of Technology, pp. 78–93 (2004)
18. Billet, O., Ding, J.: Overview of Cryptanalysis Techniques in Multivariate Public Key Cryptography. In: Sala, M., et al. (eds.) Gröbner Bases, Coding, and Cryptography, pp. 263–284. Springer, Heidelberg (2009)

Generating Parameters for Algebraic Torus-Based Cryptosystems

Tomoko Yonemura, Yoshikazu Hanatani, Taichi Isogai,
Kenji Ohkuma, and Hirofumi Muratani

Toshiba Corporation,
1, Komukai Toshiba-cho, Saiwai-ku, Kawasaki 212-8582, Japan
{tomoko.yonemura,hirofumi.muratani}@toshiba.co.jp

Abstract. Algebraic torus-based cryptosystems are public key cryptosystems based on the discrete logarithm problem, and have compact expressions compared with those of finite field-based cryptosystems. In this paper, we propose parameter selection criteria for the algebraic torus-based cryptosystems from the viewpoints of security and efficiency. The criteria include the following conditions: consistent resistance to attacks on algebraic tori and their embedding fields, and a large degree of freedom to select parameters suitable for each implementation. An extension degree and a characteristic size of a finite field on which the algebraic tori are defined are adjustable. We also provide examples of parameters satisfying the criteria.

1 Introduction

Practical public key encryption schemes are fundamental technology in the field of network security. For instance, the RSA encryption scheme is famous and its 2048-bit public key is recommended at present. The typical public key size is increasing with the progress of computer science. The growth of the public key size poses problems for machines with small memory or narrow bandwidth. However, the public key size of the RSA encryption scheme is not compressed in keeping with security.

We consider discrete logarithm problem-based cryptosystems defined on a prime-order subgroup of a multiplicative group in a finite field. It has been proposed that public key size can be compressed safely [1–3]. To compress the public key size is to represent the subgroup with fewer bits than the size of the finite field. For instance, the recommended size of the finite field is 2048 bits, and the corresponding size of the prime-order subgroup is 224 bits [4], because the discrete logarithm problem in the finite field is easier than in a general group.

Rubin and Silverberg constructed the (de)compression map by using birational maps between algebraic tori and affine spaces [3]. The security of cryptosystems on the algebraic torus $T_n(\mathbb{F}_{p^m})$ is based on the hardness of the discrete logarithm problem in it and the embedding field $\mathbb{F}_{(p^m)^n}$.

S.-H. Heng, R.N. Wright, and B.-M. Goi (Eds.): CANS 2010, LNCS 6467, pp. 156–168, 2010.
© Springer-Verlag Berlin Heidelberg 2010

1.1   Previous Work

Tables 1 and 2 show the parameters given in previous research. Let F be a minimal embedding field, and G be a subgroup on which cryptosystems are defined. The relation $G \subset T_n(\mathbb{F}_{p^m}) \subset F^{\times}$ holds. Let $|\bullet|$ denote the bit size of $\bullet$. The parameters with $n = 30$ shown by van Dijk et al. [5]. Gower proposed the parameter searching method and some examples with n=6, 30, and 210 [6]. Granger et al. used the algebraic torus T_6 defined over the extension field $\mathbb{F}_{3^m}$ in order to improve the Duursma-Lee method that computes the Tate pairing [7]. We can take parameters that also follow from all constructions of the pairing-friendly elliptic curves.

Table 1. Parameters with 2048-bit $|F|$ **Table 2.** Parameters with 1024-bit $|F|$

author(s)	n	$|F|$	$|G|$	m	$\lceil \log p \rceil$
van Dijk et al. [5]	30	1920	200	1	64
Granger et al. [7]	6	1830	305	193	2
Granger et al. [7]	6	2268	378	239	2

author(s)	n	$|F|$	$|G|$	m	$\lceil \log p \rceil$
van Dijk et al. [5]	30	960	160	1	32
Gower [6]	6	1024	160	1	171
Gower [6]	30	1024	160	1	35
Gower [6]	210	1024	160	1	6
Granger et al. [7]	6	906	150	97	2
Granger et al. [7]	6	1548	258	163	2

There are two problems concerning the above parameters: security and efficiency respectively. We consider the former problem. Some parameters are weak against the Joux-Lercier function field sieve [8]. Joux and Lercier pointed out that the complexity of the function field sieve over F is smaller than Pollard's ρ over G in the case of 1024-bit F regarded as a degree 30 extension field. Similarly, the function field sieve over F regarded as a degree m extension of $\mathbb{F}_{3^n}$ could be easier. Therefore, the parameters in Tables 1 and 2 don't survive except $(n, |F|, |G|, m, \lceil \log p \rceil) = (30, 1920, 200, 1, 64)$ in Table 1 and $(6, 1024, 160, 1, 171)$ in Table 2.

We consider the latter problem. In general, arithmetical operations are inefficient when characteristic p is unsuitable for the word size (e.g. 32 bits or 64 bits). If m is equal to 1, then $\lceil \log p \rceil$ is often larger than the typical word size. The pairing-friendly elliptic curves [9] are constructed with $m = 1$ except for the MNT curves [10]. Therefore, most of the parameters following from the pairing-friendly curves are inefficient. Selecting primitive polynomials is another point. Although cyclotomic polynomials were used in the previous research [3, 11], it is difficult to find suitable ones for the variable extension degree m.

1.2   Our Contributions

In this paper, we propose parameter selection criteria for the algebraic tori. The criteria consist of four security criteria and four efficiency criteria. Parameters include (n, m, p) of $T_n(\mathbb{F}_{p^m})$, the order q of the group G, and primitive polynomials of the embedding field $\mathbb{F}_{(p^m)^n}$. We also provide examples of parameters satisfying the criteria. For security, the parameters must resist all known attacks consistently. In view of Hitt's indication [12], $T_n(\mathbb{F}_{p^m})$ can be covered by a sub-field of $\mathbb{F}_{(p^m)^n}$, and its extension degree is not divisible by m. Therefore, G must not be covered in any proper subfields of $\mathbb{F}_{(p^m)^n}$. In order to resist attacks on algebraic tori and their embedding fields, the complexity of the index calculus in F, and the Granger-Vercauteren method [7] in $T_n(\mathbb{F}_{p^m})$, and Pollard's ρ method in G must be sufficiently large.

For efficiency, we choose the extension field $\mathbb{F}_{p^m}$ on which the algebraic tori are defined, binomials for primitive polynomials, and the algebraic tori equal to the group G. The binomials facilitate efficient arithmetic in the extension field, and they also have a large degree of freedom to decide a constant term for various m. Therefore, the extension degree m and the size of characteristic p are adjustable. In order to define cryptosystems on G, the prime-order algebraic tori themselves are easy to deal with.

The remainder of this paper is organized as follows: In section 2, we explain fundamental concepts used in the following discussion. In section 3, we propose the parameter selection criteria. In section 4, we provide examples of parameters satisfying the above criteria. In section 5, we estimate calculation costs.

2   Preliminaries and Notation

Let p and q be primes, and n and m be positive integers. Let $\mathbb{F}_{p^m}$ be a finite field of an order p^m. $\mathbb{F}_{p^m}^{\times}$ denotes a multiplicative group of $\mathbb{F}_{p^m}$. Let G be a group on which cryptosystems are defined of an order q. Let F be a minimal finite field in which G is covered. The Greek letters α, β, γ, δ denote elements of $\mathbb{F}_{(p^m)^3}$. The bold letters $\mathbf{c}$, $\mathbf{d}$, $\mathbf{w}$ denote elements of $\mathbb{F}_{p^m}$. Let λ_I and λ_P be security parameters for the index calculus and Pollard's ρ method respectively.

2.1   Cyclotomic Polynomials

Definition 1. *Let n be a positive integer. μ is the Möbius function. The n-th cyclotomic polynomial $\Phi_n(x)$ is defined by*

$$\Phi_n(x) = \prod_{d|n}(x^d - 1)^{\mu(n/d)} \ . \tag{1}$$

Theorem 1. *Let p be a prime.*

(a) If a is a positive integer not divisible by p, then $\Phi_{ap}(x)\Phi_a(x) = \Phi_a(x^p)$.
(b) If b is a positive integer divisible by p, then $\Phi_{bp}(x) = \Phi_b(x^p)$.

2.2   Algebraic Tori

Definition 2. *Let n be a positive integer. $N_{\mathbb{F}_{(p^m)^n}/f}$ is a norm map to a field f. An algebraic torus T_n over $\mathbb{F}_{p^m}$ is defined by*

$$T_n(\mathbb{F}_{p^m}) = \bigcap_{\mathbb{F}_{p^m} \subset f \subsetneq \mathbb{F}_{(p^m)^n}} \mathrm{Ker}\left[N_{\mathbb{F}_{(p^m)^n}/f}\right] \ . \tag{2}$$

Theorem 2. (a) $\#T_n(\mathbb{F}_{p^m}) = \Phi_n(p^m)$.
(b) If $h \in T_n(\mathbb{F}_{p^m})$ has a prime order not dividing n, then $h \notin \mathbb{F}_{(p^m)^d}$ for any $d|n$ with $d < n$.

Proof. (a) Note that f can be $\mathbb{F}_{(p^m)^d}$ for any $d|n$ with $d < n$. See also [3].
(b) Prime q denotes the order of h. Since $q \nmid n$, $X^n - 1$ has no repeated roots in the algebraic closure of $\mathbb{F}_q$. See also [13].   □

If the extension degree m is greater than 1, Theorem 2 (b) is insufficient to prove $T_n(\mathbb{F}_{p^m})$ is not covered by any proper subfields of $\mathbb{F}_{(p^m)^n}$. In view of Hitt's indication [12], $T_n(\mathbb{F}_{p^m})$ can be covered by a proper subfield that is not the form $\mathbb{F}_{(p^m)^d}$. We discuss this point in section 3.1.

2.3 Representations

We choose $n = 6$ for a good compression ratio and less transformation costs. The finite field $\mathbb{F}_{(p^m)^6}$ is constructed as follows:

$$\begin{cases} \mathbb{F}_{(p^m)^6} = \mathbb{F}_{(p^m)^3}[x]/f_2(x), \ f_2(x) = x^2 - \delta, \ \delta \in \mathbb{F}_{(p^m)^3}^{\times}, \\ \mathbb{F}_{(p^m)^3} = \mathbb{F}_{p^m}[y]/f_3(y), \ f_3(y) = y^3 - w, \ w \in \mathbb{F}_{p^m}^{\times}, \\ \mathbb{F}_{p^m} = \mathbb{F}_p[z]/g_m(z), \ g_m(z) = z^m - s, \ s \in \mathbb{F}_p^{\times}. \end{cases} \tag{3}$$

Projective representation. For all $\alpha + \beta x \in \mathbb{F}_{(p^m)^6}^{\times}$, the $\{(p^m)^3 - 1\}$-th power of the element is an element of $T_2(\mathbb{F}_{(p^m)^3})$, where $x^{(p^m)^3} = -x$ is led from the condition that $f_2(x)$ is irreducible. Since $T_6(\mathbb{F}_{p^m}) \subset T_2(\mathbb{F}_{(p^m)^3})$, elements of $T_6(\mathbb{F}_{p^m})$ are represented as eq. (4).

$$T_6(\mathbb{F}_{p^m}) = \left\{ \frac{\alpha - \beta x}{\alpha + \beta x} \,\middle|\, \alpha, \beta \in \mathbb{F}_{(p^m)^3}, (\alpha, \beta) \neq (0,0), \left(\frac{\alpha - \beta x}{\alpha + \beta x} \right)^{\Phi_6(p^m)} = 1 \right\} \tag{4}$$

In the representation of $T_2(\mathbb{F}_{(p^m)^3})$, the element corresponding to (α, β) is equivalent to the element corresponding to $(\lambda\alpha, \lambda\beta)$ for any $\lambda \in \mathbb{F}_{(p^m)^3}^{\times}$. So, this representation can be called the projective representation [14].

Affine representation. Let $\delta = \mathbf{d}$ be an element of $\mathbb{F}_{p^m}^{\times}$. Let $-\mathbf{d}/3$ be a quadratic nonresidue in $\mathbb{F}_{p^m}$. The elements of $T_6(\mathbb{F}_{p^m})\backslash\{1\}$ are also represented as eq. (5). This representation is called the affine representation.

$$T_6(\mathbb{F}_{p^m})\backslash\{1\} = \left\{ \frac{\mathbf{c}_0\mathbf{c}_1 + \mathbf{c}_1^2 y + (\mathbf{c}_0^2 + \mathbf{d}/3)\mathbf{w}^{-1}y^2 - \mathbf{c}_1 x}{\mathbf{c}_0\mathbf{c}_1 + \mathbf{c}_1^2 y + (\mathbf{c}_0^2 + \mathbf{d}/3)\mathbf{w}^{-1}y^2 + \mathbf{c}_1 x} \,\middle|\, \mathbf{c}_0 \in \mathbb{F}_{p^m}, \mathbf{c}_1 \in \mathbb{F}_{p^m}^{\times} \right\} \tag{5}$$

In the affine representation, for all $(\mathbf{c}_0, \mathbf{c}_1) \in \mathbb{F}_{p^m} \times \mathbb{F}_{p^m}^{\times}$, there is a corresponding element of $T_6(\mathbb{F}_{p^m})\backslash\{1\}$. This property is suitable for using the affine representation in cryptosystems. For instance, checking an element in $T_6(\mathbb{F}_{p^m})\backslash\{1\}$ is simple. We just check $\mathbf{c}_0 \in \mathbb{F}_{p^m}$ and $\mathbf{c}_1 \in \mathbb{F}_{p^m}^{\times}$; it is unnecessary to calculate the $\Phi_6(p^m)$-th power of the element. In addition, an arbitrary bit string is mapped to an element of $T_6(\mathbb{F}_{p^m})\backslash\{1\}$ by using $\{0,1\}^l \to \mathbb{F}_{p^m}$ and $\{0,1\}^l \to \mathbb{F}_{p^m}^{\times}$.

Eq. (5) is obtained from solving the condition $\left(\frac{\alpha-\beta x}{\alpha+\beta x}\right)^{\Phi_6(p^m)} = 1$ of eq. (4) by substituting $\gamma = \alpha/\beta$. The detail is in subsection 5.2. The suggested affine representation have only one exception point and the constraint of the characteristic p is not hard. There are two different affine representations. Granger et al. also solved the condition [7]. They limited characteristic $p = 3$. The other method of obtaining the affine representation was employed by Rubin and Silverberg [3, 15]. They used a given solution to describe the other solutions. The corresponding element is another exceptional point of the affine representation.

3   Parameter Selection Criteria

In this section, we propose eight criteria. Criteria 1, 2, 3 and 4 are security criteria. Criteria 5, 6, 7 and 8 are efficiency criteria.

Parameters include (n, m, p) of $T_n(\mathbb{F}_{p^m})$, the order q of the group G, and primitive polynomials of the embedding field $\mathbb{F}_{(p^m)^n}$. Where, we choose $n = 6$ for a good compression ratio and lower transformation costs.

3.1   Security

First, in the case $G \subset T_n(\mathbb{F}_{p^m})$, $F = \mathbb{F}_{(p^m)^n}$ must hold. In other words, G must not be covered in any proper subfields of $\mathbb{F}_{(p^m)^n}$. Criterion 1 is a sufficient but not necessary condition for the above statement. Second, the complexity of the index calculus in F, and the Granger-Vercauteren method in $T_n(\mathbb{F}_{p^m})$, and Pollard's ρ method in G must be sufficiently large. Criteria 2, 3 and 4 are sufficient and necessary conditions for the above statements, respectively.

Criterion 1. *Let q be the order of G. $G \subset T_{6m}(\mathbb{F}_p)$ and $q \nmid 6m$ hold.*

If G is a prime-order subgroup of $T_{nm}(\mathbb{F}_p)$ and $q \nmid nm$, then G is not a subgroup of $T_j(\mathbb{F}_p)$ with $j|nm$ and $j \neq nm$. We consider the condition $q \nmid nm$. In accordance with Theorem 2 (b), if $q \nmid nm$ then $X^{nm} - 1$ has no repeated roots in the algebraic closure of $\mathbb{F}_q$. $X^{nm} - 1 = \prod_{j|nm} \Phi_j(X)$ by the definition of the cyclic polynomials. If $G \subset T_{nm}(\mathbb{F}_p)$, then $q|\Phi_{nm}(p)$. So we have $\Phi_{nm}(p) \equiv 0 \bmod q$. Since $X^{nm} - 1$ has no repeated roots, $\Phi_j(p) \not\equiv 0 \bmod q$ for any j with $j|nm$ and $j \neq nm$. Therefore, subgroups of $T_j(\mathbb{F}_p)$ do not have the order divided by q for any j with satisfying $j|nm$ and $j \neq nm$.

We note what happens when Criterion 1 does not hold. A security level of such cryptosystem may not reach the expected security level. Because a minimal embedding field F can be smaller than $\mathbb{F}_{(p^m)^n}$. We show the following Fact 1 as an example of this indication.

Fact 1. *Let m_n be the maximal divisor of m such that all its prime factors are also prime factors of n. If $m_n \neq m$, then $T_{nm_n}(\mathbb{F}_p) \subsetneq T_n(\mathbb{F}_{p^m})$ is covered in $\mathbb{F}_{p^{nm_n}} \subsetneq \mathbb{F}_{(p^m)^n}$.*

Proof. By using Theorem 1, $\Phi_n(p^m)$ is factorized as $\prod_{d|\frac{m}{m_n}} \Phi_{nm_nd}(p)$. Therefore, the corresponding algebraic torus is decomposed as eq. (6).

$$T_n(\mathbb{F}_{p^m}) = \times_{d|\frac{m}{m_n}} T_{nm_nd}(\mathbb{F}_p) \qquad (6)$$

If $m_n \neq m$, then $T_{nm_n}(\mathbb{F}_p)$ in the right-hand side of eq. (6) is smaller than $T_n(\mathbb{F}_{p^m})$ in the left-hand side of eq. (6).   □

Where, $\frac{m}{m_n}$ is the maximal divisor of m such that each its prime factor is not prime factor of n. For instance, if $n = 6$ and $m = 60$ then we obtain $m_n = 12$ and $\frac{m}{m_n} = 5$.

Criterion 2. *The size of F is larger than the security parameter λ_I. When an extension degree l' and an order p' of a corresponding base field satisfy $\log p' < O(\sqrt{l'} \log l')$, the complexity of the index calculus should be larger than the security parameter λ_P.*

There are two efficient methods of calculating the index calculus: the number field sieve method [16] in a prime field and the function field sieve method [17] in a finite field with a large extension degree. Complexity is $L_p[1/3, (64/9)^{1/3}]$ and $L_{p^l}[1/3, (32/9)^{1/3}]$ by using eq. (7), respectively.

$$L_{p^l}[a, b] = \exp((b + o(1))(\log p^l)^a (\log \log p^l)^{1-a}) \qquad (7)$$

Both of the above methods have the same order of subexponential complexity. The constant term of the function field sieve method is smaller than that of the number field sieve method. The security parameter $\lambda_I = 2048, 1024$ is defined by the former for safety.

In the case of a medium-size characteristic, variations of the number field sieve method and the function field sieve method have been proposed [8, 18]. Therefore, the complexity of the index calculus for $\mathbb{F}_{p^l}$ is $L_{p^l}[1/3]$ for all characteristic sizes. When the constant part of the varied function field sieve is smaller than the original, we should check that the complexity of the index calculus satisfies the required security. For $\mathbb{F}_{p^l}$, the variation of the function field sieve is applied to the field as $\mathbb{F}_{p'^{l'}}$ with a divisor l' of l, where $p' = p^{l/l'}$. The variation is efficient when $\log p' < O(\sqrt{l'} \log l')$ holds. If $\log p'$ and $\sqrt{l'} \log l'$ are balanced, then $l' \sim 50$ for $\log p'^{l'} = 2048$, and $l' \sim 35$ for $\log p'^{l'} = 1024$. When the sieving step is fast owing to a good implementation, we should check the complexity of the linear algebra step that is estimated at $O(p'^{2D}) = L_{p^l}[1/3]$. The parameter D is fixed by a parameter α with $p' = L_{p^l}[1/3, \alpha]$.

Criterion 3. *When $\log p \sim 3m \log(3m)$ holds, the complexity of the T_2 method is larger than the security parameter λ_P. When $\log p \sim 2m \log(2m) + 12m - 3 \log m$ holds, the complexity of the T_6 method is larger than λ_P.*

Granger and Vercauteren proposed two algorithms [7]. One is applied to $T_2(\mathbb{F}_{p^{3m}})$, the other is applied to $T_6(\mathbb{F}_{p^m})$. The complexity of the T_2 method is $O((3m)!p((3m)^3 + m^2 \log p) + (3m)^3 p^2)$ and the complexity of the T_6 method is $O((2m)!p(2^{12m} + 3^{2m} \log p) + m^3 p^2)$. In the above complexity, if the first term and the third term are balanced, then the second term is negligible. The complexity becomes $O(L_{p^m}[1/2])$.

Criterion 4. *The size of G is larger than the security parameter λ_P.*

There are only exponential time algorithms for solving the discrete logarithm problem in the general group. For instance, the complexity of Pollard's ρ method is $O(\sqrt{q})$, where q is the order of G.

The computational complexity of these attacks for the parameters listed in section 4 is given in the same section.

3.2   Efficiency

First, we choose the extension field $\mathbb{F}_{p^m}$ on which the algebraic tori are defined, and binomials for primitive polynomials. The binomials must be irreducible. Criteria 5, 6 and 7 are sufficient and necessary conditions for the above statement. Second, we choose the algebraic tori equal to G. When we choose $n = 6$, $T_6(\mathbb{F}_{p^m})$ must be equal to G. Criterion 8 is a sufficient and necessary condition for the above statement.

Criterion 5. *Let m' be a prime factor of m. Both of eq. (8) hold.*

$$\begin{cases} \forall m'\ m'|(p-1) \wedge s^{(p-1)/m'} \neq 1 \\ 4|(p-1)\ if\ 4|m \end{cases} \tag{8}$$

Criterion 6. $3|(p^m - 1) \wedge w^{(p^m - 1)/3} \neq 1$ *holds.*

Criterion 7. $p \neq 2 \wedge \delta^{(p^{3m} - 1)/2} \neq 1$ *holds.*

Criteria 5 and 6 are transformed from Theorem 3 by using Fact 2.

Theorem 3. *Let $t > 1$ be an integer and $a \in \mathbb{F}_{q'}^\times$. Then the binomial $f_t(w) = w^t - a$ is irreducible in $\mathbb{F}_{q'}[w]$ if and only if eq. (9) is satisfied:*

$$\begin{cases} \forall m'\ m'|ord(a) \wedge m' \nmid \frac{q'-1}{ord(a)} \\ 4|(q'-1)\ if\ 4|m\ . \end{cases} \tag{9}$$

m' denotes prime factor of t, and $ord(a)$ denotes the order of a in $\mathbb{F}_{q'}$.

Proof. See chapter 3 section 5 of [19].   □

Fact 2. *Let m' be a prime, and $a \in \mathbb{F}_{q'}^\times$. $ord(a)$ is the order of a in $\mathbb{F}_{q'}^\times$. Then eq. (10) holds. And also, $ord(a) \nmid \frac{q'-1}{m'}$ is equivalent to $a^{(q'-1)/m'} \neq 1$.*

$$m'|ord(a) \wedge m' \nmid \frac{q'-1}{ord(a)} \Leftrightarrow m'|(q'-1) \wedge ord(a) \nmid \frac{q'-1}{m'} \tag{10}$$

Proof. $q' - 1$, ord(a) and m' are represented as the following eq. (11). Let p_i be a prime integer. e_i and e'_i are nonnegative integers. Since ord(a)$|(q' - 1)$, then $e'_i \leq e_i$ for all i.

$$\begin{cases} (q' - 1) = \prod_i p_i^{e_i} \\ \text{ord}(a) = \prod_i p_i^{e'_i} \\ m' = p_j \end{cases} \tag{11}$$

In the left-hand side of eq. (10), $m'|$ord(a) and $m' \, / \frac{q'-1}{\text{ord}(a)}$ mean $1 \leq e'_j$ and $e_j - e'_j < 1$. It leads to $e_j - 1 < e'_j \leq e_j$. Since e_j and e'_j are integers, then $e_j = e'_j$. Therefore, we obtain $1 \leq e'_j = e_j$. The right-hand side of eq. (10) means $1 \leq e_j$ and $e_j - 1 < e'_j$. It leads to $e_j = e'_j$. Therefore, we obtain $1 \leq e_j = e'_j$. The converse is proved in the same way. $\qquad\square$

Criterion 7 is transformed from Theorem 3 by using Fact 2, and $2|(p^{3m} - 1)$ is replaced by $p \neq 2$.

Criterion 8. *Let a and b be nonnegative integers. $m = 2^a 3^b$ holds. And, $q = \Phi_6(p^m) = \Phi_{6m}(p)$ is prime.*

If $G = T_6(\mathbb{F}_{p^m})$, then $q = \Phi_6(p^m) = \prod_{d|\frac{m}{m_6}} \Phi_{6m_6 d}(p)$. The condition $\frac{m}{m_6} = 1$ means that m has no prime factor except 2 and 3.

- In the case of $\frac{m}{m_6} = 1$, $q = \Phi_6(p^m) = \Phi_{6m_6}(p)$.
- In the case of $\frac{m}{m_6} \neq 1$, if $q = \Phi_6(p^m)$ is assumed, then there exist d with $\Phi_{6m_6 d}(p) = q$, and $\Phi_{6m_6 d'}(p) = 1$ for all d' satisfying with $d'|\frac{m}{m_6}$ and $d' \neq d$. We consider $d = 1$. Since $\Phi_{6m_6}(p) = p^{2m_6} - p^{m_6} + 1$, then $\Phi_{6m_6}(p) \neq 1$ with $1 < p$, so we obtain $\Phi_{6m_6}(p) = q$. However, $\Phi_{6m_6}(p) = q$ conflicts with the assumption $q = \Phi_6(p^m)$. The reason is as follow: $\Phi_{6m_6}(p) = \Phi_6(p^m)$ leads to $p^m + p^{m_6} - 1 = 0$ when $m \neq m_6$, but $p^m + p^{m_6} - 1 = 0$ is inconsistent with $1 < p$ and $1 \leq m_6 < m$. Therefore the assumption is denied. $q \neq \Phi_6(p^m)$ in this case.

We note what happens when Criterion 8 does not hold. A non-prime q does not match many cryptosystems. Although a proper subgroup G of the algebraic torus may be preferred to the algebraic torus itself from the viewpoint of size, there was not a good representation of the subgroup.

4   Suggested Parameters

We find the parameters $(m, p, s, \mathbf{w}, \delta)$ with the following steps:

1. obtaining possible m,
2. deciding (m, p),
3. deciding $(s, \mathbf{w}, \delta)$.

Inputs are the security parameters $(\lambda_I, \lambda_P) = (2048, 224), (1024, 160)$. In the case of the prime-order algebraic torus, $\log q \sim \lambda_I/3$. Then, the second part of

Table 3. $(\lambda_I, \lambda_P) = (2048, 224)$

n	$\lvert F\rvert$	$\lvert G\rvert$	m	$\lceil \log p \rceil$	p	s	$\mathbf{w}$	$\mathbf{d}$
6	2048	683	6	57	∘	3	z	$1+z$
6	2048	683	12	29	365292517	5	z	z

∘ = 133432608300027847

Table 4. $(\lambda_I, \lambda_P) = (1024, 160)$

n	$\lvert F\rvert$	$\lvert G\rvert$	m	$\lceil \log p \rceil$	p	s	$\mathbf{w}$	$\mathbf{d}$
6	1024	342	8	22	2643241	7	2	z

Criterion 1 " $q \nmid 6m$ " and Criterion 4 " $\lambda_P \lesssim \log q$ " are always satisfied with the above security parameters. The suggested parameters are shown in Tables 3 and 4. Note that $\delta = \mathbf{d} \in \mathbb{F}_{p^m}^{\times}$ for constructing the affine representation. We explain details of the above steps as follows.

At the first step, we obtain possible m by using Criterion 2 " $\lambda_I \sim 6m \log p$ ", the first part of Criterion 7 " $p \neq 2$ ", and Criterion 8 " $m = 2^a 3^b$ ". The upper bound of m is led by $1 < \log p \sim \lambda_I/(6m)$. The lower bound of m is led by smaller characteristic p than the typical word size (e.g. 32 bits or 64 bits). We check the second part of Criterion 2. When a divisor l' of $6m$ is around 50 (or 35) for $\lambda_I = 2048$ (or 1024 respectively), $D = 1$ in the complexity of the index calculus $O(p'^{2D})$. We estimate that:

- $l' = 48, 54$ for $\lambda_I = 2048$, the complexity $\log (p')^2 \sim 85, 76$.
- $l' = 36$ for $\lambda_I = 2048$, the complexity $\log (p')^2 \sim 144 > \lambda_P = 112$.
- $l' = 32, 36$ for $\lambda_I = 1024$, the complexity $\log (p')^2 \sim 64, 57$.
- $l' = 24$ for $\lambda_I = 1024$, the complexity $\log (p')^2 \sim 85 > \lambda_P = 80$.

Therefore, we obtain $m = 6, 12$ for $\lambda_I = 2048$, and $m = 8$ for $\lambda_I = 1024$. We also check the first and the second part of Criterion 3. The complexity becomes subexponential when $m = 6$ for $\lambda_I = 2048$.

- $m = 6$ for $\lambda_I = 2048$, the complexity $\log (3m)^3 p^2 \sim 127 > \lambda_P = 112$.

Therefore, possible m are unaffected by the Granger-Vercauteren method.

At the second step, we decide (m, p) by using the second part of Criterion 8 " $q = \Phi_{6m}(p)$ ", the first part and the third part of Criterion 5 " $m'|(p - 1)$ " and " $4|(p - 1)$ if $4|m$ ", and the first part of Criterion 6 " $3|(p^m - 1)$ ".

At the third step, we decide $(s, \mathbf{w}, \delta)$ by using the second part of Criterion 5 " $s^{(p-1)/m'} \neq 1$ ", the second part of Criterion 6 " $\mathbf{w}^{(p^m - 1)/3} \neq 1$ ", and the second part of Criterion 7 " $\delta^{(p^{3m} - 1)/2} \neq 1$ ".

We estimate the number of available parameters satisfying the above constraints. Whenever (m, p) is fixed, $(s, \mathbf{w}, \mathbf{d})$ still have a great degree of freedom. We are interested in the number of (m, p). When p satisfies $\lfloor 6m \log p \rfloor = \lambda_I$ exactly, 656 different p for $(\lambda_I, m) = (2048, 12)$, many more p for $(2048, 6)$, and 36 different p for $(1024, 6)$.

5   Computation Costs

Costs of operations and transformations are shown in Tables 5 and 6. Where I_3, M_3, F_3 are a cost of inversion, multiplication, the Frobenius exponentiation

Table 5. Costs of operations

Operation	Cost
Mult	$3M_3 + 3B = 18M + 9B$
Frob	$2F_3 + B = 6F + 7B$

Table 6. Cost of transformations

Map	Cost
P2A $I_3 + M_3 = I + 21M + 6F + 12B$	
A2P	$M + 2S + B$

in $\mathbb{F}_{(p^m)^3}$, respectively. I, M, S, F are a cost of inversion, multiplication, square, the Frobenius exponentiation in $\mathbb{F}_{p^m}$, respectively. B is a cost of multiplication between unknown element and the constant $\mathbf{w}, \mathbf{d}$ in $\mathbb{F}_{p^m}$.

Although the extension degree m and the size of characteristic p are adjustable, our result is similar to the case of cyclotomic polynomials [11] and the case of characteristic 3 [7].

5.1   Arithmetical Operations

Multiplication. The primitive polynomial $f_2(x) = x^2 - \delta$ leads to $x^2 \equiv \delta$ (mod $f_2(x)$). The multiplication between $(\alpha, \beta) \in T_6(\mathbb{F}_{p^m})$ and $(\alpha', \beta') \in T_6(\mathbb{F}_{p^m})$ implies the multiplication between $\frac{\alpha - \beta x}{\alpha + \beta x} \in T_6(\mathbb{F}_{p^m})$ and $\frac{\alpha' - \beta' x}{\alpha' + \beta' x} \in T_6(\mathbb{F}_{p^m})$, and becomes $(\alpha'', \beta'') \in T_6(\mathbb{F}_{p^m})$ by

$$(\alpha \pm \beta x)(\alpha' \pm \beta' x) = (\alpha\alpha' + \delta\beta\beta') \pm (\beta\alpha' + \alpha\beta')x = \alpha'' \pm \beta'' x. \tag{12}$$

Eq. (12) is equal to the multiplication in the finite field $\mathbb{F}_{(p^m)^6}$. We can compute that with $3M_3 + 3B$ by using the Karatsuba method, which performs better than classical multiplication when the cost of the multiplication in the subfield is higher than that of the addition in the subfield.

The Frobenius exponentiation. $x^2 \equiv \delta$ (mod $f_2(x)$) leads $x^{p^l} = \delta^{(p^l-1)/2}x$. The p^l-th power of $(\alpha, \beta) \in T_6(\mathbb{F}_{p^m})$ implies the p^l-th power of $\frac{\alpha - \beta x}{\alpha + \beta x} \in T_6(\mathbb{F}_{p^m})$, and becomes $(\alpha', \beta') \in T_6(\mathbb{F}_{p^m})$ by

$$(\alpha \pm \beta x)^{p^l} = \alpha^{p^l} \pm \beta^{p^l} \delta^{(p^l-1)/2}x = \alpha' \pm \beta' x. \tag{13}$$

Eq. (13) is equal to the Frobenius exponentiation in the finite field $\mathbb{F}_{(p^m)^6}$.

5.2   Transformations

From the projective into the affine (P2A). We consider the case that $\delta = \mathbf{d} \in \mathbb{F}_{p^m}^\times$. If $\beta \neq 0$, by substituting $\gamma = \alpha/\beta$ to the condition of eq. (4), we have the following eq. (14).

$$\gamma^{p^{2m}+p^m} + \gamma^{p^m+1} + \gamma^{p^{2m}+1} = -\mathbf{d} \tag{14}$$

By substituting $\gamma = \mathbf{c}_0 + \mathbf{c}_1 y + \mathbf{c}_2 y^2$, $\mathbf{c}_i \in \mathbb{F}_{p^m}$ to eq. (14), we obtain

$$\mathbf{c}_1\mathbf{c}_2\mathbf{w} = \mathbf{c}_0^2 + \mathbf{d}/3. \tag{15}$$

$\mathbf{c}_2$ can be calculated from $\mathbf{c}_0$ and $\mathbf{c}_1$ in the case of $\mathbf{c}_1 \neq 0$. If $m = 2^a 3^b$, then $\mathbf{c}_1 \neq 0$ and $\mathbf{c}_2 \neq 0$ is proved. We show $-\mathbf{d}/3$ is a quadratic nonresidue in $\mathbb{F}_{p^m}$ in order to prove that. See appendix A.

From the affine into the projective (A2P). $c_0 c_1 + c_1^2 y + (c_0^2 + d \cdot 3^{-1}) w^{-1} y^2$ and c_1 are substituted to α and β, respectively.

6   Concluding Remarks

In this paper, we produced the parameter selection criteria from the viewpoints of security and efficiency. There are an improvement of the index calculus for medium characteristic, subexponential time algorithm for solving the discrete logarithm problem in algebraic tori, and the fact the algebraic torus $T_n(\mathbb{F}_{p^m})$ can be covered by a subfield of $\mathbb{F}_{(p^m)^n}$. We established criteria in order to resist attacks consistently. We also made the extension degree m and the size of characteristic p adjustable. We provide examples of parameters satisfying the criteria in the case that characteristics are shorter than the word size. Our analysis confirmed that the element of the algebraic torus T_6 in the affine representation has good properties: small size and simple use. Certain issues are left as subjects for future work, including investigating the performance of an actual implementation, optimizing torus-based public key schemes, and selecting more efficient parameters such that G is the minimum.

References

1. Smith, P., Skinner, C.: A Public-key Cryptosystem and a Digital Signature Based on the Lucas Function Analogue to Discrete Logarithms. In: Safavi-Naini, R., Pieprzyk, J.P. (eds.) ASIACRYPT 1994. LNCS, vol. 917, pp. 357–364. Springer, Heidelberg (1995)
2. Lenstra, A.K., Verheul, E.R.: The XTR Public Key System. In: Bellare, M. (ed.) CRYPTO 2000. LNCS, vol. 1880, pp. 1–19. Springer, Heidelberg (2000)
3. Rubin, K., Silverberg, A.: Torus-based Cryptography. In: Boneh, D. (ed.) CRYPTO 2003. LNCS, vol. 2729, pp. 349–365. Springer, Heidelberg (2003)
4. Barker, E., Barker, W., Burr, W., Polk, W., Smid, M.: Recommendation for Key Management - Part 1: Genaral (Revised). Special Publication 800/57, NIST (2007)
5. van Dijk, M., Granger, R., Page, D., Rubin, K., Silverberg, A., Stam, M., Woodruff, D.: Practical Cryptography in High Dimensional Tori. In: Cramer, R. (ed.) EUROCRYPT 2005. LNCS, vol. 3494, pp. 234–250. Springer, Heidelberg (2005)
6. Gower, J.E.: Prime Order Primitive Subgroups in Torus-based Cryptography. Cryptology ePrint Archive, Report 2006/466 (2006)
7. Granger, R., Vercauteren, F.: On the Discrete Logarithm Problem on Algebraic Tori. In: Shoup, V. (ed.) CRYPTO 2005. LNCS, vol. 3621, pp. 66–85. Springer, Heidelberg (2005)
8. Joux, A., Lercier, R.: The Function Field Sieve in the Medium Prime Case. In: Vaudenay, S. (ed.) EUROCRYPT 2006. LNCS, vol. 4004, pp. 254–270. Springer, Heidelberg (2006)
9. Freeman, D., Scott, M., Teske, E.: A Taxonomy of Pairing-Friendly Elliptic Curves. Journal of Cryptology 23(2), 224–280 (2010)
10. Miyaji, A., Nakabayashi, M., Takano, S.: New Explicit Conditions of Elliptic Curve Traces for FR-Reduction. IEICE Transactions on Fundamentals of Electronics, Communications and Computer Sciences E84-A(5), 1234–1243 (2001)

11. Granger, R., Page, D., Stam, M.: A Comparison of CEILIDH and XTR. In: Buell, D.A. (ed.) ANTS 2004. LNCS, vol. 3076, pp. 235–249. Springer, Heidelberg (2004)
12. Hitt, L.: On the Minimal Embedding Field. In: Takagi, T., Okamoto, T., Okamoto, E., Okamoto, T. (eds.) Pairing 2007. LNCS, vol. 4575, pp. 294–301. Springer, Heidelberg (2007)
13. Bosma, W., Hutton, J., Verheul, E.R.: Looking beyond XTR. In: Zheng, Y. (ed.) ASIACRYPT 2002. LNCS, vol. 2501, pp. 321–332. Springer, Heidelberg (2002)
14. Galbraith, S.: Disguising Tori and Elliptic Curves. Cryptology ePrint Archive, Report 2006/248 (2006)
15. Rubin, K., Silverberg, A.: Compression in Finite Fields and Torus-based Cryptography. SIAM Jour. on Computing 37(5), 1401–1428 (2008)
16. Gordon, D.: Discrete Logarithms in GF (p) Using the Number Field Sieve. SIAM Jour. on Discrete Math. 6, 124–138 (1993)
17. Adleman, L.M.: The Function Field Sieve. In: Huang, M.-D.A., Adleman, L.M. (eds.) ANTS 1994. LNCS, vol. 877, pp. 108–121. Springer, Heidelberg (1994)
18. Joux, A., Lercier, R., Smart, N.P., Vercauteren, F.: The Number Field Sieve in the Medium Prime Case. In: Dwork, C. (ed.) CRYPTO 2006. LNCS, vol. 4117, pp. 326–344. Springer, Heidelberg (2006)
19. Lidl, R., Niederreiter, H.: Finite Fields, 2nd edn. Encyclopedia of Mathematics and its Applications, vol. 20. Cambridge University Press, Cambridge (1997)

A   Quadratic Nonresidue in $\mathbb{F}_{p^m}$

Fact 3. *Let a and b be nonnegative integers, p be a prime, and $m = 2^a 3^b$. $g_m(z) = z^m - s$, $f_3(y) = y^3 - \mathbf{w}$ and $f_2(x) = x^2 - \delta$ are irreducible in $\mathbb{F}_p[z]$, $\mathbb{F}_{p^m}[y]$ and $\mathbb{F}_{(p^m)^3}[x]$, respectively. If $\delta = \mathbf{d} \in \mathbb{F}_{p^m}^{\times}$, $-\mathbf{d}/3$ is a quadratic nonresidue in $\mathbb{F}_{p^m}$.*

Proof. -3 is a quadratic residue and $\mathbf{d}$ is a quadratic nonresidue. Then $-\mathbf{d}/3$ is a quadratic nonresidue in $\mathbb{F}_{p^m}$.

- Since $f_3(y)$ is irreducible, $3|(p^m - 1)$ holds and $3 \nmid p$. If $3|m$, then $3|(p - 1)$ because $g_m(z)$ is irreducible. The fact that -3 is a quadratic residue in $\mathbb{F}_{p^m}$ is proved by using Fact 4 (a) and Fact 5 (a). If $2|m$ and $3|(p - 1)$, the fact is proved in the above way. If $2|m$ and $3|(p - 2)$, the fact is proved by using Fact 4 (b) and Fact 5 (b).
- Since $f_2(x)$ is irreducible, $\mathbf{d}^{(p^m - 1)/2} \neq 1$. $\mathbf{d}$ is a quadratic nonresidue in $\mathbb{F}_{p^m}$ straightforwardly. $\square$

Fact 4. *Let p be a prime.*

(a) *If $3|(p - 1)$, -3 is a quadratic residue in $\mathbb{F}_p$.*
(b) *If $3|(p - 2)$, -3 is a quadratic nonresidue in $\mathbb{F}_p$.*

Proof. We use the Legendre symbol $\left(\frac{a}{b}\right)$, the first supplementary law, and the reciprocity law.

$$\left(\frac{-3}{p}\right) = \left(\frac{-1}{p}\right)\left(\frac{3}{p}\right) = (-1)^{\frac{p-1}{2}}(-1)^{\frac{p-1}{2}}\left(\frac{p \bmod 3}{3}\right) = \left(\frac{p \bmod 3}{3}\right) \qquad (16)$$

If $p \bmod 3 = 1$, then -3 is a quadratic residue in $\mathbb{F}_p$. Else if $p \bmod 3 = 2$, then -3 is a quadratic nonresidue in $\mathbb{F}_p$. $\square$

Fact 5. *Let m be a positive integer, p be a prime.*

(a) If a is a quadratic residue in $\mathbb{F}_p$, then a is a quadratic residue in $\mathbb{F}_{p^m}$.

(b) If a is a quadratic nonresidue in $\mathbb{F}_p$, a is a quadratic residue in $\mathbb{F}_{p^m}$ in the case of $2 \mid m$, and a is a quadratic nonresidue in $\mathbb{F}_{p^m}$ in the case of $2 \nmid m$.

Proof. (a) If $a^{(p-1)/2} = 1$, then $a^{\frac{p^m-1}{2}} = a^{\frac{p-1}{2}(p^{m-1}+\cdots+p+1)} = 1$.

(b) If $a^{(p-1)/2} = -1$, then $a^{\frac{p^m-1}{2}} = (-1)^{p^{m-1}+\cdots+p+1} = (-1)^m$.   □

Analysis of the MQQ Public Key Cryptosystem

Jean-Charles Faugère[2], Rune Steinsmo Ødegård[1,*],
Ludovic Perret[2], and Danilo Gligoroski[3]

[1] Centre for Quantifiable Quality of Service in Communication Systems at the
Norwegian University of Science and Technology in Trondheim, Norway
rune.odegard@q2s.ntnu.no
[2] SALSA Project - INRIA (Centre Paris-Rocquencourt)
UPMC, Univ Paris 06 - CNRS, UMR 7606, LIP6
104, avenue du Président Kennedy 75016 Paris, France
jean-charles.faugere@inria.fr, ludovic.perret@lip6.fr
[3] Department of Telematics at the Norwegian University of Science and Technology
in Trondheim, Norway
danilog@item.ntnu.no

Abstract. MQQ is a multivariate public key cryptosystem (MPKC)
based on multivariate quadratic quasigroups and a special transform
called "*Dobbertin transformation*" [17]. The security of MQQ, as well
as any MPKC, reduces to the difficulty of solving a non-linear system of
equations easily derived from the public key. In [26], it has been observed
that that the algebraic systems obtained are much easier to solve that
random non-linear systems of the same size. In this paper we go one
step further in the analysis of MQQ. We explain why systems arising
in MQQ are so easy to solve in practice. To do so, we consider the so-
called the degree of regularity; which is the exponent in the complexity
of a Gröbner basis computation. For MQQ systems, we show that this
degree is bounded from above by a small constant. This is due to the
fact that the complexity of solving the MQQ system is the minimum
complexity of solving just one quasigroup block or solving the Dobbertin
transformation. Furthermore, we show that the degree of regularity of
the Dobbertin transformation is bounded from above by the same constant as the bound observed on MQQ system. We then investigate the
strength of a tweaked MQQ system where the input of the Dobbertin
transformation is replaced with random linear equations. It appears that
the degree of regularity of this tweaked system varies both with the size
of the quasigroups and the number of variables. We conclude that if a
suitable replacement for the Dobbertin transformation is found, MQQ
can possibly be made strong enough to resist pure Gröbner attacks for
adequate choices of quasigroup size and number of variables.

Keywords: multivariate cryptography, Gröbner bases , public-key, multivariate quadratic quasigroups, algebraic cryptanalysis.

* Rune Steinsmo Ødegård was visiting the SALSA team at LIP6 during the research
of this paper.

S.-H. Heng, R.N. Wright, and B.-M. Goi (Eds.): CANS 2010, LNCS 6467, pp. 169–183, 2010.
© Springer-Verlag Berlin Heidelberg 2010

1   Introduction

The use of polynomial systems in cryptography dates back to the mid eighties with the design of Matsumoto and Imai [25], later followed by numerous other proposals. Two excellent surveys on the current state of proposals for multivariate asymmetric cryptosystems has been made by Wolf and Preneel [33] as well as Billet and Ding [6]. Basically the current proposals can be classified into four main categories, some of which combine features from several categories: Matsumoto-Imai like schemes [28,30], Oil and Vinegar like schemes [29,20], Stepwise Triangular Schemes [31,18] and Polly Cracker Schemes [11]. In addition Gligoroski et al. has proposed a fifth class of trapdoor functions based on multivariate quadratic quasigroups [17].

As pointed out in [6], it appears that most multivariate public-key cryptosystems (MPKC) suffer from obvious to less obvious weaknesses. Some attacks are specific and focus on one particular variation and breaks it due to specific properties. One example is the attack of Kipnis and Shamir against the Oil and Vinegar scheme [21]. However, most attacks use general purpose algorithms that solve multivariate system of equations. Generic algorithms to solve this problem are exponential in the worst case, and solving random system of algebraic equations is also known to be difficult (i.e. exponential) in the average case. However, in the case of multivariate public-key schemes the designer has to embed some kind of trapdoor function to enable efficient decryption and signing. To achieve this, the public-key equations are constructed from a highly structured system of equations. Although the structure is hidden, it can be exploited for instance via differential or Gröbner basis based techniques.

Using Gröbner basis [8] is a well established and general method for solving polynomial systems of equations. The complexity of a Gröbner basis computation is exponential in the degree of regularity, which is the maximum degree of polynomials occurring during the computation [4]. The first published attack on multivariate public-key cryptosystems using Gröbner basis is the attack by Patarin on the Matsumoto-Imai scheme [27]. In this paper Patarin explains exactly why one is able to solve the system by using Gröbner bases. The key aspect is that there exists bilinear equations relating the input and output of the system [6]. This low degree relation between the input and the output means that only polynomials of a low degree will appear during the computation of the Gröbner basis. Consequently, the complexity of solving the system is bounded by this low degree.

Another multivariate cryptosystem which has been broken by Gröbner bases cryptanalysis is the MQQ public key block cipher [17]. The cipher was broken both by Gröbner bases and MutantXL independently in [26]. Given a ciphertext encrypted using the public key, the authors of [26] were able to compute the corresponding plaintext. However, the paper did not theoretically explain why the algebraic systems of MQQ are easy to solve in practice. In this paper we explain exactly why the MQQ cryptosystem is susceptible to algebraic cryptanalysis. This is of course interesting from a cryptanalysis perspective, but also from a design perspective. If we want to construct strong multivariate cryptographic schemes we must understand why the weak schemes have been broken.

1.1 Organisation of the Paper

This paper is organized as follows. In Section 2 we give an introduction to multivariate quadratic quasigroups. After that we describe the MQQ public key cryptosystem. In Section 3 we give a short introduction to the theory of Gröbner bases and reiterate the generic complexity of computing such bases. In Section 4 we show that the degree of regularity of MQQ systems is bounded from above by a small constant. We then explain this charcteristic by looking at the shape of the inner system. In Section 5 we further elaborate on the weaknesses of MQQ, and investigate if some tweaks can make the system stronger. Finally, Section 6 concludes the paper.

2 Description of the MQQ Public Key Cryptosystem

In this section we give a description of the multivariate quadratic quasigroup public key cryptosystem [17]. The system is based on previous work by Gligoroski and Markovski who introduced the use of quasigroup string processing in cryptography [23,24].

2.1 Multivariate Quadratic Quasigroups

We first introduce the key building block of the MQQ PKC, namely multivariate quadratic quasigroups. For a detailed introduction to quasigroups in general, we refer the interested reader to [32].

Definition 1. *A quasigroup is a set Q together with a binary operation $*$ such that for all $a, b \in Q$ the equations $\ell * a = b$ and $a * r = b$ have unique solutions ℓ and r in Q respectively. A quasigroup is said to be of order n if there are n elements in the set Q.*

Let $(Q, *)$ be a quasigroup of order 2^d, and β be a bijection from the quasigroup to the set of binary strings of length d, i.e

$$\begin{aligned} \beta : Q &\to GF(2^d) \\ a &\mapsto (x_1, \dots, x_d) \end{aligned} \tag{1}$$

Given such a bijection, we can naturally define a vector valued Boolean function

$$\begin{aligned} *_{vv} : GF(2^d) \times GF(2^d) &\to GF(2^d) \\ (\beta(a), \beta(b)) &\mapsto \beta(a * b) \end{aligned} \tag{2}$$

Now let $\beta(a * b) = (x_1, \dots, x_d) *_{vv} (x_{d+1}, \dots, x_{2d}) = (z_1, \dots, z_d)$. Note that each z_i can be regarded as a $2d$-ary Boolean function $z_i = f_i(x_1, \dots, x_{2d})$, where each $f_i : GF(2^d) \to GF(2)$ is determined by $*$. This gives us the following lemma [17].

Lemma 1. *For every quasigroup $(Q, *)$ of order 2^d and for each bijection β : $Q \to GF(2^d)$ there is a unique vector valued Boolean function $*_{vv}$ and d uniquely determined $2d$-ary Boolean functions $f_1, f_2, \ldots, f_d$ such that for each $a, b, c \in Q$:*

$$a * b = c$$
$$\Updownarrow \qquad\qquad (3)$$
$$(x_1, \ldots, x_d) *_{vv} (x_{d+1}, \ldots, x_{2d}) = (f_1(x_1, \ldots, x_{2d}), \ldots, f_d(x_1, \ldots, x_{2d})).$$

This leads to the following definition for multivariate quadratic quasigroups.

Definition 2. *([17]) Let $(Q, *)$ be a quasigroup of order 2^d, and let $f_1, \ldots, f_d$ be the uniquely determined Boolean functions under some bijection β. We say that the quasigroup is a multivariate quadratic quasigroup (MQQ) of type $Quad_{d-k} Lin_k$ (under β) if exactly $d - k$ of the corresponding polynomials f_i are of degree 2 and k of them are of degree 1, where $0 \le k \le d$.*

Gligoroski et al. [17] mention that quadratic terms might cancel each other. By this we mean that some linear transformation of $(f_i)_{1 \le i \le n}$ might result in polynomials where the number of linear polynomials is larger than k, while the number of quadratic polynomials is less than $d - k$. Later Chen et al. [9] have shown that this is more common than previously expected. In their paper they generalizes the definition of MQQ above to a family which is invariant by linear transformations, namely:

Definition 3. *Let $(Q, *)$ be a quasigroup of order 2^d, and let $f_1, \ldots, f_d$ be the unique Boolean functions under some bijection β. We say that the quasigroup is a multivariate quadratic quasigroup (MQQ) of strict type $Quad_{d-k} Lin_k$ (under β), denoted by $Quad_{d-k}^s Lin_k^s$, if there are at most $d - k$ quadratic polynomials in $(f_i)_{1 \le i \le d}$ whose linear combination do not result in a linear form.*

Chen et al. also improved Theorem 2 from [17] which gives a sufficient condition for a quasigroup to be MQQ. We restate this result below.

Theorem 1. *Let $\mathbf{A}_1 = [f_{ij}]_{d \times d}$ and $\mathbf{A}_2 = [g_{ij}]_{d \times d}$ be two $d \times d$ matrices of linear Boolean expressions with respect to $x_1, \ldots, x_d$ and $x_{d+1}, \ldots, x_{2d}$ respectively. Let $\mathbf{c}$ be a binary column vector of d elements. If $det(\mathbf{A}_1) = det(\mathbf{A}_2) = 1$ and*

$$\mathbf{A}_1 \cdot (x_{d+1}, \ldots, x_{2d})^T + (x_1, \ldots, x_d)^T = \mathbf{A}_2 \cdot (x_1, \ldots, x_d)^T + (x_{d+1}, \ldots, x_{2s})^T, \quad (4)$$

then the vector valued Boolean operation $(x_1, \ldots, x_d) *_{vv} (x_{d+1}, \ldots, x_{2d}) =$

$$\mathbf{B}_1 \mathbf{A}_1 \cdot (x_{d+1}, \ldots, x_{2d})^T + \mathbf{B}_2 \cdot (x_1, \ldots, x_d)^T + \mathbf{c} \quad (5)$$

defines a quasigroup $(Q, *)$ of order 2^d which is MQQ for any two non-singular Boolean matrices $\mathbf{B}_1$ and $\mathbf{B}_2$

In addition Chen et al. [9] proved that no MQQ as in Theorem 1 can be of strict type $Quad_d^s Lin_0^s$. This result uncovered a possible weakness in [17] as the proposed scheme used 6 quasigroups of type $Quad_5 Lin_0$.

Notice that the vector valued Boolean function defining the MQQ in Theorem 1 have no terms of the form $x_i x_j$ with $i, j \leq d$ or $i, j > d$. This means that if we set the first or the last half of the variables to a constant, we end up with only linear terms in the MQQ. It is still an open question if there exists MQQ that are not as in Theorem 1.

The MQQs used in this paper have been produced using the algorithm provided in Appendix A. The algorithm is based on the paper [9], and produces MQQs that are more suitable for encryption since they are guaranteed to be of strict type $\text{Quad}_{d-k}^s \text{Lin}_k^s$ for $0 < k \leq d$.

2.2 The Dobbertin Bijection

In addition to MQQs, [17] also uses a bijection introduced by Dobbertin in [12]. Dobbertin proved that the following function, in addition to being multivariate quadratic over $GF(2)$, is a bijection in $GF(2^{2r+1})$:

$$
\begin{aligned}
D_r : GF(2^{2r+1}) &\to \quad GF(2^{2r+1}) \\
x \quad &\mapsto x^{2^{r+1}+1} + x^3 + x
\end{aligned}
\tag{6}
$$

2.3 A Public Key Cryptosystem Based on MQQ

We are now ready to describe the public key cryptosystem presented by Gligoroski et al. in [17]. Let $N = nd$ be the desired number of variables $(x_1, \ldots, x_N)$, and let $\{*_{vv}^1, \ldots, *_{vv}^k\}$ be a collection of MQQs of size 2^d represented as $2d$-ary vector valued Boolean functions. The public key is constructed as follows.

Algorithm. *MQQ public key construction*

1. Set $\mathbf{X} = [x_1, \ldots, x_N]^T$. Randomly generate an $N \times N$ non-singular Boolean matrix $\mathbf{S}$, and compute $\mathbf{X} \leftarrow \mathbf{S} \cdot \mathbf{X}$.
2. Randomly choose a n-tuple $I = \{i_1, \ldots, i_n\}$, where $i_j \in \{1, \ldots, k\}$. The tuple I will decide which MQQ, $*_{vv}^{i_j}$, to use at each point of the quasigroup transformation.
3. Represent $\mathbf{X}$ as a collection of vectors of length d, $\mathbf{X} = [X_1, \ldots, X_n]^T$. Compute $\mathbf{Y} = [Y_1, \ldots, Y_n]^T$ where $Y_1 = X_1, Y_2 = X_1 *_{vv}^{i_1} X_2$, and $Y_{j+1} = X_j *_{vv}^{i_j} X_{j+1}$ for $j = 1, \ldots, n-1$.
4. Set $\mathbf{Z}$ to be the vector of all the linear terms of $Y_1, \ldots, Y_n$. Here Y_1 will be all linear terms, while each Y_j has between 1 and k linear terms depending on the type $\text{Quad}_{d-k}^s \text{Lin}_k^s$ of MQQ used. Transform $\mathbf{Z}$ with one or more Dobbertin bijections of appropriate size. For example if $\mathbf{Z}$ is of size 27 we can use one Dobbertin bijection of dimension 27, three of dimension 9, or any other combination summing up to 27. Finally, set $\mathbf{W} \leftarrow \text{Dob}(\mathbf{Z})$.
5. Replace the linear terms of $\mathbf{Y} = [Y_1, \ldots, Y_n]^T$ with the terms in $\mathbf{W}$. Randomly generate an $N \times N$ non-singular Boolean matrix $\mathbf{T}$, and compute $\mathbf{Y} \leftarrow \mathbf{T} \cdot \mathbf{Y}$
6. **return** the public key $\mathbf{Y}$. The private key is $\mathbf{S}, \mathbf{T}, \{*_{vv}^1, \ldots, *_{vv}^k\}$ and I.

3   Gröbner Bases

This section introduces the concept of Gröbner bases as well as a complexity bound to compute such bases. We refer to (for instance) [10] for basic definitions, and a more detailed description of the concepts.

Let $\mathbb{K}$ be a field and $\mathbb{K}[x_1, \ldots, x_N]$ be the polynomial ring over $\mathbb{K}$ in the variables $x_1, \ldots, x_N$. Recall that a *monomial* in a collection of variables is a product $x^\alpha = x_1^{\alpha_1} \cdots x_N^{\alpha_N}$ where $\alpha_i \geq 0$. Let $>$ be an admissible *monomial order* on $\mathbb{K}[x_1, \ldots, x_N]$. The most common example of such ordering is the *lexicographical order* where $x^\alpha > x^\beta$ if in the difference $\alpha - \beta \in \mathbb{Z}^N$, the leftmost nonzero entry is positive. Another frequently encountered order is the *graded reverse lexicographical* order where $x^\alpha > x^\beta$ iff $\sum_i \alpha_i > \sum_i \beta_i$ or $\sum_i \alpha_i = \sum_i \beta_i$ and in the difference $\alpha - \beta \in \mathbb{Z}^N$ the rightmost nonzero entry is negative. For different monomial orderings Gröbner bases hold specific theoretical properties and show different practical behaviors. Given a monomial order $>$, the *leading term* of a polynomial $f = \sum_\alpha c_\alpha x^\alpha$, denoted $LT_>(f)$, is the product $c_\alpha x^\alpha$ where x^α is the largest monomial appearing in f in the ordering $>$.

Definition 4. *([10]) Fix a monomial order $>$ on $\mathbb{K}[x_1, \ldots, x_N]$, and let $I \subset \mathbb{K}[x_1, \ldots, x_N]$ be an ideal. A Gröbner basis for I (with respect to $>$) is a finite collection of polynomials $G = \{g_1, \ldots, g_t\} \subset I$ with the property that for every nonzero $f \in I$, $LT_>(f)$ is divisible by $LT_>(g_i)$ for some i.*

Let

$$f_1(x_1, \ldots, x_N) = \cdots = f_m(x_1, \ldots, x_N) = 0 \tag{7}$$

by a system of m polynomials in N unknowns over the field $\mathbb{K}$. The set of solutions in $\mathbb{K}$, which is the *algebraic variety*, is defined as

$$V = \{(z_1, \ldots, z_N) \in k | f_i(z_1, \ldots, z_N) = 0 \forall 1 \leq i \leq m\} \tag{8}$$

In our case we are interested in the solutions of the MQQ system, which are defined over $GF(2)$.

Proposition 1. *([15]) Let G be a Gröbner basis of $[f_1, \ldots, f_m, x_1^2 - x_1, \ldots, x_N^2 - x_N]$. Then the following holds:*

1. $V = \emptyset$ *(no solution) iff $G = [1]$.*
2. V *has exactly one solution iff $G = [x_1 - a_1, \ldots, x_N - a_N]$ where $a_i \in GF(2)$. Then $(a_1, \ldots, a_N)$ is the solution in $GF(2)$ of the algebraic system.*

It is clear that as we are solving systems over $GF(2)$ we have to add the field equations $x_i^2 = x_i$ for $i = 1, \ldots, N$. This means that we have to compute Gröbner bases of $m + N$ polynomials and N variables. This is quite helpful, since the more equations you have, the more able you are to compute Gröbner bases [15].

3.1   Complexity of Computing Gröbner Bases

Historically, the concept of Gröbner bases, together with an algorithm for computing them, was introduced by Bruno Buchberger in his PhD-thesis [8]. Buchberger's algorithm is implemented in many computer algebra systems. However, in the last decade, more efficient algorithms for computing Gröbner bases have been proposed. Most notable are Jean-Charles Faugère's F_4[13] and F_5 [14] algorithms. In this paper we have used the magma [22] 2.16-1 implementation of the F_4 algorithm on a 4 core Intel Xeon 2.93GHz computer with 128GB of memory.

The complexity of computing a Gröbner basis of an ideal I depends on the maximum degree of the polynomials appearing during the computation. This degree, called *degree of regularity*, is the key parameter for understanding the complexity of a Gröbner basis computation [4]. Indeed, the complexity of the computation is polynomial in the degree of regularity D_{reg}, more precisely the complexity is:

$$\mathcal{O}(N^{\omega D_{reg}}), \tag{9}$$

which basically correspond to the complexity of reducing a matrix of size $\approx N^{D_{reg}}$. Here $2 < \omega \leq 3$ is the "linear algebra constant", and N the number of variables of the system. Note that D_{reg} is also a function of N, where the relation between D_{reg} and N depends on the specific system of equations. This relation is well understood for regular (and semi-regular) systems of equations [1,4,2,5]. However, as soon as the system has some kind of structure, this degree is much more difficult to predict. In some particular cases, it is actually possible to bound the degree of regularity (see the works done on HFE [15,19]). But this is a hard task in general.

As already pointed out, the degree of regularity is abnormally small for algebraic systems occuring in MQQ. This fact explains the weakness observed in [26]. In this paper, we go one step further in the security analysis by explaining why the degree of regularity is so small for MQQ.

Note that the degree of regularity is related to the ideal $I = \langle f_1, \ldots, f_m \rangle$ and not the equations $f_1, \ldots, f_m$ themselves. In particular, for any non-singular matrix T, the degree of regularity of $[f'_1, \ldots, f'_m]^t = T \cdot [f_1, \ldots, f_m]^t$ is similar to the degree of regularity of $[f_1, \ldots, f_m]$. More generally, we can assume that this degree is generically (i.e. with high probability) invariant for a random (invertible) linear change of variables, and an (invertible) combination of the polynomials. These are exactly the transformations performed to mask the MQQ structure. Note that such a hypothesis has already been used for instance in [19].

4   Why MQQ Is Susceptible to Algebraic Cryptanalysis

In [26], MQQ systems with up to 160 variables was broken usin MutantXL (the same result has aslo been obtained independently with F_4). The most important point made by [26] is that the degree of regularity is bounded from above by 3. This is much lower than a random system of quadratic equations where the degree of regularity increases linearly with the number of variables N. Indeed, for a random system it holds that D_{reg} is asymptotically equivalent to $\frac{N}{11.114}$ [2]. The authors of [26] observed that this low degree is due to the occurrence of many new low degree relations during the computation of a Gröbner basis. In Section 4.2, we will explain in detail how the very structure of the MQQ system results in the apparance of the low degree relations. First, however, we will show that same upper bound on the degree of regularity is obtained using the improved quasigroups described in Section 2.1.

4.1   Experimental Results on MQQ

To test how the complexity of Gröbner bases computation of MQQ systems is related to the number of variables, we constructed MQQ systems in $30, 60, 120$ and 180 variables following the procedure described in Section 2.3. In this construction we used 17 MQQs of strict type $\text{Quad}_8^s\text{Lin}_2^s$ and Dobbertin bijections over different extension fields of dimension 7 and 9 respectively. We then tried to compute the plaintext given a ciphertext encrypted with the public key. The results of this test are presented in Table 1. From the table we see that the

Table 1. Results for MQQ-(30,60,120,180). Computed with magma 2.16-1's implementation of the F_4 algorithm on a Intel Xeon 2.93GHz quad core computer with 128GB of memory.

Variables	D_{reg}	Solving Time (s)	Memory (b)
30	3	0,06	15,50
60	3	1,69	156,47
120	3	379,27	4662,00
180	3	4136,31	28630,00

degree of regularity does not increase with the number of variables, but remains constant at 3. This means breaking the MQQ system is only polynomial in the number of variables. Once again, this is not the behaviour of a random system of equations, for which the degree of regularity increases linearly with the number of variables, and the solving time therefore increases exponentially. We explain the reason of such difference in the next section.

4.2   Shape of the MQQ System

The non-random behavior described above can be explained by considering the shape of the "unmasked" MQQ system. By unmasked we mean the MQQ system without the linear transformations S and T. As already explained in Section 3.1, the maximum degree of the polynomials occurring in the computation of a Gröbner basis is invariant under the linear transformation S and T.

In Figure 1 we show which variables appear in each equation for an unmasked MQQ system of 60 variables. The staircase shape comes from the cascading use of quasigroups, while the three blocks of equations at the bottom are from the Dobbertin bijection of size 7. Obviously, a random multivariate system would use all 60 variables in all equations. For this instance of MQQ, only $\frac{1}{3}$ of the variables are used in each quasigroup and about $\frac{2}{3}$ is used in each block of the Dobbertin transformation.

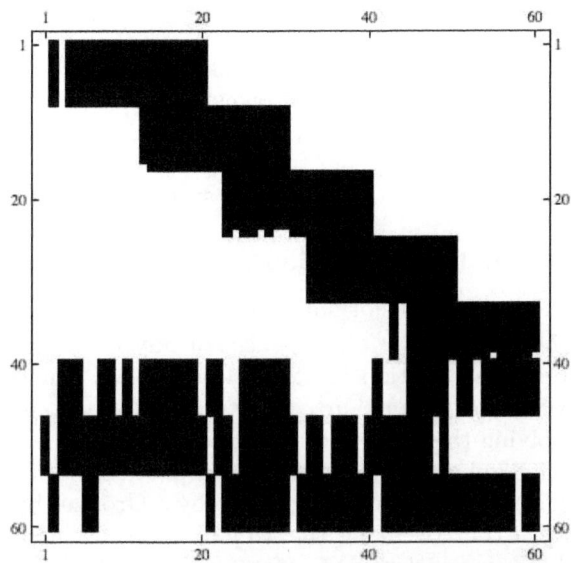

Fig. 1. Shape of 60 variable MQQ public key system without the use of S and T transformations. The black color means that the corresponding variables is used in the equation. The system was constructed using 4 MQQs of type $\mathrm{Quad}_8^s\mathrm{Lin}_2^s$, one MQQ of type $\mathrm{Quad}_7^s\mathrm{Lin}_3^s$, and 3 Dobbertin bijections defined over 3 different extension fields of dimension 7.

Now assume that the Gröbner basis algorithm somewhere during the calculation has found the solution for one of the quasigroup blocks $Y_j = X_j *_{vv}^{ij} X_{j+1}$. Due to the cascading structure of the MQQ system, the variables of X_j are used in the block $Y_{j-1} = X_{j-1} *_{vv}^{ij-1} X_j$ and the variables of X_{j+1} are used in the block $Y_{j+1} = X_{j+1} *_{vv}^{ij+1} X_{j+2}$. In Section 2.1 we showed that if we set the first or the last half of the variables of an MQQ to constant, all equations become linear. This means that if we have solved the block Y_j, the equations of the blocks Y_{j-1} and Y_{j+1} becomes linear. The blocks Y_{j-1} and Y_{j+1} can then be solved easily. This gives a solution for the variables X_{j-1} and X_{j+2}, which again makes the equations in the blocks Y_{j-2} and Y_{j+2} linear. Continuing like this we have rapidly solved the whole system.

Similarly, assume the Gröbner basis has solved the Dobbertin blocks at some step. This gives us the solution to all the variables in X_1 which makes the first quasigroup block $Y_1 = X_1 *_{vv}^{i1} X_2$ linear. Solving this gives us the first half of the equations of the block Y_2 and so on. As a conclusion, solving a MQQ system is reduced to either solving just one block of quasigroup equations, or solving the Dobbertin transformation. The security of solving an MQQ system is therefore the minimum complexity between solving the Dobbertin transformation or one MQQ block.

5   Weaknesses of MQQ

The goal of this part is to determine the weakest part of the system; the Dobbertin transformation or the quasigroup transformation. We first look closer at the Dobbertin block of equations. Since these equations constitutes a square system of equations, we expect them to be easier to solve then the quasigroup block of equations, which is an undetermined system of equations.

5.1   The Dobbertin Transformation

Recall that the Dobbertin transformation is a bijection over $GF(2^{2r+1})$ defined by the function $D_r(x) = x^{2^{r+1}+1} + x^3 + x$. For any r, we can view this function as $2r + 1$ Boolean equations in $2r + 1$ variables. Using magma 2.16-1's implementation of the F_4 algorithm[1], we experimentally computed the degree of regularity for solving this system of equations for $r = 2, \ldots, 22$. We observed that the degree of regularity was 3 for all computed instances. Therefore the Dobbertin transformation can be easily solved by a Gröbner basis computation. In addition we learn that tweaking the MQQ system by increasing the size of the extension field, over which the transformation is defined, will have no effect on strengthening the system.

Proving mathematically (if true) that the degree of regularity of $D_r(x)$ is constant at 3 for all r is difficult. We can, however, explain why the degree of regularity is low for all practical r. Let $\mathbb{K} = \mathbb{F}_q$ be a field of q elements, and let $\mathbb{L}$ be an extension of degree n over $\mathbb{K}$. Recall that an HFE polynomial f is a low-degree polynomial over $\mathbb{L}$ with the following shape:

$$f(x) = \sum_{\substack{0 \leq i,j \leq n \\ q^i+q^j \leq d}} a_{i,j} x^{q^i+q^j} + \sum_{\substack{0 \leq k \leq n \\ q^k \leq d}} b_k x^{q^k} + c, \tag{10}$$

where $a_{i,j}, b_k$ and c all lie in $\mathbb{L}$. The maximum degree d of the polynomial has to be chosen such that factorization over $\mathbb{L}$ is efficient [7]. Setting $q = 2$ and $n = 2r + 1$ we notice that the Dobbertin transformation is actually an HFE polynomial, $D_r(x) = x^{2^{r+1}+2^0} + x^{2^1+2^0} + x^{2^0}$. This is very helpful since a lot of work has been done on the degree of regularity for Gröbner basis compuation of HFE polynomials [15,7]. Indeed, it has been proved that the degree of regularity for HFE polynomial of degree d is bounded from above by $\log_2(d)$ [15,16]. For Dobbertin's transformation this means the degree of regularity is bounded from above by $r + 1$ at least.

However, since the coefficients of the Dobbertin transformation all lie in $GF(2)$, we can give an even tighter bound on the degree of regularity. Similarly to the weak-key polynomials in [7], the Dobbertin transformation commutes with the Frobenius automorphism and its iterates $F_i(x) : x \mapsto x^{2^i}$ for $0 \leq i \leq n$, namely

$$D_r \circ F_i(x) = F_i \circ D_r(x). \tag{11}$$

---

[1] The computer used was 4 processor Intel Xeon 2.93GHz computer with 128GB of memory.

Table 2. Effects of quasigroup size and the Dobbertin transformation on the observed degree of regularity for different MQQ. D_{reg} is the observed degree of regularity of normal MQQ systems, while D^*_{reg} is the observed degree of regularity for the same system where the input to Dobbertin has been replaced with random linear equations.

Variables	Quasigroup size	Quasigroups type	Dobbertin	D_{reg}	D^*_{reg}
30	2^5	4 $\text{Quad}_3^s\text{Lin}_2^s$ and 1 $\text{Quad}_2^s\text{Lin}_3^s$	7,9	3	3
	2^{10}	2 $\text{Quad}_8^s\text{Lin}_2^s$	7,7	3	4
40	2^5	5 $\text{Quad}_3^s\text{Lin}_2^s$ and 2 $\text{Quad}_2^s\text{Lin}_3^s$	7,7,7	3	4
	2^{10}	3 $\text{Quad}_8^s\text{Lin}_2^s$	7,9	3	4
	2^{20}	1 $\text{Quad}_{17}^s\text{Lin}_3^s$	7,7,9	3	4
50	2^5	9 $\text{Quad}_3^s\text{Lin}_2^s$	7,7,9	3	3
	2^{10}	4 $\text{Quad}_8^s\text{Lin}_2^s$	9,9	3	4
60	2^5	11 $\text{Quad}_3^s\text{Lin}_2^s$	9,9,9	3	3
	2^{10}	4 $\text{Quad}_8^s\text{Lin}_2^s$ and 1 $\text{Quad}_7^s\text{Lin}_3^s$	7,7,7	3	5
	2^{20}	1 $\text{Quad}_{18}^s\text{Lin}_2^s$ and 1 $\text{Quad}_{17}^s\text{Lin}_3^s$	7,9,9	3	5

Thus $D_r(x) = 0$ implies that $F_i \circ D_r(x) = 0$. This means for each i we can add the $2r + 1$ equations over $GF(2)$ corresponding to the equation $D_r \circ F_i(x) = 0$ over $GF(2^{2r+1})$ to the ideal. However, many of these equations are similar. Actually, we have that F_i and F_j are similar if and only if $gcd(i, 2r + 1) = gcd(j, 2r + 1)$ [7]. Worst case scenario is when $2r + 1$ is prime. The Frobenius automorphism then gives us (only) $2(2r + 1)$ equations in $2r + 1$ variables. From [3] we have the following formula for the degree of regularity for a random system of multivariate equations over $GF(2)$ when the number of equations m is a multiple of the number of variables N. For $m = N(k + o(1))$ with $k > 1/4$ the degree of regularity is

$$\frac{D_{\text{reg}}}{N} = \frac{1}{2} - k + \frac{1}{2}\sqrt{2k^2 - 10k - 1 + 2(k + 2)\sqrt{k(k + 2)}} + o(1). \quad (12)$$

Setting $k = 2$ we get $D_{\text{reg}} = -\frac{3}{2} + \frac{1}{2}\sqrt{-13 + 16\sqrt{2}} \cdot (2r + 1) \approx 0.051404 \cdot (2r + 1) = 0.102808 \cdot r + 0.051404$. Note that the degree of regularity can not be smaller then 3. This means we have $max(3, 0.102808 \cdot r + 0.051404)$ as an upper bound for a *random* multivariate system with the same number of equations and variables as the Dobbertin transformation. This provides a good indication that the degree of regularity for Dobbertin (which is not random at all) should be small, as observed in the experiments, and even smaller than a regular HFE polynomial.

5.2 The Quasigroup Transformation

To get an idea how strong the quasigroup transformation is, we performed some experiments where we replaced the input of the Dobbertin transformation by random linear equations. This means that solving a Dobbertin transformation block will no longer make all the equations in the first quasigroup transformation linear. The result of our experiment on this special MQQ system where the linear equations are perfectly masked is listed in Table 2. Note that the degree of regularity of 5 is still too small to prevent Gröbner bases attacks. What is important is how the degree of regularity increases when we increase different parameters. From the table it appears that both the quasigroup size and the number of variables have an effect on the degree of regularity. This tells us that if we replace the Dobbertin transformation with a stronger function, the MQQ system can possibly be made strong enough to resist pure Gröbner attacsk for adequate choices of quasigroup size and number of variables.

6   Conclusion

We further explained the results of [26] by showing that the degree of regularity for MQQ systems are bounded from above by a small constant. Therefore even MQQ systems with large number of variables can easily be broken with Gröbner bases cryptanalysis. The main result of this paper is an explanation of the underlying reason for this abnormal degree of regularity. We demonstrated how the complexity of solving MQQ systems with Gröbner bases is equal to the minimum of the complexity of solving the Dobbertin transformation and the complexity of solving one MQQ block. Furthermore, our experimental data showed that the degree of regularity for solving the Dobbertin transformation is bounded from above by 3, the same as the bound on the MQQ system. These experimental results were also explained mathematically. A natural interpretation of the results of our investigation is that the Dobbertin transformation employed is a serious weakness in the MQQ system.

From a design point of view, we also showed that if Dobbertin's transformation is replaced with an ideal function – which perfectly hides the linear parts of the system – the degree of regularity varies with the size of the quasigroups and the number of variables. We conclude that if a suitable replacement for Dobbertin's transformation is found, MQQ can possibly be made strong enough to resist pure Gröbner attacsk for adequate choices of quasigroup size and number of variables. This remains an interesting open problem.

References

1. Bardet, M.: Étude des systèmes algébriques surdéterminés. Applications aux codes correcteurs et à la cryptographie. PhD thesis, Université de Paris VI (2004)
2. Bardet, M., Faugère, J.-C., Salvy, B.: Complexity study of Gröbner basis computation. Technical report, INRIA (2002), http://www.inria.fr/rrrt/rr-5049.html
3. Bardet, M., Faugère, J.-C., Salvy, B.: Complexity of Gröbner basis computation for semi-regular overdetermined sequences over F2 with solutions in F2. Technical report, Institut national de recherche en informatique et en automatique (2003)
4. Bardet, M., Faugère, J.-C., Salvy, B.: On the complexity of Gröbner basis computation of semi-regular overdetermined algebraic equations. In: Proc. International Conference on Polynomial System Solving (ICPSS), pp. 71–75 (2004)
5. Bardet, M., Faugère, J.-C., Salvy, B., Yang, B.-Y.: Asymptotic behaviour of the degree of regularity of semi-regular polynomial systems. In: Proc. of MEGA 2005, Eighth International Symposium on Effective Methods in Algebraic Geometry (2005)

6. Billet, O., Ding, J.: Overview of cryptanalysis techniques in multivariate public key cryptography. In: Sala, M., Mora, T., Perret, L., Sakata, S., Traverso, C. (eds.) Gröbner Bases, Coding and Cryptography, pp. 263–283. Springer, Heidelberg (2009)
7. Bouillaguet, C., Fouque, P.-A., Joux, A., Treger, J.: A family of weak keys in hfe (and the corresponding practical key-recovery). Cryptology ePrint Archive, Report 2009/619 (2009)
8. Buchberger, B.: Ein Algorithmus zum Auffinden der Basiselemente des Restklassenringes nach einem nulldimensionalen Polynomideal. PhD thesis, Leopold-Franzens University (1965)
9. Chen, Y., Knapskog, S.J., Gligoroski, D.: Multivariate Quadratic Quasigroups (MQQ): Construction, Bounds and Complexity. Submitted to ISIT 2010 (2010)
10. Cox, D., Little, J., O'Shea, D.: Using Algebraix Geometry. Springer, Heidelberg (2005)
11. Levy dit Vehel, F., Marinari, M.G., Perret, L., Traverso, C.: A survey on polly cracker system. In: Sala, M., Mora, T., Perret, L., Sakata, S., Traverso, C. (eds.) Gröbner Bases, Coding and Cryptography, pp. 263–283. Springer, Heidelberg (2009)
12. Dobbertin, H.: One-to-one highly nonlinear power functions on $GF(2^n)$. Appl. Algebra Eng. Commun. Comput. 9(2), 139–152 (1998)
13. Faugère, J.-C.: A new efficient algorithm for computing Gröbner bases (F_4). Journal of Pure and Applied Algebra 139(1-3), 61–88 (1999)
14. Faugère, J.-C.: A new efficient algorithm for computing Gröbner bases without reduction to zero (F_5). In: Proceedings of the 2002 International Symposium on Symbolic and Algebraic Computation. ACM, New York (2002)
15. Faugère, J.-C., Joux, A.: Algebraic cryptanalysis of Hidden Field Equation (HFE) cryptosystems using Gröbner bases. In: Boneh, D. (ed.) CRYPTO 2003. LNCS, vol. 2729, pp. 44–60. Springer, Heidelberg (2003)
16. Fouque, P.-A., Macario-Rat, G., Stern, J.: Key recovery on hidden monomial multivariate schemes. In: Smart, N.P. (ed.) EUROCRYPT 2008. LNCS, vol. 4965, pp. 19–30. Springer, Heidelberg (2008)
17. Gligoroski, D., Markovski, S., Knapskog, S.J.: Multivariate quadratic trapdoor functions based on multivariate quadratic quasigroups. In: Proceedings of the American Conference on Applied Mathematics, MATH 2008, Stevens Point, Wisconsin, USA, pp. 44–49. World Scientific and Engineering Academy and Society (WSEAS), Singapore (2008)
18. Goubin, L., Courtois, N.T., Cp, S.: Cryptanalysis of the TTM cryptosystem. In: Okamoto, T. (ed.) ASIACRYPT 2000. LNCS, vol. 1976, pp. 44–57. Springer, Heidelberg (2000)
19. Granboulan, L., Joux, A., Stern, J.: Inverting HFE is quasipolynomial. In: Dwork, C. (ed.) CRYPTO 2006. LNCS, vol. 4117, pp. 345–356. Springer, Heidelberg (2006)
20. Kipnis, A., Hotzvim, H.S.H., Patarin, J., Goubin, L.: Unbalanced oil and vinegar signature schemes. In: Stern, J. (ed.) EUROCRYPT 1999. LNCS, vol. 1592, pp. 206–222. Springer, Heidelberg (1999)
21. Kipnis, A., Shamir, A.: Cryptanalysis of the oil & vinegar signature scheme. In: Krawczyk, H. (ed.) CRYPTO 1998. LNCS, vol. 1462, pp. 257–266. Springer, Heidelberg (1998)
22. MAGMA. High performance software for algebra, number theory, and geometry — a large commercial software package, http://magma.maths.usyd.edu.au

23. Markovski, S.: Quasigroup string processing and applications in cryptography. In: Proc. 1st Inter. Conf. Mathematics and Informatics for Industry MII 2003, Thessaloniki, April 14-16, pp. 278–290 (2003)

24. Markovski, S., Gligoroski, D., Bakeva, V.: Quasigroup string processing. In: Part 1, Contributions, Sec. Math. Tech. Sci., MANU, XX, pp. 1–2 (1999)

25. Matsumoto, T., Imai, H.: Public quadratic polynomial-tuples for efficient signature-verification and message-encryption. In: Günther, C.G. (ed.) EUROCRYPT 1988. LNCS, vol. 330, pp. 419–453. Springer, Heidelberg (1988)

26. Mohamed, M.S., Ding, J., Buchmann, J., Werner, F.: Algebraic attack on the MQQ public key cryptosystem. In: Garay, J.A., Miyaji, A., Otsuka, A. (eds.) CANS 2009. LNCS, vol. 5888, pp. 392–401. Springer, Heidelberg (2009)

27. Patarin, J.: Cryptanalysis of the Matsumoto and Imai public key scheme of Eurocrypt 1988. In: Coppersmith, D. (ed.) CRYPTO 1995. LNCS, vol. 963, pp. 248–261. Springer, Heidelberg (1995)

28. Patarin, J.: Hidden field equations (hfe) and isomorphisms of polynomials (ip): two new families of asymmetric algorithms. In: Maurer, U.M. (ed.) EUROCRYPT 1996. LNCS, vol. 1070, pp. 33–48. Springer, Heidelberg (1996)

29. Patarin, J.: The oil & vinegar signature scheme (1997)

30. Patarin, J., Goubin, L., Courtois, N.T.: $C * -+$ and HM: Variations around two schemes of T. Matsumoto and H. Imai. In: Ohta, K., Pei, D. (eds.) ASIACRYPT 1998. LNCS, vol. 1514, pp. 35–49. Springer, Heidelberg (1998)

31. Shamir, A.: Efficient signature schemes based on birational permutations. In: Stinson, D.R. (ed.) CRYPTO 1993. LNCS, vol. 773, pp. 1–12. Springer, Heidelberg (1994)

32. Smith, J.D.H.: An introduction to quasigroups and their representations. Chapman & Hall/CRC, Boca Raton (2007)

33. Wolf, C., Preneel, B.: Taxonomy of public key schemes based on the problem of multivariate quadratic equations. Cryptology ePrint Archive, Report 2005/077 (2005)

A   Algorithm for Generating Random MQQ

In this section we present the pseudo-code for how the MQQs used in this paper have been generated. The code was implemented in magma.

Algorithm. *MQQ algorithm*

```
1.   n ←{size of quasigroup}
2.   L ←{number of linear terms}
3.   if L ≤ 2
4.      then Q = n
5.      else  Q = n − L
6.   CorrectDeg ←True
7.   while CorrectDeg
8.       do A1 ←IdentityMatrix(n) (∗ The identity matrix of size n ∗)
9.          X1 ←[x₁,...,xₙ]ᵀ
10.         X2 ←[xₙ₊₁,...,x₂ₙ]ᵀ
11.         for i ←1 to Q
12.            do for j ←i + 1 to n
13.               do for k ←i + 1 to (n)
14.                  r ∈_R {0,1} (∗ random element from the set
                     {0,1} ∗)
15.                  A1_(i,j) = A1_(i,j) + r ∗ X1_k
16.         B ←RandomNonSingularBooleanMatrix(n) (∗ Random non singular
            Boolean matrix of size n ∗)
17.         C ←RandomBooleanVector(n) (∗ Random Boolean vector of size
            n ∗)
18.         A1 ←B ∗ A1
19.         X1 ←B ∗ X1 + C
20.         L1 ←RandomNonSingularBooleanMatrix(n) (∗ Random non singu-
            lar Boolean matrix of size n ∗)
21.         L2 ←RandomNonSingularBooleanMatrix(n) (∗ Random non singu-
            lar Boolean matrix of size n ∗)
22.         A1 ←LinTrans(A1, L1) (∗ Lineary transform the indeterminates of
            A1 according to L1 ∗)
23.         X1 ←LinTrans(X1, L1) (∗ Lineary transform the indeterminates of
            X1 according to L1 ∗)
24.         X2 ←LinTrans(X2, L2) (∗ Lineary transform the indeterminates of
            X2 according to L2 ∗)
25.         MQQ ←A1 ∗ X2 + X1
26.         GBMQQ ←Gröbner(MQQ,2) (∗ The truncated Gröbnerbasis of de-
            gree 2 under graded reverse lexicographical ordering. ∗)
27.         Deg ←{number of linear terms in GBMQQ}
28.         if Deg= L
29.            then CorrectDeg ←False
30.  return GBMQQ
```

Efficient Scalar Multiplications for Elliptic Curve Cryptosystems Using Mixed Coordinates Strategy and Direct Computations

Roghaie Mahdavi and Abolghasem Saiadian

Department of Electronic Engineering, Amirkabir University of Technology, Tehran, Iran
{Samaneh5628,eeas335}@aut.ac.ir

Abstract. Scalar multiplication is the heart of elliptic curve cryptosystems. Several techniques have been proposed for efficient scalar multiplication. Mixed coordinate strategy is a useful technique for implementing efficient scalar multiplication. It splits a scalar multiplication into a few parts, and performs each part in the best coordinate. Also, the running time of scalar multiplication can be reduced by applying direct computations in the evaluation stage. This technique directly computes points of the form $2P + Q$ from points P and Q on the elliptic curve. In this paper, we apply mixed coordinate strategy and direct computations to various scalar multiplication algorithms such as binary method, NAF and window NAF methods, MOF and window MOF methods to find the best combinations of mixed coordinates strategy and direct computations for scalar multiplication with respect to the computational costs and memory consumption.

Keywords: Elliptic curve cryptosystem, Scalar multiplication, Coordinate system, Mixed coordinates strategy, Direct computations.

1 Introduction

Elliptic Curve Cryptosystem (ECC) was first proposed by Koblitz [1] and Miler [2] independently in 1985. Security of ECC is derived from hardness of the discrete logarithm problem on the additive group of points on an elliptic curve over a finite field [3].

The benefits of ECC, when compared with other traditional public key cryptosystems such as RSA in the same security level, include: shorter key length, higher speed, lower power consumption. These are especially useful for mobile and wireless devices which typically have limited computational resources and bandwidth.

The main operation in ECC is scalar multiplication that is $K.P$ where K is an integer, and P is an elliptic curve point. The common method for scalar multiplication is performed by iterative additions (ECADD) and doubling (ECDBL) on an elliptic curve according to bits k_i of binary representation K. Various methods have been proposed for the efficient computation of $K.P$ by reducing EC-operations (ECADD, ECDBL). One method can be made by taking different binary representations of the multiplier K such as non-adjacent form (NAF) [3], window non-adjacent forms such as (wNAF) [3], NAF+SW [3] and Frac-wNAF [3]. The other signed binary

S.-H. Heng, R.N. Wright, and B.-M. Goi (Eds.): CANS 2010, LNCS 6467, pp. 184–198, 2010.

representations of K is Mutual Opposite Form (MOF) [4], wMOF [5] and Frac-wMOF [6]. The conversion of MOF representations of an integer is highly flexible, because it can be made from left-to-right and right-to-left and it is more efficient in memory-constraint devices such as smart cards.

The running time of scalar multiplication is computed based on two levels of complexity, elliptic curve operations and finite filed operations. Performing fast addition and doubling on an elliptic curve is crucial for efficient scalar multiplication. Computation time of EC-operation depends on the coordinate system adapted. EC-operation in affine coordinate involves inversion, which is particularly a very costly operation over prime fields. To avoid inversion for prime filed, various coordinate systems have been proposed such as Projective coordinate [7], Jacobian coordinate [7], modified Jacobian coordinate, and Chudnovsky Jacobian coordinate [7]. Using these coordinates, we can remove inversions from the EC-operations at the cost of increase in the other simpler filed operations.

Computation cost of an ECADD and ECDBL are different for various coordinates. Some coordinate systems such as modified Jacobian coordinate have faster doubling than the other coordinate systems and some coordinate systems such as Chudnovsky Jacobian coordinate have faster addition. One possible way for an efficient scalar multiplication is to switch the coordinate systems in the computation of ECADD and ECDUB which is called "mixed coordinate strategy" [8]. Another approach for efficient scalar multiplication is direct computation [9]. In recent years, several algorithms have been proposed for direct computation of points of the form $2P + Q$ where P and Q are on the elliptic curve.

In this paper, we use both mixed coordinate strategy and direct computations to get efficient scalar multiplication. We compute the running time of various scalar multiplication algorithms with mixed coordinate strategy and combination of mixed coordinate strategy and direct computations to discover the best strategy to compute scalar multiplication from running time and memory consumption point of view.

The rest of the paper organized as follow, a brief overview of scalar multiplication algorithms is presented in Section 2. In Section 3, we discuss about various coordinate systems and mixed coordinate strategy, and we calculate the running time of scalar multiplication algorithms with mixed coordinate strategy according to computational cost. Direct computations are discussed in Section 4. In this section, we also calculate the running time of scalar multiplication algorithms with combination of mixed coordinate strategy and direct computations with respect to computational cost. In Section 5, we compare both strategies to find the optimal strategy to obtain efficient scalar multiplication and finally, conclusions are drawn.

2   Scalar Multiplication

Scalar multiplication is the heart of elliptic curve cryptosystems. The speed of scalar multiplication plays an important role in the efficiency of these systems. Several algorithms have been proposed to compute scalar multiplication in an efficient way. They increase speed of scalar multiplication by reducing ECADD and ECDBL operations. Some of them are discussed in the following subsection. They are studied from the running time perspective of scalar multiplication based on EC-operations. For the number of elliptic curve additions and doublings, the time needed to perform an addition will be denoted A and time of doubling D.

2.1 Binary Method

The most common method for performing scalar multiplication is the binary method which computes $K.P$ according to bits k_i where $K = \sum_{i=0}^{d-1} k_i 2^i$, $k_i \in \{0,1\}$. Algorithm 1 computes scalar multiplication with the binary method.

Algorithm 1. Binary Scalar Multiplication

Input: Affine point P, positive integer K with binary representation $(k_{d-1}, \dots, k_0)_2$

Output: $Q = K.P$

$\quad Q \leftarrow P$

$\quad$ **For** $i = d - 2$ **to** 0

$\quad\quad$ **If** $k_i \neq 0$

$\quad\quad\quad Q \leftarrow 2Q + P$

$\quad\quad$ **Else**

$\quad\quad\quad Q \leftarrow 2Q$

$\quad$ **Return** Q

The cost of multiplication depends on the bit length of the binary representation of K and the number of non-zero digits that is called the Hamming weight of scalar multiplication. On average, Hamming weight of binary representation of k is $\frac{1}{2}$, so this method requires $(d-1)$ doubling and $(\frac{d-1}{2})$ adding. The efficiency of binary method may be enhanced if pre-computation is allowed. In this case each non-zero bit k_i is not restricted to be 1, but is an element of a suitable digit set $\mathcal{T}$ of integers. We call $K = \sum_{i=0}^{d-1} k_i 2^i$ a $\mathcal{T}$-representation if $k_i \in \mathcal{T} \cup \{0\}$ hold for each $0 \leq i \leq d$. In this technique, in the pre-computation stage the points $k_i P$ for $k_i \in \mathcal{T}$ are pre-computed and stored. In the recoding stage, the scalar is rewritten to a $\mathcal{T}$-representation and in the evaluation stage eventually the scalar multiplication is performed. The most established technique for generating $\mathcal{T}$-representation is window method [6]. In the window method with width w, successively, w consecutive bits of the binary scalar are scanned and replaced by a table-entry corresponding. The Hamming weight of window method is $\frac{1}{w+1}$, therefore the computational cost of this method is $(\frac{d-1}{w+1} - 1)$ elliptic curve addition and $(d-1)$ elliptic curve doubling operations. In the window method there is a trade-off between the running time of scalar multiplication and memory consumption. We have to store 2^{w-1} of $k_i P$s in the pre-computation stage but the Hamming weight is reduced to $\frac{1}{w+1}$ which causes faster scalar multiplication.

2.2 NAF Method

The most popular signed binary representation is NAF (non-adjacent form) that the integer K is represented as $K = \sum_{i=0}^{d-1} k_i 2^i$ where $k_i \in \{-1,0,1\}$. In NAF representation, among any two adjacent digits, at most one is non-zero. Fortunately, the NAF of K is at most one digit longer than the binary representation. The Hamming weight of NAF representation is 1/3 and it has a lower Hamming weight than the binary representation. Hence we can save 1/6 of additions required for evaluating $K.P$, but the number of doubling is not affected. To reduce the number of additions, wNAF representation have been proposed with a pre-computed table of size 2^{w-2} points. Algorithm 2 computes scalar multiplication from Left-to-Right window NAF method. In Table 2.1, we have summarized computational cost of NAF and window NAF methods according to EC-operations. The terms $\# \mathcal{T}$ and $\# EC$ are computational costs of pre-computation and evaluation stages according to EC-operations, respectively.

Algorithm 2. Left-to-Right window NAF method

Input: Affine point P, positive integer K with $NAF_w(K) = (k_{d-1}, \dots, k_0)_2$

Output: $Q = K.P$

Pre-computation:

    Set $P_1 \leftarrow P$, $P_2 \leftarrow 2P$

    For each $i \in \{3,5,\dots,2^{w^-}-1\}$

    $P_i \leftarrow P_{i-2} + P_2$

Evaluation:

    $Q \leftarrow P_{k_{d-1}}$

    For $i = d - 2$ **to** 0

      If $k_i > 0$

        $Q \leftarrow 2Q + P_{k_i}$

      Else if $k_i < 0$

        $Q \leftarrow 2Q - P_{k_i}$

      Else $Q \leftarrow 2Q$

  Return Q

Table 2.1. Comparison of Non-Zero Density and EC-operations Cost NAF Methods

Scheme	1/N.Z. Density	# τ	# EC
NAF[3]	3	—	$(\frac{d-3}{3})A + (d-1)D$
wNAF[3]	$w+1$	$D + (2^{w-2}-1)A$	$(\frac{d}{w+1}-1)A + (d-1)D$
NAF+SW[3]	$v(w) = w + \frac{4}{3} - \frac{(-1)^w}{3.2^{w-2}}$	$D + \left(\frac{2^w+(-1)^{w+1}}{3}-1\right)A$	$(\frac{d}{v(w)}-1)A + (d-1)D$
Frac- wNAF[3]	$v(w,m) = w + \frac{m+1}{2^w} + 2$	$D + \left(\frac{2^w+m+1}{2}-1\right)A$	$(\frac{d}{v(w,m)}-1)A + (d-1)D$

According to Table 2.1, the running time of window NAF algorithms is reduced by decreasing the number of additions with some pre-computation points. The disadvantage of the NAF methods is that they can be computed only from the least significant bit, that is, right-to-left; however, when used with memory constraint devices such as smart cards; left-to-right recoding schemes are by far more valuable. We have to compute the recoding and store them before starting left-to-right evaluation stage. Hence, it requires additional n-bit memory for the right-to-left exponent recoding.

2.3   MOF Method

The first left-to-right recoding algorithm was proposed by Joye and Yen[13] in 2000. In CRYPTO 2004, Okeya[5] proposed a new efficient left-to-right recoding scheme called Mutual Opposite Form(MOF) with these properties that:

- Signed adjacent non-zero bits (without considering 0 bits) are opposite.
- Most non-zero bit and the least non-zero bit are 1 and -1, respectively.
- All the positive integers with d bit binary string can be represented by unique (d+1) bit MOF.

The d bit binary string K can be converted to a signed binary string by computing $\mu = 2k - k$, where '(-)' stands for a bitwise subtraction. Algorithm 3 efficiently converts the binary string K to MOF from most significant bit.

Algorithm 3. Computing the Left-to-Right MOF from Binary Representation of K

Input: Binary string $K = (k_{d-1}, \dots, k_0)$

Output: MOF of $K = (mk_d, \dots, mk_0)$

   $mk_d \leftarrow k_{d-1}$

   For $i = d-1$ **to** 1 **do**

      $mk_i \leftarrow k_{i-1} - k_i$

   $mk_0 \leftarrow -k_0$

Return $(mk_d, \dots, mk_0)$

Hamming weight of MOF representation is $\frac{1}{2}$. To reduce the hamming weight of MOF representation, we can use the window methods. In Table 2.2; we summarized computational costs of MOF and window MOF scalar multiplication algorithms according to EC-operations.

Table 2.2. Comparison of Non-Zero Density and EC-operations Cost MOF Methods

Scheme	1/N.Z. Density	# τ	# EC
MOF[4]	2	—	$(\frac{d}{2}) A + (d - 1)D$
wMOF[5]	$w + 1$	$D + (2^{w-2} - 1)A$	$(\frac{d}{w+1} - 1) A + (d - 1)D$
Frac- wMOF[6]	$w + 1$	$D + (2^{w-2} - 1)A$	$(\frac{d}{w+1} - 1) A + (d - 1)D$

3   Elliptic Curve Coordinate Systems

An elliptic curve can be represented using several coordinate systems. For each system, the computation cost of elliptic curve addition and doubling operations are different. For efficient curve scalar multiplication, choosing a coordinate system with fast addition and doubling operations is an important factor.

Let $E: y^2 = x^3 + ax + b$, where $a, b \in F(p)$, be an elliptic curve over $F(p)$. Let $P = (x_1, y_1)$ and $Q = (x_2, y_2)$ be two points on E. The addition formula for affine coordinate is mentioned in Table 3.1. We denote the computational time of an elliptic curve addition operation with $t(P + Q)$ and an elliptic curve doubling operation with $t(2P)$. We also denote a field squaring, a filed multiplication and a field inversion by S, M, I; respectively. EC-operations in affine coordinate involve inversion, which is a very costly operation. To avoid inversion, Projective coordinate systems have been proposed. Using these coordinates, we can remove inversions from EC-operation with increasing the simpler field operations.

Table 3.1. Addition Formula in Affine Coordinate

$P \neq Q$	$P_1 = P_2$
$h = \dfrac{y_2 - y_1}{x_2 - x_1}$	$h = \dfrac{3x_1^2 + a}{2y_1}$
$x = h^2 - x_2 - x_1$	$x = h^2 - x_2 - x_1$
$y = h(x_1 - x) - y_1$	$y = h(x_1 - x) - y_1$
Cost: $t(P + Q) = I + 2M + S$	Cost: $t(2P) = I + 2M + 2S$

In Projective coordinate systems, the formula for point arithmetic can be obtained from the formula for affine coordinate by the substitutions $x = \frac{X}{Z^c}$ and $y = \frac{Y}{Z^e}$. The appropriateness of using coordinates is strongly determined by the ratio $\frac{I}{M}$.

Table 3.2. Computation Cost of Addition and Doubling Operations in Projective Coordinate Systems

Coordinate	Transform	Field-Cost		M-Cost	
		$t(P+Q)$	$t(2\,P)$	$t(P+Q)$	$t(2\,P)$
Affine	(X,Y)	I +2 M + S	I + 2 M + 2S	32.8 M	33.6 M
Projective	$(X/Z, Y/Z)$	12M + 2S	7M + 5S	13.6 M	11M
Jacobian	$(X/Z^2, Y/Z^3)$	11M + 5S	1M + 8S	15 M	7.4M
modified Jacobian	(X,Y,Z,aZ^4)	11M + 7S	3M + 5S	16.6M	7M
Chudnovsky Jacobian	(X,Y,Z,Z^2,Z^3)	11M + 3S	6M + 4S	13.4M	9.2M

Projective coordinate systems are proposed for ratio $\dfrac{I}{M} \geq 10$ satisfying. In Table 3.2, we summarized the computation cost of elliptic curve point additions in Projective coordinate systems according to field-cost and M-cost. We used the newest algorithms proposed for addition and doubling operations in Projective coordinate systems [16]. We assumed that S = 0.8 M and I = 30M.

The key observation is that point addition operations in projective coordinates can be done only using field multiplication, with no inversion required. The cost of eliminating inversion is an increased number of field multiplications.

3.1 Mixed Coordinate Strategy

Coordinate system has a vital role to speed up scalar multiplication in elliptic curve cryptosystems. In previous section, the computation cost of scalar multiplication algorithms were computed according to EC-operations. To speed up scalar multiplication, it is important to choose coordinate systems with fast addition operation. To compute elliptic curve addition operations, coordinate systems have different computational costs. Some coordinate systems have faster addition operation than the other and some have faster doubling operation. Therefore, a possible way for an efficient scalar multiplication is to switch coordinate system in each stage of the scalar multiplication algorithms. Cohn, Miyaji and Ono [8] proposed this strategy, which is called "mixed coordinate strategy". In recent years, for efficient scalar multiplication, several algorithms have been proposed that would correctly add two points in different coordinate systems or take a point in one coordinate system and return its double in another. In Table 3.3, we mentioned some of these algorithms and their computational costs. We denote by A, J, J^m and J^c the affine coordinates, the Jacobian coordinate, the modified Jacobian coordinates and the Chudnovsky Jacobian coordinates, respectively.

3.2 Computational Costs of Scalar Multiplications with Mixed Coordinate Strategy

We first describe some notation for representing mixed coordinate in this paper. The notation "$2C^1 = C^{2}$" or (C^1, C^2) represents a mixed doubling of $2P$ in coordinate system C^2 from P in coordinate system C^1. The notation "$C^1 + C^2 = C^3$ " or (C^1, C^2, C^3) represents a mixed addition $P + Q$ in coordinate system C^3 from P and Q

in coordinate systems C^1 and C^2, respectively. In this section, we compute the computational cost of various scalar multiplication algorithms such as binary and sign binary algorithms, window NAF and window MOF algorithms by mixed addition (C^1, C^2, C^3) and mixed doubling (C^1, C^2). For example, we will study the required steps to compute scalar multiplication in binary and window NAF methods in the details. Then, we will use of Algorithm1 and Algorithm2 for binary and window NAF methods, respectively. In both methods, we set base point P in KP in affine coordinate to obtain faster mixed additions.

Table 3.3. Mixed Addition and Doubling Algorithms

Computation	Computation Cost
$t(2A = J^c)$	$4M + 3S$
$t(2A = J)$	$1M + 5S$
$t(2J = J^c)$	$5M + 7S$
$t(2J^c = J)$	$4M + 5S$
$t(2A = J^m)$	$2M + 5S$
$t(J + A = J)$	$7M + 4S$
$t(J^c + A = J)$	$7M + 2S$
$t(J + J^c = J)$	$11M + 3S$
$t(J^c + J^c = J)$	$10M + 2S$
$t(A + A = J^c)$	$5M + 3S$
$t(J + A = J^c)$	$8M + 5S$
$t(J^c + A = J^c)$	$8M + 3S$
$t(J + J = J^c)$	$12M + 6S$
$t(J + J^c = J^c)$	$12M + 4S$
$t(J^m + A = J^m)$	$7M + 6S$

Table 3.4. Direct Computation for $2P + Q$[8]

Computation	Computation Cost
$t(2A + A = A)$	$I + 9M + 2S$
$t(2A + A = J)$	$11M + 3S$
$t(2A + J = J)$	$15M + 4S$
$t(2J + A = J)$	$13M + 5S$
$t(2J + J = J)$	$17M + 6S$
$t(2J^c + J^c = J)$	$15M + 4S$
$t(2J^c + A = J)$	$12M + 4S$
$t(2J^c + J = J)$	$16M + 5S$
$t(2J + J^c = J)$	$16M + 5S$
$t(2J + A = J^m)$	$14M + 7S$
$t(2J^c + A = J^m)$	$13M + 6S$
$t(2J^c + J^c = J^m)$	$16M + 6S$

To perform scalar multiplication in the binary method, we have proposed some combinations of mixed doubling (C^1, C^2) and mixed additions in the form of (C^1, A, C^3). In Table 3.5, we computed the computational cost of these combinations according to the field cost and M-cost. On average, the combination (J, A, J) for mixed addition operation and (J, J) for doubling is an optimal combination that speeds up the scalar multiplication.

Table 3.5. Combinations of Mixed Addition and Doubling to Compute Scalar Multiplication

	combinations	$t(C^1, C^2, C^3)$		$t(C^1, C^2)$	
		Field cost	M-cost	Field cost	M-cost
1	$(J, A, J), (J, J)$	$7M + 4S$	$10.2M$	$1M + 8S$	$7.4M$
2	$(J^m, A, J^m), (J^m, J^m)$	$7M + 6S$	$11.8M$	$3M + 5S$	$7M$
3	$(J^c, A, J^c), (J^c, J^c)$	$8M + 3S$	$10.4M$	$6M + 4S$	$9.2M$

The computational cost of the binary algorithm according to EC-operation is $\frac{(d-1)}{2}A + (d-1)D$ that d is the bit length of scalar multiplier K in KP. For simplicity in calculation of computational cost, we assumed that $k_{d-2} = 0$. Since the base point P is in affine coordinate, for doubling operation, we have (A,J) at first and (J,J) otherwise. In the following, we compute the time of computation of scalar multiplication for binary algorithm according to Field-cost and M-cost by substituting computational cost of algorithms according to Table 3.3. We denote TC as the time of computation for scalar multiplication. The final result in window NAF method with this strategy is in Jacobian form, so $(J \rightarrow A)$ is needed to return the result in affine form with computational cost of $I + 3M + S$.

$$TC = t(A,J) + (d-2)t(J,J) + \frac{(d-1)}{2}t(J,A,J) + t(J \rightarrow A) \tag{3.1}$$
$$= I + (4.5d - 11.5)M + (10d - 4)S$$
$$= (12.5d + 15.3)M$$

We also used this combination for signed binary methods, such as NAF and MOF. In table 3.6, we computed the computational cost of scalar multiplication for binary methods with this strategy.

To perform window NAF method, there are two stages to compute scalar multiplication, pre-computed and evaluation stages. At first, it is necessary to compute some pre-computation points in pre-computed stage. The computational cost of window NAF algorithm in pre-computed stage is $D + (2^{w-2} - 1)A$ .Since the computational cost in pre-computed stage depends on ECADD, it is efficient to choose a coordinate system with the lowest computation cost for addition operations. According to Table 2.2, Chudnovsky Jacobian coordinate has the lowest computational cost for elliptic curve addition operation. Hence, for efficient scalar multiplication, we choose this system and represent pre-computation points in Chudnovsky Jacoobian coordinate system. Since the base point is an affine coordinate, we have a mixed doubling (A,J^c) at first and after that for $w \geq 3$ we must compute (J^c,A,J^c) then (J^c,J^c,J^c) otherwise.The time of computation in pre-computed stage is computed by the following formula:

$$TC_1 = t(A,J^c) + t(J^c,A,J^c) + (2^{w-2} - 2)t(J^c,J^c,J^c) \tag{3.2}$$
$$= (12 + 11(2^{w-2} - 2))M + (11 + 3(2^{w-2} - 2))S$$
$$= (16.8 + 13.4(2^{w-2} - 2))M$$

In evaluation stage, since the pre-computed points are in Chudnovsky Jacobian coordinate, the optimal combination to compute scalar multiplication in a fast way is (J,J^c,J) for addition operations and (J,J) for doubling operations. But first, we must compute (A,J). When window representation of scalar multiplier K is none zero, $k_i = \mp1$ happened with probability around $\frac{1}{2^{w-2}}$, we will have (J,A,J) with probability $\frac{1}{2^{w-2}}$ .The time of computation in evaluation stage is computed by the following formula:

$$TC_2 = t(A,J) + (d-1)t(J,J) + \frac{1}{2^{w-2}}\left(\frac{d}{w+1}-1\right)t(J,A,J) + \frac{(2^{w-2}-1)}{2^{w-2}}\left(\frac{d}{w+1}-1\right)t(J,J^c,J) \quad (3.3)$$

$$= \left(d + \frac{(11(2^{w-2}-1)+7)}{2^{w-2}}\left(\frac{d}{w+1}-1\right)\right)M + \left(8d-3+\frac{(3(2^{w-2}-1))+4)}{2^{w-2}}\left(\frac{d}{w+1}-1\right)\right)S$$

$$= \left(7.4d - 2.4 + \frac{(13.4(2^{w-2}-1)+10.2)}{2^{w-2}}\left(\frac{d}{w+1}-1\right)\right)M$$

The total time of computation to compute scalar multiplication in window NAF according to M-cost is computed in Equation (3.4). At the end, we return the representation of the point in affine coordinate with the computational cost of $I + 3M + S$.

$$TC_3 = TC_1 + TC_2 + t(J \rightarrow A) \qquad (3.4)$$

$$= \Bigg(48.68 + 7.4d + 13.4(2^{w-2}-2)$$

$$+ \frac{(13.4(2^{w-2}-1)+10.2)}{2^{w-2}}\left(\frac{d}{w+1}-1\right)\Bigg)M$$

In table 3.7, the computational costs of window NAF methods are presented. Since the density of representation of window MOFs is identical to that of the window NAF, the computational costs of window MOFs are the same as the window NAF and we do not report them again. In Table 3.7, the term $\# Pre$ is the number of pre-computation points.

Table 3.6. Computation Cost of Binary Methods with Mixed Coordinates Strategy

Scheme	Field-Cost	M-Cost
Binary method	$I + (4.5d - 1.5)M + (10d - 6)S$	$(12.5d + 22.7)M$
NAF[3]	$I + (3.33d - 5)M + (9.33d - 14)S$	$(10.794d + 13.8)M$
MOF[4]	$I + (4.5d + 2)M + (10d - 8)S$	$(12.5d + 25.6)M$

Table 3.7. Computation Cost of Window NAF Methods with Mixed Coordinates Strategy

w	wNAF		NAF+SW		Frac- wNAF	
	M-Cost	# Pre	M-Cost	# Pre	M-Cost	# Pre
3	$(10.35 d + 29)M$	2	$(9.898d + 43.47)M$	3	$(9.788d + 81.27)M$	6
4	$(10.08d + 55)M$	4	$(9.742d + 68.88)M$	5	$(9.518d + 134.8)M$	10
5	$(9.566d + 108.2)M$	8	$(9.409d + 148.6)M$	11	$(9.267d + 241.9)M$	18
6	$(9.282d + 215.2)M$	16	$(9.186d + 283.3)M$	21	$(9.055d + 456.3)M$	34
7	$(9.064d + 429.5)M$	32	$(8.988d + 577)M$	43	$(8.88d + 885.1)M$	66

4 Mixed Coordinate Strategy and Direct Computation Algorithms

Direct computation is another approach to get efficient scalar multiplications. In recent years, several algorithms have been proposed for direct computation of points of the form $2P + Q$ directly. Since such points are repeatedly computed in the evaluation stage, direct computation algorithms can be effective for efficient scalar multiplication. We can use mixed coordinate strategy in direct computations and improve the computational cost of these algorithms. In table 3.4, some of these algorithms and their computational costs are reported.

4.1 Computational Costs of Scalar Multiplications with Mixed Coordinate Strategy and Direct Computation

In Section 3.2, the computational cost of scalar multiplication is computed by mixed coordinate strategy. In this section, we combine mixed coordinates with direct computation to get more efficient scalar multiplications. For direct computation, $2P + Q$, we use the notation "$2C^1 + C^2 = C^3$" or $(2C^1, C^2, C^3)$ representing a direct computation $2P + Q$ in coordinate system C^3 from P and Q in coordinate systems C^1 and C^2, respectively.

At first, to perform scalar multiplication in binary method we choose some combinations of mixed doubling (C^1, C^2) and direct computations in the form of $(2C^1, A, C^3)$. In Table 4.1, we computed the computational cost of these combinations according to field cost and M-cost. On average, we can see that the combination $(2J, A, J)$ for direct computation and (J, J) for doubling operations are optimal combinations to speed up the scalar multiplication process.

Table 4.1. Computation Cost of Some Combinations of Direct Computations and Doubling Operation

	combinations	$t(C^1, C^2, C^3)$		$t(C^1, C^2)$	
		Field cost	M-cost	Field cost	M-cost
1	$(2J, A, J)$, (J, J)	$13M + 5S$	$17M$	$1M + 8S$	$7.4M$
2	$(2J, A, J^m)$, (J^m, J)	$14M + 7S$	$19.6M$	$3M + 4S$	$6.2M$
3	(J^c, A, J^m) , (J^m, J^c)	$13M + 6S$	$17.8M$	$4M + 5S$	$8M$
4	(J^c, A, J) , (J, J^c)	$12M + 4S$	$15.2M$	$5M + 7S$	$10.6M$

For binary method, Equation (4.1) computes the time required for computation algorithm. For simplicity, we assumed that $k_{d-2} = 0$.

$$TC = t(A, J) + \frac{(d-3)}{2} t(J, J) + \frac{(d-1)}{2} t(2J, A, J) + t(J \rightarrow A) \qquad (4.1)$$
$$= I + (7d - 7)M + (6.5d - 9.5)S$$
$$= (12.2d + 15.4)M$$

For signed binary methods such as NAF and MOF, we use this combination too. In table 4.2, we computed the computational cost of scalar multiplication for binary methods by the combination of mixed coordinate strategy and direct computation.

In window methods, the operations in pre-computed stage are the same. For the evaluation stage, the optimal combinations to compute scalar multiplication are $(2J, A, J)$

for direct computation and (J,J) for doubling operation. In bellow, we computed the computational cost of window NAF in evaluation stage according to Field-cost and M-cost.

$$TC_2 = t(A,J) + \left(d - \frac{d}{w+1} - 1\right)t(J,J) + \frac{1}{2^{w-2}}\left(\frac{d}{w+1} - 1\right)t(2J,A,J) + \frac{(2^{w-2}-1)}{2^{w-2}}\left(\frac{d}{w+1} - 1\right)t(2J,J^c,J) \quad (4.2)$$

$$= \left(1 + \left(d - \frac{d}{w+1} - 1\right) + \frac{(16(2^{w-2}-1)+13)}{2^{w-2}}\left(\frac{d}{w+1} - 1\right)\right)M + \left(5d + 8\left(d - \frac{d}{w+1} - 1\right) + \frac{(5(2^{w-2}-1)+5)}{2^{w-2}}\left(\frac{d}{w+1} - 1\right)\right)S$$

$$= \left(5 + 7.4\left(d - \frac{d}{w+1} - 1\right) + \frac{(20(2^{w-2}-1)+17)}{2^{w-2}}\left(\frac{d}{w+1} - 1\right)\right)M$$

The total time to compute window NAF according to M-cost is shown in the following equation:

$$TC_3 = TC_1 + TC_2 + t(J \to A) \quad (4.3)$$

$$= \left(55.6 + 7.4\left(d - \frac{d}{w+1} - 1\right) + 13.4(2^{w-2} - 2)\right.$$

$$\left. + \frac{(20(2^{w-2} - 1) + 17)}{2^{w-2}}\left(\frac{d}{w+1} - 1\right)\right)M$$

The output of window NAF and window MOFs are the same and so we do not mention them. In Table 4.3, we computed computational cost of window NAF methods with this strategy for several window lengths.

Table 4.2. Computational Cost of Binary Methods with Direct Computation Strategy

Scheme	Field-Cost	M-Cost
Binary method	$I + (7d - 7)M + (6.5d - 7)S$	$(12.2d + 17.4)M$
NAF[3]	$I + (5d - 13)M + (7d - 7)S$	$(10.6d + 14.4)M$
MOF[4]	$I + (7d - 1)M + (6.5d - 8.5)S$	$(12.2d + 22.2)M$

Table 4.3. Computational Cost of Window NAF Methods with Direct Computation Strategy

w	wNAF		NAF+SW		Frac- wNAF	
	M-Cost	# Pre	M-Cost	# Pre	M-Cost	# Pre
3	$(10.175d + 29.7)M$	2	$(9.754d + 43.6)M$	3	$(9.643d + 82.05)M$	6
4	$(9.77d + 55.75)M$	4	$(9.57d + 69.6)M$	5	$(9.392d + 135.55)M$	10
5	$(9.43d + 109)M$	8	$(9.29d + 149.35)M$	11	$(9.154d + 242.68)M$	18
6	$(9.11d + 216)M$	16	$(9.081d + 283.1)M$	21	$(8.956d + 457.04)M$	34
7	$(8.863d + 430.3)M$	32	$(8.842d + 577.8)M$	43	$(8.448d + 885.83)M$	66

5 Comparison

In Sections 3 and 4, we calculated the running time of scalar multiplication based on computational time of field multiplication and bits number of multiplier K. In Section 3, we applied mixed coordinate strategy to compute the running time of scalar

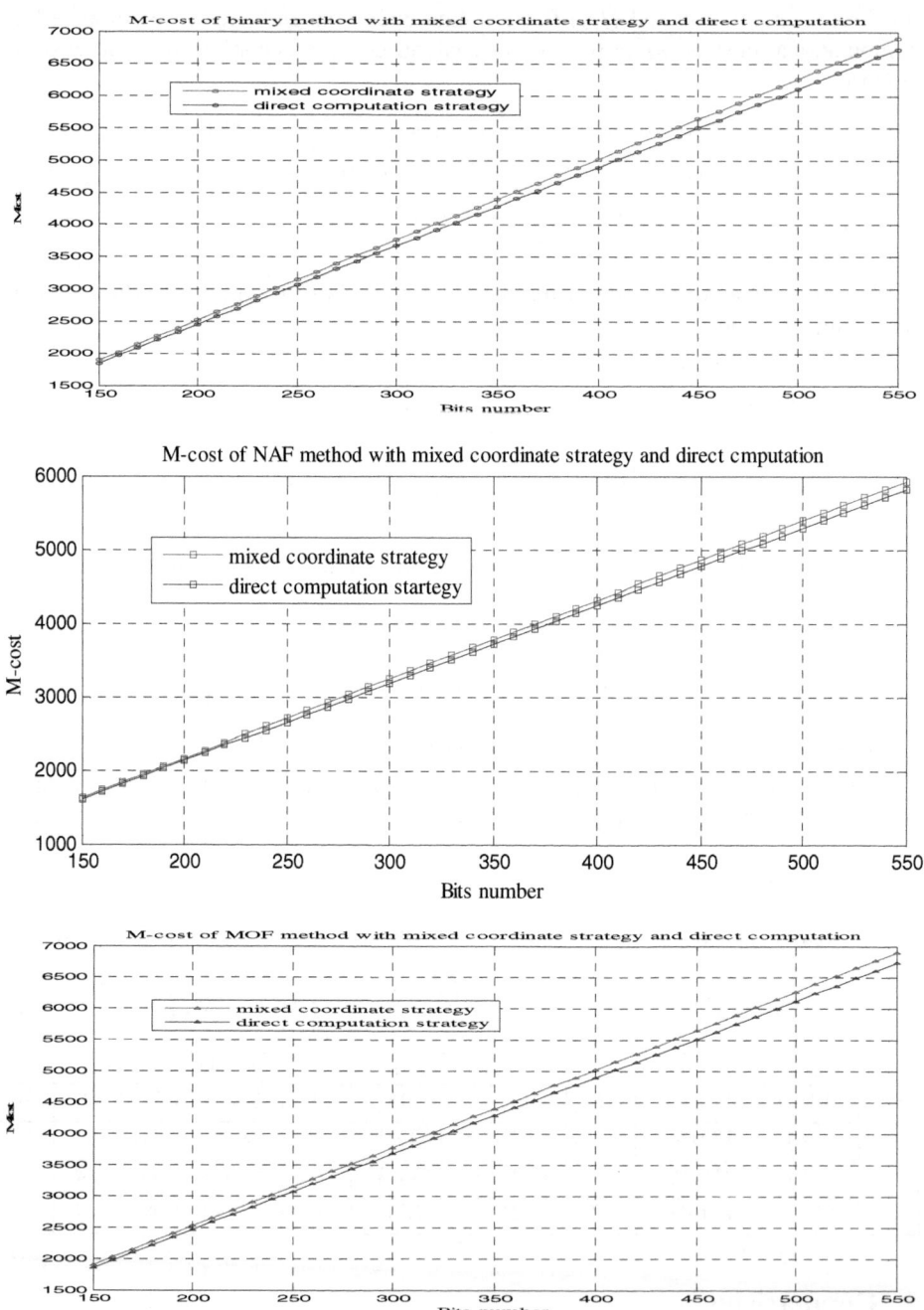

Fig. 5.1. Comparison between computation cost of mixed coordinate strategy and combination of mixed coordinate strategy and direct computation in binary and signed binary methods

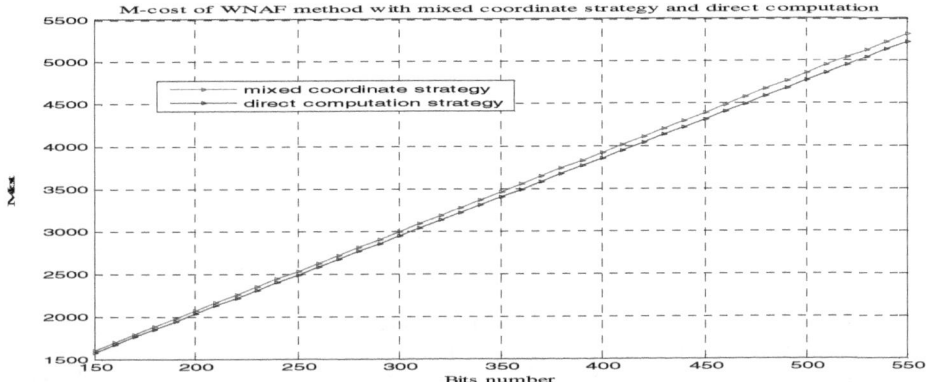

Fig. 5.2. comparison between computation cost of mixed coordinate strategy and combination of mixed coordinate strategy and direct computation in wNAF method

multiplication algorithms. In Tables 3.6 and 3.7, we calculated running time of binary methods and window NAFs, respectively. In window NAFs methods, there is a trade-off between the running time of scalar multiplication and memory consumption. The running time of scalar multiplication can be reduced by doing some pre-computation points in pre-computed stage, but it requires an additional memory space. Window NAF and window MOF methods have the same output but the advantage of window MOFs to widow NAF is in flexibility of conversion an integer to window representation. It can be done from left-to-right and right-to-left and it is more efficient in memory-constraint devices such as smart cards. In Section 4, we used of combinations of mixed coordinate strategy and direct computations to compute running time of scalar multiplications. Figure 5.1 and Figure 5.2 show the running time of binary and signed binary methods and wNAF method for various bits number for mixed coordinate strategy and combination of mixed coordinate strategy and direct computation. In Figure 5.2, we assume that window length is 6.

In Figure 5.1 and Figure 5.2, we can see that the running time of scalar multiplication using direct computations is lower than the strategy of mixed coordinate alone. To speed up scalar multiplication, the use of direct computation algorithms $(2C^1, C^2, C^3)$ is more efficient instead of computing (C^1, C^2, C^3) and (C^1, C^2) separately and finally adding them.

6 Conclusion

In this paper, we compared computational costs of two strategy mixed coordinate strategy and direct computations strategy. We applied both strategies in binary and signed binary methods and window NAFs algorithms. We observed that direct computation with mixed coordinate strategy was more efficient and had lower computation cost to compute scalar multiplication. Combination of mixed coordinates strategy and direct computations is a useful technique to speed up scalar multiplication in all methods.

References

[1] Koblitz, N.: Elliptic curve cryptosystem. Mathematics of Computation 48, 203–209 (1987)

[2] Miler, V.S.: Use of elliptic curves in cryptography. In: Williams, H.C. (ed.) CRYPTO 1985. LNCS, vol. 218, pp. 417–426. Springer, Heidelberg (1986)

[3] Hankerson, D., Menezes, A., Vanstone, S.: Guide to Elliptic Curve Cryptography. Springer, New York (2004)

[4] Morain, F., Olivos, J.: Speed up the computations on an elliptic curve using addition-subtraction. chains. RAIRO Theoretical Informatics and Applications 24, 531–543 (1990)

[5] Okeya, K.: Signed binary representations revisited. In: Franklin, M. (ed.) CRYPTO 2004. LNCS, vol. 3152, pp. 123–139. Springer, Heidelberg (2004)

[6] Schmidt-Samoa, K., Semay, O., Takagi, T.: Analysis of Fractional Window Recoding Methods and Their Application to Elliptic curve cryptosystems. IEEE Transactions on Computers 55, 48–57 (2006)

[7] IEEE Standard 1363-2000, IEEE Standard Specifications for Public Key Cryptography. IEEE Computer Society, Los Alamitos (August 29, 2000)

[8] Cohen, H., Miyaji, A., Ono, T.: Efficient elliptic curve exponentiation using mixed coordinates. In: Ohta, K., Pei, D. (eds.) ASIACRYPT 1998. LNCS, vol. 1514, pp. 51–65. Springer, Heidelberg (1998)

[9] Guajardo, J., Paar, C.: Efficient algorithms for elliptic curve cryptosystems. In: Kaliski Jr., B.S. (ed.) CRYPTO 1997. LNCS, vol. 1294, pp. 342–356. Springer, Heidelberg (1997)

[10] Balasubramaniam, P., Karthikeyan, E.: Elliptic curve scalar multiplication algorithm using complementary recoding. Applied Mathematics and Computation 190, 51–56 (2007)

[11] Cilardo, A., Coppolino, L., Mazzocca, N., Romano, L.: Elliptic Curve Cryptography Engineering. Proceeding of the IEEE 94, 395–406 (2006)

[12] Lauter, K.: The advantages of elliptic curve cryptography for wireless security. IEEE Wireless Commun. 11(1), 62–67 (2004)

[13] Joye, K., Yen, S.: Optimal left-to-right binary signed digit recoding. IEEE Transactions on Computers 49, 740–748 (2000)

[14] Li, Z., Higgins, J., Clement, M.: Performance of Finite Field Arithmetic in an Elliptic Curve Cryptosystem. IEEE, New York (2001); 0-7695-1315-8/01

[15] Adachi, D., Gamou, M., Hirata, T.: Efficient scalar multiplication on elliptic curve using direct computations. IEICE Trans. Fundamentals J88(1), 54–61 (2005) (in Japanese)

[16] http://www.hyperelliptic.org/EFD

Cryptography Meets Hardware: Selected Topics of Hardware-Based Cryptography (Invited Talk)

Ahmad-Reza Sadeghi

Technical University Darmstadt, System Security Lab, Germany
ahmad.sadeghi@cased.de

Modern cryptography provides a variety of methods and protocols that allow different entities to collaborate securely without mutual trust, and hence constitutes the basic technology for a wide range of security and privacy critical applications. However, even the most basic cryptographic functionalities such as commitments, oblivious transfer, or set intersection require computationally expensive public key cryptography when implemented in software only, and their secure universal composition cannot be achieved without additional setup assumptions.

A recent line of research aims to incorporate tamper-proof (tamper-resistant) hardware tokens (e.g., smartcards) into cryptographic schemes benefiting both, theory and practice: Theoreticians aim at founding cryptography on tamper-proof hardware to overcome known, or show new impossibility results, e.g., in established security frameworks such as Universal Composability (UC); practitioners aim to substantially improve the performance of cryptographic protocols (e.g., use hardware instead of computationally expensive public-key operations or fully homomorphic encryption), to reduce or prevent physical information leakage of cryptographic operations (through side-channel attacks), or to realize Trusted Computing functionality in hardware allowing to link software to the underlying hardware (e.g., as proposed by the Trusted Computing Group, TCG).

In this talk we consider selected topics and aspects of hardware-based cryptographic protocols: Motivated by applications we discuss different trust models and security goals. We then focus on secure two-party computation, often called Secure Function Evaluation (SFE), allowing two untrusting parties to jointly compute an arbitrary function on their respective private inputs while revealing no information beyond the outcome. Although in the past SFE was widely believed to be inefficient, the rapidly growing speed of computing devices and communication networks along with algorithmic improvements and the automatic generation and optimizations of SFE protocols has made them usable in practical application scenarios. We show how the complexity of SFE protocols can be reduced through using tamper-proof hardware tokens where the token is *not* fully trusted, i.e., only by one (the issuer) but not both parties. In this context we also evaluate the practical performance of one-time programs (OTPs), a functionality which can be evaluated exactly once in an insecure environment

S.-H. Heng, R.N. Wright, and B.-M. Goi (Eds.): CANS 2010, LNCS 6467, pp. 199–200, 2010.
© Springer-Verlag Berlin Heidelberg 2010

without leaking any information about the security critical data. In future work token-based SFE protocols can be combined with cryptographic compilers such as the most recent TASTY (Tool for Automating Secure Two-partY computation) compiler.

The tamper-proofness assumption is critical in real world implementations given the fact that physical side channel attacks on cryptographic implementations and devices have become crucial today. More concretely, in practice, computation and memory leak information about the secret values, and hence, the conventional algorithmic provable security (black box approach) is not sufficient. In this context a recent line of research considers physical primitives called Physically Unclonable Functions (PUFs). They represent a promising new technology that allows to store secrets in a *tamper-evident* and unclonable manner, and enjoy their security from the unique physical structures at deep submicron level. Here the assumption is that the same randomness/key cannot be extracted when the device is physically tampered with. We discuss some recent approaches towards combining and binding algorithmic properties of cryptographic schemes with physical structure of the underlying hardware by means of PUFs.

Towards a Cryptographic Treatment of Publish/Subscribe Systems

Tsz Hon Yuen, Willy Susilo, and Yi Mu

Center for Computer and Information Security Research
School of Computer Science and Software Engineering
University of Wollongong, Australia
{thy738,wsusilo,ymu}@uow.edu.au

Abstract. Publish/subscribe mechanism is a typical many-to-many messaging paradigm when multiple applications want to receive the same message or when a group of applications want to notify each other. Nonetheless, there exist only a few works that deal with this topic formally, in particular addressing their security issues. Although security issues and requirements for content-based publish/subscribe systems have been partially addressed by Wang *et al.*, there are *no* formal definition for all of these security requirements in the literature. As a result, most of the existing schemes do not have any security proof and there is no way to justify whether those schemes are really secure or not in practice. Furthermore, there is no comprehensive scheme that satisfies the most essential security requirements at the same time. In this paper, *for the first time in the literature*, we introduce the security model for all security requirements of content-based publish/subscribe systems. We then exhibit a new publish/subscriber system that fulfills most of the security requirements. Furthermore, we also provide a comprehensive proof for our concrete construction according to the new model.

1 Introduction

Publish/subscribe (pub/sub) is an efficient communication infrastructure that supports dynamic, many-to-many data dissemination in a distributed environment. It allows decoupled messaging between: (1) *subscribers*, having *subscriptions* to the interested information, and (2) *publishers*, providing *notifications* for the information they provide. This kind of many-to-many communication is being more and more popular in social networking websites.

All pub/sub technologies use subject or topic names as the loosely coupled link between publishers and subscriber systems. Publishers produce messages on a particular subject or topic name and subscribers receive those messages by registering interest in the subject name either explicitly or through some broader subscription scheme using wildcards. Subscribers and publishers are loosely coupled by a network of *brokers* that route the notifications to the interested subscribers. Pub/sub allows subscribing applications to select messages that these applications receive by topic (as specified by the publishing application) or by content (by specifying filters). The latter is usually referred to as

S.-H. Heng, R.N. Wright, and B.-M. Goi (Eds.): CANS 2010, LNCS 6467, pp. 201–220, 2010.

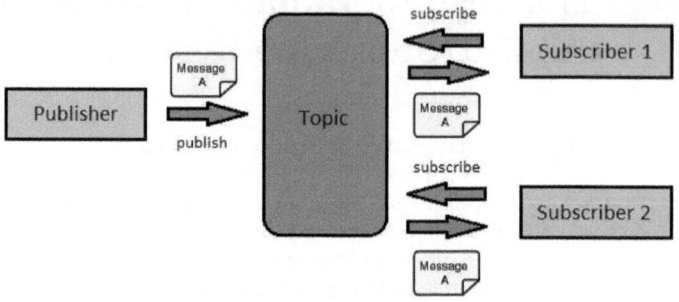

Fig. 1. Publish/subscribe system

the content-based pub/sub systems (CBPS). Any messages addressed to a topic are delivered to all the topic's subscribers. Every subscriber receives a copy of each message. Information is automatically *pushed* to subscribing applications without them having to *pull* or request it. In short, pub/sub topologies publish messages directly to the bus or network, and these topologies are known as *shared bus*-based solutions.

The current proposed or existing pub/sub systems tend to focus on the performance, scalability and expressiveness issues of the mechanism. Security issues and requirements are firstly addressed by Wang *et al.* [11]. The main issues include authentication, integrity and anonymity, which can usually be achieved by minor modification to the existing approaches. On the other hand, confidentiality is considered more difficult to achieve. Therefore, we refer Wang *et al.*'s work as addressing the security of pub/sub network *partially*.

Our Contributions. Publish/subscribe systems are important to the future social networking services. However, there is no formal security model for the pub/sub systems. Wang *et al.* [11] proposed some security issues and requirements, without defining a formal model. Nikander and Giannis [6] points out the difficulty of modeling pub/sub systems using traditional send/receive paradigm. They only give a general model to reflect the multicast nature of the pub/sub systems, without concerning the security requirements. As a result, most of the existing schemes do not have any security theorem or proof. To the best of the author's knowledge, only Raicius and Rosenblum [8] proposed the security model for confidentiality and they proved the confidentiality of their scheme. A complete security theorem and proof is essential to analysis the security level of a CBPS protocol. Moreover, a complete security model is needed to identify the security requirements and the attacker's capability. Therefore, in this paper, we propose a formal security model for all security requirements for the CBPS.

Secondly, Wang *et al.* [11] suggested some possible solutions for each security requirements that they proposed. However, it is not clear that if these methods can work together under the *same* threat model. Moreover, some methods are out-of-band solutions and are handled independently of the pub/sub infrastructure. Additionally, most of the existing CBPS schemes enabling confidentiality do not consider authenticity and integrity simultaneously. In this paper, we propose a comprehensive CBPS scheme which fulfills most security requirements concurrently. We prove the security of our scheme under the new security model.

In this paper, *for the first time in the literature*, we provide a formal cryptographic treatment of Content-based pub/sub systems (CBPS). Our model can therefore be used to analyze the security of any CBPS system. Furthermore, we also provide a concrete construction of CBPS that satisfies our model. We provide a security proof for our scheme that our scheme is secure under our proposed model.

2 Publish/Subscribe Systems and Their Security Models

In this section we first give the definition of publish/subscribe system. After that, we describe the security requirements and security models for CBPS. It is the *first* comprehensive security model of pub/sub system. Without a formal security model, we cannot analysis the concrete security of any pub/sub system.

2.1 Publish/Subscribe Systems

A publish/subscribe system is a system with interactions between four parties:

- Publishers *notify* the brokers for the information they provide in the pub/sub system. They do not know who will obtain the information.
- Subscribers *subscribe* to the interested information. They only receive the information which matches their subscription.
- Brokers *match* the subscription and the notification by the subscribers and the publishers. The broker network will route and forward the packets to the matching subscribers. Sometimes they are further categorized into:
  - Intermediate brokers. They only route packets within the broker network.
  - Border brokers. They act as a link between the broker network and the other parties in the pub/sub network.
  - Publisher hosting brokers. They are a kind of border brokers that connect between the broker network and the publishers.
  - Subscriber hosting brokers. They are a kind of border brokers that connect between the broker network and the subscribers.
- Managers maintains and coordinates the keys used within the pub/sub system. According to the basic concept of pub/sub system, the publisher does not know the public keys of subscribers. Therefore, the publisher cannot encrypt using subscribers' public keys. In order to provide confidentiality, managers are needed to act as the target of the encryption scheme. If confidentiality is not considered in the pub/sub system, then this party can be ignored. Some papers [8,4] assume that the publishers and the subscribers have a pre-shared session key. They do not concern about how the managers help to share the session key within the pub/sub system. The managers are called *key distribution center* in [9], *accounting server* in [3] and *secure administrator* in [14].

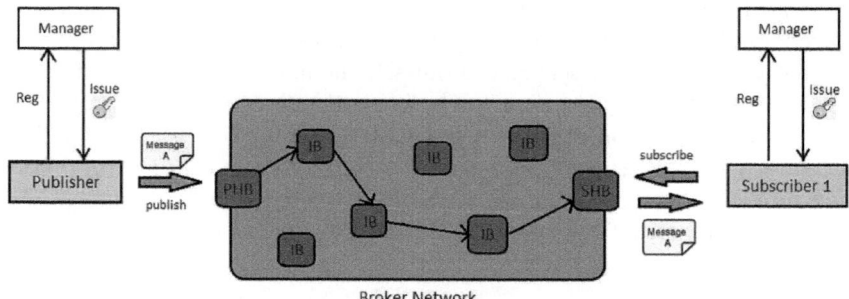

Fig. 2. Brokers in publish/subscribe system. PHB stands for Publisher Hosting Brokers, SHB stands for Subscriber Hosting Brokers and IB stands for Intermediate Brokers. PHB and SHB are both border brokers.

A content-based publish/subscribe system consists of ten algorithms defined as follows:

- Setup(1^λ): On input a security parameter 1^λ, it outputs the system parameter param and the manager's secret key msk.
- KeyGen(param): On input the system parameter param, it outputs a secret key and a public key. It can be further divided into publisher's KeyGen$_p$ to generate publisher secret key psk and public key ppk; subscriber's KeyGen$_s$ to generate subscriber secret key ssk and public key spk; and broker's KeyGen$_b$ to generate broker secret key bsk and public key bpk.
- RegP(param, $filter$, psk), IssueP(param, msk, ppk): The interactive algorithms RegP and IssueP are run by the publisher and the manager respectively, where param is the system parameter, $filter$ is the filter set by the publisher, (psk, ppk) is the publisher's secret key, public key pairs and msk is the manager's secret key. RegP first sends the notification to IssueP and IssueP returns a publisher key K_p to RegP.
- RegS(param, sub, ssk), IssueS(param, msk, spk): The interactive algorithms RegS and IssueS are run by the subscriber and the manager respectively, where param is the system parameter, sub is the subscription by the subscriber, (ssk, spk) is the subscriber's secret key, public key pairs and msk is the manager's secret key. RegS first sends the subscription to IssueS and IssueS returns a subscriber key K_s to RegS.
- Pub(param, m, psk, K_p): On input (param, m, psk, K_p) where param is the system parameter, m is the message, psk is the publisher's secret key and K_p is the publisher key for some $filter$, the publisher outputs a notification n^1 to the broker network.

---

[1] We use the term "notification" instead of "ciphertext" for a few reasons. Firstly, part of the information sent by the publisher may not be encrypted, such as the keyword of the message, to facilitate routing. Secondly, the subscription by the subscriber may also be encrypted as another ciphertext in the system.

- Sub(param, sub, ssk, bpk): On input (param, sub, ssk, bpk) where param is the system parameter, sub is the subscription, ssk is the subscriber's secret key and bpk is the (subscriber hosting) broker public key, the subscriber outputs a ciphertext for subscription C_{sub} to the broker network.
- Match(param, n, C_{sub}, bsk): On input (param, n, C_{sub}) where param is the system parameter, n is the notification, C_{sub} is the subscription ciphertext and bsk is the (subscriber hosting) broker secret key, the broker first outputs spk for matching the subscription by spk, 0 for not match, $\perp_p$ for invalid notification or $\perp_s$ for invalid subscription.
- Retrieve(param, n, K_s): On input (param, n, K_s) where param is the system parameter, n is the notification and K_s is the subscriber key, the subscriber outputs a pair (m, ppk) or $\perp$ for invalid, where m is the message and ppk is the publisher's public key.

We define that the subscription condition sub, the filter condition $filter$ and also the notification n (except the encrypted part) are in the format of $(name, op, value)$. For example, it can be $(price, =, 10)$ or $(age, \geq, 30)$. We define the symbol $n \subseteq_s F$ as the notification n satisfies the boolean relationship in F. In the CBPS system, we require that the publisher's notification n should satisfy the publisher's filter ($n \subseteq_s filter$). Otherwise, the publisher does not have a valid publisher key K_p for n. Similarly, we require that the notification n should satisfy the subscription sub ($n \subseteq_s sub$). Otherwise, the subscriber does not have a valid subscriber key K_s for decryption of n. For example if

$$sub = \langle((code, =, ABC) \text{ AND } (date, <, \text{May } 30 \ 2008)) \text{ OR } (code, =, DEF)\rangle,$$
$$filter = \langle(code, =, ABC)\rangle,$$
$$n = \langle(code, =, ABC) \text{ AND } (price, =, ??)$$
$$\text{AND } (time, =, 14:20) \text{ AND } (date, =, \text{May } 12 \ 2008)\rangle,$$

then $n \subseteq_s sub$ and $n \subseteq_s filter$.

2.2   Correctness

The content-based publish/subscribe system has two types of correctness: matching correctness and retrieval correctness. Matching correctness means that an honest broker can always correctly match a valid notification to a subscriber with a valid satisfying subscription. Retrieval correctness means that an honest subscriber can obtain the message if the notification which matches his subscription criteria.

Formally, they are defined as follows:

- Matching Correctness. We require that

$$\text{Match}(\text{param}, n, \text{Sub}(\text{param}, sub, ssk, bpk), bsk) = spk,$$

where $(ssk, spk) \leftarrow \text{KeyGen}(\text{param}), (bsk, bpk) \leftarrow \text{KeyGen}(\text{param}), n \leftarrow \text{Pub}(\text{param}, m, psk, \text{RegP}(\text{param}, filter))$, and $n \subseteq_s sub$.

- Retrieval Correctness. We require that

$$\text{Retrieve}(\text{param}, n, \text{RegS}(\text{param}, sub, ssk)) = (m, ppk),$$

where $(psk, ppk) \leftarrow \text{KeyGen}_p(\text{param}), (ssk, spk) \leftarrow \text{KeyGen}_s(\text{param}), n \leftarrow$ $\text{Pub}(\text{param}, m, psk, \text{RegP}(\text{param}, filter, psk)), n \subseteq_s sub$ and $n \subseteq_s filter$.

2.3 Trust Model

There are three types of trust regarding the underlying broker network:

1. *A complete trust to the broker network.* The adversary is not given any information within the broker network.
2. *A trust to the border brokers only.* The adversary can access any information in the intermediate brokers, but not the border brokers.
3. *Untrusted broker network.* The adversary can access any information in the the broker network.

Notice that the trust we discuss here is whether the brokers' keys (if any) and data accessed by the brokers are available to the adversary. We always assume that the brokers honestly route the packets. If not, the subscribers may never receive any packets.

According to different trust level of the broker network, the broker public keys and secret keys can be used as the input in the Pub, Sub or Match protocol.

2.4 Confidentiality

The publish/subscribe system has three types of confidentiality: information confidentiality, subscription confidentiality and publisher confidentiality as discussed by Wang *et al.* [11].

In some cases, part of the information can be known by the brokers to facilitate routing (e.g. stock code, update date in a pub-sub stock quote application) while part of the information must be kept secret from untrusted brokers (e.g. stock price, percentage change). Therefore we consider the confidentiality for the secret information instead of the whole document to be sent.

Information confidentiality means that the secret information in the notification should not be known by the untrusted brokers and all outsiders. Subscription confidentiality means that the secret information in the subscription should not be known by the untrusted brokers, publishers, other subscribers and all outsiders. Publisher confidentiality means that the secret information in the notification should not be known by the non-subscribers of that notification. It includes the untrusted brokers and all outsiders. Therefore publisher confidentiality implies information confidentiality.

We note that our model for confidentiality only involves one trusted manager only. In real system, there may be many managers. Our model can be modified for multiple managers. We give the current confidentiality model of one manager for simplicity.

Publisher Confidentiality. We describe the publisher confidentiality for the secret information in the pub/sub network. The indistinguishability game is formally defined as follows:

1. The challenger runs $(\text{param}, msk) \leftarrow \text{Setup}(1^\lambda)$ and $(bsk, bpk) \leftarrow \text{KeyGen}_b(\text{param})$. The challenger gives the public parameters param and the secret/public key pairs of the untrusted brokers to the adversary $\mathcal{A}$. The manager's secret key msk is unknown to $\mathcal{A}$.
2. $\mathcal{A}$ is allowed to query the following oracles:
   - IssueS Oracle: On input the subscription sub and the subscriber's public key spk, it runs the $\text{IssueS}(\text{param}, msk, spk)$ protocol and interacts with the $\text{RegS}(\text{param}, sub, \cdot)$ run by $\mathcal{A}$. The oracle outputs the subscriber key K_s from IssueS.
   - IssueP Oracle: On input the publisher's filter $filter$ and the publisher's public key ppk, it runs the $\text{IssueP}(\text{param}, msk, ppk)$ protocol and interacts with the $\text{RegP}(\text{param}, filter, \cdot)$ run by $\mathcal{A}$. The oracle outputs the publisher key K_p from IssueP.
   - Retrieval Oracle: On input (n, sub) where n is the notification and sub is the subscription, the oracle first runs $(ssk, spk) \leftarrow \text{KeyGen}_s(\text{param})$. Then the oracle runs both $\text{IssueS}(\text{param}, msk, spk)$ and $\text{RegS}(\text{param}, sub, ssk)$ by itself and obtains K_s. Finally, it outputs the secret information $(m, ppk)/\bot \leftarrow \text{Retrieve}(\text{param}, n, K_s)$.
3. $\mathcal{A}$ sends two messages m_0^* and m_1^* from the message space, a publisher secret key psk^{*2} and a filter $filter^*$ to the challenger. The messages m_0^* and m_1^* are only different in the part of the secret information. The challenger encrypts m_b^* as $n_b^* \leftarrow \text{Pub}(\text{param}, m_b^*, psk^*, \text{RegP}(\text{param}, filter^*, psk^*))$. There should be no subscription sub queried to the IssueS Oracle, such that $n_b^* \subseteq_s sub$. The challenger picks a bit $b \in \{0, 1\}$ and sends the notification n_b^* to $\mathcal{A}$.
4. $\mathcal{A}$ is allowed to query the oracles, with the exception that no subscription sub queried to the IssueS Oracle, such that $n_b^* \subseteq_s sub$; and n^* should not be queried to the Retrieval Oracle.
5. Finally $\mathcal{A}$ output his guess b'.

The advantage of $\mathcal{A}$ in the game is $|\Pr[b' = b] - \frac{1}{2}|$.

Definition 1. *A CBPS scheme is $(\epsilon, t, q_s, q_p, q_r)$-publisher confidential against chosen ciphertext attack if there is no t-time adversary with q_s queries to the IssueS oracle, q_p queries to the IssueP oracle and q_r queries to the retrieval oracle has an advantage over ϵ in the game.*

Information Confidentiality. Due to the similarity of the definition between information confidentiality and publisher confidentiality, we can define the indistinguishability game of publisher confidentiality same as the one of information confidentiality without query to the IssueS Oracle.

---

[2] Our publisher confidentiality is a strong model since the publisher secret key of the challenge notification is chosen by the adversary. It is possible to define a weaker model where the adversary is only given the publisher public key.

Definition 2. *A CBPS scheme is (ϵ, t, q_p, q_r)-information confidential against chosen ciphertext attack if there is no t-time adversary with q_p queries to the IssueP oracle and q_r queries to the retrieval oracle has an advantage over ϵ in the game.*

Notice that for both publisher and information confidentiality, we say that a system is *selectively* secure if we require the adversary commits to the challenge filter $filter^*$ at the beginning of the game.

Subscription Confidentiality. We describe the confidentiality for the subscription in the pub/sub system. The subscribers may want their subscriptions to be confidential against the broker network[3]. Then the brokers need to match the "encrypted subscriptions" with the notifications[4]. The indistinguishability game is defined as follows:

1. The challenger runs $(\mathsf{param}, msk) \leftarrow \mathsf{Setup}(1^\lambda)$ and $(bsk, bpk) \leftarrow \mathsf{KeyGen}_b(\mathsf{param})$. The challenger gives the public parameters param and the secret/public key pairs of the untrusted brokers to the adversary $\mathcal{A}$. The manager's secret key msk is unknown to $\mathcal{A}$.
2. $\mathcal{A}$ is allowed to query the IssueS Oracle, IssueP Oracle and Retrieval Oracle defined in the publisher confidentiality game.
3. $\mathcal{A}$ sends two subscription sub_0^* and sub_1^* and the subscriber secret key ssk^*, where sub_0^* and sub_1^* have never been queried to the IssueS Oracle. The challenger picks a bit $b \in \{0, 1\}$ and computes $C_{sub}^* \leftarrow \mathsf{Sub}(\mathsf{param}, sub_b^*, ssk^*, bpk)$. He sends the resulting ciphertext C_{sub}^* to $\mathcal{A}$.
4. $\mathcal{A}$ is allowed to query the oracles, with the exception that no subscription sub_0^* and sub_1^* are queried to the IssueS Oracle.
5. Finally $\mathcal{A}$ output his guess b'.

The advantage of $\mathcal{A}$ in the game is $|\Pr[b' = b] - \frac{1}{2}|$.

Definition 3. *A CBPS scheme is $(\epsilon, t, q_s, q_p, q_r)$-subscription confidential against chosen ciphertext attack if there is no t-time adversary with q_s queries to the IssueS oracle, q_p queries to the IssueP oracle and q_r queries to the retrieval oracle has an advantage over ϵ in the game.*

Notice that all of the above definitions for confidentiality is against chosen ciphertext attack (CCA). If we do not allow any query to the retrieval oracle, then the above confidentiality definition is reduced to against chosen plaintext attack (CPA).

2.5 Unforgeability

We describe the unforgeability in the pub/sub system. It provides authentication and integrity for the pub/sub system. Wang *et al.* [11] mentioned

---

[3] For example, an investor may not want other people to know which stock price he has subscribed, since it may leak information of which stock he may buy.

[4] Public key Encryption with Keyword Search (PEKS)[2] can be one of the method to solve this dilemma. We will explain in details in the full version of the paper.

that authentication (end-to-end and point-to-point), information integrity, subscription integrity and service integrity are important security requirements for the pub/sub system. We use the standard notion of unforgeability for digital signature to cover the authentication and integrity requirements.

Information Unforgeability. Information unforgeability means that the subscriber believes that the notification is produced by the publisher and is not altered in the broker network. The game for information unforgeability is formally defined as follows:

1. The challenger runs (param, msk) $\leftarrow$ Setup(1^λ), (bsk, bpk) $\leftarrow$ KeyGen$_b$(param) and (psk, ppk) $\leftarrow$ KeyGen$_p$(param). The challenger gives the public parameters param, the manager's secret key msk, the secret/public key pairs of the untrusted brokers and the publisher public key ppk to the adversary $\mathcal{A}$. The publisher's secret key psk is unknown to $\mathcal{A}$.
2. $\mathcal{A}$ is allowed to query the Pub Oracle: On input the message m and the publisher filter $filter$, the oracle first runs both IssueP(param, msk, ppk) and RegP(param, $filter, psk$) by itself and obtains K_p. Then, it outputs the notification $n \leftarrow$ Pub(param, m, psk, K_p).
3. $\mathcal{A}$ returns a message m^*, a notification n^* for a subscription sub^*.

$\mathcal{A}$ wins the game if (m^*, ppk) $\leftarrow$ Retrieve(param, n^*, K_s), where K_s is the output of RegS(param, sub^*, ssk)) interacting with IssueS(param, msk, spk), n^* was not the output of Pub Oracle query with input m^* and (ssk, spk) $\leftarrow$ KeyGen$_s$(param).

Definition 4. *A CBPS scheme is (ϵ, t, q_p)-information unforgeable against chosen message attack if there is no t-time adversary winning the above game with probability at least ϵ with q_p queries to the Pub oracle.*

Subscription Unforgeability. Subscription unforgeability means that the broker believes that the subscription is produced by the subscriber and is not altered in the broker network. The game for subscription unforgeability is formally defined as follows:

1. The challenger runs (param, msk) $\leftarrow$ Setup(1^λ), (bsk, bpk) $\leftarrow$ KeyGen$_b$(param) and (ssk, spk) $\leftarrow$ KeyGen$_s$(param). The challenger gives the public parameters param, the manager's secret key msk, the secret/public key pairs of the untrusted brokers and the subscriber's public key spk to the adversary $\mathcal{A}$. The subscriber's secret key ssk is unknown to $\mathcal{A}$.
2. $\mathcal{A}$ is allowed to query the Sub Oracle: On input the subscription sub, the oracle first runs both IssueS(param, msk, spk) and RegS(param, sub, ssk) by itself and obtains K_s. Then, it outputs the subscription ciphertext $C_{sub} \leftarrow$ Sub(param, sub, ssk, bpk).
3. $\mathcal{A}$ returns a subscription ciphertext C^*_{sub} and a notification n^*.

$\mathcal{A}$ wins the game if $spk \leftarrow$ Match(param, n^*, C^*_{sub}, bsk) and C^*_{sub} was not the output of Sub Oracle query.

Definition 5. *A CBPS scheme is* (ϵ, t, q_s)*-subscription unforgeable against chosen message attack if there is no t-time adversary winning the above game with probability at least* ϵ *with* q_s *queries to the Sub oracle.*

Service Unforgeability. Service unforgeability means that the broker believes that the notification is produced by the publisher and is not altered in the previous broker network. It ensures that once malicious faults arises at the infrastructure level, it could be detected by the next broker. Information unforgeability provides end-to-end authentication of the publisher, while service unforgeability provides authentication of the publisher to every point in the network. It minimizes the damage by a malicious broker who insert bogus notifications into the pub/sub network. The game for information unforgeability is formally defined as follows:

1. The challenger runs $(\mathsf{param}, msk) \leftarrow \mathsf{Setup}(1^\lambda)$, $(bsk, bpk) \leftarrow \mathsf{KeyGen}_b(\mathsf{param})$ and $(psk, ppk) \leftarrow \mathsf{KeyGen}_p(\mathsf{param})$. The challenger gives the public parameters param, the manager's secret key msk, the secret/public key pairs of the untrusted brokers and the publisher public key ppk to the adversary $\mathcal{A}$. The publisher's secret key psk is unknown to $\mathcal{A}$.
2. $\mathcal{A}$ is allowed to query the Pub Oracle: On input the message m and the publisher filter $filter$, the oracle first runs both $\mathsf{IssueP}(\mathsf{param}, msk, ppk)$ and $\mathsf{RegP}(\mathsf{param}, filter, psk)$ by itself and obtains K_p. Then, it outputs the the notification $n \leftarrow \mathsf{Pub}(\mathsf{param}, m, psk, K_p)$.
3. $\mathcal{A}$ returns a notification n^*, a subscription ciphertext C^*_{sub}, a subscriber's public key spk^* and the corresponding subscriber key K^*_s.

$\mathcal{A}$ wins the game if $(m^*, ppk) \leftarrow \mathsf{Retrieve}(\mathsf{param}, n^*, K^*_s)$, $spk^* \leftarrow \mathsf{Match}(\mathsf{param}, n^*, C^*_{sub}, bsk)$ and n^* was not the output of any Pub Oracle query.

Definition 6. *A CBPS scheme is* (ϵ, t, q_p)*-service unforgeable against chosen message attack if there is no t-time adversary winning the above game with probability at least* ϵ *with* q_p *queries to the Pub oracle.*

2.6   Anonymity

The anonymity in the pub/sub system is different for the publishers and subscribers. We will consider two cases separately. The trust model for anonymity is different from confidentiality and unforgeability, since the border brokers directly connecting to the publisher and subscriber must know who is communicating with them. Therefore the border brokers must be trusted for anonymity. To be more specific, publisher hosting broker is trusted for publisher anonymity; and subscriber hosting broker is trusted for subscriber anonymity.

Publisher Anonymity. The anonymity for the publisher means the publisher remains anonymous when he sends a notification. Only the legitimate subscribers can know the identity of the publisher (for authentication purpose). The publisher anonymity game is formally defined as follows:

1. The challenger runs $(\text{param}, msk) \leftarrow \text{Setup}(1^\lambda)$ and $(bsk, bpk) \leftarrow \text{KeyGen}_b(\text{param})$. The challenger gives the public parameters param and the secret/public key pairs of the untrusted brokers to the adversary $\mathcal{A}$. The manager's secret key msk is unknown to $\mathcal{A}$.
2. $\mathcal{A}$ is allowed to query the IssueS Oracle, IssueP Oracle and Retrieval Oracle defined in the publisher confidentiality game.
3. $\mathcal{A}$ sends two publisher key pairs (ppk_0^*, psk_0^*) and (ppk_1^*, psk_1^*), a message m^* and a filter $filter^*$ to the challenger. The challenger computes $n_b^* \leftarrow \text{Pub}(\text{param}, m^*, psk_b^*, \text{RegP}(\text{param}, filter^*, psk_b^*))$. There should be no subscription sub queried to the IssueS Oracle, such that $n_b^* \subseteq_s sub$, no matter $b = 0$ or 1. He picks a bit $b \in \{0, 1\}$ and sends the notification n_b^* to $\mathcal{A}$.
4. $\mathcal{A}$ is allowed to query the oracles, with the exception that no subscription sub queried to the IssueS Oracle, such that $n_b^* \subseteq_s sub$; and n_b^* should not be queried to the Retrieval Oracle.
5. Finally $\mathcal{A}$ output his guess b'.

The advantage of $\mathcal{A}$ in the game is $|\Pr[b' = b] - \frac{1}{2}|$.

Definition 7. *A CBPS scheme is $(\epsilon, t, q_s, q_p, q_r)$-publisher anonymous against chosen ciphertext attack if there is no t-time adversary with q_s queries to the IssueS oracle, q_p queries to the IssueP oracle and q_r queries to the retrieval oracle has an advantage over ϵ in the game.*

Subscriber Anonymity. The anonymity for the subscriber means that the subscriber remains anonymous when he sends a subscription. The subscriber anonymity game is formally defined as follows:

1. The challenger runs $(\text{param}, msk) \leftarrow \text{Setup}(1^\lambda)$ and $(bsk, bpk) \leftarrow \text{KeyGen}_b(\text{param})$. The challenger gives the public parameters param, the manager's secret key msk and the subscriber hosting broker's public key bpk to the adversary $\mathcal{A}$. The subscriber hosting broker's secret key bsk is unknown to $\mathcal{A}$.
2. $\mathcal{A}$ is allowed to query the following oracles: Match Oracle: On input (n, C_{sub}) where n is the notification and C_{sub} is the subscription ciphertext to bpk, it outputs the matching result: spk, 0, $\perp_s$ and/or $\perp_p$ which is the output from Match$(\text{param}, n, C_{sub}, bsk)$.
3. $\mathcal{A}$ sends two publisher key pairs (spk_0^*, ssk_0^*) and (spk_1^*, ssk_1^*), a subscription sub^* to the challenger. The challenger picks a bit $b \in \{0, 1\}$ and computes $C_{sub}^* \leftarrow \text{Sub}(\text{param}, sub^*, ssk_b^*, bpk)$. He sends the subscription ciphertext C_{sub}^* to $\mathcal{A}$.
4. $\mathcal{A}$ is allowed to query the oracles, with the exception that no subscription ciphertext C_{sub}^* queried to the Match Oracle.
5. Finally $\mathcal{A}$ output his guess b'.

The advantage of $\mathcal{A}$ in the game is $|\Pr[b' = b] - \frac{1}{2}|$.

Definition 8. *A CBPS scheme is (ϵ, t, q_m)-subscriber anonymous against chosen ciphertext attack if there is no t-time adversary with q_m queries to the Match oracle has an advantage over ϵ in the game.*

Notice that subscription anonymity may contradict the accountability requirement in [11]. In commercial pub/sub applications, publishers may want to charge subscribers for the information they provide. If the charge is time basis, subscribers pay when they get the subscription key for a period of time from the manager. Each independent subscription can still be anonymous and our current subscription anonymity model can still be used. However if the charge is per notification basis, subscribers' identities must be revealed for accountability and auditability purposes. The security model need to be changed, such that a publisher needs to know whose subscription matches his notification.

3 Our Construction

In this section, we will construct some CBPS protocols and we will prove their security against the security model defined in the previous section. We first describe the main idea of the scheme. We review the relevant cryptographic background and then we show the basic construction. Our basic construction satisfies most security requirements in confidentiality and unforgeability. However, our construction does not satisfy all the security requirements mentioned in previous section. Finally, we demonstrate that our basic scheme can be further extended to satisfy the other security requirements.

3.1 Main Idea of Our Basic Scheme

The main weakness of the existing CBPS protocols is that they do not have any proof of security. Furthermore, some of them only consider either confidentiality or authenticity. A secure CBPS protocol should have security proofs for both confidentiality and unforgeability. To provide information confidentiality and information unforgeability at the same time, we use an approach commonly used in signcryption schemes. It means that the randomness used in the signature and the encryption are the same. It ensures that the signature and the encryption protocol are run by the same party. An adversary cannot use the ciphertext from a legitimate user and append the adversary's signature to it; nor use the signature from a legitimate user and append a ciphertext computed by the adversary.

To facilitate routing while providing confidentiality in the pub/sub system, we employ the approach that only the part of the document containing the secret information is encrypted. For example, in a pub-sub stock quote application, a publisher (the bank) provides stock quote to subscribers (the bank's customers). The stock price is encrypted while the stock name is not. Therefore the document can be routed to subscribers who are interested in a particular stock.

Unforgeability. The challenge of encrypting the partial document is how the brokers authenticate the document without knowing the plaintext. Refer to the previous example, an obvious solution is to sign on the stock name and the encrypted stock price. However, a signature on the encrypted stock price does not guarantee the authenticity of the stock price. A more complicated solution in [3] is to encrypt the stock price and the signature of the stock price. After that the stock name and the whole ciphertext is signed again.

In this paper, we use a simpler approach by sanitizable signatures [5]. A sanitizable signature scheme allows one to verify a signature even when part to the original message is not known. Therefore, we can use compute a sanitizable signature to the whole document and encrypt the stock price. The brokers only needs to verify the signature for the part of the stock name. For the subscribers, the same signature is verified against the whole document after decryption. By the property of sanitizable signatures, it is difficult to obtain any information on the sanitized messages from the sanitizable signature. Therefore authenticity is preserved while having confidentiality in the (untrusted) broker network. Our scheme uses the sanitizable signatures by Suzuki *et al.* [10].

Confidentiality. The challenge of confidentiality is that how publishers can restrict the access of the secret information. By the loose coupling property of the pub/sub network, publishers do not know who are going to subscribe the notifications. Hence publishers have no public key to encrypt the secret information. Some schemes ([8,4]) assume that publishers and subscribers share a symmetric key. However, it contradicts the very first assumption of decoupling of publishers and subscribers. These schemes are only suitable for private pub/sub systems over public networks. An internet-scale, dynamic pub/sub network with a universe of publishers and subscribers are unlikely to share a symmetric key. Another possible solution ([7]) for confidentiality is through access control to the broker network. Encryption and decryption is performed by the border brokers and therefore trust must be placed upon them. If the broker network is not trusted, it is difficult for the publisher to find a suitable public key for encryption (brokers are not trusted and subscribers are not known).

Our scheme uses the Ciphertext-Policy Attribute-Based Encryption (CP-ABE) by Waters [12] to solve this problem. In CP-ABE, attributes are used to describe the user's (subscriber's) credentials and the encrypting party (publisher) can encrypt the message according to some formulas over these credentials. Therefore, publishers can encrypts the secret information by some attributes that describe the information. Subscribers can request keys from the manager about the attributes that they are interested in.

3.2   Cryptographic Backgrounds

We present a brief revision on groups with efficiently computable bilinear maps and then review some number theoretic assumptions. After that, we review the definition of access structures and relevant backgrounds on Linear Secret Sharing

Schemes (LSSS), sanitizable signatures and CP-ABE. They are extensively used in our concrete construction of pub/sub system.

Pairings and Intractability Assumptions. Let $\mathbb{G}$ and $\mathbb{G}_T$ be two multiplicative cyclic groups of prime order p. Let g be a generator of $\mathbb{G}$.

Definition 9. *A map $\hat{e} : \mathbb{G} \times \mathbb{G} \to \mathbb{G}_T$ is called a bilinear map if, for all $u, v \in \mathbb{G}$ and $a, b \in \mathbb{Z}_p$, we have $\hat{e}(v^a, v^b) = \hat{e}(u, v)^{ab}$, and $\hat{e}(g, g) \neq 1$.*

Definition 10 (CDH). *The Computational Diffie-Hellman problem is that, given g, $g^x, g^y \in \mathbb{G}$ for unknown $x, y \in \mathbb{Z}_p$, to compute g^{xy}.*

We say that the (ϵ, t)-CDH assumption holds if no t-time algorithm has the non-negligible probability ϵ in solving the CDH problem.

Definition 11 (DBDH). *The Decisional Bilinear Diffie-Hellman problem is that, given $(g, g^a, g^b, g^c) \in \mathbb{G}$ and $T \in \mathbb{G}_T$ for unknown $a, b, c \in \mathbb{Z}_p$, to distinguish if $T = \hat{e}(g, g)^{abc}$ or T is a random element in $\mathbb{G}_T$.*

We say that the (ϵ, t)-DBDH assumption holds if no t-time algorithm has the non-negligible probability ϵ minus half in solving the DBDH problem.

Definition 12 (decisional q-BDHE). *The decisional q-Bilinear Diffie-Hellman Exponent problem is that, given $(g, g^a, g^{a^2}, \ldots, g^{a^q}, g^{a^{q+2}}, \ldots, g^{a^{2q}}, g^s) \in \mathbb{G}$ and $T \in \mathbb{G}_T$ for unknown $a, s \in \mathbb{Z}_p$, to distinguish if $T = \hat{e}(g, g)^{a^{q+1}s}$ or T is a random element in $\mathbb{G}_T$.*

We say that the (ϵ, t)-decisional q-BDHE assumption holds if no t-time algorithm has the non-negligible probability ϵ minus half in solving the decisional q-BDHE problem.

Access Structures. We first review the definition of access structure in [1].

Definition 13 (Access Structure [1]). *Let $\{P_1, P_2, \ldots, P_n\}$ be a set of parties. A collection $\mathbb{A} \subseteq 2^{\{P_1, P_2, \ldots, P_n\}}$ is monotone if $\forall B, C$: if $B \in \mathbb{A}$ and $B \subseteq C$ then $C \in \mathbb{A}$. An* access structure *(resp. monotone access structure) is a collection (resp. monotone collection) $\mathbb{A}$ of non-empty subsets of $\{P_1, P_2, \ldots, P_n\}$. The sets in $\mathbb{A}$ are called the* authorized sets, *and the sets not in $\mathbb{A}$ are called the* unauthorized sets.

In our context, the role of parties is taken by the attributes, which is equivalent to the subscription condition. Thus the access structure $\mathbb{A}$ will contain the authorized set of attributes (subscription condition). We restrict our attention to monotone access structure. From now on, unless stated otherwise, by an access structure we mean a monotone access structure.

Linear Secret Sharing Schemes. We adapt the definition of Linear Secret Sharing Schemes (LSSS) in [1].

Definition 14 (LSSS [1]). *A secret sharing scheme Π over a set of parties $\mathcal{P}$ is called* linear *over $\mathbb{Z}_p$ if:*

1. *The shares for each party form a vector over $\mathbb{Z}_p$.*
2. *There exists a matrix M called the share-generating matrix for Π. The matrix M has ℓ rows and n columns. For $i = 1, \ldots, \ell$, the i-th row of M we let the function ρ defined the party labeling row i as $\rho(i)$. When we consider the column vector $v = (s, r_2, \ldots, r_n)$, where $s \in \mathbb{Z}_p$ is the secret to be shared, and $r_2, \ldots, r_n \in \mathbb{Z}_p$ are randomly chosen, then Mv is the vector of ℓ shares of the secret s according to Π. The share $(Mv)_i$ belongs to party $\rho(i)$.*

Beimel [1] showed that every LSSS enjoys the *linear reconstruction* property, defined as follows: Suppose that Π is an LSSS for the access structure $\mathbb{A}$. Let $S \in \mathbb{A}$ be any authorized set, and let $I \subset \{1, \ldots, \ell\}$ be defined as $I = \{i : \rho(i) \in S\}$. Then there exist constants $\{\omega_i \in \mathbb{Z}_p\}_{i \in I}$ such that, if $\{\lambda_i\}$ are valid shares of any secret s according to Π, then $\sum_{i \in I} \omega_i \lambda_i = s$. Furthermore, Beimel [1] showed that these constants $\{\omega_i\}$ can be found in time polynomial in the size of the share-generating matrix M.

Sanitizable Signatures. A digital signature prohibits any alteration of the original message once it is signed. It protects the signer against the message forgery. Nevertheless, it also prevents the message from being process further legitimately as well, which sometimes is actually desirable.

A typical example of sanitizable signature includes the case when the government wants to release some *partial* information in an officially signed document. In this particular case, a government officer may want to delete some sensitive information such as personal information or national secrets. In order to avoid the process of having the message to be signed again (since the original signer may not be available at that time), a sanitizable signature can be used to sign the document at the first place; and the sensitive information can be sanitized *prior to* the release of the signature. The major goal of sanitizable signature is to protect the confidentiality of part of the document while ensuring the integrity of the document.

Ciphertext-Policy Attribute-based Encryption. During encryption, the data provider can express how he wants to share data in the encryption algorithm. In traditional public key encryption, the data provider uses the recipient's public key to encrypt, such that the data is shared with the intended recipient only (by decryption).

In Ciphertext-Policy Attribute-based Encryption (CP-ABE), the recipient is ascribed a secret key associated with a set of string called "attributes". The data provider will provide a formula over these attributes, describing how he wants to share the data. The recipient can correctly decrypt a ciphertext encrypted with a formula only if his secret key associates with attributes which satisfy the formula.

216   T.H. Yuen, W. Susilo, and Y. Mu

3.3   The Basic Scheme

We use the sanitizable signature scheme by Suzuki et al. [10] and CP-ABE scheme by Waters [12]. Some input parameters described in §2.1 are omitted here when they are not used in the basic scheme. Denote $\oplus$ as the bit-wise XOR function.

- **Setup.** On input 1^λ, it picks the pairing $\hat{e} : \mathbb{G} \times \mathbb{G} \to \mathbb{G}_T$ and generators $g, g_1 \in \mathbb{G}$ and collision resistant hash functions $H_1 : \{0,1\}^* \to \mathbb{G}$ and $H_2 : \{0,1\}^* \to \mathbb{G}$. It chooses a random exponent $\alpha \in \mathbb{Z}_p$. The manager secret key is g^α. It outputs the system parameter param $= \{g, g_1, \hat{e}(g,g)^\alpha, \hat{e}, H_1, H_2\}$. Let (Sig, Vfy) be a secure signature scheme.
- **KeyGen.** On input the system parameter param, the publisher randomly picks his secret key $x_p \leftarrow \mathbb{Z}_p$. He outputs his public key $y_p = g^{x_p}$. The subscriber randomly picks his secret key $x_s \leftarrow \mathbb{Z}_p$. He outputs his public key $y_s = g^{x_s}$.
- **RegP, IssueP.** The publisher chooses the filter as an LSSS access structure (M, ρ). We limit ρ to be an injective function, that is an attribute is associated with at most one row of an $\ell \times n$ matrix M. It is the same as the publisher key K_p and therefore he does not need to interact with the manager.
- **RegS, IssueS.** On input the subscription as a set of attributes S, the subscriber sends it to the manager. the manager with master secret key g^α first chooses a random $t \in \mathbb{Z}_p$. He creates the subscriber key as

$$K = g^\alpha g_1^t, \quad L = g^t, \quad \forall x \in S \ K_x = H_1(x)^t$$

The manager sends the subscriber key $K_s = (K, L, \{K_x : \forall x \in S\})$ to the subscriber.
- **Pub.** On input $(\text{param}, m, x_p, K_p)$ where param is the system parameter, m is the message, x_p is the publisher's secret key and $K_p = (M, \rho)$ is the publisher key.
  The publisher first chooses a random vector $\boldsymbol{v} = (s, y_2, \ldots, y_n) \in \mathbb{Z}_p^n$. For $i = 1, \ldots, \ell$, he calculates $\lambda_i = \boldsymbol{v} \cdot M_i$, where M_i is the vector corresponding to the i-th row of M. The publisher then chooses random $r_1, r_2 \in \mathbb{G}$ and $s \in \mathbb{Z}_p$. He computes

$$w_1 = H_2(m||r_1), \quad w_2 = H_2(M||\rho||r_2), \quad C = \hat{e}(g,g)^{\alpha s} \oplus (m, \sigma_1, r_1),$$
$$C' = g^s, \quad C_1 = g_1^{\lambda_1} H_1(\rho(1))^{-s}, \quad \ldots, \quad C_\ell = g_1^{\lambda_\ell} H_1(\rho(\ell))^{-s},$$
$$\sigma_1 = w_1^{x_p}, \quad w_3 = H_2(w_1||w_2||C'||C||C_1||\ldots||C_\ell), \quad \sigma_2 = (w_2 w_3)^{x_p}.$$

  The notification is published as $n = (C', r_2, w_1, \sigma_2, C, C_1, \ldots, C_\ell, M, \rho)^5$.
- **Sub.** On input (param, x_s) where param is the system parameter and x_s is the subscriber's secret key, the subscriber signs the subscription attributes S by $\sigma_s = \text{Sig}(x_s, S)$. He sends $C_{sub} = (S, \sigma_s, y_s)$ to the broker network.

---

[5] (σ_1, σ_2) can be viewed as the sanitizable signature part of the notification. Even without the knowledge of m, the broker can check the validity of σ_2 in the Match algorithm to ensure that the notification is authenticated.

– Match. On input $(\mathsf{param}, n, C_{sub})$ where param is the system parameter, $n = (C', r_2, w_1, \sigma_2, C, C_1, \ldots, C_\ell, M, \rho)$ is the notification from publisher y_p and $C_{sub} = (S, \sigma_s, y_s)$ is the subscription, the broker first computes

$$w_2 = H_2(M||\rho||r_2), \quad w_3 = H_2(w_1||w_2||C'||C||C_1||\ldots||C_\ell).$$

If $\hat{e}(\sigma_2, g_2) \neq \hat{e}(w_2 w_3, y_p)$, the broker outputs $\perp_p$. If $\mathsf{Vfy}(y_s, S, \sigma_s) = 0$, the broker also outputs $\perp_s$.

Otherwise, when S satisfies the access structure (M, ρ), the broker forwards the notification to the subscriber and outputs y_p. If S does not satisfy, the broker outputs 0.

– Retrieve. On input (param, n, K_s) where param is the system parameter, $n = (C', r_2, w_1, \sigma_2, C, C_1, \ldots, C_\ell, M, \rho)$ is the notification and $K_s = (K, L, \{K_x : \forall x \in S\})$ is the subscriber key, suppose that S satisfies the access structure (M, ρ). The subscriber finds the set $I = \{i : \rho(i) \in S\}$. Let $\{\omega \in \mathbb{Z}_p\}_{i \in I}$ be a set of constants such that if $\{\lambda_i\}$ are valid shares of any secret s according to M, then $\sum_{i \in I} \omega_i \lambda_i = s$. Then he computes

$$\frac{\hat{e}(C', K)}{(\prod_{i \in I}(\hat{e}(C_i, L)\hat{e}(C', K_{\rho(i)}))^{\omega_i})} = \frac{\hat{e}(g, g)^{\alpha s}\hat{e}(g, g_1)^{st}}{(\prod_{i \in I}\hat{e}(g, g_1)^{t\lambda_i \omega_i})} = \hat{e}(g, g)^{\alpha s}.$$

The subscriber obtains $(m, \sigma_1, r_1) = \hat{e}(g, g)^{\alpha s} \oplus C$. He computes

$$w_1 = H_2(m||r_1), \quad w_2 = H_2(M||\rho||r_2), \quad w_3 = H_2(w_1||w_2||C'||C||C_1||\cdots||C_\ell).$$

If $\hat{e}(\sigma_1 \sigma_2, g) \neq \hat{e}(w_1 w_2 w_3, y_p)$, the subscriber outputs $\perp$. Otherwise, he outputs a pair (m, ppk), where m is the message and ppk is the publisher's public key.

3.4 Security

Our basic scheme has publisher confidentiality, information confidentiality, information unforgeability, subscription unforgeability and service unforgeability.

Theorem 1. *Suppose the (ϵ, t')-decisional q-BDHE assumption holds. Then our basic scheme is (ϵ, t, q_s, q_p)-selectively publisher confidential against chosen plaintext attack in the random oracle model, with a challenge filter (M^*, ρ^*) and*

$$t' = t + (q_s + q_h)O(n^*(\tau_m + \tau_e)),$$

where M^* is of size $\ell^* \times n^*$ and $n^* \leq q$, q_h is the number of query to the H_1 oracle, τ_m and τ_e are the time for a multiplication and an exponentiation in $\mathbb{G}$, respectively.

As discussed in §2.1, publisher confidentiality implies information confidential. Therefore our basic scheme is also selectively secure for information confidentiality against the chosen plaintext attack. However, we can give a direct proof without selective model and use a weaker assumption.

Theorem 2. *Suppose the (ϵ, t')-DBDH assumption holds. Then our basic scheme is (ϵ, t, q_p)-information confidential against chosen plaintext attack in the random oracle model, with $t' = t + q_h O(\tau_e)$, where q_h is the number of query to the H_1 oracle and τ_e is the time for an exponentiation in $\mathbb{G}$.*

Theorem 3. *Suppose the (ϵ', t')-CDH assumption holds. Then our basic scheme is (ϵ, t, q_p)-information unforgeable against chosen message attack in the random oracle model, where*

$$\epsilon' \geq \epsilon(\frac{3}{q_h} - \frac{3}{q_h^2} + \frac{1}{q_h^3}), \qquad t' = t + (q_p + q_h)O(\tau_m + \tau_e),$$

where τ_m and τ_e are the time for a multiplication and an exponentiation in $\mathbb{G}$, respectively; and q_h is the number of query to the H_2 oracle.

Theorem 4. *Suppose that (Sig, Vfy) is EUF-CMA secure. Then no poly-time adversary can break the subscription unforgeability.*

Theorem 5. *Suppose the (ϵ', t')-CDH assumption holds. Then our basic scheme is (ϵ, t, q_p)-service unforgeable against chosen message attack in the random oracle model, where*

$$\epsilon' \geq \epsilon(\frac{2}{q_h} - \frac{1}{q_h^2}), \qquad t' = t + (q_p + q_h)O(\tau_m + \tau_e),$$

where τ_m and τ_e are the time for a multiplication and an exponentiation in $\mathbb{G}$, respectively; and q_h is the number of query to the H_2 oracle.

Theorem 4 is straightforward by the construction of our basic scheme. The security proofs of other theorems are given in the full version of the paper due to the space limit. We also discuss some possible extensions that can be applied to our basic scheme in the full version of the paper.

4   Related Works

In this section we compare our basic CBPS scheme and the extension with the existing CBPS schemes providing confidentiality. The result of the comparison can be found in Table 1.

The scheme of Li, Lu and Shi [4] and Srivatsa and Liu [9] use prefix-preserving tree structure for information and subscription confidentiality as well as efficient matching. However, if the adversary have a large number of matching notification and subscription pairs, then the adversary may obtain some information about the prefix in the notification and subscription. Therefore they are only secure if the adversary knows a few notification and subscription pairs. They are not secure in our security model.

Khurana [3] proposed a CBPS scheme with a threshold key sharing scheme such that t out of n managers are responsible to generates keys for subscribers and publishers. It reduces the trust to a single manager. However, the group of

Table 1. Comparison of pub/sub schemes providing confidentiality. For confidentiality (Conf), I stands for information confidentiality, S stands for subscription confidentiality and P stands for publisher confidentiality. For unforgeability (Unf), I stands for information unforgeability, S stands for subscription unforgeability and V stands for service unforgeability. For anonymity (Anon), S stands for subscriber anonymity. A small letter means that it is secure in a weaker security model in the original cited paper only. BB. stands for border brokers. For * in the table, it will be explained in §4.

Scheme	Conf	Unf	Anon	Pre-Shared Key	Proof	Trust
Li et al. [4]	i, s	-	-	Yes	No	No
Khurana [3]	I	I, V	-	No	No	*
Zhao and Sturman [14]	I	i, s	-	No	No	BB.
Raicius and Rosenblum [8]	I, S	i, s	-	Yes	Yes	No
Srivatsa and Liu [9]	i, p, s	-	-	No	No	No
Pesonen et al. [7]	I, S	*	-	No	No	BB.
Zhang et al. [13]	i, s	-	-	No	No	No
Our Basic Scheme	I, P	I, S, V	-	No	Yes	No
Our Extension (in full version of the paper)	I, P, S	I, S, V	s	No	Yes	No

n managers must help to calculate the notification when the notification travels from a broker to another. It greatly increases the workload of the managers.

Zhao and Sturman [14] placed a complete trust to the border brokers in their CBPS scheme. Encryption is performed between border brokers. Publishers and subscribers access pub/sub system through the access control list. Information confidentiality and authenticity is protected by this access control. The scheme is not secure in our unforgeability model.

Pesonen, Eyers and Bacon [7] also placed a trust to the border brokers in their CBPS scheme. Publishers and subscribers access pub/sub system through the access control list. Information and subscription confidentiality are protected by this access control. Since authenticated encryption is used, integrity is also protected. However, the scheme is not secure in our unforgeability model.

Raicius and Rosenblum [8] proposed the first CBPS scheme with proof of information and subscription confidentiality. It comes with the cost of publishers and subscribers having a pre-shared key. The unforgeability of the scheme is also protected by this pre-shared key, since encryption cannot be performed without the symmetric key. The scheme is not secure in our unforgeability model.

Zhang et al. [13] proposed a CBPS scheme using a new mechanism called information foiling. The publishers and subscribers generate a set of fake messages to hide the authentic message. Their new algorithm does not fit into our model since their confidentiality is in a probabilistic sense.

5   Conclusion

In this paper, we introduced the *first* security model for all security requirements of CBPS. We proposed a new CBPS scheme that fulfills most of the security requirements concurrently. We proved its security according to our new model.

References

1. Beimel, A.: Secure schemes for secret sharing and key distribution. Ph.D. thesis, Department of Computer Science, Israel Institute of Technology (1996)
2. Boneh, D., Crescenzo, G.D., Ostrovsky, R., Persiano, G.: Public key encryption with keyword search. In: Cachin, C., Camenisch, J. (eds.) EUROCRYPT 2004. LNCS, vol. 3027, pp. 506–522. Springer, Heidelberg (2004)
3. Khurana, H.: Scalable security and accounting services for content-based publish/subscribe systems. In: Haddad, H., Liebrock, L.M., Omicini, A., Wainwright, R.L. (eds.) SAC 2005, pp. 801–807. ACM, New York (2005)
4. Li, J., Lu, C., Shi, W.: An efficient scheme for preserving confidentiality in content-based publish-subscribe systems, Tech. Rep. GIT-CC-04-01, Georgia Institute of Technology (2004)
5. Miyazaki, K., Susaki, S., Iwamura, M., Matsumoto, T., Sasaki, R., Yoshiura, H.: Digital documents sanitizing problem. IEICE Technical Report ISEC2003-20, 61–67 (2003)
6. Nikander, P., Giannis, M.F.: Towards understanding pure publish/subscribe cryptographic protocols. In: 16th International Workshop on Security Protocols (2008)
7. Pesonen, L.I.W., Eyers, D.M., Bacon, J.: Encryption-enforced access control in dynamic multi-domain publish/subscribe networks. In: DEBS 2007. ACM International Conference Proceeding Series, vol. 233, pp. 104–115. ACM, New York (2007)
8. Raiciu, C., Rosenblum, D.S.: Enabling confidentiality in content-based publish/subscribe infrastructures. In: Securecomm 2006. IEEE, Los Alamitos (2006)
9. Srivatsa, M., Liu, L.: Secure event dissemination in publish-subscribe networks. In: ICDCS 2007, p. 22. IEEE Computer Society, Los Alamitos (2007)
10. Suzuki, M., Isshiki, T., Tanaka, K.: Sanitizable signature with secret information. In: Symposium on Cryptography and Information Security, 4A1-2 (2006)
11. Wang, C., Carzaniga, A., Evans, D., Wolf, A.L.: Security issues and requirements for Internet-scale publish-subscribe systems. In: HICSS 2002. IEEE Computer Society, Los Alamitos (2002)
12. Waters, B.: Ciphertext-policy attribute-based encryption: An expressive, efficient, and provably secure realization. Cryptology ePrint Archive, Report 2008/290 (2008), http://eprint.iacr.org/
13. Zhang, H., Sharma, A., Chen, H., Jiang, G., Meng, X., Yoshihira, K.: Enabling information confidentiality in publish/subscribe overlay services. In: ICC 2008, pp. 5624–5628. IEEE, Los Alamitos (2008)
14. Zhao, Y., Sturman, D.C.: Dynamic access control in a content-based publish/subscribe system with delivery guarantees. In: ICDCS 2006, p. 60. IEEE Computer Society, Los Alamitos (2006)

STE3D-CAP: Stereoscopic 3D CAPTCHA

Willy Susilo[1,*], Yang-Wai Chow[2], and Hua-Yu Zhou[2]

[1] Centre for Computer and Information Security Research
[2] Centre for Multimedia and Information Processing
School of Computer Science and Software Engineering
University of Wollongong, Australia
{wsusilo,caseyc,hz285}@uow.edu.au

Abstract. We present STE3D-CAP (pronounced as "steed-cap" /'stidkæp/)[1], a text-based CAPTCHA that is built from stereoscopic 3D images. This is a completely new direction in CAPTCHA techniques. Our idea is to incorporate stereoscopic 3D images in order to present the CAPTCHA challenge in 3D, which will be easy for humans to read (as the text stands out in the 3D scene) but hard for computers. The main idea is to produce a stereo pair, two images of the distorted 3D text objects generated from two different camera/eye viewpoints, that are presented to a human user's left and right eyes, respectively. When the two images are supplied to hardware capable of displaying stereoscopic 3D images, the resulting CAPTCHA can easily be solved by humans, as the text will appear to stand out from the rest of the scene, but computers will not be able to solve them easily. As per the usual practice, the text in the produced images will be distorted (e.g. translated, scaled, warped) and overlapped but additionally the depth of the 3D text objects in the stereoscopic images will add a degree of complexity to the CAPTCHA and make it harder for CAPTCHA attacks (due to positive and negative parallax in the stereo pair). We demonstrate that the existing attacks on STE3D-CAP will fail with an overwhelming probability and that we can increase our CAPTCHA's resistance to segmentation attacks whilst maintaining usability. We also note that our technique is applicable to other stereoscopic approaches, such as anaglyph.

1   Introduction

The invention of CAPTCHAs (Completely Automated Public Turing test to tell Computers and Humans Apart)[2] was put forth by von Ahn et al. in 2003 [26]. CAPTCHAs are designed to be simple problems that can be quickly solved by humans, but are difficult for computers to solve. After von Ahn et al.'s seminal work, hundreds of design variants have appeared either in practice or in the literature. CAPTCHAs have quickly gained popularity over the past few years, since they are used to prevent exploitations by bots and automated scripts in public web services, which are rapidly increasing. Essentially, CAPTCHAs are challenge response tests that have become almost ubiquitous

---

* This work is supported by ARC Future Fellowship FT0991397.
[1] The name is inspired by a working mount (horse) especially for warfare.
[2] The term CAPTCHAs have also been known as Human Interaction Proofs (HIPs) [6].

S.-H. Heng, R.N. Wright, and B.-M. Goi (Eds.): CANS 2010, LNCS 6467, pp. 221–240, 2010.

222 W. Susilo, Y.-W. Chow, and H.-Y. Zhou

on the World Wide Web to determine whether a user is a human or a computer. Using CAPTCHAs, services can distinguish legitimate users from computer bots while requiring minimal effort by the human user. Many companies currently employ the use of CAPTCHAs to protect their services against email spam, as well as to prevent fraud and denial of service attacks in online registrations, ticket/event reservations, online voting, chat rooms, weblogs, etc. [8].

To date, there exist three main types of CAPTCHAs: [30]

- Text-based CAPTCHAs: typically obtained by selecting a sequence of letters, rendering them, distorting the image and adding some noise;
- Image-based CAPTCHAs: typically ask users to conduct an image recognition task; and
- Sound-based CAPTCHAs (or audio CAPTCHAs): typically require users to solve a speech recognition task.

Among the three families, text-based CAPTCHAs are the most popular since they are simple, small and easy to design and implement. Therefore, they have been widely used in major web sites such as Google, Yahoo and Microsoft, since they are very intuitive to users world-wide, in addition to having good potential to provide strong security [30]. Theoretically, challenges as short as five characters are robust against random guessing[3], namely $62^5 \approx 912$ million possible five-character challenges that comprising case-insensitive letters and digits. Nevertheless, computing efforts, such as Optical Character Recognitions (OCR) or segmentation techniques, have been found very successful to achieve human-like accuracy [10,29,6]. Generally, text-based CAPTCHAs have universally suffered from a property that making them hard for computers also implies making them hard for humans [10].

Image-based CAPTCHAs were first described using labelled photographs by Chew and Tygar [9], which rely on Google Image Search [15]. This work was known to be unsuccessful due to Google's method of inferring photo contents based on surrounding descriptive text [10]. A more recent example in this family is Asirra [10], which is an image-based CAPTCHAs proposed in ACM CCS 2007 that uses images from Petfinder.com database. The security of Asirra relies on the problem of distinguishing images of cats and dogs, which is a task that is trivial for humans. Unfortunately, Golle demonstrated an attack based on machine learning to produce a classifier with 82.7% accuracy in telling apart the images of cats and dogs used in Asirra [13].

Audio CAPTCHAs were introduced to provide an alternative for those who are unable to use visual CAPTCHAs. Nevertheless, a recent study by Bigham and Cavender demonstrated that existing audio CAPTCHAs are clearly more difficult and time-consuming to complete as compared to visual CAPTCHAs for both blind and sighted users [3]. They also questioned how audio CAPTCHAs could be created that are easier for humans to solve while still addressing the improved automatic techniques for defeating them and posed it as an open problem for future research [3].

Elson et al. [10] classified CAPTCHAs into two different classes. In a Class I CAPTCHA, a secret value, which is merely a random number, is fed into a publicly

---

[3] We assume that random guessing is taken over upper and lowercase letters plus the digits.

known algorithm to produce a challenge, which is analogous to a public-key cryptosystem. In a Class II CAPTCHA, two inputs are required namely a secret value and a secret high-entropy database, which is somewhat analogous to a one-time-pad cryptosystem. One of the challenges in building a secure Class II CAPTCHA is populating the database with a sufficiently large set of classified and high-entropy entries [10].

Combining the two classifications above, it is clear that the most desirable way to construct CAPTCHAs is to employ Class I text-based CAPTCHAs, as they are small and easy to design and implement. The key challenge is how to enlarge the gap between human and non-human success rates whereas the resulting CAPTCHAs will be tolerated by users. Elson et al. [10] criticized text-based CAPTCHAs and commented that they must be intolerable to users or else the resulting CAPTCHAs would be too easy to break. In addition, Jakobsson further argues that in its current incarnation, CAPTCHAs may be nearing the end of its useful life [16]. He also contends that the current trend is such that strengthening CAPTCHAs to withstand increasingly powerful automated attacks also results in them becoming increasingly difficult for human users to solve. This will hurt user tolerance and at some point this trend will simply make CAPTCHAs too hard for people to use [16]. Therefore, it is an interesting open question to produce a Class I text-based CAPTCHA.

Our Contributions. In this paper, we present STE3D-CAP, a text-based CAPTCHA that is built from stereoscopic 3D images, as shown in Figure 1.

Fig. 1. STE3D-CAP without appropriate stereoscopic viewing equipment

STE3D-CAP is easy for human users, as it can be solved by humans who are equipped with an "appropriate" 3D visualization device (eg. Stereoscopic 3D display and glasses). The stereoscopic 3D images that represent left and right views are rendered by a graphics system on the server side and the resulting images are sent to the client system. The client system's hardware is therefore only provided with the two images and has to display these images to the user in stereo. We implemented our STE3D-CAP technique using an NVIDIA 3D Vision kit on a compatible NVIDIA graphics card and an Alienware^{TM}OptX AW2310, 3D capable, 120Hz monitor. We observe that all the known attacks, such as segmentation attacks [29] or pattern recognition analysis, will not be successful in analyzing STE3D-CAP.

The main drawback of STE3D-CAP is that it relies on the required stereoscopic display hardware that the user must own. Nevertheless, with the recent surge in the popularity of 3D movies and 3D games, LCD monitors and 3D TVs capable of displaying stereoscopic 3D images are becoming more and more common place nowadays (such as [12]). Further evidence of this can also be observed from the fact that the cost of 3D capable devices have become significantly cheaper in the past few years[4]. Our goal is to exploit advances in the lastest technology in order to breathe new life into current CAPTCHA techniques which are susceptible to novel attacks, by strengthening text-based CAPTCHAs against such attacks whilst maintaining human usability.

1.1 Related Work

After the seminal concept of a CAPTCHA was introduced by von Ahn [26], many design variations of CAPTCHAs have been proposed and used. The text-based CAPTCHAs are very popular due to its simplicity. Unfortunately, computer attacks have been found against most of the existing text-based CAPTCHAs. Simard et al. demonstrated the use of Optical Character Recognition (OCR) to obtain human-like accuracy, as long as the letters can be segmented reliably [24]. Mori and Malik showed that von Ahn et al.'s original GIMPY CAPTCHA [25] can be solved automatically 92% of the time [19].

Reading text-based CAPTCHA challenges typically consists of a segmentation challenge and a recognition challenge [6]. The segmentation challenge involves the identification of character locations in the right order, whereas the recognition challenge is in recognizing individual characters. The difficulty of these challenges can be increased using various techniques like cluttering the foreground and background, distorting individual characters, etc. Research has shown that computers are extremely successful in recognizing individual characters, even characters that are highly distorted [7,29]. This therefore suggests that the challenge for text-based CAPTCHAs is to design CAPTCHAs that are resistant to segmentation yet human readable.

In 2005, Microsoft published a text-based CAPTCHA, which was designed to be segmentation resistant [6]. The resulting CAPTCHA had been widely deployed in many Microsoft's online services, such as Hotmail, MSN and Windows Live for years. Unfortunately, a low-cost attack with success rate higher than 90% was developed by Yan and Ahmad [29], which demonstrated that this carefully designed CAPTCHA is vulnerable to novel, but simple attacks.

Usability issues of text-based CAPTCHAs have also been investigated by Yan and Ahmad [30]. The four common distortion on text-based CAPTCHAs that will not make them difficult for human users to recognize them are as follows [30].

- Translation: characters can be moved up or down and left or right by an amount;
- Rotation: characters can be turned either in the clockwise or counter clockwise direction;
- Scaling: characters can be stretched or compressed in either the x-direction or y-direction;
- Warp: elastic deformation of CAPTCHA images at different scales.

The length of text strings used in CAPTCHAs also plays an important role to their security. Some schemes choose to use a fixed length, and they turned out to be insecure.

---

[4] A set of NVIDIA [21] capable devices nowadays are around US$ 500, which is significantly lower compared to the cost last year.

For example, the Microsoft's CAPTCHAs use 8 characters in their challenge [6]. It turns out that the segmentation attack can be done easier knowing this fact [29]. On the other hand, Google's CAPTCHAs incorporate a different number of characters in each challenge, although their security has not been rigorously tested [30].

The use of color also plays an important role in CAPTCHA design. In terms of usability, color is good for drawing a user's visual attention. The incorporation of color can also make the CAPTCHA challenge more appealing, it can potentially aid in the recognition and comprehension of the text and it potentially makes the CAPTCHA seem less intrusive in the context of the application [6,30]. Color schemes can potentially increase a CAPTCHA's security against some attacks, e.g. OCR software attacks which are poor at recognising text in colored images. However as highlighted in [30], if used inappropriately color may add little or nothing to the security of a CAPTCHA, but at the same time can significantly reduce the usability of the CAPTCHA. The misuse of color can make the text in a CAPTCHA very difficult to read even for people with normal vision. In fact, color can have a negative effect on the security of a CAPTCHA, as it might make it easy for a computer to distinguish the important text from the background/foreground clutter, etc. As such, care must be taken when using color in CAPTCHAs.

The idea of representing text-based CAPTCHAs with 3D text objects has been used in [22]. Their CAPTCHA, which is called Teabag 3D, is obtained by making a picture in 3D with text objects. Unfortunately, it can be seen clearly that the location of the text objects can easily be distinguished due to the somewhat regular pattern in the surrounding regions, and therefore these CAPTCHAs would be breakable by using something like a simple segmentation technique [29].

There have also been a number of recent approaches to representing CAPTCHA challenge based on 3D models. A 3D object matching CAPTCHA challenge was introduced in [31]. This is an image-based CAPTCHA approach where users are presented with images of 3D models, which are rendered from different angles using Lambertian lighting, and are required to select matching 3D models from a set of images. However as pointed out in [23], this approach is susceptible to attacks using basic computer vision techniques. Ross et al. [23] introduced 'Sketcha', an image-based CAPTCHA approach based on images of line drawings which are rendered from 3D models. In Sketcha, users are presented with a set of randomly oriented line drawings, and are required to rotate each image until all the images are upright. Mitra et al. [18] proposed a technique of generating 'emergence images' by rendering extremely abstract representations of 3D objects models placed in a 3D environment. Their approach is based on 'emergence' which is the unique human ability to perceive objects from seemingly meaningless patches in an image. However when the image is viewed as a whole, a human can perceive the form of the main subject which pops out from the clutter [18].

The idea of CAPTCHAs has also been turned into another useful purpose, namely to help to digitize old printed material from books that computerized optical character recognition failed to recognize. This technique is known as reCAPTCHA [27]. Interestingly, it has reported that this method can transcribe text with a word accuracy exceeding 99.99% matching the guarantee of professional human transcriber [27].

2   CAPTCHA Revisited

Formally, CAPTCHAs have been defined by von Ahn et al. [26] as follows.

"A CAPTCHA is a cryptographic protocol whose underlying hardness assumption is based on an Artificial Intelligence problem."

When the underlying Artificial Intelligence (AI) problem is useful, a CAPTCHA implies an important situation, namely either the CAPTCHA is broken and there is a way to differentiate humans from computers, or the CAPTCHA is broken and a useful AI problem is solved [26].

2.1   Definitions and Notation

The following definitions and notation are adapted and simplified from [26]. Intuitively, a CAPTCHA is a test V where most humans have success close to 1, while it is hard to write a computer program that has overwhelming probability of success over V. That means, any program that has high probability of success over V can be used to solve a hard AI problem. In the following, let C be a probability distribution. If $P(\cdot)$ is a probabilistic program, let $P_r(\cdot)$ denote the deterministic program that results when P uses random coins r.

Definition 1. [26] A test V is said to be (α, β)-human executable if at least an α portion of the human population has success probability greater than β over V.

Definition 2. [26] An *AI problem* is a triple $\mathcal{P} = (S, D, f)$ where S is a set of problem instances, D is a probability distribution over S and $f : S \rightarrow \{0,1\}^*$ answers the problem instances. Let $\delta \in (0,1]$. For $\alpha > 0$ fraction of the humans H, we require $Pr_{x \leftarrow D}[H(x) = f(x)] > \delta$.

Definition 3. [26] An AI problem $\mathcal{P}$ is said to be (ψ, τ)-solved if there exists a program $\mathcal{A}$ that runs in time for at most τ on any input from S, such that

$$Pr_{x \leftarrow D,r}[\mathcal{A}_r(x) = f(x)] \geq \psi.$$

Definition 4. [26] An (α, β, η)-CAPTCHA is a test V that is (α, β)-human executable and if there exists $\mathcal{B}$ that has success probability greater than η over V to solve a (ψ, τ)-hard AI problem $\mathcal{P}$, then $\mathcal{B}$ is a (ψ, τ) solution to $\mathcal{P}$.

Definition 5. An (α, β, η)-CAPTCHA is *secure* iff there exists no program $\mathcal{B}$ such that

$$Pr_{x \leftarrow D,r}[\mathcal{B}_r(x) = f(x)] \geq \eta$$

for the underlying AI problem $\mathcal{P}$.

3   Review on 3D Stereoscopy

Stereoscopy relates to the perception of depth in the human visual system that arises from the horizontal separation of our eyes by the interocular distance (distance between the eyes) [5]. In real life, this results in our visual cortex being presented with two slightly different views of the world. When viewing a 3D scene, binocular disparity refers to the difference in the images that are projected onto the left and right eye retinas, then onto the visual cortex [17]. The human visual system perceives the sensation of depth through a process known as stereopsis, by using binocular disparity to obtain depth cues from the 2D images that are projected onto the retinas.

Though a variety of 3D display devices have been developed over the years, in this study we only concern ourselves with stereo pair based technologies. A stereo pair is a set of two images, one created for the left eye and the other for the right eye. Stereo pair based technologies simulate binocular disparity based on the presentation of the different images to each of the viewer's eyes independently [17]. By synthetically creating and presenting two correctly generated images of the left and right views of a scene, the visual cortex will fuse the images as it does in normal viewing to give rise to the sense of depth [4].

However, it is important to note that there are a variety of other depth cues that the human brain can infer from a 2D image, e.g. perspective (objects further away from the viewer look smaller), occlusion (where closer objects block objects that are further away), shading and shadows, etc. Depth cues are generally additive, in other words the more the better. Therefore, it is important for the depth cues to be consistent and to avoid conflicting depth cues in the generation of the stereo pair [5,17].

3.1   Stereo Pair Generation

To generate the stereo pair, two camera/eye viewpoints are used to create the left and right images by horizontally displacing the cameras by an appropriate eye separation. Stereo pairs that are not created correctly will make viewing very uncomfortable or the brain might not even fuse the images at all resulting in the viewer seeing two separate images. Therefore a number of factors have to be considered in practice when attempting to generate the stereo pair, so as to not overwhelm the visual system. For example, eye separation that is set to be too large results in a condition known as hyperstereo, and although this exaggerates the stereo effect the brain might find it hard to fuse the images. Another consideration particularly relevant to our study where we attempt to clutter the CAPTCHA with noise, is that if the frequency of the noise is too high, there will essentially be little matching visual information for the brain to resolve between the images in the stereo pair [4].

Parallax refers to the signed distance on the projection plane between the projected positions of a point in the stereo pair. Parallax is a function of the depth of a point in eye space [11]. A point in space that is projected onto the projection plane can be classified as having one of three relationships: zero parallax, positive parallax and negative parallax. Note that these refer to the horizontal distance on the projection plane as the vertical parallax should always be zero, otherwise the user will generally suffer from uncomfortable physical symptoms from misaligned cameras. While the amount that can be tolerated will vary from viewer to viewer, adverse side effects include headaches, eye strain, and in severe cases even nausea [17].

Zero parallax occurs when the projected point is on the projection plane. The pixel position of the projected point is exactly the same position on both left and right images. This is depicted in Figure 2(a) from a top-down view. As illustrated in Figure 2(b), positive parallax occurs when the projected point is located behind the projection plane.

In this case, the pixel position of the projected point is located on the right in the right image and on the left in the left image. To the observer, the point appears at a depth 'into' the screen. The maximum possible positive parallax is equal to the eye separation and arises when the point is located at infinity. Figure 2(c) depicts negative parallax which occurs when the projected point is located in front of the projection plane. When this happens, the pixel position of the projected point is located of the left in the right image and on the right in the left image. The observer perceives the point as coming 'out' of the screen [11].

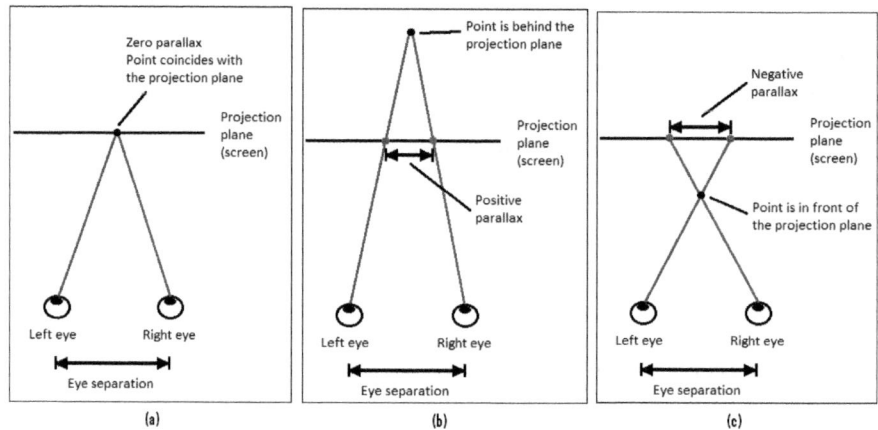

Fig. 2. Parallax

3.2 Stereoscopic 3D Display Technologies

Stereo pair based technologies require a method of ensuring that the left eye only sees the left eye's image and the right eye only sees the image for the right eye. There are a variety of methods that have been developed to achieve this. Here we highlight a number of stereo pair based technologies of relevance to our study, a comprehensive overview of 3D display technologies can be found in [17].

A common technique used in a number of stereoscopic display devices is to alternate the display of left and right views on a single display. These techniques require the viewer to use equipment such as viewing glasses to prevent the left eye from seeing the right view and vice versa. These can either be active or passive viewing glasses. Active systems employ blocking lenses which synchronize with the display to alternately cause the left and right eye lenses to become opaque, thereby blocking the respective eye's view. On passive systems, the display device produces polarized light where left and right eye images are polarized in orthogonal directions. The viewing glasses for these systems have similarly polarized lenses for each eye that only allows through light that is polarized along an axis parallel to the respective eye [17]. Either of these systems is suitable for STE3D-CAP.

Unlike the previous approaches, in the anaglyph method the viewer is not presented with alternate left and right views independently. Rather, both views are presented to the viewer simultaneously on a single image, where the left and right eye views are color encoded using two colors. The viewer has to wear glasses with red/green filters (or similar red/cyan, red/blue, etc.) which filter out colors with certain frequencies for each eye [5]. Some of the drawbacks of this approach include the lack of representation of the full range of color, and this approach typically suffers from a lot of cross-talk (this means that a portion of the view intended for one eye is visible to the other eye, resulting in what is known as 'ghosting'). However, this presents a low-cost solution.

There are other stereoscopic display technologies that do not require special viewing glasses, these devices are called autostereoscopic. These devices use a variety of approaches such as lenticular sheets and parallax barriers, which are designed to focus and redirect light to different viewing regions causing the viewer to perceive a different image for each eye. Such display devices are increasingly gaining popularity and a number of companies have recently announced the launching of their glass-free 3D devices [20]. STE3D-CAP can be used on any of these systems as long as they are stereo pair based.

4 Design and Implementation of STE3D-CAP

In this section, we will describe the design of our new CAPTCHA, stereoscopic 3D CAPTCHA, or STE3D-CAP for short. To describe it precisely, we will proceed with presenting the underlying AI problem and then commence with the detail of our design and implementation.

STE3D-CAP is a CAPTCHA that are built using the stereoscopic 3D technology. The idea is to present CAPTCHA text as 3D objects that will be easily identified by humans (who are equipped with the proper equipment), but they are hard to be analyzed by machines. For the implementation of our idea, we use existing NVIDIA technology that requires us to supply two images (2D) that represent the 3D object for the left and right eyes, resp. Then, the NVIDIA card will render the two images and present the 3D objects.

STE3D-CAP have several attractive features:

- Humans can solve it quickly (§4.1).
- Computers cannot solve it easily (§4.1).
- STE3D-CAP is easy to generate as it is a text-based CAPTCHA.
- STE3D-CAP uses the latest technology.
- STE3D-CAP uses 3D, and hence, more noise can be added in the 3D scene while the resulting CAPTCHA is still usable.
- STE3D-CAP is a variable length CAPTCHA, which are more difficult to defeat.
- STE3D-CAP is built in a 3D environment, and therefore more distortion can be added to the CAPTCHA (eg. negative/positive parallax).

Nevertheless, STE3D-CAP has several disadvantages:

- STE3D-CAP requires special-type of equipment, namely equipment to display 3D[5].

---

[5] Even though specialized 3D displays are preferable, for practical applications, the anaglyph approach can be used. Moreover, the requirement for dedicated hardware in new CAPTCHA techniques has never been an issue (e.g. physical CAPTCHAs) [14].

- STE3D-CAP challenges may require more screen space than traditional text-based CAPTCHAs.
- Like virtually all other CAPTCHAs, STE3D-CAP is not accessible to those with visual impairments including those who are stereo-blind.

It should be noted that although STE3D-CAP is currently text-based, it can easily be extended to use models of other 3D objects instead of only 3D text. Nevertheless, we choose to focus on using a text-based CAPTCHA approach for reasons outlined in [8]; namely, that text characters were designed by humans for humans, humans have been trained to recognize characters since childhood, text-based CAPTCHA tasks are easily understood by users without much instruction and that each character has a corresponding key on the keyboard which gives rise to many possible input combinations.

An interesting approach to confuse segmentation attacks would be to randomly interleave 3D models among the 3D text. A human user would clearly be able to distinguish between the text and the random objects. However, it would make segmentation and recognition harder for a computer which cannot easily differentiate between the text and objects.

In addition, we use random character strings in STE3D-CAP rather than dictionary words. While the use of words from a dictionary will probably make the text in STE3D-CAP easier to perceive and has implications on the security, we avoid this as it unfairly disadvantages people unfamiliar with the chosen language.

4.1   New AI Problem Family

In this section, we introduce a family of AI problems that will be used to build our CAPTCHA, STE3D-CAP. An image is defined as an $h \times w$ matrix (where h stands for height and w stands for width), whose entries are pixels. A pixel is defined as a triplet (R, G, B), where $0 \leq R, G, B \leq M$, for a constant M.

Let $\mathcal{I}_{2d}$ be a distribution on images, $\mathcal{I}_{3d}$ be a distribution on stereoscopic 3D images and $\mathcal{T}$ be a distribution on stereoscopic image transformations, that include rotation, scaling, translation and warp. Let Ω be a distribution on noise frequency, and Υ be a distribution on erosion factors. Let $\mathcal{C} : \mathcal{I}_{3d} \times \Omega \times \Upsilon \to \mathcal{I}_{3d}$ be a distribution of clutter functions. A clutter function is a function that accepts a 3D image, a noise frequency $\in \Omega$ and an erosion factor $\in \Upsilon$ and outputs a cluttered 3D image. Let $|A|$ denote the cardinality of A.

Let $\Delta : |\mathcal{I}_{3d}| \to \mathcal{I}_{3d}$ be a lookup function that maps an index in $|\mathcal{I}_{3d}|$ and output a stereoscopic 3D image in $\mathcal{I}_{3d}$. Let $\vartheta : \mathcal{I}_{2d} \times \mathcal{I}_{2d} \to \mathcal{I}_{3d}$ be a function that maps two images (left and right images) to a stereoscopic image. Subsequently, let $\vartheta^{-1} : \mathcal{I}_{3d} \to \mathcal{I}_{2d} \times \mathcal{I}_{2d}$ be a function that given a stereoscopic image, outputs two images that represent left and right images, resp. Subsequently, we also assume that ϑ is publicly available. For simplicity, we denote the left and right images with superscript L and R, resp.

For clarify, for the rest of this paper, we will use **Roman boldface** characters to denote elements of $\mathcal{I}_{3d}$, while Sans Serif characters to denote elements of $\mathcal{I}_{2d}$.

Problem Family ($\mathcal{P}$STE3D-CAP).
Consider the following experiment.

1. Randomly select $i \in |\mathcal{I}_{3d}|$.
2. Compute $\mathbf{i} \leftarrow \Delta(i)$.
3. Select a transformation $t \leftarrow \mathcal{T}$.
4. Compute $\bar{\mathbf{i}} \leftarrow t(\mathbf{i})$.
5. Select a clutter function $c \leftarrow \mathcal{C}$.
6. Compute $\mathbf{j} \leftarrow c(\bar{\mathbf{i}}, \omega, \upsilon)$, where $\omega \in \Omega$ and $\upsilon \in \Upsilon$ are selected randomly.
7. Output $\vartheta^{-1}(\mathbf{j})$.

The output of the experiment is $(j^L, j^R) \leftarrow \vartheta^{-1}(\mathbf{j})$, where $(j^L, j^R) \in \mathcal{I}_{2d} \times \mathcal{I}_{2d}$.

$\mathcal{P}$STE3D-CAP is to write a program that takes $(j^L, j^R) \in \mathcal{I}_{2d} \times \mathcal{I}_{2d}$ as input and outputs $i \in |\mathcal{I}_{3d}|$, assuming the program has precise knowledge of $\mathcal{T}, \mathcal{C}$ and $\mathcal{I}_{2d}$.

More formally, let

$$S_{\mathcal{I}_{2d}, \mathcal{T}, \mathcal{C}} = \{\vartheta^{-1}\left(c\left(t(\Delta(i)), \omega, \upsilon\right)\right) = (j^L, j^R) : t \leftarrow \mathcal{T},$$
$$c \leftarrow \mathcal{C}, \omega \in \Omega, \upsilon \in \Upsilon, (j^L, j^R) \in \mathcal{I}_{2d} \times \mathcal{I}_{2d}\}$$

Let $D_{\mathcal{I}_{2d}, \mathcal{T}, \mathcal{C}}$ be the distribution of $S_{\mathcal{I}_{2d}, \mathcal{T}, \mathcal{C}}$ that are obtained from executing the above experiment, and

$$f_{\mathcal{I}_{2d}, \mathcal{T}, \mathcal{C}} : S_{\mathcal{I}_{2d}, \mathcal{T}, \mathcal{C}} \rightarrow |\mathcal{I}_{3d}|$$

such that $f_{\mathcal{I}_{2d}, \mathcal{T}, \mathcal{C}} = i, i \in |\mathcal{I}_{3d}|$. Then,

$$\mathcal{P}\text{STE3D-CAP} = (S_{\mathcal{I}_{2d}, \mathcal{T}, \mathcal{C}}, D_{\mathcal{I}_{2d}, \mathcal{T}, \mathcal{C}}, f_{\mathcal{I}_{2d}, \mathcal{T}, \mathcal{C}}).$$

Hard Problem in $\mathcal{P}$STE3D-CAP.
We believe that $\mathcal{P}$STE3D-CAP contains a hard problem. Given $\mathcal{P}$STE3D-CAP $= (S_{\mathcal{I}_{2d}, \mathcal{T}, \mathcal{C}}, D_{\mathcal{I}_{2d}, \mathcal{T}, \mathcal{C}}, f_{\mathcal{I}_{2d}, \mathcal{T}, \mathcal{C}})$, for any program $\mathcal{B}$,

$$Pr_{x \leftarrow D_{\mathcal{I}_{2d}, \mathcal{T}, \mathcal{C}}, r} [\mathcal{B}_r(x) = f(x)] < \eta.$$

Based on this hard problem, we can construct a secure (α, β, η)-CAPTCHA.

Theorem 1. A secure (α, β, η)-CAPTCHA can be constructed from $\mathcal{P}$STE3D-CAP as defined above.

Proof. We will provide the proof in two stages. First, we show that (α, β, η)-CAPTCHA is (α, β)-human executable. Then, we show that (α, β, η)-CAPTCHA is hard for a computer to solve. We also show an instantiation of our proof.

Given $\mathcal{P}$STE3D-CAP, humans can get the instance $(j^L, j^R) \leftarrow \vartheta^{-1}(\mathbf{j})$, where $(j^L, j^R) \in \mathcal{I}_{2d} \times \mathcal{I}_{2d}$. Then, by executing $\vartheta(j^L, j^R)$ humans can easily see the instance of the problem and output i. We should justify that in practice, $\vartheta(\cdot)$ is implemented in a 3D hardware capable of displaying 3D, such as an NVIDIA card [21]. Executing $\vartheta(j^L, j^R)$ means that humans will be able to see the objects provided clearly using the required equipments (such as 3D glasses), and humans can output i easily. Hence, (α, β, η)-CAPTCHA is (α, β)-human executable.

However, given $\mathcal{P}_{\mathsf{STE3D\text{-}CAP}}$, machines cannot output i. Although $\vartheta(\mathsf{j}^L, \mathsf{j}^R)$ is available publicly, machines cannot "view" the 3D representation of i and hence, cannot output i easily. The best way to analyze the problem is by processing $(\mathsf{j}^L, \mathsf{j}^R)$ directly, which will not help machines to identify i. Hence, $Pr_{x \leftarrow D_{\mathcal{I}_{2d}, \mathcal{T}, c}, r}[\mathcal{B}_r(x) = f(x)] < \eta$, for any $\mathcal{B}$.

An in-depth security analysis on $\mathcal{P}_{\mathsf{STE3D\text{-}CAP}}$ will be provided in §5.

4.2 Design Principles of STE3D-CAP

1. Differences between left and right images Intuitively, the first design principle requires that the "difference" between the left and the right images must be sufficiently noisy to ensure that segmentation attacks will fail. The idea is to ensure that **j** will be clearly visible for humans to identify i, while machines observing $(\mathsf{j}^L, \mathsf{j}^R)$ cannot deduce i. Note that j^L is the 2D version of the image from the left eye's perspective and j^R is from the right eye's perspective. While the 3D version of i in both j^L and j^R must exist, the noise in both images can be made vary since the noise visible from the left eye maybe blocked and invisible from the right eye, and vice versa. Suppose we define that $\mathsf{j}^L = \bar{t}(\mathsf{j}^R) + \delta$, where $\mathcal{T}$ is a distribution on 2D image transformation (eg. translation) and $\bar{t} \leftarrow \mathcal{T}$. We require δ to be sufficiently noisy to deter against segmentation attacks. Formally, we have the following theorem.

Theorem 2. An (α, β, η)-CAPTCHA constructed from $\mathcal{P}_{\mathsf{STE3D\text{-}CAP}}$ is secure against segmentation attacks, *iff* for $\delta = \mathsf{j}^L - \bar{t}(\mathsf{j}^R)$, where $\mathcal{T}$ is a distribution on 2D image transformation and $\bar{t} \leftarrow \mathcal{T}$, we require that there exists no program $\mathcal{B}$ such that

$$Pr_{x \leftarrow D_{\mathsf{STE3D\text{-}CAP}_{\mathcal{I}_{2d}, \mathcal{T}, c}}, r}[\mathcal{B}_r(x) = i] \geq \eta.$$

2. Human Factors Although we vary a number of parameters in STE3D-CAP in order to make it less predictable and harder for computers to perform automated attacks, a number of human factors issues had to be kept in mind. This is because our aim is for a human user to be able to use STE3D-CAP comfortably.

In terms of stereoscopic viewing, it is generally easier on the eyes to view objects that are at screen depth or objects that appear into the screen, even though this does not result in the pop out of screen effect. Caution must be exercised for objects in front of the screen as parallax diverges quickly to negative infinity for objects closer to the eyes [11]. Focusing on objects that are positioned too close to the eyes forces the viewer's eyes to cross at a point in front of the screen, which can be very strenuous on the eyes. The focal length refers to the distance at which objects in the scene will appear to be at zero parallax. In general, objects should not be positioned closer than half the focal length, in other words negative parallax should not exceed the eye separation [4].

In addition, eye separation must also be kept within a reasonable range. Ideally, eye separation should be made as large as possible. This way the images for the left and right views will be rendered from very different angles and some information present in one image might be missing from the other and vice versa, as shown in Figure 3(b). Therefore, information will only be complete if both images are viewed together. However, it must be kept in mind that too large an eye separation will lead to hyperstereo

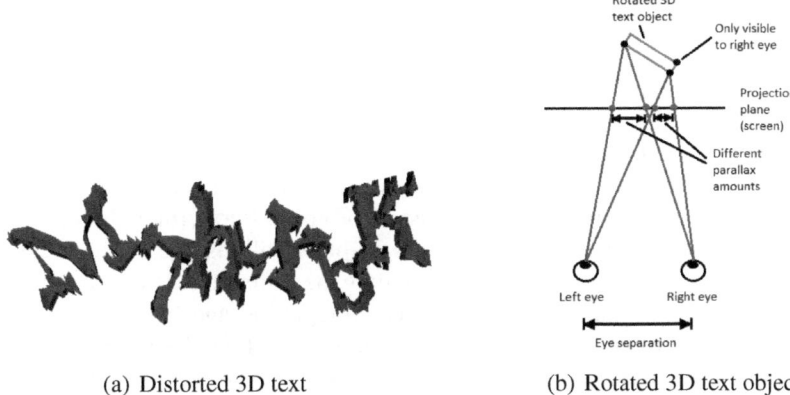

(a) Distorted 3D text   (b) Rotated 3D text object

Fig. 3.

which is uncomfortable for users, whereas a value which is too small results in hypostereo. In hypostereo, the user does not really perceive a 3D effect and furthermore the images in the stereo pair will not differ by much.

4.3   Implementation

We implemented STE3D-CAP by using a graphics system to render models of 3D text objects against foreground and background clutter. To develop and view STE3D-CAP, we used an NVIDIA 3D Vision kit with an Alienware™OptX AW2310 3D monitor. To avoid user eye strain, perspective camera properties, view frustums, etc. were set up using the general rules of thumb. For quantitative guidance, please refer to [4] and [11].

Customizable vertex and fragment shaders were used to give rise to random vertex perturbations and erosion effects. The vertices of the text models were randomly perturbed in 3D to distort the text. Figure 3(a) gives a depiction of this for the left eye. Please note that all STE3D-CAP images in this paper were generated using the same character string as shown in Figure 3(a) for comparison sake.

Individual 3D text objects also undergo random 3D transformations. Unlike conventional 2D CAPTCHAs, we can add more variation to the transformations as we are now dealing with 3D. Translation is not merely left/right and up/down but can also be varied with respect to depth from screen. Similarly rotation is not restricted to being clockwise and counter clockwise, as we can also rotate the 3D text objects about the vertical axis as well as the horizontal axis pointing to the right. Rotation must be limited to be within a certain range, otherwise the text may not be readable if slanted at angles which are too steep. We chose a conservative rotation range of between +/- 20 degrees for the rotation axes parallel to the projection plane and +/- 45 degrees clockwise/counter clockwise. By rotating the text objects the parallax of the projected points on the object will also be different in screen space. This is illustrated in Figure 3(b). We also randomly scale the 3D text objects to alter their size.

To increase the difficulty of segmentation attacks we adopted the "crowding characters together" method (letting characters touch or overlap with each other) which is suggested to be segmentation resistant [30], and we also attempted to clutter the scene with noise. Our implementation allows us to adjust the frequency of the noise. However, as noted in Section 3.1 high frequency noise makes it difficult for the brain to correlate matching visual information and subsequently makes it hard to fuse the stereo pair. Also, if the noise is too fine, it will be easy to differentiate the text from the noise by simply removing small individual clusters of noise. Furthermore, completely random noise that appears in one image but not the other will also be hard to fuse, and at the same time easy to filter out by simply finding the differences between the stereo pair.

Instead, in our implementation we use foreground and background surfaces with randomly perturbed vertices and eroded surface sections base on a 3D Perlin noise function. The surfaces' vertices were perturbed to avoid completely smooth surfaces which will be easy to filter out between stereo pair images. We can also adjust the scale and amount of erosion. In addition, we made the foreground clutter slightly translucent for usability reasons, so that the text behind it will not be completely obscured. In this manner, the clutter does not appear as random noise, but rather from the user's point of view looks like two eroded surfaces, and they can perceive the text amidst the surfaces. Figure 4[6] shows an example of such a stereo pair. It can be seen that one cannot use the images individually to complete the CAPTCHA challenge.

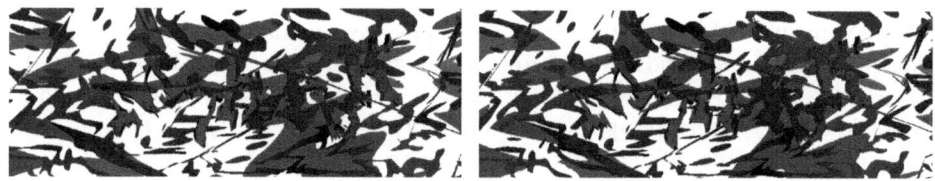

Fig. 4. Example of a STE3D-CAP stereo pair

At the moment, color is used in STE3D-CAP merely from a usability standpoint to improve the attractiveness of the CAPTCHA rather than for any particular security reason. This is because even though it will be much easier to see the text if it was highlighted with a different color from the clutter, this would also make it very easy to filter out the clutter by just separating the text based on color. Furthermore, if we introduced a lot of random colors to improve security, it would make the stereo pair very hard to fuse. Moreover, an automated attack could simply convert the images to greyscale and attempt to threshold the intensity levels rather than dealing with the color. The colors in STE3D-CAP were deliberately made to overlap with the clutter to make it harder for automated attacks whilst still being usable.

Despite STE3D-CAP being rendered in 3D, confusing character combinations still had to be avoided as highlighted in [30]. For example, a distorted 'vv' might look like a 'w', etc.

---

[6] Though we do not recommend this, it is possible to see the 3D text by crossing one's eyes à la magic eye images.

5   Security of **STE3D-CAP**

Before describing the security of **STE3D-CAP**, it is useful to review the threat model associated with CAPTCHAs, as CAPTCHAs are an unusual area of security where we are not trying to provide absolute assurances, but rather to merely slow down attackers. CAPTCHAs are considered to be "successful" if they force an automated attack to cost more than approximately 30 seconds worth of a human's time in part of the world where labor is cheap [10]. It is generally acceptable if CAPTCHAs can admit bots with less than 1/10,000 probability [25].

When considering the security of **STE3D-CAP**, we provide the adversary with (j^L, j^R) instead of j, due to the following reason. First, although the input for humans is j, machines cannot view 3D objects like humans. Therefore, it would be easier for machines to be provided with (j^L, j^R) instead. Second, the function φ is available publicly. This function will transform $j \leftarrow \varphi(j^L, j^R)$. This assumption is very reasonable as the implementation is on the client's system. Therefore, the machine adversary can also make use of this function whenever it is deemed necessary. Third, since the 3D view is generated by the client's machine, the two images (j^L, j^R) will need to be sent to the client's machine. Therefore, even though the view that is shown in the client's machine is j, it is reasonable to assume that the adversary can capture both (j^L, j^R) by observing the TCP/IP packets.

5.1   Single Image Attacks

Any of the existing CAPTCHA attacks can be attempted on the left and right images of **STE3D-CAP** individually. However, unlike existing 2D text-based CAPTCHA approaches which cannot be overly cluttered in order to maintain usability (which makes them more susceptible to segmentation attacks), it is possible to increase the foreground and background clutter in **STE3D-CAP**. This will increase **STE3D-CAP**'s security against segmentation attacks, while at the same time when **STE3D-CAP** is viewed in 3D, the viewer can still perceive the text and differentiate this against the clutter. Figure 5(a) was obtained by passing the left image through a Sobel edge detection filter. While it highlights the edges in the image, it does not give rise to much useful information.

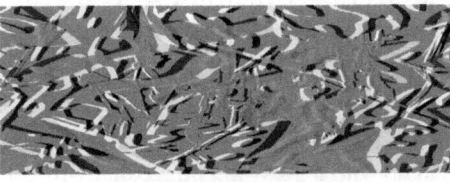

(a) Edge detection image   (b) Difference image

Fig. 5.

5.2   2D Image Difference Attacks

Unlike the existing attack models in the literature, we introduce a new type of attack namely 2D image difference attacks. In this type of attack, an adversary who is given a pair of 2D images, (j^L, j^R) will first find the difference between these two. This relies on the fact that $j^L = \bar{t}(j^R) + \delta$, where $\bar{\mathcal{T}}$ is a distribution on 2D image transformation (eg. translation) and $\bar{t} \leftarrow \bar{\mathcal{T}}$. The adversary will need to find the appropriate $\bar{t}'$. Then, subsequently, the adversary will try to eliminate δ, which is $\delta = j^L - \bar{t}(j^R)$. By eliminating δ, the leftover image will then be analyzed using the existing segmentation techniques, such as [29], to identify the segments and break the CAPTCHAs. To demonstrate this, Figure 5(b) depicts the difference image[8] between left and right views. Sections in white are in the left image but not in the right, whereas black represents sections in the right image but not in the left, and grey shows overlapping sections. It can be seen that little useful information can be gathered from the image. Figure 6 is the anaglyph version of Figure 1. For usability and ease of use of **STE3D-CAP**, one can view Figure 6 using a low cost red-cyan anaglyph glasses. Note however that the anaglyph version will look slightly different compared to when using appropriate stereoscopic devices, as one can see greater variation in the depth of the characters using the latter approach.

Fig. 6. Anaglyph

Theorem 3. An (α, β, η)-CAPTCHA constructed from $\mathcal{P}_{STE3D\text{-}CAP}$ as defined above is secure against 2D image differences attacks.

Proof. Our (α, β, η)-CAPTCHA has been designed according to the first design principle in §4.2. This means δ has been chosen such that it will be sufficient to deter against segmentation attacks. Hence, an adversary launching 2D image differences attacks will end up with a noisy 2D image that will not represent mere the object i. Therefore, image segmentation attacks on the newly developed image will not be able to extract i.

---

[7] We note that this action can be done trivially by comparing the two images (j^L and j^R).

[8] This image is obtained by taking the left image minus the right image, and scaled between black and white.

Subsequently, we will also obtain the following theorem.

Theorem 4. An (α, β, η)-CAPTCHA constructed from $\mathcal{P}_{\text{STE3D-CAP}}$ as defined above is secure against segmentation attacks.

5.3 Brute Force Attacks

A straightforward attack on STE3D-CAP is brute force. In this attack, the adversary will just provide a random solution to challenges until one succeeds. This means, given a STE3D-CAP challenge, (j^L, j^R), the adversary will find a random solution for it. Note that STE3D-CAP are variable length CAPTCHAs. Hence, the adversary has no knowledge on the length of the challenge. Suppose the length of the challenge is λ, and there are 62 possible characters comprising lower and upper case letters and digits. Then, the chance of successful brute force attacks is $\frac{1}{62^\lambda}$. In practice, this chance can be considered as negligible, especially when CAPTCHAs are combined with techniques such as token bucket algorithms [10] to combat denial-of-service attacks.

Theorem 5. An (α, β, η)-CAPTCHA constructed from $\mathcal{P}_{\text{STE3D-CAP}}$ as defined above is secure against brute force attacks.

5.4 3D Reconstruction Attacks

While one may be able to approximate the reconstruction of the 3D scene from the stereo pair, this still leaves the problem of how to separate the 3D text from the 3D clutter which is difficult due to the different parallax. In addition, because of the characters are touching/overlapping this still leaves the problem of segmenting to individual characters. In short, even if one can successfully remove the clutter this will reduce to the difficulty of segmentation attacks which forms the basis of security for existing 2D CAPTCHA approaches. Furthermore, human visual perception of 3D scenes is still an open research problem that cannot easily be modeled.

6 Applications

The incorporation of STE3D-CAP in an application requires that the application be usable with a stereoscopic 3D display. Two current areas are increasingly moving toward the use of 3D displays; namely, 3D games and 3D movies. STE3D-CAP can be included seamlessly into applications like 3D Massively Multiplayer Online Games (MMOGs). These are online games that support multiple players who interact in the same shared virtual space. Many of these games are already being developed or modified to cater for stereoscopic 3D displays. An example is the popular World of Warcraft™[2]. The use of CAPTCHAs in these applications will help deter the use of bots to gain an unfairly advantage over other players and ruin the fun for other players [14].

The number of applications that use stereoscopic 3D displays will certainly increase with more and more companies currently developing and producing glass-free 3D display devices [20]. A number of web pages already contain anaglyph flash and java applets, while others provide 3D images and videos [1]. Web browser plugins are currently being developed to be able to display stereoscopic 3D images and videos on web pages [28], and this will certainly become more and more widespread. In that respect, stereoscopic 3D CAPTCHAs is anticipated to be the way of the future. While stereoscopic displays have yet to proliferate, the existing solution is to adopt the low-cost anaglyph approach as a drop-in replacement for current CAPTCHAs on web pages.

7   Conclusion and Further Work

In this paper we presented a new stereoscopic 3D CAPTCHA, called STE3D-CAP, which attempts to overcome limitations with existing 2D approaches. We demonstrated that STE3D-CAP is resistant against the existing 2D CAPTCHA attacks. Our approach has opened a new research direction to incorporate CAPTCHA challenges in 3D scenes.

Our approach also gives rise to the possibility of producing animated 3D CAPTCHAs where either the camera's viewpoint is translated in 3D or the scene is moved with respect to the camera. The differences between animated frames will give rise to 3D depth perception of the text in the scene via motion parallax, where objects at a distance appear to move slower compared to objects what are close to the viewer. Furthermore, as the camera moves from one position to another, the 3D text which might have been obscured in one frame will become visible in another frame. In order to break this CAPTCHA, one would have to somehow correlate the content between frames whilst attempting to separate the 3D text from the background and foreground clutter, which is not an easy task. This approach will work on standard displays and can be incorporated into web pages as animated Graphics Interchange Format (GIF) images.

References

1. Anaglyph flash gallery, http://www.3dmix.com/eng/flash-gallery/
2. Activision Blizzard. World of Warcraft, http://www.worldofwarcraft.com/
3. Bigham, J.P., Cavender, A.C.: Evaluating existing audio CAPTCHAs and an interface optimized for non-visual use. In: Proceedings of the 27th International Conference on Human Factors in Computing Systems, pp. 1829–1838 (2009)
4. Bourke, P.: Calculating stereo pairs, http://local.wasp.uwa.edu.au/~pbourke/miscellaneous/stereographics/stereorender/
5. Bourke, P., Morse, P.: Stereoscopy: Theory and Practice. In: Workshop at the 13th International Conference on Virtual Systems and Multimedia, VSMM 2007 (2007), http://local.wasp.uwa.edu.au/~pbourke/papers/vsmm2007/stereoscopy_workshop.pdf
6. Chellapilla, K., Larson, K., Simard, P., Czerwinski, M.: Building Segmentation Based Human-friendly Human Interaction Proofs. In: Baird, H.S., Lopresti, D.P. (eds.) HIP 2005. LNCS, vol. 3517, pp. 1–26. Springer, Heidelberg (2005)
7. Chellapilla, K., Larson, K., Simard, P., Czerwinski, M.: Computers beat humans at single character recognition in reading based human interaction proofs. In: 2nd Conference on Email and Anti-Spam (2005)

8. Chellapilla, K., Larson, K., Simard, P., Czerwinski, M.: Designing human friendly human interaction proofs (HIPs). In: Proceedings of the SIGCHI Conference on Human Factors in Computing Systems, pp. 711–720 (2005)
9. Chew, M., Tygar, J.D.: Image Recognition CAPTCHAs. In: Zhang, K., Zheng, Y. (eds.) ISC 2004. LNCS, vol. 3225, pp. 268–279. Springer, Heidelberg (2004)
10. Elson, J., Douceur, J.R., Howell, J., Saul, J.: Asirra: A CAPTCHA that Exploits Interest-Aligned Manual Image Categorization. In: Proceedings of the 14th ACM Conference on Computer and Communications Security (ACM CCS 2007), Conference on Computer and Communications Security, pp. 366–374 (2007)
11. Gateau, S.: Th. In: and Out: Making Games Play Right with Stereoscopic 3D Technologies. NVIDIA presentation, Game Developers Conference (2009)
12. Gizmodo: Sony plans to introduce 3D LCD television by the end (2010),
    `http://gizmodo.com/5350607/`
    `sony-plans-to-introduce-3d-lcd-television-by-end-of-2010`
13. Golle, P.: Machine Learning Attacks Against the Asirra CAPTCHA. In: Proceedings of the 14th ACM Conference on Computer and Communications Security (ACM CCS 2008), Conference on Computer and Communications Security, pp. 535–542 (2008)
14. Golle, P., Ducheneaut, N.: Preventing bots from playing online games. Computers in Entertainment (CIE) 3(3), 3 (2005)
15. Google Images, `http://images.google.com`
16. Jakobsson, M.: Captcha-free throttling. In: Proceedings of the 2nd ACM Workshop on Security and Artificial Intelligence, pp. 15–22 (2009)
17. McAllister, D.: 3D Displays. Wiley Encyclopedia on Imaging, pp. 1327–1344 (2002),
    `http://research.csc.ncsu.edu/stereographics/wiley.pdf`
18. Mitra, N.J., Chu, H.-K., Lee, T.-Y., Wolf, L., Yeshurun, H., Cohen-Or, D.: Emerging images. ACM Transactions on Graphics (TOG) 28(5) (December 2009)
19. Mori, G., Malik, J.: Recognizing objects in adversarial clutter: Breaking a visual CAPTCHA. In: Conference on Computer Vision and Pattern Recognition (CVPR 2003), pp. 134–144 (2003)
20. Murph, D.: Intel shows off glasses-free 3D demo (2010),
    `http://www.engadget.com/2010/01/10/intel-shows-`
    `off-glasses-free-3d-demo-now-this-is-more-like-it/`
21. NVIDIA. NVIDIA 3D Vision,
    `http://www.nvidia.com/object/3D_Vision_Main.html`
22. OCR Research Team. Teabag 3D Revolution,
    `http://www.ocr-research.org.ua/teabag.html`
23. Ross, S., Chen, T.L.: The Effects of Promoting Patient Access To Medical Records. Journal of American Medical Informatics Association 10, 129–138 (2003)
24. Simard, P., Steinkraus, D., Platt, J.C.: Best practices for convolutional neural networks applied to visual document analysis. In: International Conference on Document Analysis and Recognition, pp. 958–962 (2003)
25. von Ahn, L., Blum, M., Hopper, N.J., Langford, J.: The CAPTCHA web page,
    `http://www.captcha.net`
26. von Ahn, L., Blum, M., Hopper, N.J., Langford, J.: CAPTCHA: Using hard AI problems for security. In: Biham, E. (ed.) EUROCRYPT 2003. LNCS, vol. 2656, pp. 294–311. Springer, Heidelberg (2003)
27. von Ahn, L., Maurer, B., McMillen, C., Abraham, D., Blum, M.: reCAPTCHA: Human-Based Character Recognition via Web Security Measures. Science 321(5895), 1465–1468 (2008)

28. VREX. DepthCharge V3 Browser Plug-In,
    http://www.vrex.com/depthcharge/
29. Yan, J., Ahmad, A.S.E.: A Low-cost Attack on a Microsoft CAPTCHA. In: Proceedings of
    the 14th ACM Conference on Computer and Communications Security (ACM CCS 2008),
    Conference on Computer and Communications Security, pp. 543–554 (2008)
30. Yan, J., Ahmad, A.S.E.: Usability of CAPTCHAs - Or Usability issues in CAPTCHA design.
    In: Symposium on Usable Privacy and Security (SOUPS) 2008, pp. 44–52 (2008)
31. YUNiTi, http://www.yuniti.com

TRIOB: A Trusted Virtual Computing Environment Based on Remote I/O Binding Mechanism

Haifeng Fang[1,2], Hui Wang[3], Yiqiang Zhao[1], Yuzhong Sun[1], and Zhiyong Liu[1]

[1] Key Laboratory of Computer System and Architecture, Institute of Computing Technology, Chinese Academy of Sciences
[2] Graduate University of Chinese Academy of Sciences
[3] High Performance Computer Research Center, Institute of Computing Technology, Chinese Academy of Sciences
{fanghaifeng,wanghui}@ncic.ac.cn, {zhaoyiqiang,yuzhongsun,zyliu}@ict.ac.cn

Abstract. When visiting cloud computing platforms, users are very concerned about the security of their personal data. Current cloud computing platforms have not provided a virtual computing environment which is fully trusted by users. Meanwhile, the management domain of cloud computing platform is subject to malicious attacks, which can seriously affect the trustworthiness of the virtual computing environment. This paper presents a new approach to build a trusted virtual computing environment in data centers. By means of three innovative technologies, the user's data can be remotely stored into trusted storage resources, the user's virtual computing environment is isolated, and the user can automatically detect the rootkit attacks against the cloud computing management domain. We design and implement a Xen-based prototype system called TRIOB. This manuscript presents the design, implementation, and evaluation of TRIOB, with a focus on rootkits detection.

Keywords: virtual computing environment, remote I/O binding, virtual machine isolation, rootkits detection.

1   Introduction

Cloud computing has become an increasingly popular computing solution. Using VMM technologies such as Xen[1], and VMware, in the cloud computing platform people can dynamically create a large number of virtual machines (VMs) to meet customers' requirements. For example, through IaaS[2] services (like Amazon's EC2[15]), users can get their own exclusive VMs for running a number of security-sensitive applications. However, from the users' point of view, data security and privacy is still an unmet concern, which is currently the major obstacle to the development of cloud computing[3].

Presently, Xen, an open source virtualization system software, is widely used in data centers. In Xen system, the virtual machine (VM) is called Domain. In order to provide a management entrance for system administrator, after Xen

S.-H. Heng, R.N. Wright, and B.-M. Goi (Eds.): CANS 2010, LNCS 6467, pp. 241–260, 2010.

started, it immediately creates a unique privilege VM (called Domain0) to run
the VM management program. Domain0 provides interface for the administrator to create, delete, start and stop multiple user VMs (called DomainU), and
allocate resources to the users' VMs. In Xen-based virtual computing environment, the trustworthiness of DomainU depends not only on its internal software
configuration, but also on the internal software in Domain0. This is decided by
the device driver model of Xen (as illustrated in Fig. 1)[1].

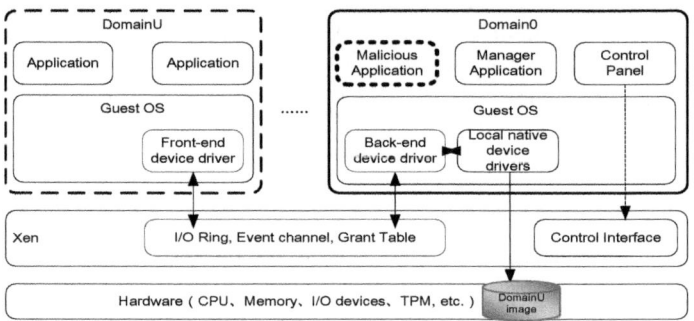

Fig. 1. The architecture of Xen and splitting device driver model

However, there are still some security weaknesses in Xen-based virtual computing environment as follows.

First, the code size of the whole trusted computing base (TCB) of the virtual
computing environment keeps increasing. To solve this problem, Xen introduces
a new domain called Isolated Driver Domain (IDD)[29]. In this way, Domain0 is
split away, having some device drivers in Domain0 encapsulated in a separated
IDD. This not only reduces the code size of the guest OS kernel in Domain0,
but also improves the device driver's reliability. However, in the current Xen
architecture, it is very common that cloud providers still take Domain0 as the
IDD. Besides, the operating system codes related to the management applications inside Domain0 could not be completely separated away. Consequently, we
need to dynamically monitor them to ensure their integrity.

Second, the assumption that Domain0 of each server in data centers is trusted
may not always be true. There exist several external or internal weakness in Do-
main0, e.g., wrong configurations and software bugs, which can compromise the
trustworthiness of Domain0. It is common in the current virtualized environment
to have a management console (like XenCenter, HyperVM) that manages DomainUs through the control interface (like XenAPI) in Domain0. While these consoles
help the administrators to manage the machines, they open new vulnerabilities,
e.g., the XenAPI HTTP interface has a cross-site scripting vulnerability[31]. Compromising a management console allows an attacker to control all the VMs managed by it. For instance, on June 8th 2009, 100,000 hosted websites were affected
by a zero-day SQL injection hole in the HyperVM 2.09[32]. With this attack, the
intruders gained the root privileges to perform malicious operations (like installing
rootkits) in Domain0.

Finally, malicious applications may be able to affect the data integrity of DomainU through Domain0. Most of the management tasks are executed in Domain0, including the I/O processing, Domain management, etc. After DomainU is created, Domain0 can read and write the memory space that belongs to DomainU through one management interface (xc_map_foreign_range). Unfortunately, malicious applications can also run in Domain0, as we mentioned. Once these applications control the interface, it will seriously affect the integrity of DomainU. To address this issue, Xen developers introduced one new concept called DomB[13]. With this new addition, the interface can only be called by the DomB in which there only deployed some special programs. Although the interface can be protected by the special Domain, sometimes DomainU need to close down the xc_map_foreign_range interface for security purpose, but it can not do this currently. Therefore, we should re-arrange the privileges amongst Xen, Domain0 and DomainU to meet the users'security requirements.

Currently, through the virtual trusted platform model (TPM)[8], we can establish the trust connection between Domains belonging to different management fields. But the TPM technology can only ensure the Domains' configuration integrity at startup[9][21], and it cannot guarantee the runtime integrity of the processes running in the Domains when they are attacked by malicious programs like rootkits. Existing research work [16][17][18] mainly enhance the trustworthiness of DomainU by means of "out-of-the-box" monitoring technologies. However, because the trustworthiness of DomainU also depends on Domain0, we must ensure that Domain0 is in a trusted running state[34][35][36]. Therefore, no matter how Domain0 was split, the integrity of the processes in Domain0 needs to be monitored, otherwise DomainU will be in an untrustworthy state.

Based on the above analyses, we designed and implemented a new trusted virtual computing environment (called TRIOB) for data center with some new trust-enhancing techniques. In this virtual computing environment, users can specify the resource configuration and the running mode of DomainU in which users' applications are deployed. In detail, by means of the new remote I/O binding technique, DomainU can be bound with the remote storage resources located in the users' physical machines outsides cloud computing platform. During users' applications are running, DomainU can be in a new trusted running mode to absolutely isolate from other DomainUs and Domain0 in data center. Meanwhile, the integrity of the Domain0 is measured dynamically and remotely verified by users.

We have tested the TRIOB prototype system. Our experimental results show that compared to the traditional remote I/O binding technique, the TRIOB's performance is improved. At the same time, under the introduction of the trust-enhancing mechanisms, the TRIOB system can timely detect the rootkit attacks against the Domain0, while the overhead remains in an acceptable range for users.

The rest of this paper is organized as follows: Section 2 describes the background information. Section 3 gives a specific application scenario, and discusses the existing challenges. In Section 4, we give the architecture of prototype system and its work principle. Section 5 introduces an implementation of the prototype system in detail. Section 6 is the prototype evaluation. Some related works are introduced in Section 7. Finally concluding remarks are presented in Section 8.

2 Background

2.1 Splitting Device Driver Model

The unique splitting device driver model (illustrated in Fig. 1)[1] is the core feature of Xen. Applications in DomainU access I/O devices through the front-end device driver which provides virtual I/O devices for DomainU, such as virtual block device. Whenever these applications produce I/O operations, the front-end device driver receives these I/O requests. Then these I/O requests are forwarded to the back-end device driver in Domain0 through the communication mechanisms (I/O ring, event channel, grant table) between VMs. After the back-end device driver receives these I/O requests, it checks whether the requests are legitimate and then calls the local native device drivers to perform real I/O operations. When the I/O operations are completed, the back-end device driver will notice the front-end device driver which reports to the applications in DomainU.

With the splitting device driver model, DomainU just works with virtual I/O devices, and it does not need to concern with the physical I/O devices' type and location. For example, a VM can be configured with some virtual CPUs and virtual memory which derive from the local physical node in data center, and virtual storage devices which derive from other non-local physical nodes' storage devices. In this situation, the VM can use local resources for running applications, and transparently store data into the other remote physical nodes. However, Xen currently only supports this idea indirectly based on traditional technologies such as NFS, NBD[10][11], and it does not consider the relevant trustworthiness issues that may arise because of introduction of some remote I/O binding in the current model.

2.2 Kernel-Level Rootkit Attacks against Domain0

To ensure the trustworthiness of Domain0, we need to analyze the possible attacks against Domain0. Currently, the guest OS running in Domain0 is Xeno-Linux (Linux's modified version running on Xen). Various rootkit attacks, especially the kernel-level ones, are the most serious threats for Linux operating system[19]. Unfortunately, due to the back-end device drivers locating within the Linux kernel in Domain0, the kernel-level rootkit attacks against Domain0 will also affect DomainU.

In general, rootkits are collections of programs that enable attackers who have gained administrative control of a host to modify the host's software, usually causing it to hide their presence from the host's genuine administrators[27]. Linux-oriented kernel-level rootkit usually enters into the kernel space through accessing /dev/kmem (or /dev/mem), or it is loaded into kernel space as a kernel module (LKM). After the rootkit enters into the kernel space, it can make damages as follows. 1) Modifying the kernel text code, such as the virtual file system, the device drivers; 2) Modifying the kernel's critical data structures such as the system call table and the interrupt table. From users' point of view, it is necessary to monitor the integrity of the kernel code, the key kernel data structures and other critical kernel memory space in Domain0, so that users can timely make awareness of the exception in Domain0 and take effective countermeasures.

3   Scenario and Challenges

Typically, a trusted computing environment should enable users to know its software configuration and runtime state. In addition, while the computing environment is under damage, it should provide the mechanism to protect users' sensitive data[4].

To build such a Xen-based trusted virtual computing environment, we suppose that the users need to get a VM from the IaaS service in cloud computing platform. But, they do not trust the cloud computing platform, especially its internal storage resources. Meanwhile, they only want to lease the virtual computing environment (that is DomainU in Xen) to run some security-sensitive applications. In this scenario, users can customize the lent DomainU with a special recourse configuration in which its virtual storage device is bound with the users' storage resources. From user's point of view, the actual storage resource remotely locates in user's local physical machine.

At first, we assume that Xen is trustworthy because its code size is very small and the Domain in running state has almost no power to destroy Xen[30]. Note that this assumption is consistent with that of many other VMM-based security research efforts [5][14]. Then, the virtual computing environment for user needs to undergo two stages as follows.

(1) Starting DomainU. User needs to specify the DomainU's resource configuration (like the location information of the remote storage resources) through the service interface of the cloud computing platform. The information will be sent to Domain0 which is responsible for creating the corresponding DomainU. When starting DomainU, user needs to get the configuration metric information of DomainU, Domain0 and Xen to determine whether he can establish a trusted connection with them. By means of TPM technology, we can achieve this goal in that we do not focus on this stage in this paper.

(2) Running user's applications in DomainU. These applications are installed in user's storage resources, so user can believe that DomainU is trustworthy. When these applications access data from user's remote storage devices, the I/O requests in DomainU will be forwarded to the guest OS kernel in Domain0. However, currently user cannot control Domain0 from DomainU. According to the preceding analysis, at this stage, Domain0 should be excluded from the TCB.

Based on the above analysis, the Xen-based virtual computing environment is trustworthy only if it has two basic properties: trusted storage resources and secure runtime computing environment (we call it TRIOB-DomainU). To achieve these goals, we need to overcome the following challenges.

First, how to bind TRIOB-DomainU with the remote user's storage resources? By using NFS technology, we can mount the remote storage image file onto Domain0. But this solution is not secure because Domain0 can directly access the content inside the mounted image file. Meanwhile, it would be better that the remote I/O binding mechanism can match with the splitting device driver model and facilitate with TRIOB-DomainU for trusty purpose. Thus, we need to design a new secure remote I/O binding mechanism for TRIOB-DomainU, by which users only need to provide the location information of the remote image file.

Second, how to allow user to control the runtime state of TRIOB-DomainU? DomainU should be absolutely isolated from other VMs, so that the user believes his applications running in a secure space. As mentioned before, we need to re-divide the privileges amongst Xen, Domain0 and the DomainU. For example, after DomainU is created, user can close up the xc_map_foreign_range interface to prohibit Domain0 from mapping the memory of the DomainU. Most importantly, DomainU should own the power of monitoring some memory regions of Domain0 to check its integrity. At the same time, we should not allow attacker to damage Domain0 by means of the privileges of the DomainU. That is to say, DomainU can only specify the location and type of mapped memory area in Domain0, and can only obtain the integrity information related to the memory area. On the other hand, the memory area that can be read by DomainU must be verified by Xen so that some valuable contents in Domain0 are protected by Xen. The most important thing is that the proposed mechanism needs to ensure that only the memory area that mapped to the DomainU can be accessed by the user.

Finally, how to allow user to be aware of the exceptions, e.g., the attacks against VM's kernel, in cloud computing platform? According the security requirements of TRIOB-DomainU, user hopes to perceive Domain0 kernel's exception in time and make the appropriate treatments. Meanwhile, in data center there are many DomainUs serving to different users and the users would take care of different security issues. It is obvious that someone needs to dynamically monitor and measure the memory of Domain0 to check its integrity. But, who should be responsible for this task? Is TRIOB-DomainU or Xen? We believe that Xen is better because in Xen space we can easily monitor any memory area belonging to different Domains including Domain0. Furthermore, who is responsible for the verification task? For the special scenario above, it is the user who should do this work. Especially, user needs to determine whether the I/O data produced from TRIOB-DomainU could be stored into user's storages based on the runtime state of Domain0.

To address the above challenges, we propose a new approach to construct a trusted virtual computing environment and we call it TRIOB system. Its architecture is shown in Fig. 2.

4   Architecture of TRIOB

The TRIOB system consists of two parts, the computing environment running in data center and the storage resource locating in the user's physical node. In

Fig. 2, the left side is the computing environment (TRIOB-DomainU) which provides a trusted running environment for user's applications, and the right side is the user's physical machine which provides reliable storage resources. In TRIOB-DomainU, the storage resources are some VM image files (that is, DomainU's image). From the functionality viewpoint, the TRIOB system includes three subsystems, that is, the remote I/O subsystem, the mode-control subsystem and the dynamic monitoring and measurement subsystem. The three subsystems can help overcome the challenges respectively.

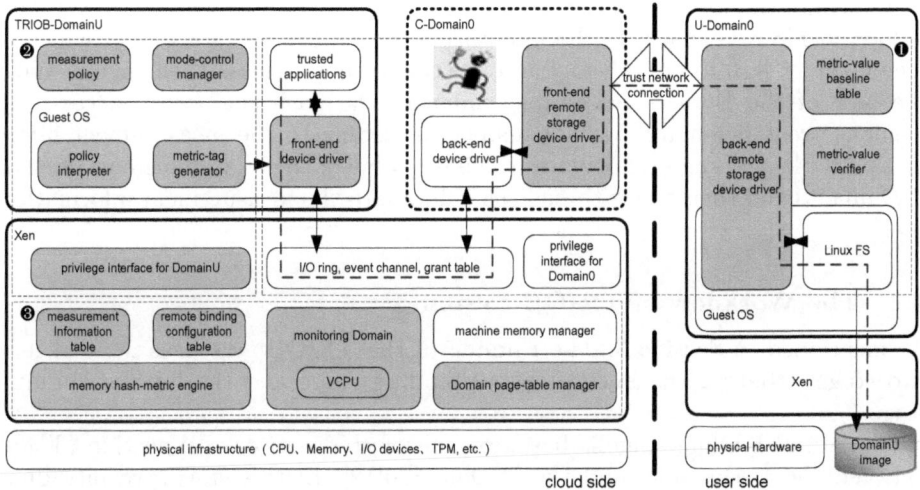

Fig. 2. Architecture of TRIOB system (①remote I/O subsystem; ②mode-control subsystem; ③dynamic monitoring and measurement subsystem)

The remote I/O subsystem consists of the front-end remote storage device driver, the back-end remote storage device driver, and the associated remote verification modules. The front-end remote storage device driver is responsible for transferring the I/O requests produced by applications to the back-end remote storage device driver through network. If TRIOB-DomainU is in the trusted running mode, each I/O request will be attached with the tag which contains the metric-hash value of C-Domain0. In the user side, when the back-end remote storage device driver receives the I/O requests, it will first send the tag to the metric-value verifier who is responsible for the verification based on the metric-value baseline. Only these I/O data belonging to trusted I/O requests can be stored into the user's local storage resource, that is, the DomainU's image. To enhance the security of the remote I/O channel, we can encrypt the corresponding I/O data in TRIOB-DomainU and decrypt them in user side.

The mode-control subsystem consists of the mode-control manager, the measurement policy, the policy semantic interpreter, the metric-tag generator and the privilege interface for DomainU. When security-sensitive applications are running in TRIOB-DomainU, the mode-control manager can switch DomainU into the trusted mode through calling the privilege interface. The user can specify the policy what content of kernel memory in C-Domain0 to be measured and how to measure. Then the policy interpreter is responsible for translating the policy and registering this information into the measurement information table and the remote binding configuration table, respectively. The privilege interfaces for DomainU mainly include running mode interface, measurement interface, memory mapping interface, etc.

The dynamic monitoring and measurement subsystem consists of the monitoring Domain, the memory hash engine, the domain page-table manager, etc. The monitoring Domain is a special Domain which is started together with Xen and hidden inside Xen space. In the monitoring Domain, there is a VCPU which is scheduled by Xen to periodically activate the memory metric-hash engine. With the help of the Domain page-table manager, the hash engine can access any memory areas belonging to C-Domain0. According to the measurement information table, the engine calculates the hash value of the related memory area of C-Domain0 and then stores these hash values into the measurement information table.

4.1   The Workflow of TRIOB System

As mentioned in Section 3, The running of the TRIOB system is divided into two stages, that is, the initial trusted binding stage and the trusted running stage.

The initial binding stage is based on the trusted network connection (TNC) between the C-Domain0 and the U-Domain0[12]. In this stage, we introduce a protocol to ensure the trusted exclusive connection between the TRIOB-DomainU and remote storage resources. The protocol is listed as follows.

1) *U-Domain0 sends message to C-Domain0. The message contains image configuration, U-Domain0's public encryption key (Kg);*
2) *C-Domain0 stores the message into both configuration file for DomainU and the remote binding configuration table in Xen space;*
3) *According to the remote binding configuration table, Xen generates a random symmetric encryption key (Ku) used to encrypt I/O requests;*
4) *C-Domain0 creates and starts TRIOB-DomainU based on the configuration file;*
5) *When C-Domain0 loads the guest OS in TRIOB-DomainU, Xen checks whether the guest OS matches with the requirements based on the remote binding configuration table. For example, the version is correct or not;*
6) *The Monitoring-Domain in Xen measures the hash value (Mu) of the guest OS kernel in TRIOB-DomainU and stores Mu into the measurement information table in Xen;*
7) *TRIOB-DomainU initializes the virtual storage device, gets Ku from the remote binding configuration table, and gets Mu from the measurement information table;*
8) *TRIOB-DomainU sends message to U-Domain0. The message contains the command of binding image, EKg (Ku), Mu, image configuration, etc.;*

9) *U-Domain0 gets the message. Then it decrypts **Ku** using its private key (**Kp**), and checks whether **Mu** matches with the user's requirement;*

10) *If all above steps are passed through, U-Domain0 completes the binding process.*

It can be proved that the binding protocol is a one-to-one exclusive binding protocol. First, we ensure the one-to-one trusted connection between the two Domain0s by means of TPM/TNC specifications. Secondly, Xen is responsible for verifying the loaded guest OS kernel in TRIOB-DomainU, and generating **Ku** which cannot be obtained by C-Domain0. Furthermore, in the end of the binding process, user can verify the configuration of TRIOB-DomainU.

In the trusted running stage, the main task is to ensure TRIOB-DomainU to be in an absolutely close state, so that the I/O data produced by applications are securely stored into remote storage resources.

According to the users' requirements, TRIOB-DomainU can work in two modes, namely, the normal mode and the trusted mode. In the normal mode, TRIOB-DomainU is an ordinary DomainU which can be controlled by C-Domain0 as usual. In the trusted mode, C-Domain0 cannot control TRIOB-DomainU. In addition, the I/O data is attached with the metric-hash tag. In the trusted mode, C-Domain0 cannot transparently read or write the memory of TRIOB-DomainU so that TRIOB-DomainU is isolated from C-Domain0 and other DomainUs.

In the trusted mode, when the users run security-sensitive applications in TRIOB-DomainU, the processing steps of TRIOB are listed at follows:

1) *User logs on to TRIOB-DomainU by means of some remote access protocol software such as ssh, then configures the measurement policy;*

2) *The policy interpreter translates the measurement policy and registers this information into the measurement information table in Xen;*

3) *User switches TRIOB-DomainU into the trusted mode;*

4) *User starts the trusted applications;*

5) *The VCPU in monitoring Domain is periodically scheduled, then the VCPU activates the memory metric-hash engine which calculates hash of the memory areas in C-Domain0 according to the measurement information table;*

6) *The applications produce I/O operations which are received by the front-end device driver;*

7) *The metric-tag generator gets the metric information from the measurement information table to generate the metric-tag, then it sends the tag to the front-end device driver;*

8) *The front-end device driver attaches the tag onto the block I/O requests and then forwards the block I/O requests to the back-end device driver;*

9) *The front-end remote storage device driver receives these I/O requests and further transfers them to back-end remote storage device driver;*

10) *The back-end remote storage device driver receives these I/O requests. Then it separates the metric-tag and I/O data from these I/O requests, and sends them to the metric-value verifier and Linux file system, respectively;*

11) *The metric-value verifier checks the metric-tag based on the metric-value baseline table;*
12) *If the metric value matches with the metric-value baseline, the I/O data can be stored into the DomainU image; otherwise, the metric-value verifier will report an error alarm.*

In the normal mode, user can run some ordinary applications in TRIOB-DomainU. We do not elaborate the scenario in this paper.

5   Implementation

Currently, we have implemented the TRIOB prototype based on Xen 3.1 and Linux 2.6.18. Our implementation is described in this section.

5.1   Trusted Remote I/O Processing

In Xen, the front-end virtual block device driver in DomainU is "blkfront" module and the back-end virtual block device driver is "blktap" module[16]. To solve the first challenge, we firstly expand the blktap module and add two new modules (that is, remote-blkfront and remote-blkback) so that the block I/O request can be transferred to the remote storage through network. At the same time, we modify the blkfront module so that the metric-hash tag can be attached onto per block I/O request. Fig. 3 shows the detail internal structure.

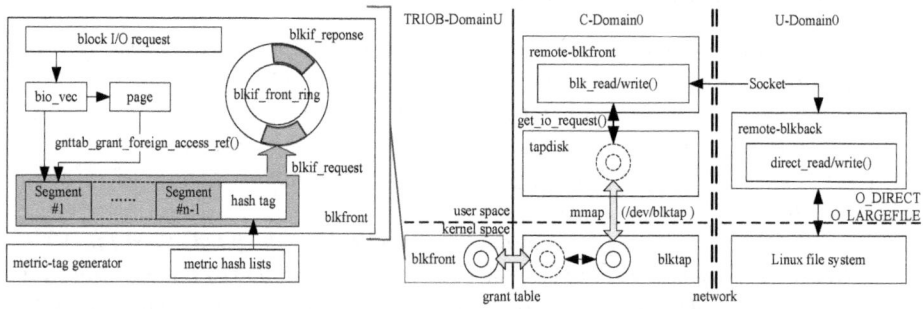

Fig. 3. Transfer the I/O requests attached with tag to remote storage in user's physical machine

As shown in the right part of Fig. 3, the user space part (tapdisk) of blktap maps the I/O ring memory to the user space through mmap interface. In the new remote-blkfront module, *blk_read* calls the *get_io_request* function to get the I/O requests from the user space I/O ring. In the other side, the new remote-blkback module, which has a socket connection with the remote-blkfront module, receives the I/O requests. The remote-blkback module can read or write the image with O_DIRECT and O_LARGEFILE operational mode. According to the blkif_request in I/O ring, the remote-blkback can quickly locate the block data in the image so as to ensure the I/O processing efficiency.

The I/O ring is a data structure shared between Domain0 and DomainU which contains the I/O requests. For virtual storage device, the I/O ring is called blkif_front_ring. When the blkfront module receives the block I/O request, it will interpret the request and re-organize it. In detail, it will translate the bio_vec into the blkif_request which encapsulates some segments. Amongst these segments, we select the last segment to save the metric hash value (called hash tag). The last segment is made by metric-tag generator which will get the corresponding metric hash values of some kernel memory areas in C-Domain0 through the privilege hypercall for TRIOB-DomainU.

5.2 The Privilege Interface for DomainU and Policy Interpreter

To solve the second challenge, firstly, we add a new hypercall which can only be called by TRIOB-DomainU. Through passing different command parameter for the hypercall, user can own some privileges as follows.

Command	Functions
DOMU_switch_to_trusted_mode	TRIOB-DomainU in trusted mode
DOMU_switch_to_normal_mode	TRIOB-DomainU in normal mode
MONITOR_checking_register	Registering metric-related policy
MONITOR_checking_get_digests	Getting metric-hash value

Secondly, we modify the xc_map_foreign_range interface. There is a lot of work based on this interface, and the most famous one is XenAccess[16]. In the TRIOB system, we modify the interface to ensure that, when TRIOB-DomainU is in the trusted mode, the DomainU can map the physical memory area in Domain0 into its virtual address space, at the same time prohibits C-Domain0 to do the same thing.

The xc_map_foreign_range interface triggers Xen's memory management functions by accessing the character driver (/proc/privcmd). We modify the ioctl function's subroutine (IOCTL PRIVCMD MMAP, IOCTL PRIVCMD MMAP-BATCH) and add the privilege checking logic. The Domain can access the interface only if it matches one of two conditions 1) the mapping Domain is the initial Domain (like Domain0), and the mapped Domain is running in the normal mode; 2) the mapping Domain is not the initial Domain (like TRIOB-DomainU), and it is running in the trusted mode.

XenAccess currently can only run in Domain0. Based on the modified xc_map foreign_range interface and with some minor patches, XenAccess can also run in DomainU. In the TRIOB system, we rename the modified XenAccess as policy interpreter (as shown in Fig. 4).

The policy interpreter gets the physical-to-machine (P2M) table of Domain0 by accessing the modified xc_map_foreign_range interface, and calculates the starting machine address of the kernel page table of Domain0 by referring to the P2M table. According to the measurement policy, the policy interpreter reads the "System.map" file to look up the corresponding starting virtual address of the measured memory areas in Domain0. Then by means of the P2M table and the kernel page table, it can look up the corresponding page table entries (PTE) from which it can get the starting machine address (start mfn) of the measured memory areas, and then it registers these addresses into the measurement information table by calling the new hypercall with "MONITOR_checking_register" command.

5.3   The Memory Hash-Metric Engine, the Domain Page-Table Manager and the Monitoring Domain

The memory hash-metric engine, the Domain page-table manager and the monitoring Domain are all implemented in Xen space for two reasons. On the one hand, due to the trustworthiness of Xen, we can ensure these components are trust. On the other hand, these components need to access any memory areas belonging to C-Domain0, which is easy to do in Xen space. All of the three components cooperate together to solve the third challenge.

The memory hash-metric engine is implemented as a set of hash functions which can calculate the digest of the machine memory area. Meanwhile, we choose the MD5 hash algorithm to calculate the digest. When it is activated, the engine computes the hash of the machine memory area specified in the measurement information table and saves the hash value into the table again.

Through periodically activating the engine, the functions within the engine can be dynamically executed in Xen space. However, currently Xen does not support the concept like kernel thread in Linux OS. We found that there is an idle_domain within Xen space. The idle_domain manages all of the idle_vcpu, and these idle_vcpus and the VCPUs belonging to all other Domains are scheduled by the Xen scheduler. When the idle_vcpu is scheduled, some function codes in Xen space will run on this VCPU. Based on this idea, we introduce a new Domain (called monitoring Domain) in Xen space. The monitoring Domain has three features as follows. First, its data structure is not embodied in global domain management list, so that it is hidden in Xen space. Secondly, it shares the page table (idle_pg_table) with Xen so that it can call all of functions within Xen. Finally, it only creates one VCPU which is similar to the idle_vcpu, but the VCPU's priority is the same with the general VCPU so that it can timely activate the memory hash-metric engine.

In the measurement information table, there are some items related to machine memory regions for which the engine needs to calculate the corresponding hash value. However, the engine cannot access them directly in Xen space currently.

On the 32-bit x86 physical platform, the virtual address space of Xen in non-PAE mode covers only 64MB memory, which means that Xen can only access a maximum 64MB of machine memory for one time. To access the machine memory belonging to C-Domain0, we add a simple Domain memory mapping module (called Domain page-table manager). Now, we illustrate the way how to access the kernel machine memory of C-Domain0 as follow (Fig. 4 gives the details).

First, as mentioned in Section 5.2, we can get PTE or the starting machine address (start mfn) of the kernel memory region in C-Domain0 and the length of the kernel memory area, from the measurement information table.

Then, we allocate one-page memory from Xen heap space by calling the *alloc_xenheap_page* function. The purpose of allocating the page is to temporary use its virtual address region occupied by the page. By the *virt_to_mfn* macro, we can translate the page's starting virtual address into the corresponding starting machine page frame number (mfn). Meanwhile, by traversing the kernel page table (idle_pg_table) of Xen, we get the corresponding PTE of the page.

Finally, the Domain page-table manager temporarily replaces the PTE for Xen heap page memory with the PTE for kernel memory in C-Domain0 and refreshes TLB. Then the memory remapping work is completed. From now on the engine can directly access the kernel memory in C-Domain0. If the measured memory area is larger than one page, we can repeat the process for several times. We do not tell the way how to recover the original mapping in this paper.

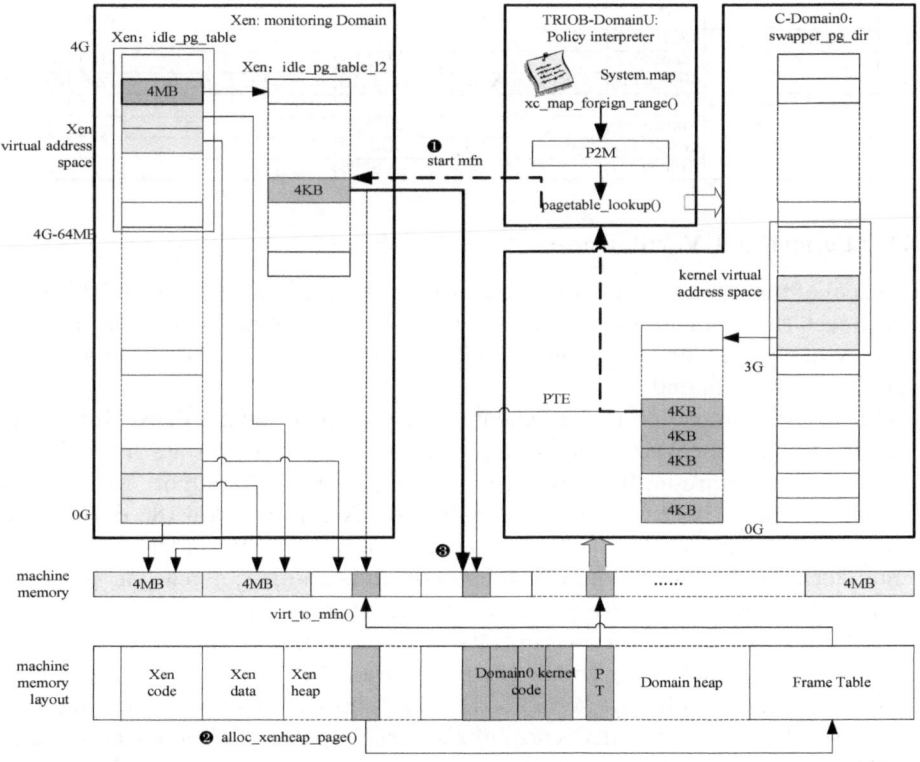

Fig. 4. Accessing the memory in C-Domain0 from Xen space (①getting the start machine page number (start mfn) and the PTE of machine memory belonging to kernel code; ②allocating the temporary machine memory from Xen heap space; ③replacing the PTE belonging to Xen heap page with the PTE belonging to kernel machine page)

6   Evaluation

Currently, our research group has built a virtual computing platform (called TRainbow[12]) for data center. TRainbow provides the IaaS service and TRIOB is its key component. According to the previous scenario, we get a high performance physical node from the TRainbow platform for the cloud side. Meanwhile, we also choose one lower performance node for the user side. The two machines are connected via Gigabit Ethernet network. The detail configuration of the experiment environment is shown in Table 1.

Table 1. Experiment environment configuration

		Cloud side	User side
Physical Machine	CPU	Intel Xcon E5410 2.33GHz	AMD Athlon2200+ 1.800GHz
	Core	8	2
	Cache	6144 KB	256 KB
	Memory	16GB	1GB
	Network	Gigabit Ethernet Controller	Gigabit Ethernet Controller
	Disk	SCSI 2*512GB	IDE 36GB
	VMM	Xen 3.1 (modified)	Xen 3.1
Domain0	VCPU	1	1
	Memory	512MB	512MB
	Guest OS	FC 6 XenoLinux 2.6.18	FC 6 XenoLinux 2.6.18
TRIOB-DomainU	VCPU	1	
	Memory	256MB	
	Guest OS	XenoLinux 2.6.18(modified)	
	Image	8GB	

6.1   Functional Verification

The goal of functional testing experiment is to check whether the TRIOB system can detect the change of the integrity when the kernel in C-Domain0 is damaged by attacker. Currently, the system focuses on rootkits attacks against the XenoLinux in Domain0.

Our system prototype is implemented on the XenoLinux 2.6.18. During the experiment, we discovered that many of the well-known rootkits listed in Table 2, such as adore-ng 0.56, lvtes, override, phalanx-b6, cannot, without additional changes, be compiled or installed on XenoLinux 2.6.18. Besides, we cannot find other suitable rootkits. Through deeply analyzing the principle of Linux kernel-level rootkits and Xen, we conclude the major reasons as follow. First, the "/dev/mem" driver's *mmap* interface is re-implemented in Xen, as a result, traditional mem-type rootkits (like phalanx-b6) cannot easily locate the memory area belonging to the kernel. Second, Xen is the only piece of code running in ring 0, while the kernel of Domains runs in less privileged ring (ring 1 in case of x86_32). So, in XenoLinux's kernel space, traditional kernel-level rootkits cannot directly execute the privilege operations such as setting control registers. Third, most LKM-type rootkits are based on the symbols exported by the kernel, for example, the sys_call_table, as a way to hook its own code to it. But, since Linux version 2.6, some critical kernel symbols are not exported any longer and even protected in read-only memory area, which improves the system hardening process.

Therefore, we port these rootkits to the XenoLinux 2.6.18. For mem-type rootkits we resort to the C-Domain0's "System.map" to find the locations of the attacked memory area[33]. Based on this approach, we successfully install phalanx-b6. For LKM-type rootkits we need to bypass the memory protection. Traditional kernel-level rootkits can directly clear the WP bit of CR0, while XenoLinux's kernel prevents it from happening. However, there is a hypercall (HYPERVISOR update_va_mapping) by which rootkits can make the read-only memory area to be writable. Based on the hypercall, we successfully install those LKM-type rootkits (like adore-ng 0.56, lvtes, override). Meanwhile, we also implement a typical kernel-level rootkit (called hack_open) for XenoLinux 2.6.18. The rootkit is a kernel module which can modify the system call table and some other kernel memory areas. (The key function based on this hypercall is shown as follow.)

```
static int make_syscall_table_writable(unsigned long va)
{
        pte_t *pte;
        int rc = 0;
        pte = virt_to_ptep(va);
        rc = HYPERVISOR_update_va_mapping(
                (unsigned long)va, pte_mkwrite(*pte), 0);
        if (rc)
        {
                xen_l1_entry_update(pte, pte_mkwrite(*pte));
                return rc;
        }
        return rc;
}
```

Table 2 lists the public representative kernel rootkits which can be detected by the TRIOB prototype.

Table 2. Public representative kernel rootkits for Linux 2.6 and detection results

rootkit name	kernel version	loading mode (type)	Can run on XenoLinux 2.6.18?	Can be detected?	integrity verification report	
					modify syscall_table	modify kernel text
adore-ng-0.56	2.6.16	LKM	√ (ported)	√		√
lvtes	2.6.3	LKM	√ (ported)	√	√	
mood-nt	2.6.16	kmem	√	√	√	
override	2.6.14	LKM	√ (ported)	√	√	
phalanx-b6	2.6.14	mem	√ (ported)	√		√
suckit2priv	2.6.x	kmem	√	√	√	
hack_open	2.6.18	LKM	√	√	√	√

6.2   Performance Evaluation

To further assess the validity of the system, we measure its time performance index. We deploy a few typical application workloads into TRIOB-DomainU, and test theirs total execution time by using "time" command under different I/O binding mode. The results are shown in Table 3.

Table 3. The execution time of different application workloads under three binding modes

kernel-build			emacs			bzip2		
	Time(s)	% NFS		Time(s)	% NFS		Time(s)	% NFS
NFS	456.16		NFS	7.85		NFS	140.34	
RIOB	371.59	-18.54	RIOB	7.23	-7.90	RIOB	116.64	-16.89
TRIOB	523.57	14.78	TRIOB	8.09	3.06	TRIOB	155.24	10.62

We run three different typical application workloads under three different modes. Kernel-build is a typical I/O intensive and computation intensive application workload, and opening a large file with Emacs and then immediately closing it up is cache-sensitive, and decompressing the compressed Linux kernel source package file by bzip2 is a typical computation intensive workload.

RIOB means the DomainU (in the normal mode) bound with remote storage in the user's physical machine by our new I/O binding technology, and TRIOB represents the situation in the trusted mode. NFS means the DomainU (in the normal mode) bound with remote storage using NFS technology (version 4). As depicted in Table 3, the system's performance in the normal mode (RIOB) is better than NFS. Especially for I/O intensive workloads, the performance increases up about 19% than NFS. In the trusted mode (TRIOB) the performance lowers down about 15% than NFS under the I/O intensive workload, the overhead is significant but still in an acceptable range for users.

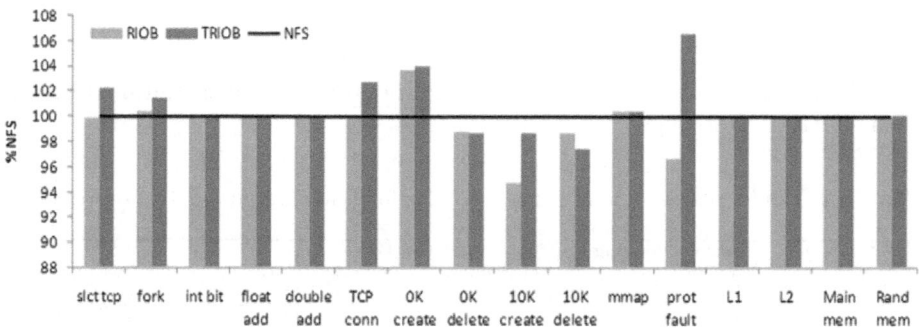

Fig. 5. Related lmbench results

In order to reduce the overhead, we run the **lmbench** benchmark suite in TRIOB-DomainU to help for system performance analysis. As shown in Fig.5, the overhead of TRIOB includes three parts, that is, process creation (slct tcp, fork), inter-process communication (TCP conn) and file system (file create, delete, protection fault). All of these factors are related to the "trusted I/O processing" subsystem, especially the metric-hash tag attached onto per block I/O request. In our current implementation, the tag occupied the location belonging to original block segment, which leads the times of I/O data transmission and the total transmission time to be extended. The time of accessing meta-data (related to "file create, protection fault") is critical to the overhead. For example, during compiling the Linux kernel, there will be a large number of small temporary files to be created. The time of "0KB file create" becomes longer which results in the total compile time longer. We will deal with this problem in future work.

7   Related Work

VPFS[20] proposes a virtual private file system built on L4. Its implementation is divided into two parts. The front-end part provides a secure computing environment for the applications, and the back-end part reuses the traditional file system in an untrusted computing environment. As the untrusted part controls the storage resources, VPFS modifies the traditional file system with encryption functions to ensure the trustworthiness of the whole system. In contrary, in TRIOB the users can directly control the storage resources so as to avoid the additional encryption overhead. Storage Capsules[21] provides a trusted computing environment based on VM hosted on PC. The VM can run in a trusted mode (disconnecting the network) to protect user's data from attacks. Similarly, TRIOB has a trusted mode, but it focuses on the cloud computing platform. Software-Privacy Preserving Platform[22], Overshadow[23] both assume the guest OS is not trustworthy, and they both provide similar mechanism to bypass the guest OS for trust application. These works only focus on ensuring the isolation amongst the applications within the same VM, while TRIOB assumes that the guest OS in Domain0 is not trustworthy, and provides a trusted virtual machine in cloud computing platform to protect users' applications and data.

Remote I/O Binding. Collective[24] provides a remote I/O binding technology by which user's notebook can access the virtual storage resources in cloud computing platform. In contrary, in TRIOB a user can bind the VM in cloud computing platform with the virtual storage resources in the user's physical machine. Netchannel[25] provides a remote back-end I/O device driver for DomainU to help for the VM migration. In the TRIOB system, the back-end remote storage device driver adopts similar technology with the I/O requests verification feature. NFS, NBD, iSCSI[10][11] and other traditional remote I/O technologies are widely used in data centers. They are implemented in virtual file system layer or block device driver layer of traditional OS. Due to the splitting device driver model, there are redundant layers in the implementation when they are used in virtual computing environment. In the TRIOB system the implementation of remote I/O binding makes better of the features of the splitting device driver model to bypass some layers, and the performance is better than these traditional technologies.

Dynamic Measurement. To enforce the trustworthiness of the virtual computing environment, researchers have introduced a series of trust-enhancing mechanisms into data center[5][7][9]. In this field, many works are based on the TCG's TPM technology. In Terra framework[5], VMs are divided into two types, that is, gray-box and white-box. In gray-box VM, users can determine the VM's trustworthiness by the way of remote verification based on TPM. However, the TPM-based measurement can only ensure a VM's integrity at startup. To further ensure the integrity of the computing platform at running time, researchers have bought up with many technologies related to dynamic measurement and verification of the integrity. Copilot[27] provides a dynamic memory measurement mechanism by means of a memory monitoring co-processor installed on the mainboard. In this system, the hardware can hash the memory area containing kernel text and other key components through direct memory access (DMA). In the TRIOB system, it does the similar work in Xen without adding special hardware. Pioneer[28] provides a dynamic trust root which is a running program. To verify the program's trustworthiness, pioneer periodically monitors the integrity and response time of the running program. Similarly, TRIOB dynamically monitors the integrity of Domain0 so as to determine the trustworthiness of the I/O data.

VM Monitoring. As virtualization technology is gradually mature, Tal Garfinkel and Mondel Rosenblum[14] propose the idea of VM introspection (VMI), an approach to intrusion detection which co-locates an IDS on the same machine as the host it is monitoring and leverages a VMM to isolate the IDS from the monitored host. In VMwatcher[18], the IDS system running in Domain0 monitors the memory in DomainU through xc_map_foreign_range interface. It can deduce the processes information in DomainU based on the data structure of task and sends this information to the anti-virus software. Lares[6] provides an active monitoring framework based on virtual machine architecture in which some hooks are inserted into the critical path in the kernel of the monitored VM. In TRIOB, we don't need to modify the kernel in the monitored VM. As for full virtual machine, by observing the related hardware behaves (like CR3 changing, TLB flush) of the process running in VM, Antfarm[17] can transparently guess the processes information. Being different from all these works, in TRIOB we propose a new function for VMI technology, that is, the monitored virtual machine can be a virtual machine (like Domain0) with higher privileges.

8   Conclusion

We introduced a new approach to build a trusted virtual computing environment, called TRIOB, that is geared towards data center protection and security. In TRIOB there are three key technical contributions to meet users' data security requirements. Firstly, user's data can be securely stored into theirs own storage resources through the trusted remote I/O binding technology. Second, the virtual computing environment for user is absolutely isolated from other VMs in cloud platform leveraging the trusted mode-control technique. This can ensure the integrity of user's computing environment. Finally, through the novel dynamic monitoring technique, user can timely detect the attacks against the administrator's computing environment. Dynamic monitoring can ensure the integrity of the remote I/O binding channel. We implemented a prototype system based on Xen and we quantified its security and performance properties. Our experiments show that the TRIOB system can achieve those goals above and the overhead is in an acceptable range for users.

Acknowledgments. This work was supported in part by the National High-Tech Research and Development Program (863) of China under grants 2009AA

01Z141 and 2009AA01Z151, the projects of National Science Foundation of China (NSFC) under grant 90718040, and the National Grand Fundamental Research Program (973) of China under grant No.2007CB310805. We would like to thank Angelos Stavrou and the anonymous reviewers for their comments.

References

1. Barham, P., Dragovic, B., Fraser, K., et al.: Xen and the Art of Virtualization. In: Proc. of the 19th ACM Symp. on Operating Systems Principles 2003, pp. 164–177 (2003)
2. Huizenga, G.: Cloud Computing: Coming out of the fog. In: Proceedings of the Linux Symposium 2008, vol. 1, pp. 197–210 (2008)
3. Armbrust, M., Fox, A., et al.: Above the Clouds: A Berkeley View of Cloud. Technical Report No. UCB/EECS-2009-28 (2009)
4. Kaufman, L.M.: Data Security in the World of Cloud Computing. IEEE Security and Privacy 7(4), 61–64 (2009)
5. Garfinkel, T., et al.: Terra: A Virtual Machine-Based Platform for Trusted Computing. In: Pro. of the 19th ACM Symp. on Operating Systems Principles 2003, pp. 193–206 (2003)
6. Payne, B.D., Carbone, M., Sharif, M., Lee, W.: Lares: An Architecture for Secure Active Monitoring Using Virtualization. In: Proceedings of the IEEE Symposium on Security and Privacy 2008, pp. 233–247 (2008)
7. Berger, S., et al.: TVDc: managing security in the trusted virtual datacenter. ACM SIGOPS Operating Systems Review 42(1), 40–47 (2008)
8. Berger, S., Cceres, R., et al.: vTPM: Virtualizing the Trusted Platform Module. In: Proc. of the 15th Conference on USENIX Security Symposium (2006)
9. Sailer, R., Zhang, X., Jaeger, T., van Doorn, L.: Design and Implementation of a TCG-based Integrity Measurement Architecture. In: Proceedings of the 13th Conference on USENIX Security Symposium (2004)
10. Tan, T., Simmonds, R., et al.: Image Management in a Virtualized Data Center. ACM SIGMETRICS Performance Evaluation Review 36(2), 4–9 (2008)
11. Warfield, A., Hand, S., Fraser, K., Deegan, T.: Facilitating the development of soft devices. In: USENIX Annual Technical Conference 2005 (2005)
12. Sun, Y., Fang, H., Song, Y., et al.: TRainbow: a new trusted virtual machine based platform. International Journal Frontiers of Computer Science in China 4(1), 47–64 (2010)
13. Murray, D.G., Milos, G., Hand, S.: Improving Xen Security through Disaggregation. In: Proc. Of the 4th ACM SIGPLAN/SIGOPS International Conference on Virtual Execution Environments 2008, pp. 151–160 (2008)
14. Garfinkel, T., Rosenblum, M.: A Virtual Machine Introspection Based Architecture for Intrusion Detection. In: Proc. of the Network and Distributed Systems Security Symposium 2003, pp. 191–206 (2003)
15. Amazon Elastic Compute Cloud (Amazon EC2), http://aws.amazon.com/ec2/
16. Payne, B., Carbone, M., Lee, W.: Secure and Flexible Monitoring of Virtual Machines. In: Computer Security Applications Conference 2007, pp. 385–397 (2007)
17. Jones, S.T., Arpaci-Dusseau, A.C., et al.: Antfarm: Tracking Processes in a Virtual Machine Environment. In: Proc. of USENIX Annual Technical Conference 2006 (2006)

18. Jiang, X., Wang, X., Xu, D.: Stealthy malware detection through vmm-based out-of-the-box semantic view reconstruction. In: Proc. of the 14th ACM Conference on Computer and Communications Security 2007, pp. 128–138 (2007)
19. Linux kernel rootkits - protecting the system's Ring-Zero,
    http://www.sans.org/reading_room/whitepapers/honors/
    linux-kernel-rootkits-protecting-systems_1500
20. Weinhold, C., Hartig, H.: VPFS: building a virtual private file system with a small trusted computing base. In: Proceedings of the 3rd ACM SIGOPS/EuroSys European Conference on Computer Systems 2008, pp. 81–93 (2008)
21. Weele, E.V., Lau, B., et al.: Protecting Confidential Data on Personal Computers with Storage Capsules. In: Proceedings of the 18th USENIX Security Symposium 2009 (2009)
22. Yang, J., Shin, K.G.: Using Hypervisor to Provide Application Data Secrecy on a Per-Page Basis. In: Pro. of the Fourth International Conference on Virtual Execution Environments 2008, pp. 71–80 (2008)
23. Chen, X., Garfinkel, T., et al.: Overshadow: a virtualization-based approach to retrofitting protection in commodity operating systems. In: Proc. of the 13th International Conference on Architectural Support for Programming Languages and Operating Systems 2008, pp. 2–13 (2008)
24. Chandra, R., Zeldovich, N., Sapuntzakis, C., Lam, M.S.: The collective: a cache-based system management architecture. In: Proc. of the 2nd Conference on Symposium on Networked Systems Design and Implementation 2005, vol. 2, pp. 259–272 (2005)
25. Kumar, S., Schwan, K.: Netchannel: a VMM-level mechanism for continuous, transparent device access during VM migration. In: Pro. of the Fourth ACM SIGPLAN/SIGOPS International Conference on Virtual Execution Environments 2008 (2008)
26. Aiken, S., Grunwald, D., Pleszkun, A.R., Willeke, J.: A performance analysis of the iSCSI protocol. IEEE MSST (2003)
27. Petroni, N.L., et al.: Copilot: a Coprocessor-based Kernel Runtime Integrity Monitor. In: Proc. of the 13th Conference on USENIX Security Symposium (2004)
28. Seshadri, A., Luk, M., Shi, E., et al.: Pioneer: Verifying Code Integrity and Enforcing Unhampered Code Execution on Legacy Systems. In: The 20th ACM Symposium on Operating Systems Principles (2005)
29. Fraser, K., et al.: Safe Hardware Access with the Xen Virtual Machine Monitor. In: Proceedings of the 1st Workshop on Operating System and Architectural Support for the on Demand IT InfraStructure, Boston, MA (2004)
30. Wojtczuk, R.: Subverting the Xen Hypervisor. Black Hat USA (2008)
31. http://cve.mitre.org/cgi-bin/cvename.cgi?name=CVE-2008-32531
32. Goodin, D.: Webhost hack wipes out data for 100,000 sites,
    http://www.theregister.co.uk/2009/06/08/webhost_attack
33. Lineberry, A.: Malicious Code Injection via /dev/mem. Black Hat 2009 (2009)
34. Milos, G., Murray, D.G.: Boxing clever with IOMMUs. In: Proceedings of the 1st ACM Workshop on Virtual Machine Security 2008, pp. 39–44 (2008)
35. Murray, D.G., Hand, S.: Privilege separation made easy: trusting small libraries not big processes. In: Proceedings of the 1st European Workshop on System Security 2008, pp. 40–46 (2008)
36. Dalton, C.I., Plaquin, D., et al.: Trusted virtual platforms: a key enabler for converged client devices. SIGOPS Oper. Syst. Rev. 43(1), 36–43 (2009)

Dynamic Group Key Exchange Revisited

Guomin Yang and Chik How Tan

Temasek Laboratories
National University of Singapore
{tslyg,tsltch}@nus.edu.sg

Abstract. In a dynamic group key exchange protocol, besides the basic group setup protocol, there are also a join protocol and a leave protocol, which allow the membership of an existing group to be changed more efficiently than rerunning the group setup protocol. The join and leave protocols should ensure that the session key is updated upon every membership change so that the subsequent sessions are protected from leaving members (*backward security*) and the previous sessions are protected from joining members (*forward security*). In this paper, we present a new security model for dynamic group key exchange. Comparing to existing models, we do a special treatment to the state information that a user may use in a sequence of setup/join/leave sessions. Our treatment gives a clear and more concise definition of session freshness for group key exchange in the dynamic setting. We also construct a new dynamic group key exchange protocol that achieves strong security and high efficiency in the standard model.

Keywords: Group Key Exchange, Dynamic Group, Insider Security, Mutual Authentication, Strong Contributiveness.

1   Introduction

Group key exchange (GKE) protocols are mechanisms by which a group of $n > 2$ parties communicate over an insecure network can generate a common secret key (usually we call this common secret key a session key as a user may have multiple key exchange sessions). A static group key exchange (SGKE) protocol consists of a long-lived key generation algorithm, and a group setup protocol which is invoked whenever a group of parties want to establish a shared key. In contrast, a dynamic group key exchange (DGKE) protocol consists of a long-lived key generation algorithm and three sub-protocols: a setup protocol, a join protocol, and a leave protocol. The join and leave protocols allow the membership of an existing group to be changed more efficiently than rerunning the group setup protocol. In this paper, we focus on the dynamic setting.

Since the goal of a GKE protocol is to establish a shared key that is known only to the group members, the main security requirement is to make sure that no information, not even a single bit, of the agreed key is leaked to any passive or active adversaries outside the group. For the special case of two-party key exchange, the problem has been extensively studied (e.g., [3, 4, 2, 13, 25] and

S.-H. Heng, R.N. Wright, and B.-M. Goi (Eds.): CANS 2010, LNCS 6467, pp. 261–277, 2010.

many more). Though it might be natural to extend the existing results for two-party key exchange to the group setting, such an extension is less trivial for the dynamic group case. For dynamic group key exchange, special attentions should be paid to the join and leave protocols: these two protocols must make sure that upon every membership change, the session key is updated so that the subsequent sessions are protected from leaving members (namely, *backward security*) and the previous sessions are protected from joining members (namely, *forward security*[1]). In other words, for the leave/join protocol, we have to assume that the leaving/joining users are the adversary, who may misbehave (i.e. deviate from the protocol specification) in one session, and try to break the security of a subsequent/previous session among other parties.

To see more clearly the differences between a two-party (or static group) protocol and a dynamic group protocol: first, in the dynamic setting, the state information of the honest parties in two sessions (e.g., one involves a joining/leaving adversary and the other the adversary wants to attack) can be related, so the definition of a fresh (or clean) session is less straightforward when session state reveal queries are allowed; secondly, for several, say t, related sessions created in a sequence of setup/join/leave events, the adversary can choose to attack one session but at the same time actively participate in the rest $t - 1$ sessions, in which the adversary may not honestly follow the protocol specification (e.g., the adversary may plant some trapdoor information in one session, and try to break the following sessions after he/she leaves the group); thirdly, when considering the notion of *forward secrecy*, a secure DGKE protocol should ensure a past session key K_i remains secure even when the adversary compromises both the long-lived keys and the state information of the subsequent join/leave sessions after session i.

For group key exchange protocols, it is also necessary to consider *insider* security. Two insider security issues need to be addressed: mutual authentication and key control resistance. Different from two-party protocols, mutual authentication for GKE protocol means when a party makes a decision "accept" in a session, he/she must be sure that all his/her partners have indeed participated in the session, in particular, a subset of protocol participants should not be able to impersonate another party. For key control resistance, we require a secure group key exchange protocol to be able to prevent a subset of users from controlling (any part) of the session key.

Related Work. The case of two-party key exchange protocols has been extensively studied in the past three decades (e.g., [16, 17, 3, 4, 2, 13, 25] and many more). A lot of efficient and provably secure two party key exchange protocols have been constructed, and many have been deployed in the real practice (e.g. SSL, SSH, IPSec).

---

[1] We remark that the notion of forward security with respect to the join protocol is different from the conventional notion of "forward secrecy". Informally, the latter means an established session key remains secure even if an adversary learns the long-lived keys of the protocol participants in the future.

Some early group key exchange protocols [12, 1, 29, 30, 21] extended two-party key exchange (e.g., the two-party Diffi-Hellman protocol) to the multi-party setting. In [9], based on the Bellare-Rogaway security model [3] for two-party key exchange, Bresson, Chevassut, Pointcheval, and Quisquater proposed the first computational security model (referred to as the BCPQ model) for (static) group key exchange protocols. The BCPQ model and its variants then became the *de facto* standard for analyzing group key exchange protocols.

In [23], Katz and Yung proposed a scalable compiler to transform any static GKE protocol secure against passive adversaries to a new protocol secure against active adversaries. They then applied their compiler to the Burmester-Desmedt protocol and obtained the first constant-round and fully-scalable static GKE protocol in the standard model.

In [8], a formal model for dynamic group key exchange was proposed, this model was later extended in [7] to consider concurrent protocol executions and strong corruption. Provably secure protocols which require linear round complexity are also proposed in [8, 7]. In [24], Lim et al. proposed another DGKE protocol provably secure in the model of Bresson et al., the protocol requires only two rounds but is proven secure in the random oracle model. In [18], Dutta et al. presented another security model for dynamic group key exchange, and a protocol with formal security proof, but neither their model nor the protocol was carefully designed, as a result the model didn't really capture forward/backward security [32] and the protocol is flawed [31]. In [26], Manulis presented a 3-round provably secure DGKE protocol (denoted by Dynamic TDH1) which uses a binary tree structure.

INSIDER SECURITY. Katz and Shin [22] provided the first formal definition of insider security for static group key exchange protocols under the Universal Composability (UC) framework [14]. They also presented a generic compiler to build UC-secure Static GKE protocols. Recently, in [19], Furukawa et al. presented a more round-efficient UC-secure Static GKE protocol which requires only two rounds, but the protocol requires each protocol participants to perform $O(n)$ pairing operations for an n-party group.

The insider security by Katz and Shin [22] considers impersonation attacks by malicious protocol participants. Later, Bresson and Manulis [11] unified this notion with key confirmation, and unknown key share resistance into their definition of Mutual Authentication. Recently, Gorantla el al. [20] further unified the Mutual Authentication definition by Bresson and Manulis with the notion of Key Compromise Impersonation resistance.

The notion of key control was first introduced by Mitchell et al. [27], which refers to attacks by which part of the protocol participants aim to control the value of the session key. Resistance to key control attacks is formalized via the notion of "contributiveness" in [6, 11] for group key exchange protocols. In [10], Bresson and Manulis presented a compiler that can transform any GKE protocols to achieve contributiveness at the cost of 2 extra communication rounds.

Our Contributions. In this paper, we present a new security model for dynamic group key exchange protocols. To define forward and backward security, we provide a special treatment to the shared state information among a sequence of setup/join/leave sessions. Our treatment gives a clear and more concise definition of session freshness than existing models. We then construct an efficient DGKE protocol and prove its security in the standard model.

2   Security Definitions

We start by some existing definitions and notations [3,5] for key exchange protocols.

Protocol Participants and Long-Lived Keys. Let $\mathcal{HU}$ denote a nonempty set of parties. Each party $U \in \mathcal{HU}$ is provided with a Long-Lived Key LLK_U that is generated by running a long-lived key generation algorithm KG. Later on, new users can still be created and added into the system by the adversary. Such an adversarial capability was first considered in [11]. Let $\mathcal{MU}$ denote the set of users added by the adversary. When being created, a long-lived key LLK_M for M is generated by the adversary with the restriction that LLK_M^{Pub}, which denotes the *public* part of LLK_M, has never been used by any other user in the system. However, we do *not* require LLK_M to be generated by honestly running the long-lived key generation algorithm KG, and the adversary may not know the secret part of LLK_M.

Instance Oracles. A party may run many instances concurrently except that an instance in a join or leave session cannot run concurrently with any of its ancestors (see footnote 2). We call each instance of a party an oracle, and we use Π_U^i ($i \in \mathbb{N}$) to denote the ith instance of party U. All the oracles within a party U share the same long-lived key LLK_U. An oracle is activated when it receives an incoming message from the network. Upon activation, the oracle performs operations by following the Setup, Join or Leave protocol. We assume that whenever a user wants to setup a new group or join an existing group, an *unused* oracle will be used.

Session and Partner IDs, State Information, and Session Keys. When an oracle Π_U^i is activated to start a protocol (Setup, Join or Leave), it learns its partner id pid_U^i (similar to [11], we let pid_U^i include the identity of U itself). At the time Π_U^i makes a decision Accept, it outputs (secret) session key k_U^i under a session id sid_U^i which is determined during the protocol execution. Since we are considering group key exchange in the dynamic setting, some state information of Π_U^i should also be saved. This information would be used if a Join or Leave event later. In this paper, we assume that, after an instance Π_U^i accepts, its state information (which includes two fields $\mathsf{pid}_{old} = \mathsf{pid}_U^i$ and $\mathsf{sid}_{old} = \mathsf{sid}_U^i$) will be passed to another unused instance Π_U^j (possibly picked by the adversary), and Π_U^j would replace Π_U^i to participate in the next Join or Leave session. Also, after Π_U^j receives the state information from Π_U^i, all the state information in Π_U^i is erased, and Π_U^j is labeled *used*.

Discussion. A straightforward way to deal with the join and leave events is to let Π_U^i keep the state information at the end of a session and become active again in the subsequent Join or Leave event. However, such an approach will introduce troubles when we later define the *freshness* of an instance, since one instance may participate in different sessions with different partners and generate different session IDs. On the other hand, as we will see later, our approach makes the security definition easy and clean. Besides, our approach is meaningful, it mimics the following space-friendly implementation in real practice: after a KE session is completed and the session key is returned to the upper layer application, the user saves the necessary state information at a safe place in the harddisk, and erases that copy of the program (which implements the DGKE protocol) and all its state information from the memory. Later, when a join or leave event occurs, the user starts a new copy of the protocol/program with the previously saved state information as the parameters to the program.

Definition 1. *A Dynamic Group Key Exchange Protocol $\mathcal{DGKE}$ consists of a Long-Lived Key generation algorithm* KG, *a group setup protocol* Setup, *a join protocol* Join, *and a leave protocol* Leave.

- KG(1^k): *On input a security parameter 1^k, the long-lived key generation algorithm provides each user with a long-lived key LLK_U.*
- Setup($\mathcal{I}$): *on input a set of user identities $\mathcal{I}$, the setup protocol creates a new multicast group $\mathcal{G} = \mathcal{I}$.*
- Join($\mathcal{G}', \mathcal{I}$): *on input an existing multicast group $\mathcal{G}'$ and a set of users $\mathcal{I}$ such that $\mathcal{I} \cap \mathcal{G}' = \emptyset$, the join protocol creates a new multicast group $\mathcal{G} = \mathcal{G}' \cup \mathcal{I}$.*
- Leave($\mathcal{G}', \mathcal{I}$): *on input an existing multicast group $\mathcal{G}'$ and a set of users $\mathcal{I} \subset \mathcal{G}'$, the leave protocol creates a new multicast group $\mathcal{G} = \mathcal{G}' \backslash \mathcal{I}$.*

An execution of $\mathcal{DGKE}$ consists of running the KG *algorithm once, and many concurrent executions of the other three protocols. We say $\mathcal{DGKE}$ is correct if, when no adversary is present, all the parties in the group $\mathcal{G}$ compute the same session key at the end of the* Setup, Join *or* Leave *protocol.*

Security Model. We consider the following game that involves all the users in the set $\mathcal{HU}$ and an adversary $\mathcal{A}$. All the users are connected via an unauthenticated network that is controlled by $\mathcal{A}$. The game is initiated by running the long-lived key generation algorithm KG to provide each user in $\mathcal{HU}$ with a long-lived key. The adversary $\mathcal{A}$ is then given $\{LLK_U^{Pub}\}_{U \in \mathcal{HU}}$ and interacts with the oracles via the queries described below.

- Register(M, LLK_M^{Pub}): This query allows the adversary $\mathcal{A}$ to create and add a new user M with long-live (public) key LLK_M^{Pub} into the system. We require that neither the user identity M nor the long-lived public key LLK_M^{Pub} has been used by any other user in the system. However, we don't require LLK_M^{Pub} to be generated by running the KG algorithm. All the activities and operations of user M will be performed by the adversary $\mathcal{A}$.

- Send(Π_U^i, msg_{in}): This query allows $\mathcal{A}$ to invoke instance Π_U^i with an incoming message msg_{in}. Upon receiving the message, Π_U^i performs operations according to the Setup, Join or Leave protocol, and generates the response. Should Π_U^i accept or reject will be made available to $\mathcal{A}$. When an oracle Π_U^i accepts, $\mathcal{A}$ chooses another unused instance Π_U^j. The state information of Π_U^j is then set to $St_U^j = St_U^i$ (this operation is assumed to be done within the user U. In particular, the adversary is unaware of the state information being passed, but the adversary may learn this information via a RevealState query (described below) to Π_U^j). Π_U^j is labeled *used* and will replace Π_U^i to participate in the subsequent Join or Leave protocol. The Setup, Join and Leave events are also activated by $\mathcal{A}$ through Send queries, as follows:
  1. When $\mathcal{A}$ wants to activate an *unused* instance Π_U^i to start the Setup protocol, it sets $msg_{in} = $ setup$\|$pid where pid is the partner id of the instance Π_U^i.
  2. When $\mathcal{A}$ wants activate an instance Π_U^i, which is either unused (i.e. U is going to join an existing group) or has been used (i.e. U is a member of the existing group), to start the Join protocol, it sets $msg_{in} = $ join$\|$pid$_{old}\|$sid$_{old}\|$pid$_{new}$ where pid$_{old}$ denotes the old group (with session id sid$_{old}$) on top of which a new group pid$_{new}$ is to be built. It is worth noting that several groups with the same set of members may exist, so the session id is necessary to uniquely identify the (old) group.
  3. Similarly, when $\mathcal{A}$ wants activate an instance Π_U^i to start the Leave protocol, it sets $msg_{in} = $ leave$\|$pid$_{new}$. Note that if Π_U^i participates in the leave protocol, it should already have the information of the existing group, so the fields pid$_{old}$ and sid$_{old}$ are not needed.
- Corrupt(U): This query allows the adversary to obtain the long-lived key LLK_U.
- RevealKey(Π_U^i): This query reveals the session key being held by the oracle Π_U^i.
- RevealState(Π_U^i): This query reveals all the state information, but not the long-lived key, currently being held by the oracle Π_U^i.
- Test(Π_U^i): This query is asked only once in the game, and is only available if oracle Π_U^i has accepted, and is *fresh* (see below). An unbiased coin b is tossed, if $b = 0$, a random value drawn from the session key space is returned; if $b = 1$, the real session key k_U^i is returned. After the Test query, the adversary can still perform those queries described above.

Oracle Freshness. An oracle Π_U^i is *fresh* if all of the following conditions hold:
1. pid$_U^i \cap \mathcal{MU} = \emptyset$;
2. No user in pid$_U^i$ is corrupted before the adversary makes a Send($\Pi_V^j, *$) query with ($V \in $ pid$_U^i \wedge$ sid$_V^j = $ sid$_U^i$);
3. No RevealState query is performed to an oracle Π_V^j with ($V \in $ pid$_U^i \wedge$ sid$_V^j = $ sid$_U^i$) or any of its ancestors[2].

---

[2] We say Π_V^i is an ancestor of Π_V^j if there exists a path ($\Pi_V^i, ..., \Pi_V^t, ...\Pi_V^j$) such that each instance in the path passes its state information to the next one.

4. No RevealKey query is performed to an oracle Π_V^j with $(V \in \mathsf{pid}_U^i \wedge \mathsf{sid}_V^j = \mathsf{sid}_U^i)$.

SK-Security. Before $\mathcal{A}$ terminates it outputs a bit b'. We say $\mathcal{A}$ wins the game if $b' = b$. We define the advantage of the adversary $\mathcal{A}$ attacking protocol $\mathcal{DGKE}$ to be

$$\mathsf{Adv}_{\mathcal{A},\mathcal{DGKE}}^{sk}(k) = |\Pr[b' = b] - \frac{1}{2}|$$

Definition 2 (SK-Security). *We say a dynamic group key exchange protocol $\mathcal{DGKE}$ is SK-secure if, for any polynomial time adversary $\mathcal{A}$, the advantage $\mathsf{Adv}_{\mathcal{A},\mathcal{DGKE}}^{sk}(k)$ is a negligible function of the security parameter k.*

Comparisons with Existing Models. In the early models by Bresson et al. [8, 7], a single instance will maintain the state information in a sequence of setup, join, and leave events. When defining the freshness, the adversary is not allowed to reveal the state information of the instance (or any of its partners) at any time. As a consequence, it does not capture the scenario that the adversary compromises the state information in the join/leave sessions which are subsequent sessions of the test session.

In the model by Dutta and Barua [18], the adversary can only passively received communication transcripts of the join and leave protocols, but in reality, the adversary is the joining/leaving user who actively participated in the join/leave session. In [32], it has been shown that the proven secure protocol in [18] doesn't provide backward security.

In the model by Manulis [26], similar to the approach by Bresson et al., a single instance will maintain the state information in a sequence of setup/join/leave events, however the adversary is allowed to reveal the state information of the instance output in the test query at some points. While in our model, due to our trick to the state information, each oracle will participate in at most one session, which makes the definition of session freshness more concise as we don't need to consider at which points the adversary is allowed to perform RevealState to an instance.

Another difference between Manulis's model and ours is that we have different meanings in backward security. In Manulis's model, backward security considers an adversary who compromises state information of the ancestors (see footnote 2) of the instances in the test session, while ours means the leaving users, who may plant some trapdoors in the previous sessions, cannot learn any information of the session key established among the remaining group of users, but our model requires none of the instances in the test session, or any of their ancestors, has been asked a RevealState query. The definition by Manulis is stronger than ours, however, such a definition may be too strong as no existing DGKE protocol (including the dynamic TDH1 by Manulis) can achieve such a security level. The reason is that in order to perform join/leave efficiently, some critical state information (such as the DH exponents) used by the join/leave protocol is related to state information of the previous session. In order to provide the backward security defined by Manulis, each instance needs to freshly generate all critical state information in each session, in which case the join/leave protocol most likely gains no advantage than the setup protocol (i.e., the protocol essentially becomes a static one).

MA-Security. We say that an instance Π_U^i is *honest* if $U \in \mathcal{HU}$ and Π_U^i honestly performs its operations according to the protocol. Below we review the definition of MA-security in [20]. The definition is a modification of the MA-security in [11] by including the notion of Key Compromise Impersonation (KCI) resistance. Recall that in a KCI attack an adversary corrupts a user U and then impersonates other (uncorrupted) users to (honest instances of) U.

Definition 3 (MA-Security). *Let $\mathcal{DGKE}$ be a correct dynamic group key exchange protocol and $\mathcal{A}'$ an adversary who is allowed to perform* **Register, Send, Corrupt, RevealKey** *and* **RevealState** *queries. We say that $\mathcal{A}'$ breaks the mutual authentication of $\mathcal{DGKE}$ if at some point during the execution of $\mathcal{DGKE}$, there exists an honest instance Π_U^i who has accepted with k_U^i and another user $V \in \mathcal{HU} \cap \mathsf{pid}_U^i$ who is uncorrupted at the time Π_U^i accepts such that*

1. *There is no instance oracle Π_V^t with $(\mathsf{pid}_V^t, \mathsf{sid}_V^t) = (\mathsf{pid}_U^i, \mathsf{sid}_U^i)$, or*
2. *There is an instance oracle Π_V^t with $(\mathsf{pid}_V^t, \mathsf{sid}_V^t) = (\mathsf{pid}_U^i, \mathsf{sid}_U^i)$ that has accepted with $k_V^t \neq k_U^i$.*

Denote $\mathsf{Adv}_{\mathcal{A}', \mathcal{DGKE}}^{ma}(k)$ the probability that $\mathcal{A}'$ breaks the mutual authentication of $\mathcal{DGKE}$. We say a dynamic group key exchange protocol $\mathcal{DGKE}$ is MA-secure if for any polynomial time adversary $\mathcal{A}'$, $\mathsf{Adv}_{\mathcal{A}', \mathcal{DGKE}}^{ma}(k)$ is a negligible function of the security parameter k.

Contributiveness. We present below the notion of contributiveness for dynamic group key exchange protocols. A group key exchange protocol secure under this notion can resist key control attacks where a subset of insiders tries to control any part of the resulting session key.

Definition 4 (Co-Security). *Let $\mathcal{DGKE}$ be a correct dynamic group key exchange protocol and $\mathcal{A}'' = (\mathcal{A}_1'', \mathcal{A}_2'')$ an adversary who is allowed to perform* **Register, Send, Corrupt, RevealKey** *and* **RevealState** *queries. $\mathcal{A}''$ runs in two stages:*

- *(Prepare.) $\mathcal{A}_1''$ performs the oracle queries and outputs a bit b, an index j, along with some state information* **St***.*
- *(Attack.) On input* **St***, $\mathcal{A}_2''$ performs the oracle queries and finally outputs (U, i).*

We say that $\mathcal{A}''$ wins if

1. *Π_U^i has terminated accepting k_U^i such that the j-th bit of k_U^i is equal to b,*
2. *Π_U^i is honest and has not started its execution in the Prepare stage,*
3. *Π_U^i has never been asked a* **RevealState** *query in the Attack phase.*

Define

$$\mathsf{Adv}_{\mathcal{A}'', \mathcal{DGKE}}^{co}(k) = \Pr[\mathcal{A}'' \text{ wins}] - \frac{1}{2}.$$

A dynamic group key exchange protocol $\mathcal{DGKE}$ is said to provide contributiveness (or Co-security) if for any polynomial time adversary $\mathcal{A}''$, $Adv^{co}_{\mathcal{A}'',\mathcal{DGKE}}(k)$ is a negligible function of the security parameter k.

Our definition of contributiveness is different from the existing definitions of *strong* contributiveness defined in [11, 20]. The later requires that the adversary cannot control the whole key, and as a result it does not capture partial-key control attacks. Below we present a protocol that achieves strong contributiveness but fail to achieve our notion of contributiveness.

U_1	U_2	U_3	U_4	U_5

$$k_i \xleftarrow{\$} \{0,1\}^\ell, x_i \xleftarrow{\$} \mathbb{Z}_q^*, \sigma_i^1 \leftarrow \mathcal{DS}.Sign(sk_i, M_i^1 \| \mathcal{I})$$

$M_1^1 = k_1 \| g^{x_1}, \sigma_1^1 \quad M_2^1 = k_2 \| g^{x_2}, \sigma_2^1 \quad M_3^1 = k_3 \| g^{x_3}, \sigma_3^1 \quad M_4^1 = k_4 \| g^{x_4}, \sigma_4^1 \quad M_5^1 = H(k_5) \| g^{x_5}, \sigma_5^1$

Broadcast Round 1

$$sid \leftarrow H(\mathcal{I} \| k_1 \| k_2 \| k_3 \| k_4 \| H(k_5)), t_i^L \leftarrow H(g^{x_{L(i)} x_i}), t_i^R \leftarrow H(g^{x_{R(i)} x_i})$$

$$T_i \leftarrow t_i^L \oplus t_i^R, \hat{T}_5 \leftarrow k_5 \oplus t_5^R, \sigma_i^2 \leftarrow \mathcal{DS}.Sign(sk_i, M_i^2)$$

$M_1^2 = T_1 \| sid, \sigma_1^2 \quad M_2^2 = T_2 \| sid, \sigma_2^2 \quad M_3^2 = T_3 \| sid, \sigma_3^2 \quad M_4^2 = T_4 \| sid, \sigma_4^2 \quad M_5^2 = \hat{T}_5 \| T_5 \| sid, \sigma_5^2$

Broadcast Round 2

Session Key $k = H(\mathcal{I} \| k_1 \| k_2 \| k_3 \| k_4 \| k_5)$

Fig. 1. A (Static) Group Key Exchange Protocol [20]. $\mathcal{I} = \{U_1, U_2, U_3, U_4, U_5\}$.

The protocol in Fig. 1 is proven secure under the strong contributiveness definition in [11, 20]. However, the protocol does not provide partial-key control resistance. In the protocol, the keying materials k_i of U_i $(1 \leq i \leq 4)$ are sent in clear at the beginning of the protocol, so the last user (i.e. U_5 in our example) can (partially) control the session key as follows: after seeing the k_i's of all the other users, U_5 repeatedly tries different values for k_5 until the session key $k = H(\mathcal{I} \| k_1 \| k_2 \| k_3 \| k_4 \| k_5)$ has a desired pattern. We can see that in order to control s bits of the final session key, the expected number of trials that U_5 needs to perform is 2^s.

A similar attack can be performed to other protocols (e.g., those in [8, 7, 24, 18, 26, 11]) where a user can repeatedly try different keying materials after seeing the keying material sent by other users.

However, in our definition of contributiveness, we don't allow the adversary to make the RevealState query. This restriction is necessary in our definition since otherwise the adversary can use the RevealState query to learn the keying material of the instance under attack, and then repeatedly choose the proper keying materials for other users. In contrast, such a restriction is not required in the strong contributiveness definition in [11, 20]. So in general, these two notions are incomparable. However, we believe partial-key control resistance may be more important in some circumstances.

It is also worth noting that partial-key control attacks have been considered in some previous work, such as in the shielded-insider privacy security notion by Desmedt et al. [15], and in the *weak* contributiveness notion by Bresson and Manulis [10]. We leave the relationship among these partial-key control resistance notions an open problem.

3 A New Dynamic Group Key Exchange Protocol

In this section, we present a new dynamic group key exchange protocol and show that it satisfies all the security definitions (i.e. SK-, MA-, and Co-Security) given in Sec. 2. Our protocol makes use of a commitment scheme $\mathcal{CMT} = (\mathsf{CMT}, \mathsf{CVF})$, a digital signature scheme $\mathcal{DS}$, and two pseudo-random function families $\hat{\mathbf{F}}$ and $\bar{\mathbf{F}}$.

3.1 Primitives

Commitment Schemes. A commitment scheme $\mathcal{CMT}$ consists of two algorithms: a commitment algorithm CMT which takes a message M to be committed as input and returns a commitment C and an opening key δ, and a deterministic verification algorithm CVF which takes C, M, δ as input and returns either 0 or 1. We say $\mathcal{CMT} = (\mathsf{CMT}, \mathsf{CVF})$ is a perfectly hiding and computationally binding commitment scheme with binding error ϵ if $\mathcal{CMT}$ achieves all the following properties.

- Consistency: for any message M

$$\Pr[(C, \delta) \xleftarrow{\$} \mathsf{CMT}(M) : \mathsf{CVF}(C, M, \delta) = 1] = 1.$$

- Perfectly Hiding: for any messages M_0 and M_1 such that $|M_0| = |M_1|$, the commitments C_0 and C_1 are identically distributed where $(C_0, \delta_0) \xleftarrow{\$} \mathsf{CMT}(M_0)$, and $(C_1, \delta_1) \xleftarrow{\$} \mathsf{CMT}(M_1)$.
- Computationally Binding: For every polynomial time algorithm $\mathcal{M}$

$$\mathrm{Adv}^{\mathrm{bind}}_{\mathcal{CMT}, \mathcal{M}} \overset{\mathrm{def}}{=} \Pr\left[\begin{array}{l} (C, (M_0, \delta_0), (M_1, \delta_1)) \leftarrow \mathcal{M}(1^k) : \\ M_0 \neq M_1 \wedge \mathsf{CVF}(C, M_0, \delta_0) = 1 \\ \wedge \mathsf{CVF}(C, M_1, \delta_1) = 1 \end{array} \right] \leq \epsilon(k).$$

In our protocol we will make use of a commitment scheme with the following additional property: for any message M, an honest execution of $\mathsf{CMT}(M)$ generates a commitment C that is uniformly distributed in the range of $\mathsf{CMT}(\cdot)$. We call such kind of commitment schemes *Uniformly Distributed* commitment schemes. A typical example of this type of commitment schemes is the Pedersen commitment scheme [28] where the computationally binding property holds under the Discrete Log assumption.

Pseudo-random Function Family. A family of efficiently computable functions $\mathbf{F} = \{\mathsf{F}_K : \mathsf{D} \to \mathsf{R} | K \in \mathsf{K}\}$ is called a pseudo-random function family, if for any polynomial time algorithm $\mathcal{A}$,

$$\mathrm{Adv}^{prf}_{\mathbf{F}, \mathcal{A}}(k) \overset{\mathrm{def}}{=} \Pr[\mathcal{A}^{\mathsf{F}_\kappa(\cdot)}(1^k) = 1] - \Pr[\mathcal{A}^{\mathsf{RF}(\cdot)}(1^k) = 1]$$

is negligible where $\kappa \xleftarrow{\$} \mathsf{K}$ and $\mathsf{RF} : \mathsf{D} \to \mathsf{R}$ is a truly random function.

Digital Signature Scheme. A digital signature scheme $\mathcal{DS}$ consists of three algorithms: a key generation algorithm $\mathcal{DS}.Kg$ that takes a security parameter 1^k as input and returns a long-lived key pair (pk, sk) where pk is public and sk is private; a signing algorithm $\mathcal{DS}.Sign$ that takes a private key sk and a message $m \in \{0,1\}^*$ as input, and returns a signature σ; and a verification algorithm $\mathcal{DS}.Ver$ that takes a public key pk, a message m and a signature σ as input, and returns a bit $b \in \{0,1\}$ indicating the validity of the signature. The consistency requirement is that for any security parameter k and any message $m \in \{0,1\}^*$,

$$\Pr[(pk, sk) \leftarrow \mathcal{DS}.Kg(1^k) : \mathcal{DS}.Ver(pk, m, \mathcal{DS}.Sign(sk, m)) = 1] = 1.$$

We say $\mathcal{DS}$ is existentially unforgeable under chosen message attacks (uf-cma), if for any polynomial time algorithm $\mathcal{F}$,

$$\mathsf{Adv}_{\mathcal{DS},\mathcal{F}}^{uf-cma}(k) \stackrel{\text{def}}{=} \Pr \left[\begin{array}{l} (pk, sk) \leftarrow \mathcal{DS}.Kg(1^k), \\ (m^*, \sigma^*) \leftarrow \mathcal{F}^{\mathcal{DS}.Sign(sk, \cdot)}(pk) : \\ \mathcal{DS}.Ver(pk, m^*, \sigma^*) = 1 \\ \wedge \; \mathcal{F} \text{ has never queried } \mathcal{DS}.Sign(sk, m^*) \end{array} \right]$$

is negligible.

3.2   The Protocol

Our protocol is an improved version of the protocol by Kim et al. [24] with the following differences: (1) the protocol in [24] is in the random oracle model while ours is in the standard model; (2) the protocol in [24] cannot achieve MA- and Co-security.

Protocol Design. The setup protocol makes use of two-party Diffie-Hellman key exchange to form a ring structure: each party generate an ephemeral public/private key pair (x_i, g^{x_i}), and generates a left key K_i^L (based on $g^{x_{i-1}x_i}$) with his left neighbor and a right key K_i^R (based on $g^{x_{i+1}x_i}$) with his right neighbor, and broadcasts $K_i^L \oplus K_i^R$. Then each group member can recover the left/right keys of all other group members due to the ring structure. One of the members also conceals its keying material using his right key, and others just send the keying materials in clear. Now only the legitimate group members can recover the concealed keying material and compute the final session key. To ensure contributiveness, we require each party to commit their keying material before receiving others' keying materials.

For the join/leave event, we let part of the existing group members to perform the same procedures as in the setup protocol, but the rest of the users do not need to run the full setup protocol, so computational and communication cost can be saved. And to ensure forward/backward security, the state information of the each user is updated using two pseudo-random functions. Below are the details of the protocol.

Let $\mathbb{G}$ denote a cyclic group of prime order q, and g is a generator of $\mathbb{G}$. Our dynamic group key exchange protocol works as follows:

- **KG:** For each user U_i inside the system, a long lived key pair $(pk_i, sk_i) \leftarrow \mathcal{DS}.Kg(1^k)$ is generated.
- **Setup** (Fig. 2): The following protocol is performed among a set $\mathcal{I} = \{U_1, U_2, ..., U_n\}$ of users.

  1. (Round 1) Each user U_i chooses $k_i \xleftarrow{\$} \{0,1\}^\ell$, $x_i \xleftarrow{\$} \mathbb{Z}_q^*$ and computes $(k_i', \delta_i) \leftarrow \mathsf{CMT}(k_i)$. U_i then broadcasts $M_i^1 = k_i' \| g^{x_i}$.

  2. (Round 2) Upon receiving all the messages M_j^1 $(j \neq i)$, each U_i computes $\mathsf{sid}_i \leftarrow k_1' \| k_2' \| ... \| k_n'$, $t_i^L \leftarrow \hat{\mathsf{F}}_{g^{x_{L(i)} x_i}}(1)^3$, $t_i^R \leftarrow \hat{\mathsf{F}}_{g^{x_{R(i)} x_i}}(1)$, $\omega_i \leftarrow t_i^L \oplus t_i^R$. U_i $(1 \leq i < n)$ sets $M_i^2 = \omega_i \| k_i \| \delta_i$ and U_n computes $T_n \leftarrow t_n^R \oplus (k_n \| \delta_n)$ and sets $M_n^2 = \omega_n \| T_n$. Each U_i then generates a signature σ_i^2 on the message $M_i^1 \| M_i^2 \| \mathcal{I} \| \mathsf{sid}_i$ and broadcast $M_i^2 \| \sigma_i^2$.

  3. (Key Computation) Upon receiving $M_j^2 \| \sigma_j^2$ $(j \neq i)$, each U_i verifies all the signatures. If the signatures are valid, U_i derives $t_{i-1}^L \leftarrow t_i^L \oplus \omega_{i-1}$, $t_{i-2}^L \leftarrow t_{i-1}^L \oplus \omega_{i-2}$, ... until $t_1^L = t_n^R$ is derived. U_i then derives $k_n \| \delta_n \leftarrow t_n^R \oplus T_n$. U_i then verifies if $\mathsf{CVF}(k_j', k_j, \delta_j) = 1$ for all $j \neq i$. If all the verifications are successful, U_i computes the session key as $\mathsf{k}_i \leftarrow \tilde{\mathsf{F}}_{\hat{k}}(1)$ where $\hat{k} \leftarrow \bigoplus_{U_j \in \mathcal{I}} k_j$.

  4. (Post Computation) Each U_i computes $h_i^L \leftarrow \hat{\mathsf{F}}_{g^{x_{L(i)} x_i}}(0)$, $h_i^R \leftarrow \hat{\mathsf{F}}_{g^{x_{R(i)} x_i}}(0)$ and $X \leftarrow \tilde{\mathsf{F}}_{\hat{k}}(0)$, saves (h_i^L, h_i^R, X) with $\mathsf{pid}_i = \mathcal{I}$ and sid_i in the memory, and erases all other state information.

U_1	U_2	U_3	U_4
	$k_i \xleftarrow{\$} \{0,1\}^\ell, (k_i', \delta_i) \leftarrow \mathsf{CMT}(k_i), x_i \xleftarrow{\$} \mathbb{Z}_q^*$		
$M_1^1 = k_1' \| g^{x_1}$	$M_2^1 = k_2' \| g^{x_2}$	$M_3^1 = k_3' \| g^{x_3}$	$M_4^1 = k_4' \| g^{x_4}$
	Broadcast Round 1		
	$\mathsf{sid}_i \leftarrow k_1' \| k_2' \| k_3' \| k_4', t_i^L \leftarrow \hat{\mathsf{F}}_{g^{x_{L(i)} x_i}}(1), t_i^R \leftarrow \hat{\mathsf{F}}_{g^{x_{R(i)} x_i}}(1), \omega_i \leftarrow t_i^L \oplus t_i^R$		
	$T_4 \leftarrow t_4^R \oplus (k_4 \| \delta_4), \sigma_i^2 \leftarrow \mathcal{DS}.Sign(sk_i, M_i^1 \| M_i^2 \| \mathcal{I} \| \mathsf{sid}_i)$		
$M_1^2 = \omega_1 \| k_1 \| \delta_1, \sigma_1^2$	$M_2^2 = \omega_2 \| k_2 \| \delta_2, \sigma_2^2$	$M_3^2 = \omega_3 \| k_3 \| \delta_3, \sigma_3^2$	$M_4^2 = \omega_4 \| T_4, \sigma_4^2$
	Broadcast Round 2		
	Session Key $\mathsf{k} = \tilde{\mathsf{F}}_{\hat{k}}(1)$ where $\hat{k} = k_1 \oplus k_2 \oplus k_3 \oplus k_4$		
	Post Computation		
	$h_i^L \leftarrow \hat{\mathsf{F}}_{g^{x_{L(i)} x_i}}(0), h_i^R \leftarrow \hat{\mathsf{F}}_{g^{x_{R(i)} x_i}}(0), X \leftarrow \tilde{\mathsf{F}}_{\hat{k}}(0)$		
h_1^L, h_1^R, X	h_2^L, h_2^R, X	h_3^L, h_3^R, X	h_4^L, h_4^R, X

Fig. 2. The Setup Protocol. $\mathcal{I} = \{U_1, U_2, U_3, U_4\}$.

---

[3] Here for simplicity we directly use the Diffie-Hellman key as the pseudo-random function key, in practice, one may need to first apply a Key Drivation Function (KDF) to the Diffie-Hellman key, and then use the output of the KDF as the pseudo-random function key.

- Join (Fig. 3): Given an old group $\mathcal{I} = \{U_1, U_2, ..., U_n\}$ where each member has state information (h_i^L, h_i^R, X) and a set of new users $\mathcal{J} = \{U_{n+1}, U_{n+2}, ..., U_{n+k}\}$, the Join protocol works as follows:

  1. (Round 1): Each U_i ($1 \le i \le n+k$) chooses $\hat{k}_i \overset{\$}{\leftarrow} \{0,1\}^\ell$ and computes $(\hat{k}_i', \hat{\delta}_i) \leftarrow \mathsf{CMT}(\hat{k}_i)$. Each U_j ($j \in \{1, n, n+1, ..., n+k\}$) also chooses $\hat{x}_j \overset{\$}{\leftarrow} \mathbb{Z}_q^*$, and U_2 sets $\hat{x}_2 = X$. Then each U_i ($1 \le i \le n+k$) broadcasts M_i^1 where $M_j^1 = \hat{k}_j' \| g^{\hat{x}_j}$ for $j \in \{1, 2, n, n+1, ..., n+k\}$ and $M_\ell^1 = \hat{k}_\ell'$ for $\ell \in \{3, ..., n-1\}$.

  2. (Round 2): Upon receiving all the messages, each U_i ($1 \le i \le n+k$) computes $\mathsf{sid}_i \leftarrow \hat{k}_1' \| \hat{k}_2' \| ... \| \hat{k}_{n+k}'$. Each U_j ($j \in \{1, 2, n, n+1, ..., n+k\}$) then computes $t_j^L \leftarrow \hat{\mathsf{F}}_{g^{\hat{x}_{L(j)} \hat{x}_j}}(1)$, $t_j^R \leftarrow \hat{\mathsf{F}}_{g^{\hat{x}_{R(j)} \hat{x}_j}}(1)$ where $\hat{x}_{L(n)} = \hat{x}_2, \hat{x}_{R(2)} = \hat{x}_n$, and $\omega_i \leftarrow t_i^L \oplus t_i^R$. Each U_ℓ ($\ell \in \{3, ..., n-1\}$) also computes $t_\ell^R = t_2^R, t_\ell^L = t_2^L$ (as U_ℓ also has X), $\omega_\ell \leftarrow t_\ell^L \oplus t_\ell^R$. U_i ($i \in \{1, 2, n+1, ..., n+k\}$) sets $M_i^2 = \omega_i \| \hat{k}_i \| \hat{\delta}_i$, U_ℓ ($3 \le \ell \le n-1$) sets $M_\ell^2 = \hat{k}_\ell \| \hat{\delta}_\ell$, and U_n computes $T_n \leftarrow t_n^R \oplus (\hat{k}_n \| \hat{\delta}_n)$ and sets $M_n^2 = \omega_n \| T_n$. Each U_i ($1 \le i \le n+k$) generates a signature σ_i^2 on the message $M_i^1 \| M_i^2 \| \mathcal{I}' \| \mathsf{sid}_i$ and broadcasts $M_i^2 \| \sigma_i^2$.

  3. (Key Computation): Each $U_i \in \mathcal{I}'$ performs the same procedures as he/she does in the Key Computation phase of the Setup protocol, except that for each U_ℓ ($3 \le \ell \le n-1$), $t_\ell^L = t_2^L$ and $t_\ell^R = t_2^R$ are used. The final session key of each U_i is computed as $\mathsf{k}_i \leftarrow \tilde{\mathsf{F}}_{\hat{k}}(1)$ where $\hat{k} \leftarrow \bigoplus_{U_j \in \mathcal{I}'} \hat{k}_j$.

  4. (Post Computation) Each U_v ($v \in \{1, n+1, ..., n+k\}$) computes $h_v^L \leftarrow \hat{\mathsf{F}}_{g^{\hat{x}_{L(v)} \hat{x}_v}}(0)$, each U_j ($j \in \{n, n+1, ..., n+k\}$) computes $h_j^R \leftarrow \hat{\mathsf{F}}_{g^{\hat{x}_{R(j)} \hat{x}_j}}(0)$, and $h_1^R, h_n^L, h_\ell^L, h_\ell^R$ ($2 \le \ell \le n-1$) remain unchanged. Each U_i ($i \in \{1, 2, ..., n+k\}$) computes $X' \leftarrow \tilde{\mathsf{F}}_{\hat{k}}(0)$, saves (h_i^L, h_i^R, X') with $\mathsf{pid}_i = \mathcal{I}'$ and sid_i in the memory, and erases all other state information.

- Leave (Fig. 4): Let $\mathcal{I} = \{U_1, U_2, ..., U_n\}$ be an existing group where each member U_i has state information (h_i^L, h_i^R, X). Let $\mathcal{I}' = \mathcal{I} \backslash \mathcal{J}$ where $\mathcal{J} = \{U_{l_1}, U_{l_2}, ..., U_{l_{n'}}\}$ denotes the set of leaving users, and $N(\mathcal{J})$ be the set of neighbors of those leaving users, i.e. $N(\mathcal{J}) = \{U_{l_1-1}, U_{l_1+1}, ..., U_{l_{n'}-1}, U_{l_{n'}+1}\}$. The Leave protocol works as follows:

  1. (Round 1): Each U_i in $\mathcal{I}'$ randomly chooses $\tilde{k}_i \in \{0,1\}^\ell$ and computes $(\tilde{k}_i', \tilde{\delta}_i) \leftarrow \mathsf{CMT}(\tilde{k}_i)$. Each U_ℓ ($\ell \in \mathcal{I}' \backslash N(\mathcal{J})$) sets $M_\ell^1 = \tilde{k}_\ell'$. Each U_j in $N(\mathcal{J})$ additionally chooses $\tilde{x}_j \overset{\$}{\leftarrow} \mathbb{Z}_q^*$, and sets $M_j^1 = \tilde{k}_j' \| g^{\tilde{x}_j}$. Each U_i in $\mathcal{I}'$ broadcasts M_i^1.

  2. (Round 2): Upon receiving all the messages, each U_i in $\mathcal{I}'$ computes sid_i as the concatenation of all the $\tilde{k}_j'$ sent by $U_j \in \mathcal{I}'$. Each pair of neighbors U_{l_j-1} and U_{l_j+1} in $N(\mathcal{J})$ generate $h_{l_j-1}^R = \hat{\mathsf{F}}_{g^{\tilde{x}_{l_j+1} \tilde{x}_{l_j-1}}}(0)$ and $h_{l_j+1}^L = \hat{\mathsf{F}}_{g^{\tilde{x}_{l_j-1} \tilde{x}_{l_j+1}}}(0)$, respectively. Each U_i in $\mathcal{I}'$ generates $t_i^L = \hat{\mathsf{F}}_{h_i^L}(1)$, $t_i^R = \hat{\mathsf{F}}_{h_i^R}(1)$, and $\omega_i \leftarrow t_i^L \oplus t_i^R$. The user $U_{l_{n'}+1}$ additionally computes $T_{l_{n'}+1} \leftarrow t_{l_{n'}+1}^R \oplus (\tilde{k}_{l_{n'}+1} \| \tilde{\delta}_{l_{n'}+1})$. Each U_i in $\mathcal{I}' \backslash \{U_{l_{n'}+1}\}$ then sets $M_i^2 = \omega_i \| \tilde{k}_i \| \tilde{\delta}_i$, $U_{l_{n'}+1}$ sets $M_{l_{n'}+1}^2 = \omega_{l_{n'}+1} \| T_{l_{n'}+1}$. Finally, each

U_1	U_2	U_3	U_4	U_5
h_1^L, h_1^R, X	h_2^L, h_2^R, X	h_3^L, h_3^R, X	h_4^L, h_4^R, X	h_5^L, h_5^R, X

$$\hat{k}_i \xleftarrow{\$} \{0,1\}^\ell, (\hat{k}_i', \hat{\delta}_i) \leftarrow \mathsf{CMT}(\hat{k}_i), \hat{x}_i \xleftarrow{\$} \mathbb{Z}_q^*, \hat{x}_2 \leftarrow X$$

$M_1^1 = \hat{k}_1' \| g^{\hat{x}_1}$	$M_2^1 = \hat{k}_2' \| g^X$	$M_3^1 = \hat{k}_3'$	$M_4^1 = \hat{k}_4' \| g^{\hat{x}_4}$	$M_5^1 = \hat{k}_5' \| g^{\hat{x}_5}$

Broadcast Round 1

$$t_i^L \leftarrow \hat{\mathsf{F}}_{g^{\hat{x}_{L(i)}\hat{x}_i}}(1), t_i^R \leftarrow \hat{\mathsf{F}}_{g^{\hat{x}_{R(i)}\hat{x}_i}}(1) \text{ for } \{U_1, U_2, U_4, U_5\}, t_3^L = t_2^L \text{ \& } t_3^R = t_2^R$$

$$\omega_i \leftarrow t_i^L \oplus t_i^R, T_4 \leftarrow t_4^R \oplus (\hat{k}_4 \| \hat{\delta}_4), \mathsf{sid}_i \leftarrow \hat{k}_1' \| \hat{k}_2' \| \hat{k}_3' \| \hat{k}_4' \| \hat{k}_5', \sigma_i^2 \leftarrow \mathcal{DS}.Sign(sk_i, M_i^1 \| M_i^2 \| \mathcal{I}' \| \mathsf{sid}_i)$$

$M_1^2 = \omega_1 \| \hat{k}_1 \| \hat{\delta}_1, \sigma_1^2$	$M_2^2 = \omega_2 \| \hat{k}_2 \| \hat{\delta}_2, \sigma_2^2$	$M_3^2 = \hat{k}_3 \| \hat{\delta}_3, \sigma_3^2$	$M_4^2 = \omega_4 \| T_4, \sigma_4^2$	$M_5^2 = \omega_5 \| \hat{k}_5 \| \hat{\delta}_5, \sigma_5^2$

Broadcast Round 2

Session Key $\mathsf{k} = \tilde{\mathsf{F}}_{\hat{k}}(1)$ where $\hat{k} \leftarrow \hat{k}_1 \oplus \hat{k}_2 \oplus \ldots \oplus \hat{k}_5$

Post Computation

$$h'^L_i \leftarrow \hat{\mathsf{F}}_{g^{\hat{x}_{L(i)}\hat{x}_i}}(0), h'^R_i \leftarrow \hat{\mathsf{F}}_{g^{\hat{x}_{R(i)}\hat{x}_i}}(0), X' \leftarrow \tilde{\mathsf{F}}_{\hat{k}}(0)$$

h'^L_1, h_1^R, X'	h_2^L, h_2^R, X'	h_3^L, h_3^R, X'	h_4^L, h'^R_4, X'	h'^L_5, h'^R_5, X'

Fig. 3. The Join Protocol. $\mathcal{I}' = \{U_1, U_2, U_3, U_4, U_5\}, \mathcal{J} = \{U_5\}$.

U_i in $\mathcal{I}'$ generates a signature σ_i^2 on the message $M_i^1 \| M_i^2 \| \mathcal{I}' \| \mathsf{sid}_i$ and broadcasts $M_i^2 \| \sigma_i^2$.

3. (Key Computation): Each $U_i \in \mathcal{I}'$ performs the same procedures as he/she does in the Key Computation phase of the Setup protocol. The final session key of each U_i is computed as $\mathsf{k}_i \leftarrow \tilde{\mathsf{F}}_{\hat{k}}(1)$ where $\hat{k} \leftarrow \bigoplus_{U_j \in \mathcal{I}'} \tilde{k}_j$.

4. (Post Computation): Each $U_i \in \mathcal{I}'$ computes $h'^L_i \leftarrow \hat{\mathsf{F}}_{h_i^L}(0), h'^R_i \leftarrow \hat{\mathsf{F}}_{h_i^R}(0), X' \leftarrow \tilde{\mathsf{F}}_{\hat{k}}(0)$, saves (h'^L_i, h'^R_i, X') with $\mathsf{pid}_i = \mathcal{I}'$ and sid_i in the memory, and erases all other state information.

3.3 Security Analysis

We prove that our protocol is secure with respect to the security definitions (i.e. SK-, MA, and Co-security) given in Sec. 2.

Decisional Diffie-Hellman (DDH) Problem: Fix a generator g of $\mathbb{G}$. The DDH assumption claims that $\{g, g^a, g^b, Z\}$ and $\{g, g^a, g^b, g^{ab}\}$ are computationally indistinguishable where a, b are randomly selected from $\mathbb{Z}_q$ and Z is a random element of $\mathbb{G}$.

Theorem 1. *The proposed dynamic group key exchange protocol is SK-secure if the DDH assumption holds in the underlying group $\mathbb{G}$, $\mathcal{DS}$ is a uf-cma secure digital signature scheme, $\mathcal{CMT}$ is a uniformly distributed perfectly hiding commitment scheme, $\hat{\boldsymbol{F}}$ and $\tilde{\boldsymbol{F}}$ are two independent pseudo-random function families.*

We prove the Theorem in three cases: (1) the Test query is made to a setup session, (2) the Test query is made to a join session, and (3) the Test query is

U_1	U_2	U_3	U_4	U_5
h_1^L, h_1^R, X	h_2^L, h_2^R, X	h_3^L, h_3^R, X	h_4^L, h_4^R, X	h_5^L, h_5^R, X

$$\tilde{k}_i \overset{\$}{\leftarrow} \{0,1\}^\ell, (\tilde{k}_i', \tilde{\delta}_i) \leftarrow \mathsf{CMT}(\tilde{k}_i), \tilde{x}_i \overset{\$}{\leftarrow} \mathbb{Z}_q^*$$

$M_1^1 = \tilde{k}_1'$	$M_2^1 = \tilde{k}_2'$	$M_3^1 = \tilde{k}_3' \| g^{\tilde{x}_3}$		$M_5^1 = \tilde{k}_5' \| g^{\tilde{x}_5}$

Broadcast Round 1

$$h_3^R \leftarrow \hat{\mathsf{F}}_{(g^{\tilde{x}_5})^{\tilde{x}_3}}(0) \qquad\qquad h_5^L \leftarrow \hat{\mathsf{F}}_{(g^{\tilde{x}_3})^{\tilde{x}_5}}(0)$$

$$\mathsf{sid}_i \leftarrow \tilde{k}_1' \| \tilde{k}_2' \| \tilde{k}_3' \| \tilde{k}_5', t_i^L \leftarrow \hat{\mathsf{F}}_{h_i^L}(1), t_i^R \leftarrow \hat{\mathsf{F}}_{h_i^R}(1), \omega_i \leftarrow t_i^L \oplus t_i^R$$

$$T_5 \leftarrow t_5^R \oplus (\tilde{k}_5 \| \tilde{\delta}_5), \sigma_i^2 \leftarrow \mathcal{DS}.Sign(sk_i, M_i^1 \| M_i^2 \| \mathcal{I}' \| \mathsf{sid}_i)$$

$M_1^2 = \omega_1 \| \tilde{k}_1 \| \tilde{\delta}_1, \sigma_1^2$	$M_2^2 = \omega_2 \| \tilde{k}_2 \| \tilde{\delta}_2, \sigma_2^2$	$M_3^2 = \omega_3 \| \tilde{k}_3 \| \tilde{\delta}_3, \sigma_3^2$		$M_5^2 = \omega_5 \| T_5, \sigma_5^2$

Broadcast Round 2

Session Key $\mathsf{k} = \tilde{\mathsf{F}}_{\hat{k}}(1)$ where $\hat{k} = \tilde{k}_1 \oplus \tilde{k}_2 \oplus \tilde{k}_3 \oplus \tilde{k}_5$

Post Computation

$$h_i'^L \leftarrow \hat{\mathsf{F}}_{h_i^L}(0), h_i'^R \leftarrow \hat{\mathsf{F}}_{h_i^R}(0), X' \leftarrow \tilde{\mathsf{F}}_{\hat{k}}(0)$$

U_1	U_2	U_3	U_4	U_5
$h_1'^L, h_1'^R, X'$	$h_2'^L, h_2'^R, X'$	$h_3'^L, h_3'^R, X'$		$h_5'^L, h_5'^R, X'$

Fig. 4. The **Leave** Protocol. $\mathcal{I}' = \{U_1, U_2, U_3, U_5\}, \mathcal{J} = \{U_4\}$.

made to a leave session. To prove (1), we just follow the same approach as other existing work does, we define a sequence of games, starting from the original SK-security game, ending with a game in which the adversary has no advantage, and show that the difference between each two consecutive games is negligible. For case (2) and (3), we use the idea that $f(0)$ and $f(1)$ "looks" random and independent to any polynomial time adversary if $f(\cdot)$ is a secure pseudo-random function, so even if the adversary see one of them, we can still replace the other with a random element.

Theorem 2. *The proposed dynamic group key exchange protocol is MA-secure if $\mathcal{DS}$ is a uf-cma secure digital signature scheme, and $\mathcal{CMT}$ is a computationally binding commitment scheme.*

Theorem 3. *The proposed dynamic group key exchange protocol is Co-secure if $\mathcal{CMT}$ is a perfectly hiding and computationally binding commitment scheme, and $\tilde{\boldsymbol{F}}$ is a pseudo-random function family.*

The detailed proofs are deferred to the full paper.

4 Conclusions and Future Work

In this paper, we presented a new security model for dynamic group key exchange (DGKE) protocols and a new definition of contributiveness which captures partial-key control attacks. Comparing to existing security models, our new model is more concise and easy to use. We also presented a new DGKE protocol that provides strong security as well as high efficiency. Some possible future work includes 1) study the relationship among the existing partial-key control resistance notions; and 2) construct a robust protocol that can handle the situation of user crash during protocol execution.

276   G. Yang and C.H. Tan

References

1. Ateniese, G., Steiner, M., Tsudik, G.: Authenticated group key agreement and friends. In: ACM Conference on Computer and Communications Security, pp. 17–26 (1998)
2. Bellare, M., Canetti, R., Krawczyk, H.: A modular approach to the design and analysis of authentication and key exchange protocols. In: Proc. 30th ACM Symp. on Theory of Computing, pp. 419–428. ACM, New York (May 1998)
3. Bellare, M., Rogaway, P.: Entity authentication and key distribution. In: Stinson, D.R. (ed.) CRYPTO 1993. LNCS, vol. 773, pp. 232–249. Springer, Heidelberg (1994)
4. Bellare, M., Rogaway, P.: Provably secure session key distribution – the three party case. In: Proc. 27th ACM Symp. on Theory of Computing, Las Vegas, pp. 57–66. ACM, New York (1995)
5. Bellare, M., Pointcheval, D., Rogaway, P.: Authenticated key exchange secure against dictionary attacks. In: Preneel, B. (ed.) EUROCRYPT 2000. LNCS, vol. 1807, pp. 139–155. Springer, Heidelberg (2000)
6. Bohli, J.-M., Gonzalez Vasco, M.I., Steinwandt, R.: Secure group key establishment revisited. Int. J. Inf. Sec. 6(4), 243–254 (2007)
7. Bresson, E., Chevassut, O., Pointcheval, D.: Dynamic group Diffie-Hellman key exchange under standard assumptions. In: Knudsen, L.R. (ed.) EUROCRYPT 2002. LNCS, vol. 2332, pp. 321–336. Springer, Heidelberg (2002)
8. Bresson, E., Chevassut, O., Pointcheval, D.: Provably authenticated group Diffie-Hellman key exchange - the dynamic case. In: Boyd, C. (ed.) ASIACRYPT 2001. LNCS, vol. 2248, pp. 290–309. Springer, Heidelberg (2001)
9. Bresson, E., Chevassut, O., Pointcheval, D., Quisquater, J.-J.: Provably authenticated group Diffie-Hellman key exchange. In: ACM Conference on Computer and Communications Security, pp. 255–264 (2001)
10. Bresson, E., Manulis, M.: Contributory group key exchange in the presence of malicious participants. IET Information Security 2(3), 85–93 (2008)
11. Bresson, E., Manulis, M.: Securing group key exchange against strong corruptions and key registration attacks. International Journal of Applied Cryptography 1(2), 91–107 (2008)
12. Burmester, M., Desmedt, Y.: A secure and efficient conference key distribution system (extended abstract). In: De Santis, A. (ed.) EUROCRYPT 1994. LNCS, vol. 950, pp. 275–286. Springer, Heidelberg (1995)
13. Canetti, R., Krawczyk, H.: Analysis of key-exchange protocols and their use for building secure channels. In: Pfitzmann, B. (ed.) EUROCRYPT 2001. LNCS, vol. 2045, pp. 453–474. Springer, Heidelberg (2001)
14. Canetti, R.: Universally composable security: A new paradigm for cryptographic protocols. Cryptology ePrint Archive, Report 2000/067 (2000), http://eprint.iacr.org/
15. Desmedt, Y., Pieprzyk, J., Steinfeld, R., Wang, H.: A non-malleable group key exchange protocol robust against active insiders. In: Katsikas, S.K., López, J., Backes, M., Gritzalis, S., Preneel, B. (eds.) ISC 2006. LNCS, vol. 4176, pp. 459–475. Springer, Heidelberg (2006)
16. Diffie, W., Hellman, M.E.: New directions in cryptography. IEEE Transactions on Information Theory 22, 644–654 (1976)
17. Diffie, W., Van Oorschot, P.C., Wiener, M.J.: Authentication and authenticated key exchanges. Designs, Codes, and Cryptography 2(2), 107–125 (1992)

18. Dutta, R., Barua, R.: Constant round dynamic group key agreement. In: Zhou, J., López, J., Deng, R.H., Bao, F. (eds.) ISC 2005. LNCS, vol. 3650, pp. 74–88. Springer, Heidelberg (2005)
19. Furukawa, J., Armknecht, F., Kurosawa, K.: A universally composable group key exchange protocol with minimum communication effort. In: Ostrovsky, R., De Prisco, R., Visconti, I. (eds.) SCN 2008. LNCS, vol. 5229, pp. 392–408. Springer, Heidelberg (2008)
20. Choudary Gorantla, M., Boyd, C., González Nieto, J.M.: Modeling key compromise impersonation attacks on group key exchange protocols. In: Jarecki, S., Tsudik, G. (eds.) PKC 2009. LNCS, vol. 5443, pp. 105–123. Springer, Heidelberg (2009)
21. Just, M., Vaudenay, S.: Authenticated multi-party key agreement. In: Kim, K.-c., Matsumoto, T. (eds.) ASIACRYPT 1996. LNCS, vol. 1163, pp. 36–49. Springer, Heidelberg (1996)
22. Katz, J., Shin, J.S.: Modeling insider attacks on group key-exchange protocols. In: ACM Conference on Computer and Communications Security, pp. 180–189 (2005)
23. Katz, J., Yung, M.: Scalable protocols for authenticated group key exchange. In: Boneh, D. (ed.) CRYPTO 2003. LNCS, vol. 2729, pp. 110–125. Springer, Heidelberg (2003)
24. Kim, H.-J., Lee, S.-M., Lee, D.H.: Constant-round authenticated group key exchange for dynamic groups. In: Lee, P.J. (ed.) ASIACRYPT 2004. LNCS, vol. 3329, pp. 245–259. Springer, Heidelberg (2004)
25. Krawczyk, H.: HMQV: A High-Performance Secure Diffie-Hellman Protocol. In: Shoup, V. (ed.) CRYPTO 2005. LNCS, vol. 3621, pp. 546–566. Springer, Heidelberg (2005)
26. Manulis, M.: Provably secure group key exchange. PhD Thesis, Ruhr University Bochum (2007), http://www.manulis.eu/phd.html
27. Mitchell, C., Ward, M., Wilson, P.: On key control in key agreement protocols. Electronics Letters 34, 980–981 (1998)
28. Pedersen, T.P.: Non-interactive and information-theoretic secure verifiable secret sharing. In: Feigenbaum, J. (ed.) CRYPTO 1991. LNCS, vol. 576, pp. 129–140. Springer, Heidelberg (1991)
29. Steer, D.G., Strawczynski, L., Diffie, W., Wiener, M.J.: A secure audio teleconference system. In: Goldwasser, S. (ed.) CRYPTO 1988. LNCS, vol. 403, pp. 520–528. Springer, Heidelberg (1988)
30. Steiner, M., Tsudik, G., Waidner, M.: Diffie-Hellman key distribution extended to group communication. In: ACM Conference on Computer and Communications Security, pp. 31–37 (1996)
31. Tan, C.-H., Yang, G.: Comment on provably secure constant round contributory group key agreement in dynamic setting. IEEE Transactions on Information Theory (to appear)
32. Teo, J.C.M., Tan, C.H., Ng, J.M.: Security analysis of provably secure constant round dynamic group key agreement. IEICE Transactions 89-A(11), 3348–3350 (2006)

Towards Practical and Secure Coercion-Resistant Electronic Elections

Roberto Araújo[1], Narjes Ben Rajeb[2], Riadh Robbana[3],
Jacques Traoré[4], and Souheib Yousfi[5]

[1] Universidade Federal do Pará, ICEN, Faculdade de Computação, Brazil
[2] LIP2, INSAT, Tunisia
[3] LIP2, Tunisia Polytechnic School, Tunisia
[4] Orange Labs, France
[5] LIP2, ENIT, Tunisia

Abstract. Coercion-resistance is the most effective property to fight coercive attacks in Internet elections. This notion was introduced by Juels, Catalano, and Jakobsson (JCJ) at WPES 2005 together with a voting protocol that satisfies such a stringent security requirement. Unfortunately, their scheme has a quadratic complexity (the overhead for tallying authorities is quadratic in the number of votes) and would therefore not be suitable for large scale elections. Based on the work of JCJ, Schweisgut proposed a more efficient scheme. In this paper, we first show that Schweisgut's scheme is insecure. In particular, we describe an attack that allows a coercer to check whether a voter followed or not his instructions. We then present a new coercion-resistant election scheme with a linear complexity that overcomes the drawbacks of these previous proposals. Our solution relies on special *anonymous credentials* and is proven secure, in the random oracle model, under the q-Strong Diffie-Hellman and Strong Decisional Diffie-Hellman Inversion assumptions.

1 Introduction

Internet elections are far from being a consensus. On one hand, many people believe that the current technology is enough for deploying such elections in large scale. On the other hand, a number of voting researchers do not recommend them nowadays. They state that Internet elections have many intrinsic problems and that these problems must be addressed before carrying out these elections in real world scenarios. Despite of the disagreement, Estonia and the city of Geneva in Switzerland have made advances towards Internet elections. They have already developed voting systems and accomplished elections over Internet. Especially, in 2007, Estonia was the first country in the world to conduct online voting in parliamentary elections.

The success of the Internet elections in Geneva and Estonia may stimulate other countries to follow them and implement Internet voting in the near future. This may be boost by the many benefits of Internet elections over the traditional ones. Voters have the possibility to vote from any convenient place including the

S.-H. Heng, R.N. Wright, and B.-M. Goi (Eds.): CANS 2010, LNCS 6467, pp. 278–297, 2010.
© Springer-Verlag Berlin Heidelberg 2010

comfort of their residences and offices. Also, Internet elections may be more attractive for voters and consequently increase voter turnout. Other benefits include a faster computation of the voting results and a possible reduction of costs.

These elections, however, have been criticized and discouraged by specialists as they have a number of problems. One of them is the fact that Internet elections are susceptible to coercion and vote-selling. Because voters are free for voting from any place they desire, coercers and vote buyers can easily influence them to vote for their candidates. Anyone may imagine a scenario where a vote buyer offers money to a voter and later observes her voting for his candidate. In order to reach a large number of voters, adversaries may even automatize these attacks. As stated by Jefferson et al. [21], "the Internet can facilitate large scale vote buying by allowing vote buyers to automate the process".

Although coercion and vote-selling may be difficult to hold in Internet elections, a number of voting protocols that mitigate these problems were proposed. Some of them deal with these problems through the property of receipt-freeness. That is, these schemes prevent voters from making or obtaining any evidence about their votes that could be transfered to adversaries via network.

In 2005 a more powerful property with regard to coercion and vote-selling was introduced by Juels, Catalano, and Jakobsson (JCJ) [23], though. The property, called coercion-resistance, takes into account that a voter cannot be able to make receipts as the receipt-free one. Also, it considers that the adversary may threat voters to abstain from voting, to reveal her private data, or to cast random votes. The coercion-resistance is the most effective property nowadays to fight coercion and vote-selling. In order to accomplish this notion, JCJ also introduced the first scheme that satisfies it.

Related Work on Coercion-resistance

The coercion-resistant scheme of Juels, Catalano, and Jakobsson (JCJ) first appeared in 2002 in the Cryptology ePrint Archive [22]. After improvements, it was effectively published in 2005 at WPES [23]. This scheme represents another step in the development of secure Internet voting systems. It was the first scheme to fight realistic attacks not well considered in previous solutions.

JCJ's proposal relies on anonymous credentials to overcome coercive attacks. The voter receives a valid credential (i.e. an alphanumeric string) in a secure way and uses it to cast her vote. When under coercion, the voter makes a fake credential and follows the instructions of the coercer. Later on, when alone, the voter votes again using her valid credential; this is the vote that will be counted in the tallying phase. The adversary is unable to distinguish between the valid and the fake credential. This scheme, however, suffers from an intrinsic drawback. As described in their paper "the overhead for tallying authorities is quadratic in the number of voters". Consequently, their solution is impractical for large scale elections.

Following JCJ's work, several coercion-resistant schemes were proposed. Clarkson et al. [15] presented a variant of Prêt-à-Voter scheme suitable for Internet voting and based on decryption mix nets. Mix nets are cryptographic techniques used to anonymize messages (e.g. votes). They perform this by permuting a set of messages and then by decrypting (or reencrypting) the permuted messages.

Schweisgut [29] and more recently Clarkson et al. [14] proposed schemes which mitigate the inefficiency problem of the JCJ's solution. The former scheme relies on decryption mix nets and on a tamper-resistant hardware, whereas the latter is a modified version of JCJ's proposal.

One of the most promising schemes based on JCJ's ideas was introduced by Smith [30]. He presented an efficient scheme with linear work factor. Weber et al. [32], however, pointed out problems of Smith's proposal and presented a protocol that combines the ideas of JCJ with a variant of Smith's mechanism. Unfortunately, the solutions of Smith and Weber et al. are not coercion-resistant as showed in [2]. The problem of Smith's scheme was also noted independently by Clarkson et al. [14].

The first practical and secure coercion-resistant scheme was given by Araújo, Foulle, and Traoré [2]. This proposal, different from the previous ones, employs special formed credentials that allows the scheme to achieve a linear work factor. It avoids the mechanism of comparisons that makes the scheme of JCJ inefficient. Unfortunately, the security of their scheme is only conjectured.

Paper Contribution and Organization

In this paper, we first show a weakness of the scheme of Schweisgut [29]. In particular, we describe an attack that allows an adversary to check whether a voter followed or not his instructions.

We then introduce a new coercion-resistant voting scheme. Our solution is practical and can be used in elections that comprehend a large number of voters. The new proposal employs some ideas similar to that presented in [2]. However, our scheme differs from the previous solution mainly in two aspects. First, we employ new anonymous credentials whose security relies on a different problem. These credentials are shorter than the credentials presented in [2] and make our proposal more efficient than the previous one. Second, while in [2] they did not prove that their scheme is coercion-resistant, we formally prove that our solution fulfills this security requirement.

This work is organized as follows: in the next section, we show that the proposal of Schweisgut is not coercion-resistant. In Section 3, we first introduce the main assumptions on which rely the security of our new voting protocol. We then present the main cryptographic building blocks used in our scheme and recall the game-based definition of coercion-resistance introduced by JCJ. We next describe our new proposal. In Section 4, we present a formal security analysis of our solution. Finally, in Section 5, we conclude this work.

2 Weaknesses on a Known Coercion-Resistant Solution

In this section we briefly describe a known coercion-resistant proposal. The scheme was given by Schweisgut [29] and aims at being more efficient than JCJ's protocol. However, we show here that Schweisgut's scheme is not coercion-resistant as claimed.

2.1 The Protocol of Schweisgut

As the original proposal of JCJ, the scheme employs anonymous credentials. These credentials identify eligible voters without revealing their identity. They also allow the voter to deceive adversaries. More specifically, the voter has a valid credential that she uses when she is not under coercion. When coerced by an adversary, the voter is able to make a fake credential and use it. As the coercer cannot distinguish between a valid and a fake credential, he cannot determine whether the voter gave him a valid credential or not.

In the scheme of Schweisgut, in particular, the voter has only two credentials. One of them is a valid one and the other is a fake one. Both credentials are stored in an observer, i.e. a tamper resistant device. Taking into account that a public generator g (among other public parameters) and that an El Gamal key pair [19] of the talliers (where T is the public key) were previously generated, the scheme is briefly described as follows:

Registration Phase. After authenticating the voter, the registration authorities (registrars) generate a random valid credential σ and encrypts it producing $E_T[\sigma]$ (where $E_X[m]$ means an El Gamal encryption of a message m computed with the public key X). They then transfer $E_T[\sigma]$ to the voter that stores it in her observer. The voter now generates a random fake credential σ', encrypts it, and stores $E_T[\sigma']$ on her observer. At the end of this phase, the registrars send a list of encrypted valid credentials through a verifiable decryption mix net (i.e. the mixes have to prove that they have correctly permuted and decrypted the tuples) and publish the mix net results.

Voting Phase. In order to vote, the voter interacts with her observer. During this, she selects two random numbers a, a', uses a to encrypt her vote v and obtains $E_T[v]$; she then employs the other random number to compute $g^{a'}$, and sends $E_T[v]$ and $g^{a'}$ to the observer. The observer now selects two fresh random numbers b and b', reencrypts $E_T[v]$ with b and obtains $E_T[v]'$, and reencrypts the encrypted valid credential $E_T[\sigma]$ (or the encrypted fake credential $E_T[\sigma']$ according to the voter intention) to obtains $E_T[\sigma]'$; it then computes $g^{a'+b'}$ and $O = [b \cdot H(g, E_T[v]', E_T[\sigma]', g^{a'+b'}) + b']$, where H is a secure hash function, and sends back to the voter: $\langle g^{a'+b'}, E_T[v]', E_T[\sigma]', O \rangle$; O is a non-malleability proof. After receiving the values from the observer, the voter computes $O' = [(a + b) \cdot H(g, E_T[v]', E_T[\sigma]', g^{a'+b'}) + (a' + b')]$ and publishes on a bulletin board the following tuple: $\langle g^{a'+b'}, E_T[v]', E_T[\sigma]', O', P \rangle$, where P is a proof that $E_T[v]'$ contains a valid vote. This is performed via an anonymous channel.

Tallying Phase. Once the voting period is finished, the talliers first verify the proof of non-malleability O' and the proof P. After excluding votes with invalid proofs, the talliers apply a plaintext equivalence test[1] [20] to identify credentials used more than once (i.e. duplicates); the talliers keeps one of the duplicates based on a policy (e.g. the last posted vote). Then, the talliers send the votes (i.e. the remaining tuples $\langle E_T[v]', E_T[\sigma]'\rangle$) through a *verifiable* decryption mix net. Finally, the resulting mixed credentials are compared with the valid credentials processed in the registration phase. A match identify a valid vote. Observe that the plaintext pairs (i.e. vote and credential) are published on a bulletin board so that anyone can verify the correctness of the protocol.

2.2   A Weakness in Schweisgut's Scheme

At first glance, the scheme of Schweisgut seems to be coercion-resistant. However, the coercer could use a simple strategy to test whether a credential is valid or not. Suppose a coercer forces the voter to reveal the encrypted credential $E_T[\sigma]$ he received from the registrars. The coercer then employs this encrypted credential to compute a new ciphertext in such a way that their underlying plaintexts satisfy a specific relation R; for example, the coercer could select a random value t and compute $E_T[t \cdot \sigma]'$ from $E_T[\sigma]$ and t (by exploiting the fact that El Gamal is malleable). He then computes the proof O' for the new encrypted credential $t \cdot \sigma$. Note that the coercer can make this proof himself without needing an observer and without knowing the plaintext σ. As the proof only involves the exponent used to encrypt the vote and the coercer makes this ciphertext, he knows the corresponding exponent and can make the proof himself. For this, he selects new fresh random numbers i, j, i', j' and computes: $O' = [(i + j) \cdot H(g, E_T[v], E_T[t \cdot \sigma]', g^{i'+j'}) + (i' + j')]$; this proof will hold as the one generated by an observer since its verification is true. The coercer then posts two votes (i.e. two tuples) on the bulletin board: one with the encrypted credential $E_T[\sigma]$ received from the voter and one containing the encrypted credential $E_T[t \cdot \sigma]'$.

In the tallying phase, after sending all tuples $\langle E_T[v]', E_T[\sigma]'\rangle$ to the decryption mix net, the talliers obtain a list L with pairs $\langle v, \sigma \rangle$; this list is published on a bulletin board. In order to verify the voter gave him a valid or an invalid credential, the coercer reads a σ in L and uses the value t to compute $t \cdot \sigma$. The coercer then search on L for a value $t \cdot \sigma$. When a match is found, the coercer verifies whether the vote corresponding to the credential σ was removed or not from the count. If this occurs, the coercer learns that the voter gave him an invalid credential and punishes her. Otherwise, the coercer can be sure that he received the correct credential and rewards the voter. If no match is found, the coercer repeats the process with another credential. Observe that, in the worst case, the complexity of this attack is roughly in $O(|L|log|L|)$ operations.

---

[1] This is a cryptographic primitive that operates on ciphertexts in a threshold cryptosystem. The input to a Plaintext Equivalence Test is a pair of ciphertexts; the output is a single bit indicating whether the corresponding plaintexts are equal or not.

3   Our New Coercion-Resistant Protocol for Internet Voting

As we showed in Section 2, the proposal of Schweisgut is not coercion-resistant as an adversary is able to distinguish between a valid and an invalid credential. In this section we introduce a new coercion-resistant scheme. Our proposal is based on JCJ's ideas and employs new credentials that prevent adversaries from checking them. The new credentials have their security based on known problems and are different from the credentials used in past coercion-resistant proposals.

3.1   Preliminaries

Notation. Let A be an algorithm. By $A(\cdot)$ we denote that A has one input (resp., by $A(\cdot, ..., \cdot)$ we denote that A has several inputs). By $y \leftarrow A(x)$, we denote that y was obtained by running A on input x. If A is deterministic, then y is unique; if A is probabilistic, then y is a random variable. If S is a finite set, then $y \leftarrow S$ denotes that y was chosen from S uniformly at random.

By $A^O(\cdot)$, we denote a Turing machine that makes queries to an oracle O.

Let b be a boolean function. By $(y \leftarrow A(x) : b(y))$, we denote the event that $b(y) = 1$ after y was generated by running A on input x. The statement $\Pr[\{x_i \leftarrow A_i(y_i)\}_{1 \leq i \leq n} : b(x_n)] = \alpha$ means that the probability that $b(x_n) = 1$ after the value x_n was obtained by running algorithms $A_1,..., A_n$ on inputs $y_1,..., y_n$ is α, where the probability is over the random choices of the probabilistic algorithms involved.

According to the standard definition, we say that a quantity $f(k)$ is *negligible* in k if for every positive integer c there is some l_c such that $f(k) < k^{-c}$ for $k > l_c$. In most cases, we use the term negligible alone to mean negligible with respect to the full set of relevant security parameters. Similarly, in saying that an algorithm has *polynomial running time*, we mean that its running time is asymptotically bounded by some polynomial in the relevant security parameters.

Complexity Assumptions. The security of our voting protocol relies on the following assumptions:

In [4], Boneh and Boyen introduced a new computational problem in *bilinear* context. However, for our purpose, we will consider this problem in the classical discrete log setting, i.e. without *bilinear map*.

q-Strong Diffie-Hellman assumption I (q-SDH-I) [4]: Let k denotes a security parameter. Let G be a group of prime order p with $2^k < p < 2^{k+1}$ and g a random generator in G. We say that the q-SDH-I assumption holds in G if for all polynomial-time adversaries $\mathcal{A}$ the advantage

$$\text{Adv}_{G,\mathcal{A}}^{q\text{-SDH-I}}(k) = \Pr[y \leftarrow Z_p^*; (c, A) \leftarrow \mathcal{A}(g, g^y, ..., g^{y^q}) : c \in Z_p \wedge A = g^{1/(y+c)}]$$

is a negligible function in k.

q-Strong Diffie-Hellman assumption II (q-SDH-II): Let k denotes a security parameter. Let G be a group of prime order p with $2^k < p < 2^{k+1}$ and g_1 and g_2 two random generators in G. We say that the q-SDH-II assumption holds in G if for all polynomial-time adversaries $\mathcal{A}$ the advantage

$$\mathrm{Adv}_{G,\mathcal{A}}^{q\text{-SDH-II}}(k) = \Pr[y \leftarrow Z_p^*; \{(x_i, r_i) \leftarrow Z_p^2; A_i = (g_1 g_2^{x_i})^{1/(y+r_i)}; B_i =$$
$$(x_i, r_i, A_i)\}_{1 \leq i \leq q-1}; B = (x, r, A) \leftarrow \mathcal{A}(g_1, g_2, g_2^y, B_1, ..., B_{q-1}) :$$
$$(x, r) \in Z_p^2 \wedge A = (g_1 g_2^x)^{1/(y+r)} \wedge \{B \neq B_i\}_{1 \leq i \leq q-1}]$$

is a negligible function in k.

Lemma 1. *If the q-SDH-I assumption holds in G then the q-SDH-II assumption holds in G*

Proof. see [17] for a proof of this Lemma.

The security of our voting protocol also relies on the Decision Diffie-Hellman assumption and on strongest variants of this assumption.

Decision Diffie-Hellman assumption (DDH) [3]: Let k denotes a security parameter. Let G be a group of prime order p with $2^k < p < 2^{k+1}$. We let $\mathrm{D_{DDH}}$ be the distribution (g, g^x, g^y, g^{xy}) in G^4 where g is a random generator in G and x, y are uniform in Z_p. We let $\mathrm{R_{DDH}}$ be the distribution (g, g^x, g^y, g^z) where g is a random generator in G and x, y, z are uniform in Z_p subject to $z \neq xy$. We say that the DDH assumption holds in G if for all polynomial-time adversaries $\mathcal{A}$ the advantage

$$\mathrm{Adv}_{G,\mathcal{A}}^{\mathrm{DDH}}(k) = |\Pr[x \leftarrow \mathrm{D_{DDH}} : \mathcal{A}(x) = 1] - \Pr[x \leftarrow \mathrm{R_{DDH}} : \mathcal{A}(x) = 1]|$$

is a negligible function in k.

The following assumption has been introduced by Camenisch et al. [8] in order to prove the security of their e-token system.

Strong Decision Diffie-Hellman Inversion assumption I (SDDHI-I) [8]: Let k denotes a security parameter. Let G be a group of prime order p with $2^k < p < 2^{k+1}$ and g a random generator in G. Let $O_a(\cdot)$ be an oracle that, on input $z \in Z_p^*$, outputs $g^{1/(a+z)}$. We say that the SDDHI-I assumption holds in G if for all polynomial-time adversaries $\mathcal{A}$, that do not query the oracle on r, the advantage

$$\mathrm{Adv}_{G,\mathcal{A}}^{\mathrm{SDDHI-I}}(k) = |\Pr[a \leftarrow Z_q^*; (r, \alpha) \leftarrow \mathcal{A}^{O_a}(g, g^a); y_0 = g^{1/(a+r)}; y_1 \leftarrow G;$$
$$b \leftarrow \{0, 1\}; b' \leftarrow \mathcal{A}^{O_a}(y_b, \alpha) : b = b'] - 1/2|$$

is a negligible function in k.

Strong Decision Diffie-Hellman Inversion assumption II (SDDHI-II):
Let k denotes a security parameter. Let G be a group of prime order p with
$2^k < p < 2^{k+1}$ and g_1 and g_2 two random generators in G. Let $O_a(\cdot)$ be an
oracle that, on input $z, t \in Z_p^*$, outputs $(g_1 g_2^t)^{1/(a+z)}$. We say that the SDDHI-
II assumption holds in G if for all polynomial-time adversaries $\mathcal{A}$, that do not
query the oracle on r, the advantage

$$\mathrm{Adv}_{G,\mathcal{A}}^{\mathrm{SDDHI\text{-}II}}(k) = |\Pr[a \leftarrow Z_q^*; \; (x, r, \alpha) \leftarrow \mathcal{A}^{O_a}(g_1, g_2, g_2^a); y_0 = (g_1 g_2^x)^{1/(a+r)};$$
$$y_1 \leftarrow G; \; b \leftarrow \{0,1\}; \; b' \leftarrow \mathcal{A}^{O_a}(y_b, \alpha) : b = b'] - 1/2|$$

is a negligible function in k.

Lemma 2. *If the SDDHI-I assumption holds in G then the DDH assumption holds in G*

Lemma 3. *If the SDDHI-I assumption holds in G then the SDDHI-II assumption holds in G*

For ease of presentation, we will call in the sequel q-SDH (respectively SDDHI)
assumption the q-SDH-II (respectively SDDHI-II) assumption.

3.2   Cryptographic Building Blocks

The new voting scheme requires a set of cryptographic primitives to ensure its
security. We describe next these primitives.

Bulletin Boards. The new scheme requires information to be publicly published so that anyone can verify them. In order to perform this, the scheme relies
on a bulletin board communication model. By using this model, the scheme allows anyone to post information on bulletin boards. However, no one can delete
or alter any information published on the board. The proposal of Cachin et al. [7]
may be used to implement the bulletin boards required here.

A Threshold Cryptosystem. Our scheme relies on a threshold version of a
semantically secure cryptosystem with homomorphic property. We require here,
though, the Modified El Gamal cryptosystem proposed by JCJ [23]. This variant,
is described as follows: let G be a cyclic group of order p where the Decision Diffie-
Hellman problem (see Boneh [3] for details) is hard. The public key is composed
of the elements $(g_1, g_2, h = g_1^{x_1} g_2^{x_2})$ with $g_1, g_2 \in G$ and the corresponding private
key is formed by $x_1, x_2 \in Z_p$. The Modified El Gamal ciphertext of a message
$m \in G$ is $(M = g_1^s, N = g_2^s, O = mh^s)$, where $s \in Z_p$ is a random number. The
message m is obtained from the ciphertext (M, N, O) by $O/(M^{x_1} N^{x_2})$. In the
threshold version, the El Gamal public key and its corresponding private key are
cooperatively generated by n parties; though, the private key is shared among
the parties. In order to decrypt a ciphertext, a minimal number of t out of n
parties is necessary. The Modified El Gamal cryptosystem is *semantically secure*

under the DDH assumption. Borrowing freely from the exposition in [23], we provide, for completeness, a sketched version of this proof. Suppose there exists a probabilistic polynomial time algorithm $\mathcal{A}$ which can break the semantic security of the Modified El Gamal cryptosystem then there exists an algorithm $\mathcal{B}$ that breaks the Decision Diffie-Hellman problem. We prove this claim by constructing $\mathcal{B}$ as follows. So assume that $\mathcal{B}$ receives on input a quadruple (g_1, g_2, h_1, h_2) from the challenger $\mathcal{C}$ of the DDH problem and has to determine whether this quadruple follows the D_{DDH} distribution or not. $\mathcal{B}$ constructs the public key for the Modified El Gamal scheme as follows. It chooses x_1 and x_2 at random, sets $h = g_1^{x_1} g_2^{x_2}$ and sends (g_1, g_2, h) to $\mathcal{A}$ as the challenge parameters of the Modified El Gamal scheme.

When $\mathcal{A}$ will come up with the two messages m_0, m_1 he wants to be challenged on, $\mathcal{B}$ will proceed as follows. It flips a random (private) bit b, and encrypts m_b as follows: $(h_1^k, h_2^k, m h_1^{k x_1} h_2^{k x_2})$ where k is a random value.

Note that if the given quadruple is a DH one then the ciphertext has the right distribution. This is because $h_1^k = g_1^{k'}$ and $h_2^k = g_2^{k'}$ for some k' and $(h_1^{x_1} h_2^{x_2})^k = h^{k'}$ (for the same k').

If on the other hand, the given quadruple is not a DH one then it is easy to check that $\mathcal{A}$ gains no information at all about the encrypted message (this is because this time to decrypt, $\mathcal{A}$ has to know the secret exponents x_1 and x_2 which remains information theoretically hidden by h). The latter property will be important in the proof that our voting protocol is coercion-resistant.

Universally Verifiable Mix Nets. In some steps of our scheme we employ mix nets to provide anonymity. This cryptographic primitive was introduced by Chaum [12] and further developed by many other authors. It performs by permuting messages, and by reencrypting or by decrypting them. Our scheme requires a re-encryption mix net based on the El Gamal cryptosystem. However, in order to reduce the trust in the mix process, the mix net should be universally verifiable. That is, after mixing messages, the mix net must prove publicly the correctness of its shuffle. The proposals of Neff [26], and Furukawa and Sako [18] are examples of universally verifiable mix nets.

Non-Interactive Zero-knowledge Proofs. The proposal we present below also requires several zero-knowledge proofs of knowledge. Zero-knowledge proofs of knowledge are interactive protocols between a verifier and a prover allowing a prover to assure the verifier his knowledge of a secret, without any leakage on it. These primitives help ensuring security in our solution. Our scheme employs a proof of knowledge of a discrete logarithm [28] to make ciphertexts plaintext aware (i.e. the party who makes the ciphertext should be aware of what he is encrypting) and so preventing the use of the El Gamal malleability by adversaries; in addition, the verifier should check that the components of the ciphertexts are of order p to prevent the attacks described in [9]. The solution, moreover, requires a protocol to prove that a ciphertext contains a vote for a valid candidate. Besides these protocols, our proposal uses the discrete logarithm equality test owing to Chaum and Pedersen [13], a protocol for proving knowledge of a representation, such as the one proposed by Okamoto [27], and a plaintext equivalence test [20].

Especially, our scheme requires a zero-knowledge proof of knowledge of the plaintext related to a M-El Gamal ciphertext $(M = g_1^s, N = g_2^s, O = mh^s)$, and a proof that this plaintext is $\neq 1$. The former proof is accomplished by first proving the knowledge of the discrete logarithms of the M-El Gamal terms M, N in the bases g_1, g_2, respectively. Then, we prove the representation of O in the bases B and h, where B denote the basis of a special formed plaintext m and h is the M-El Gamal public parameter. We finally prove that the discrete logarithm of M, N in the bases g_1, g_2 is equal to the second component of the representation of O in the bases B and h. A description of a similar proof can be found in [1]. For the latter proof, we prove that the discrete logarithm of M, N in the base g_1, g_2 is different from the discrete logarithm of O in the base h. This proof can be performed by means of the protocol of Camenisch and Shoup (see Section 5 of [10]).

These interactive proofs can also be used non-interactively (a.k.a *signatures of knowledge*) by using the Fiat-Shamir heuristic [16]. We will use in the sequel the following notation $\text{POK}[\alpha, \beta, ... : Predicate]$ to denote a non-interactive zero-knowledge proof (NIZKP) proving that the prover knows values $(\alpha, \beta, ...)$ satisfying the predicate *Predicate*. In this notation, the Greek letters will denote the secret knowledge and the others letters will denote public parameters between the prover and the verifier. For example, using this notation, $\text{POK}[\alpha : h = g^\alpha]$ will denote a proof of knowledge of the discrete logarithm of h in the base g.

3.3 Attack Model

Our coercion-resistant proposal follows the general idea presented by JCJ in their scheme. This let our scheme inherits some caracteristics from the original proposal. The security model under which our scheme relies on is similar to that one of JCJ. We take into account the following assumptions:

Limited Computational Power and Small Number of Authorities. An adversary has limited computational power and may compromise only a small number of authorities. He can force the voter to reveal any secret information that she is holding. Also, he can force her to abstain from voting or to post a random composed ballot as her vote;

Interactions with the Voter. The adversary cannot monitor or interact with the voter constantly during the whole voting process. However, he may interact with the voter occasionally during the voting;

A Registration Phase Free of Adversaries. The registration official is trustworthy and voters receive private data securely. Also, we assume that the voters communicate with the registrar via an untappable channel and without the interference of adversaries. This channel provides information-theoretical secrecy to the communication;

Anonymous Channels in the Voting Phase. The existence of some anonymous channels in the voting phase. These channels are used by the voters to post their votes and prevent adversaries from learning who sent a specific vote. In practice, voters may use computers in public places to achieve this or a mix net;

Trustworthy Voting Computers. The computers that the voters use for vote are trustworthy. We do not consider attacks where the adversary may control the voters' computers (e.g. by means of malwares) in order to obtain their votes or other private data.

Denial of Service Attacks are not considered. The scheme employs bulletin boards that receive data from anyone and hence would be susceptible to these attacks.

3.4   Formal Definitions

We will use the security model introduced by JCJ [23]. The essential properties are *correctness, verifiability,* and *coercion-resistance,* respectively abbreviated *corr, ver,* and *c-resist* in the sequel. Following [23], we will only focus on the formal security definition of the property of coercion-resistance as the two other properties (correctness and verifiability) are more classical and of less relevance to our work (see JCJ for formal definitions of these two properties).

In [23] coercion resistance centers on a kind of game between the adversary $\mathcal{A}$ and a voter targeted by the adversary for coercive attack. A coin is flipped; the outcome is represented by a bit b. If $b = 0$ then the coerced voter V_0 casts a ballot of its choice β, and provides the adversary with a false voting key (fake credential) $\widetilde{sk}$; in other words, the voter attempts to evade adversarial coercion. If $b = 1$, then the voter submits to the coercion of the adversary; she gives him her valid voting key (credential) sk and does not cast a ballot. The task of the adversary is to guess the value of the coin b, that is to determine the behavior of the voter. An election scheme ES is coercion-resistant, according to JCJ's definition [23], if for any polynomially-bounded adversary $\mathcal{A}$, any parameters n and n_C, and any probability distribution D_{n,n_C}, the quantity

$$\mathrm{Adv}_{ES,\mathcal{A}}^{c\text{-}resist} = \left| \mathbf{Succ}_{ES,\mathcal{A}}^{c\text{-}resist}(\cdot) - \mathbf{Succ}_{ES,\mathcal{A}}^{c\text{-}resist\text{-}ideal}(\cdot) \right|$$

is negligible in all security parameters for any voter function V_0. Where:

- n denotes the number of voters outside the control of the adversary
- n_C denotes the total number of candidates.
- D_{n,n_C} denotes a probability distribution that models the state of knowledge of the adversary about the intentions of honest voters
- $\mathbf{Exp}_{ES,\mathcal{A}}^{c\text{-}resist}$ represents the game between the adversary and the voter.
- $\mathbf{Exp}_{ES,\mathcal{A}}^{c\text{-}resist\text{-}ideal}$ represents an *ideal* voting experiment, between the adversary and the voter, in which the adversary never sees the bulletin board.
- $\mathbf{Succ}_{ES,\mathcal{A}}^{E}(\cdot) = \Pr[\mathbf{Exp}_{ES,\mathcal{A}}^{E}(\cdot) = '1']$

Intuitively, the definition of JCJ [23], means that in a real protocol execution, the adversary effectively learns nothing more than the election tally X. The adversary cannot learn any significant information from the protocol execution itself, even when mounting an active attack (see [23] for more details about this definition and [25,31,11] for alternative definitions in the simulation-based model and also [24] for an alternative definition in the game-based model).

3.5   Anonymous Credentials

Anonymous credentials have an important role in coercion-resistant schemes. They make possible voters to deceive adversaries when under coercion and to vote later on. In most of the coercion resistant schemes such as JCJ, a valid credential is a random string. Our scheme, however, employs a different technique of credentials that differs from past proposals. Our new credentials bear some similarities with the *membership certificates* of the group signature scheme of Boneh, Boyen, and Shacham [5].

The credentials used in our solution are presented as follows: let G be a cyclic group with prime order p where the Decision Diffie-Hellman (DDH) problem (see [3] for details) is assumed to be hard, y a secret key, (g_1, g_3) two random generators of G and (r, x) two random numbers in Z_p^*. The credential is composed of (A, r, x), where $A = (g_1 g_3^x)^{\frac{1}{y+r}}$. A credential in our system therefore corresponds to a *membership certificate* in Boneh et al's group signature scheme (we could in fact either use the original version of their membership certificates or the extended one described in Section 8 of [5]).

Due to the mathematical structure of our credentials, their security depends on two known assumptions: the *q-Strong Diffie-Hellman* and the *Strong Decisional Diffie-Hellman Inversion*. The *q-Strong Diffie-Hellman* assumption (*q*-SDH for short) ensures that even if an adversary has many genuine credentials (A_i, r_i, x_i), it is hard for him to forge a new and valid credential (A, r, x) with $(r, x) \neq (r_i, x_i)$ for all i (see Lemma 1). This assumption is known to hold for generic groups and the security of Boneh et al.'s group signature scheme also relies on it. The Strong Decisional Diffie-Hellman Inversion assumption (SDDHI), which also holds in generic group, ensures that an *active* coercer, ignoring the secret key y, cannot decide whether a triplet (A, r, x) is a valid credential or not; in other words, whether it satisfies or not the following equation: $A^{y+r} \overset{?}{=} g_1 g_3^x$. This way, a voter under coercion will generate a random value x' and give to the coercer a fake credential (A, r, x') instead of his valid credential (A, r, x). Under the SDDHI assumption, the adversary will not be able to distinguish between a fake credential and a valid one.

In a real-world scenario, our credential can be seen as containing two parts: a short one, that is x, which must be kept secret, and a longer one, that is (A, r). The first part (i.e. x) has around twenty ASCII characters (this corresponds to 160 bits, the actual secure size for the order of generic groups), so a small piece of paper and a pen are sufficient to write x down. The other part (A, r) can be stored in a device or be even sent by email to the voter without compromising (under the SDDHI assumption) the credential security.

3.6   An Overview of the Scheme

Before showing the details of our proposal, we give an intuition of the new scheme.

The protocol begins in a setup phase. In this phase, a set of authorities in cooperation generate the key materials and publish the corresponding public parameters on a bulletin board. In particular, they publish the public key of a threshold homomorphic cryptosystem. After this stage, the registration phase takes place. In order to register to vote, voters prove their identities to trustworthy registration authorities (registrars). These authorities issue for each voter a unique and valid credential. The voter uses this credential to cast her vote in the voting phase.

At time of voting, the voter makes a tuple containing her vote. The tuple contains the ciphertext of the vote and ciphertexts corresponding to a credential along with a set of non-interactive zero-knowledge proofs showing the validity of these encryptions. The voter casts his vote by sending this tuple to a bulletin board via an anonymous channel. When she wants to cast her vote, the voter makes the tuple using the credential she received in the registration phase. However, if the voter is under coercion, she makes a fake credential and may either use it to cast an invalid vote (i.e. a vote that will not be counted) or give this fake credential to an adversary. The adversary is not able to distinguish between the valid and the fake credential. The voter may vote again later on using her valid credential.

In the tallying phase, a set of talliers cooperate to compute the voting results. For this, they first verify some proofs on each tuple and discard tuples with invalid proofs. After that, the talliers check part of each tuple to detect tuples posted with the same credential (duplicates). Based on the order of postings of these tuples, the talliers keep the last posted tuple and eliminate the other ones. Now, the talliers send the remaining tuples through a verifiable mix net. From the mix net output, they begin to identify the tuples posted with valid credentials. These tuples contains the votes to be counted. The validity of the credentials is checked under the encryption by exploiting the homomorphic property of the underlying cryptosystem. After identifying the tuples with valid credentials, the authorities in cooperation decrypt the corresponding vote ciphertext and publish the voting results.

3.7   The Protocol in Details

We present now the description of our scheme in details. This description consider the building blocks and the attack model presented above.

Participants and Notation. The solution is composed of four phases: setup, registration, voting, and tallying. The setup phase is where the parameters of the voting are generated. The key pairs used in the scheme, for example, are generated in this phase. In the registration phase, the voter registers to the authorities and receives a valid credential. Afterwards, in the voting phase, the voter uses her credential to express her vote intention. Finally, in the tallying phase, the voting results are computed and published. In order to perform the steps to be described in these phases, the scheme considers three participants:

- The voter is identified by $\mathcal{B}$. She obtains a valid credential σ and is able to produce a fake credential σ'. The valid credential is used for posting a valid vote, whereas the fake one is used to deceive adversaries;
- The talliers are composed of a set of authorities and are denoted by T. They control the bulletin board, run the mix net, and compute the voting results. They share an M-El Gamal private key $\widehat{T}$ corresponding to a public key T;
- The registrars, as the talliers, are composed of more than one and are identified by R. They authenticate each eligible voter in the registration phase and issue a valid credential for her. They share a M-El Gamal private key $\widehat{R}$ associated to a public key R.

In addition to the notation above, we use the following one: BB is a bulletin board, $E_T[m]$ is a M-El Gamal encryption of a message m computed with the public key T, and $D_{\widehat{T}}[m]$ is a M-El Gamal decryption of m with the private key $\widehat{T}$.

Setup of the Voting Parameters. This phase takes place prior to the registration and is necessary for the definition of the voting parameters. In order to establish these parameters, first a cyclic group $\mathbb{G}$ with prime order p is defined. The Decision Diffie-Hellman problem must be hard in this group. After that, the authorities produces four generators $g_1, g_2, g_3, o \in \mathbb{G}$. The talliers T now collaborate to generate the public key $T = (g_1, g_2, h = g_1^{x_1} g_2^{x_2})$ of the Modified El Gamal threshold cryptosystem and its corresponding private key $\widehat{T} = (x_1, x_2)$. The resulting key $\widehat{T}$ is not known by the authorities individually. Each authority knows only a share of this key. The registrars R also cooperate to establish a public key $R = g_3^y$ and the corresponding shared private key $\widehat{R} = y$.

Registration Phase. After generating their keys, the registrars are ready to issue credentials for the voters. In order to receive a secret credential, the voter first proves to the registrars that she is eligible to vote. R then selects two random numbers $r, x \in \mathbb{Z}_p$ and computes: $A = (g_1 g_3^x)^{\frac{1}{y+r}}$ (which implies that $A^{y+r} = g_1 g_3^x$ and then that $A^y = g_1 g_3^x A^{-r}$). After that, R issues to the voter the secret credential $\sigma = (A, r, x)$.[2] The registrars might generate the credential in a threshold fashion. The communication between the voter and registars is performed through an untappable channel. Following JCJ, we assume that the majority of players in R are honest and can thus ensure that R provides the voter $\mathcal{B}$ with a valid credential. Nonetheless, it is possible for R to furnish the voter with a proof that $\sigma = (A, r, x)$ is a valid credential. To do this, R has to compute a non-interactive proof of knowledge that the discrete logarithm of $g_1 g_3^x A^{-r}$ (which should be equal to A^y if $\sigma = (A, r, x)$ is a valid credential) in the base A is equal to the discrete logarithm of $R = g_3^y$ in the base g_3. In order to prevent the

---

[2] In a variant, the value x could be jointly generated by the voter and the registrars.

voter from transferring this proof (and thus to prevent coercion), R should instead issue a *designated verifier proof* of the equality of these discrete logarithms.

Voting Phase. In order to vote, the voter with credential (A, r, x) first selects a random $s \in \mathbb{Z}_p^*$ and computes $B = A^s$ using the element A of her credential. After that, she computes the tuple: $\langle\ E_T[v], B, E_T[B^{s^{-1}}], E_T[B^{rs^{-1}}],$ $E_T[g_3^x], o^x, \Pi\ \rangle$ which is equal to $\langle E_T[v], B, E_T[A], E_T[A^r], E_T[g_3^x]\ , o^x, \Pi\rangle =$ $\langle C_1, B, C_2, C_3, C_4, o^x, \Pi\rangle$. The voter then publishes his tuple on a public bulletin board by means of an anonymous channel.

The tuple is composed of the ciphertext $E_T[v]$ that contains the voter's choice, the value B, the ciphertexts $\langle E_T[B^{s^{-1}}], E_T[B^{rs^{-1}}]\rangle$ that correspond to part of the credential σ, the ciphertext $E_T[g_3^x]$ that encrypts the public generator g_3 to the power of the part x of σ. In addition, it has a set of non-interactive zero-knowledge proofs Π. This set contains:

- (Π_1) A proof that $C_1 = E_T[v]$ encrypts a valid vote (i.e, that v represents a valid candidate choice).
- (Π_2) A proof that the voter knows the plaintext related to the ciphertext $C_2 = E_T[B^{s^{-1}}] = (M_1, N_1, O_1)$. In particular, the voter will have to prove that he knows the representation of O_1 in the bases B and h using the protocol proposed by Okamoto [27]. In other words, he will have to prove that he knows a pair (β, α) such that $O_1 = B^\beta h^\alpha$: $\Pi_2 = \text{POK}[\alpha, \beta : M_1 = g_1^\alpha \wedge N_1 = g_2^\alpha \wedge O_1 = B^\beta h^\alpha]$
- (Π_3) A proof that the voter knows the plaintext related to the ciphertext $C_3 = E_T[B^{rs^{-1}}] = (M_2, N_2, O_2)$. In particular, the voter will have to prove that he knows the representation of O_2 in the bases B and h using the protocol proposed by Okamoto [27]. In other words, he will have to prove that he knows a pair (θ, δ) such that $O_2 = B^\theta h^\delta$: $\Pi_3 = \text{POK}[\delta, \theta : M_2 = g_1^\delta \wedge N_2 = g_2^\delta \wedge O_2 = B^\theta h^\delta]$
- (Π_4) A proof that the voter knows the plaintext related to the ciphertext $C_4 = E_T[g_3^x] = (M_3, N_3, O_3)$: $\Pi_4 = \text{POK}[\lambda, \mu : M_3 = g_1^\lambda \wedge N_3 = g_2^\lambda \wedge O_3 = g_3^\mu h^\lambda]$
- (Π_5) A proof that the plaintext of $C_2 = E_T[B^{s^{-1}}]$ is different from 1, as explained in Section 3.2: $\Pi_5 = \text{POK}[\alpha, \beta : M_1 = g_1^\alpha \wedge N_1 = g_2^\alpha \wedge O_1 = B^\beta h^\alpha \wedge \beta \neq 0 \bmod p]$
- (Π_6) A proof that the voter knows the discrete logarithm of $O = o^x$ in the basis o and that it is equal to the discrete logarithm of the plaintext of $C_4 = E_T[g_3^x]$ in the basis g_3: $\Pi_6 = \text{POK}[\lambda, \mu : M_3 = g_1^\lambda \wedge N_3 = g_2^\lambda \wedge O_3 = g_3^\mu h^\lambda \wedge O = o^\mu]$

Remark: All these proofs of knowledge may be accomplished using standard techniques such as the ones mentioned in Section 3.2. As is standard practice, the challenge values for these proofs of knowledge are constructed using a call to a cryptographic hash function (the Fiat-Shamir heuristic [16]), modeled in our security analysis by a random oracle. The inputs to this cryptographic hash function for these challenges values should include $ID_{Election}$ (a random election identifier), B, $C_1 = E_T[v]$, $C_2 = E_T[B^{s^{-1}}]$, $C_3 = E_T[B^{rs^{-1}}]$, $C_4 = E_T[g_3^x]$,

$O = o^x$ and other values required for the realization of these non-interactive zero-knowledge proofs. In this way, the actual vote $C_1 = E_T[v]$ submitted by a voter will be bound to the remaining voting material. Observe that, in contrast to the scheme of JCJ that employs the plaintext equivalence test [20] to eliminate duplicates, i.e. votes posted with the same credential, our scheme uses the value o^x to perform the identification of these votes. This ensures that just one vote per credential will be processed in the tabulation phase.

Tabulation Phase. After the end of the voting period, the talliers T read all tuples $\langle E_T[v], B, E_T[B^{s^{-1}}], E_T[B^{rs^{-1}}], E_T[g_3^x], o^x, \Pi \rangle$ posted on the bulletin board and process them as follows:

1. **Checking proofs:** T checks all proofs Π_i's that compose the tuples and discards tuples with incorrect proofs. In other words, T verifies that $E_T[v]$ contains a vote for a valid candidate, checks the proof of knowledge of the plaintexts with regards to $\langle E_T[B^{s^{-1}}], E_T[B^{rs^{-1}}]\rangle$, the proof that $E_T[B^{s^{-1}}]$ does not encrypts the plaintext 1, the proof of knowledge of the plaintext related to the ciphertext $C_4 = E_T[g_3^x]$, as well as the equality of the discrete logarithm of the plaintext of $C_4 = E_T[g_3^x]$ in the basis g_3 and of o^x in the basis o. The tuples that passed the test are processed in the next step without Π and B, that is, it now contains the values $\langle E_T[v], E_T[B^{s^{-1}}], E_T[B^{rs^{-1}}], E_T[g_3^x], o^x \rangle$;

2. **Eliminating duplicates:** In order to detect and remove tuples posted with the same credential (i.e. duplicates), T compares all o^x by means of a hash table. If a duplicate is detected, T keeps the last posted tuple (based on the order of posting on the bulletin board) and removes the other ones. T processes in the next step the values $\langle E_T[v], E_T[B^{s^{-1}}], E_T[B^{rs^{-1}}], E_T[g_3^x]\rangle$;

3. **Mixing:** T now sends all tuples composed of $\langle E_T[v], E_T[B^{s^{-1}}], E_T[B^{rs^{-1}}], E_T[g_3^x]\rangle$ to a verifiable mix net. The mix net outputs a permuted and re-encrypted set of tuples $\langle E_T[v]', E_T[B^{s^{-1}}]', E_T[B^{rs^{-1}}]', E_T[g_3^x]'\rangle$, where $E_T[X]'$ means the re-encryption of $E_T[X]$.

4. **Checking credentials:** From the mixed tuples $\langle E_T[v]', E_T[B^{s^{-1}}]', E_T[B^{rs^{-1}}]', E_T[g_3^x]'\rangle$, the talliers T perform along with the registrars R the following steps (for each tuple) to identify the valid votes (that is the ones which *encrypt* valid credentials $\sigma = (A, r, x)$ satisfying the relation $A^{y+r} = g_1 g_3^x$):

   (a) By means of his secret key y, R cooperatively computes: $E_T[B^{s^{-1}}]'^y = E_T[B^{ys^{-1}}]'$. Now R uses the El Gamal homomorphic property to compute: $E_T[B^{ys^{-1}}]' \cdot E_T[B^{rs^{-1}}]' = E_T[B^{ys^{-1}+rs^{-1}}]'$

   (b) T now computes $C = E_T[B^{ys^{-1}+rs^{-1}}] g_1^{-1} g_3^{-x}]'$ from $E_T[B^{ys^{-1}+rs^{-1}}]'$, $E_T[g_3^x]'$, and the public parameter g_1. Note that if we denote by $A = B^{s^{-1}}$ then $E_T[B^{ys^{-1}+rs^{-1}}]' = E_T[A^{y+r}]'$. Hence, $C = E_T[A^{y+r} g_1^{-1} g_3^{-x}]'$.

   (c) In order to identify a valid credential, T executes a Plaintext Equivalence Test in order to determine whether C is an encryption of the plaintext 1 or not. For this, T cooperatively selects a random number $z \in \mathbb{Z}_p$ and computes C^z. T then decrypts C^z. If the decryption result is equal to 1, the credential is a valid one. Otherwise, the result will be a random number and this indicates an invalid credential.

5. **Tallying:** T discards all tuples with invalid credentials and cooperatively decrypts the value $E_T[v]'$ of the tuples with valid credentials.

4   Security Analysis

In this section, we define the security properties our scheme provides. Following JCJ [23], we will however explain why our protocol satisfies the standard security requirements i.e. correctness, democracy, verifiability and coercion-resistance.

Correctness. The purpose of the proofs Π_i's is to ensure that only ballots (tuples) which contain valid credentials will be counted. Indeed, when the talliers along with the registrars will perform the test described at Step 4 of the Tabulation Phase to determine whether the ballot contains a valid credential or not, they will compute the following ciphertext which is, using the above notation (see the **Voting Phase**), equal to $E_T[B^{y\beta}B^{\theta}g_1^{-1}g_3^{-\mu}]$. In other words this is an encryption of $B^{y\beta+\theta}g_1^{-1}g_3^{-\mu}$. If this is an encryption of 1, this means that $B^{y\beta+\theta}g_1^{-1}g_3^{-\mu} = 1$. Remember that the voter has also to prove that $\beta \neq 0$ (mod p) in the voting phase (using the proof of knowledge owing to Camenisch and Shoup [10]). Let us denote by $A = B^{\beta}$, $r = \theta/\beta$ and $x = \mu$. So $B^{y\beta+\theta}g_1^{-1}g_3^{-\mu} = 1$ can be rewritten as follows: $A^{y+r}g_1^{-1}g_3^{-x} = 1$ which is equivalent to $A^{y+r} = g_1g_3^x$. In other words, if all the zero-knowledge proofs are valid, this means that the voter knows a tuple (A, r, x) such that $A^{y+r} = g_1g_3^x$. Therefore he knows a valid credential. The proof that $\beta \neq 0$ (mod p) is crucial. Without this proof, an adversary could generate a ballot that will pass the test, without knowing a valid credential. Thanks to this proof, only ballots that encrypt a valid credential will pass this test.

The method employed in this verification ensures therefore that a valid vote cannot be identified as invalid or vice versa. In addition, no one can produce valid credentials as this would involve breaking the q-SDH assumption. Therefore, only the votes from eligible voters will appear in the final count.

Democracy. By removing duplicates in step 2, we ensure that only one vote per credential (valid or fake) is processed in the remaining steps. This is performed by comparing all values o^x and then by verifying two or more values o^x match. If this takes place, only the last posted vote is considered in the next step; the others are discard. The fact that only one valid credential is processed in the next steps ensures that one vote per eligible voter will be counted.

Universal Verifiability. The bulletin board and the NIZKPs allow anyone to verify the tuples and the work of the talliers. Each posted tuple contains a set of NIZKPs. These proofs allow anyone to verify that the tuple is well-formed. Anyone can perform this and complain to the authorities in case where tuples with invalid proofs are sent to the next step. In the same way, after processing the tuples in each step of the tallying phase, the authorities publish proofs on the bulletin board. In step 2, for example, as all the tuples are published, anyone can perform the comparisons. In addition, the mix net provide NIZKPs after processing the tuples in step 3; the authorities publish NIZKPs after verifying the credentials in step 4c; especially, the authorities prove that they have used the correct share after powering the encrypted credential to a random number.

Coercion-resistance. Our voting protocol satisfies the coercion-resistance requirement as defined in [23].

Theorem 1. *The proposed voting protocol satisfies the coercion resistance requirement, in the random oracle model, under the q-SDH and SDDHI assumptions.*

Owing to space limitations, the proof of this theorem is omitted from this extended abstract and will appear in the full version of the paper.

5 Conclusion

In this paper, we have introduced a new coercion-resistant scheme. Our solution (which has a linear work factor instead of a quadratic work factor in previous solutions) is practical and secure. It employs anonymous credentials that have a similar structure than the membership certificates in the group signature scheme of Boneh, Boyen, and Shacham [5]. These credentials have their security based on the *q-Strong Diffie-Hellman* and on the *Strong Decisional Diffie-Hellman Inversion* assumptions.

Differently from some previous schemes, the new credentials can be used in more than one election. This way, a voter does not need to issue a new credential each time a new election takes place. We have also shown that a recent coercion-resistant voting protocol is insecure.

References

1. Araújo, R.: On Remote and Voter-Verifiable Voting. PhD thesis, Technische Universität Darmstadt, Darmstadt, Germany (September 2008)
2. Araújo, R., Foulle, S., Traoré, J.: A practical and secure coercion-resistant scheme for remote elections. In: Chaum, D., Kutylowski, M., Rivest, R.L., Ryan, P.Y.A. (eds.) Frontiers of Electronic Voting, Dagstuhl Seminar Proceedings, Dagstuhl, Germany, vol. 07311, Internationales Begegnungs- und Forschungszentrum für Informatik (IBFI), Schloss Dagstuhl, Germany (2008)
3. Boneh, D.: The decision diffie-hellman problem. In: Buhler, J. (ed.) ANTS 1998. LNCS, vol. 1423, pp. 48–63. Springer, Heidelberg (1998)
4. Boneh, D., Boyen, X.: Short signatures without random oracles. In: Cachin, C., Camenisch, J.L. (eds.) EUROCRYPT 2004. LNCS, vol. 3027, pp. 56–73. Springer, Heidelberg (2004)
5. Boneh, D., Boyen, X., Shacham, H.: Short group signatures. In: Franklin, M. (ed.) CRYPTO 2004. LNCS, vol. 3152, pp. 41–55. Springer, Heidelberg (2004)
6. Brickell, E.F. (ed.): CRYPTO 1992. LNCS, vol. 740. Springer, Heidelberg (1993)

7. Cachin, C., Kursawe, K., Shoup, V.: Random oracles in constantipole: practical asynchronous byzantine agreement using cryptography (extended abstract). In: Neiger, G. (ed.) PODC, pp. 123–132. ACM, New York (2000)
8. Camenisch, J., Hohenberger, S., Kohlweiss, M., Lysyanskaya, A., Meyerovich, M.: How to win the clone wars: efficient periodic n-times anonymous authentication. In: Juels, A., Wright, R.N., De Capitani di Vimercati, S. (eds.) ACM Conference on Computer and Communications Security, pp. 201–210. ACM, New York (2006)
9. Camenisch, J., Kiayias, A., Yung, M.: On the portability of generalized schnorr proofs. In: Joux, A. (ed.) EUROCRYPT 2009. LNCS, vol. 5479, pp. 425–442. Springer, Heidelberg (2009)
10. Camenisch, J., Shoup, V.: Practical verifiable encryption and decryption of discrete logarithms. In: Boneh, D. (ed.) CRYPTO 2003. LNCS, vol. 2729, pp. 126–144. Springer, Heidelberg (2003)
11. Canetti, R., Gennaro, R.: Incoercible multiparty computation (extended abstract). In: FOCS, pp. 504–513 (1996)
12. Chaum, D.: Untraceable electronic mail, return addresses and digital pseudonyms. Communications of the ACM 24(2), 84–88 (1981)
13. Chaum, D., Pedersen, T.P.: Wallet databases with observers. In: Brickell [6], pp. 89–105
14. Clarkson, M.R., Chong, S., Myers, A.C.: Civitas: Toward a secure voting system. In: IEEE Symposium on Security and Privacy, pp. 354–368. IEEE Computer Society, Los Alamitos (2008)
15. Clarkson, M.R., Myers, A.C.: Coercion-resistant remote voting using decryption mixes. In: Workshop on Frontiers in Electronic Elections (2005)
16. Fiat, A., Shamir, A.: How to prove yourself: Practical solutions to identification and signature problems. In: Odlyzko, A.M. (ed.) CRYPTO 1986. LNCS, vol. 263, pp. 186–194. Springer, Heidelberg (1986)
17. Fuchsbauer, G., Pointcheval, D., Vergnaud, D.: Transferable constant-size fair e-cash. Cryptology ePrint Archive, Report 2009/146 (2009), http://eprint.iacr.org/
18. Furukawa, J., Sako, K.: An efficient publicly verifiable mix-net for long inputs. In: Di Crescenzo, G., Rubin, A. (eds.) FC 2006. LNCS, vol. 4107, pp. 111–125. Springer, Heidelberg (2006)
19. El Gamal, T.: A public key cryptosystem and a signature scheme based on discrete logarithms. In: Blakely, G.R., Chaum, D. (eds.) CRYPTO 1984. LNCS, vol. 196, pp. 10–18. Springer, Heidelberg (1984)
20. Jakobsson, M., Juels, A.: Mix and match: Secure function evaluation via ciphertexts. In: Okamoto, T. (ed.) ASIACRYPT 2000. LNCS, vol. 1976, pp. 162–177. Springer, Heidelberg (2000)
21. Jefferson, D., Rubin, A., Simons, B., Wagner, D.: A security analysis of the secure electronic registration and voting experiment (2004)
22. Juels, A., Catalano, D., Jakobsson, M.: Coercion-resistant electronic elections. Cryptology ePrint Archive, Report 2002/165 (2002), http://eprint.iacr.org/
23. Juels, A., Catalano, D., Jakobsson, M.: Coercion-resistant electronic elections. In: Atluri, V., De Capitani di Vimercati, S., Dingledine, R. (eds.) WPES, pp. 61–70. ACM, New York (2005)
24. Kuesters, R., Truderung, T., Vogt, A.: A game-based definition of coercion-resistance and its applications. Cryptology ePrint Archive, Report 2009/582 (2009), http://eprint.iacr.org/

25. Moran, T., Naor, M.: Receipt-free universally-verifiable voting with everlasting privacy. In: Dwork, C. (ed.) CRYPTO 2006. LNCS, vol. 4117, pp. 373–392. Springer, Heidelberg (2006)
26. Neff, C.A.: A verifiable secret shuffle and its application to e-voting. In: ACM Conference on Computer and Communications Security, pp. 116–125 (2001)
27. Okamoto, T.: Provably secure and practical identification schemes and corresponding signature schemes. In: Brickell (ed.) [6], pp. 31–53 (1992)
28. Schnorr, C.-P.: Efficient signature generation by smart cards. J. Cryptology 4(3), 161–174 (1991)
29. Schweisgut, J.: Coercion-resistant electronic elections with observer. In: Krimmer, R. (ed.) Electronic Voting. LNI, vol. 86, pp. 171–177. GI (2006)
30. Smith, W.: New cryptographic election protocol with best-known theoretical properties. In: Workshop on Frontiers in Electronic Elections (2005)
31. Unruh, D., Müller-Quade, J.: Universally composable incoercibility. Cryptology ePrint Archive, Report 2009/520 (2009), http://eprint.iacr.org/
32. Weber, S.G., Araújo, R., Buchmann, J.: On coercion-resistant electronic elections with linear work. In: 2nd Workshop on Dependability and Security in e-Government (DeSeGov 2007) at 2nd Int. Conference on Availability, Reliability and Security (ARES 2007), pp. 908–916. IEEE Computer Society, Los Alamitos (2007)

Predicate Encryption with Partial Public Keys

Carlo Blundo, Vincenzo Iovino, and Giuseppe Persiano

Dipartimento di Informatica ed Applicazioni
Università di Salerno
I-84084 Fisciano (SA), Italy
{carblu,iovino,giuper}@dia.unisa.it

Abstract. *Predicate encryption* is a new powerful cryptographic primitive which allows for fine-grained access control for encrypted data: the owner of the secret key can release partial keys, called *tokens*, that can decrypt only a specific subset of ciphertexts. More specifically, in a predicate encryption scheme, ciphertexts and tokens have attributes and a token can decrypt a ciphertext if and only if a certain predicate of the two associated attributes holds.

In this paper, ciphertext attributes are vectors x of fixed length ℓ over an alphabet Σ and token attributes, called *patterns*, are vectors y of the same length over the alphabet $\Sigma_\star = \Sigma \cup \{\star\}$. We consider the predicate $\mathsf{Match}(x, y)$ introduced by [BW06] which is true if and only if $x = \langle x_1, \ldots, x_\ell \rangle$ and $y = \langle y_1, \ldots, y_\ell \rangle$ agree in all positions i for which $y_i \neq \star$.

Various security notions are relevant for predicate encryption schemes. First of all, one wants the ciphertexts to hide its attributes (this property is called semantic security). In addition, it makes sense also to consider the property of *token security*, a security notion in which the token is required not to reveal any information on the associated pattern. It is easy to see that predicate privacy is impossible to achieve in a public-key setting. In [SSW09], the authors considered the notion of a predicate encryption scheme in the symmetric-key setting and gave the first construction with token security.

In this paper, we consider the notion of a *partial public key encryption* (as suggested in [SSW09]) in which a partial public key allows a user to generate only a subset of the ciphertexts. We give a construction which is semantically secure and in which a token does not reveal any information on the associated pattern except for the locations of the $\star$'s. The proofs of security of our construction are based on hardness assumptions in bilinear groups of *prime* order; this greatly improves the efficiency of the construction when compared to previous constructions ([SSW09]) which used groups of composite orders.

Our security proofs do not use random oracles.

1   Introduction

In a predicate encryption scheme, ciphertexts and keys have attributes and a key can decrypt a certain ciphertext if and only if a certain predicate on the

S.-H. Heng, R.N. Wright, and B.-M. Goi (Eds.): CANS 2010, LNCS 6467, pp. 298–313, 2010.

two attributes holds. In this paper, ciphertext attributes x are vectors of fixed length ℓ over an alphabet Σ and key attributes (also called *patterns*) are vectors of the same length over the alphabet $\Sigma_\star = \Sigma \cup \{\star\}$. We consider the predicate $\mathsf{Match}(x, y)$ which is true if and only if $x = \langle x_1, \ldots, x_\ell \rangle$ and $y = \langle y_1, \ldots, y_\ell \rangle$ agree in all positions i for which $y_i \neq \star$.

We are interested in two security requirements which, roughly speaking, can be described as follows. We first require that a ciphertext $\tilde{X}$ should hide all information on the associated attribute vector x (we call this notion Semantic Security). In addition, we require that a key T (also called a *token*) should hide all information on the associated pattern y (we call this notion Token Security). Formal definitions of the two security requirements are found in Section 2. We would like to stress though that Token Security is not achievable in a pure public-key scenario: given token T for an unknown pattern y an adversary could check if $\mathsf{Match}(x, y)$ holds by creating a ciphertext C for attribute vector x using the public key, and then testing T against C. We thus consider the *partial public key* model in which the key owner can decide on a policy that describes which subset of the ciphertexts can be generated. More specifically, a policy $\mathsf{Pol} = \langle \mathsf{Pol}_1, \ldots, \mathsf{Pol}_\ell \rangle$ is simply a vector of length ℓ of subsets of Σ with the following intended meaning: the public key associated with policy Pol allows to create ciphertexts with attribute vector $x = \langle x_1, \ldots, x_\ell \rangle$ iff and only for $i \in [\ell]$ we have that $x_i \in \mathsf{Pol}_i$. The private key scenario corresponds to a policy Pol with $\mathsf{Pol}_i = \emptyset$ for all i's; whereas a public key scenario corresponds to a policy with $\mathsf{Pol}_i = \Sigma$ for all i's. For example, for $\ell = 2$, $\Sigma = \{0, 1\}$, and policy $\mathsf{Pol} = \langle \{1\}, \{0, 1\} \rangle$, then public key $\mathsf{PPK}_{\mathsf{Pol}}$ associated with Pol allows to create ciphertexts with attribute vector $x = \langle 1, 0 \rangle$ but not $x = \langle 0, 1 \rangle$. In the formal definition of Token Security we thus require that an adversary is not able to distinguish between tokens with pattern y_0 or y_1 with respect to a policy Pol provided that the two patterns have the same value of the predicate Match for all attributes x that can be encrypted under policy Pol.

Previous work. The first example of predicate encryption scheme has been given by Boneh et al. [BDOP04] that introduced the concept of an encryption scheme supporting equality test. Roughly speaking, in such an encryption scheme, the owner of the public key can compute, for any message M, a token T_M that allows to test if a given ciphertext encrypts message M without obtaining any additional information. More recently, along this line of research, Goyal et al. [GPSW06] have introduced the concept of an attribute-based encryption scheme (ABE scheme). In an ABE scheme, a ciphertext is labeled with a set of attributes and private keys are associated with a predicate. A private key can decrypt a ciphertext iff the attributes of the ciphertext satisfy the predicate associated with the key. An ABE scheme can thus been seen as a special encryption scheme for which, given the key associated with a predicate P, one can test whether a given ciphertext carries a message M that satisfies predicates P without having to decrypt and without getting any additional information. The construction of [GPSW06] is very general as it supports any predicate that can be expressed as a circuit with threshold gates but attributes associated with a ciphertexts appear in clear in a ciphertext. Boneh and Waters [BW07] were the first to give predicate encryption schemes that guaranteed security of the attributes for the Match predicate and showed that this implies construction for several families of predicates including conjunctions of equality, range predicate and subset predicates. This has been subsequently extended to disjunctions, polynomial equations and inner products [KSW08]. Both constructions are based on hardness assumptions regarding bilinear groups on *composite order*. Iovino and Persiano [IP08] gave more efficient constructions based on hardness assumptions regarding bilinear group of *prime order*. Shen et al. [SSW09] were the first to consider the issue of token security and gave *private-key* predicate encryption schemes for inner product based on hardness assumptions regarding bilinear group of order product of four primes.

Our results. In this paper we give a predicate encryption scheme with partial public keys based on hardness assumptions regarding bilinear group of prime order for the Match predicate. Being able to use prime order groups greatly improves the efficiency of the resulting encryption schemes since, for the same level of security, our constructions uses groups of much smaller order. Our scheme guarantees privacy of the attributes associated with the ciphertexts (see Definition 3). In addition, we also show that tokens only reveal the positions of the $\star$-entries in the associated pattern. More precisely, for any two patterns $\boldsymbol{y}_0$ and $\boldsymbol{y}_1$ that have $\star$-entries in the same positions, no probabilistic polynomial time adversary can distinguish a token for $\boldsymbol{y}_0$ from a token for $\boldsymbol{y}_1$ better than guessing at random (see Definition 4).

2   Predicate Encryption Schemes with Partial Public Keys

In this section we present the notion of a *predicate encryption scheme with partial public keys*. Following [SSW09, KSW08], we present our definitions (and constructions in Section 4) for the case in which the ciphertexts are *predicate-only*; that is, they do not carry any message and only specify the attributes. It is straightforward to extend the definitions (and the constructions) to the case in which ciphertexts carry a message.

In the following we will denote by $[\ell]$ the set $\{1, \ldots, \ell\}$ of natural numbers. We let Σ denote an *alphabet* (that is, a finite set of symbols) and let 2^Σ denote its power set (that is, the family of all subsets of Σ). Furthermore, we let $\Sigma_\star$ denote the alphabet Σ augmented with the special symbol $\star$. Finally, we say that function $\nu : \mathbb{N} \rightarrow [0, 1]$ is *negligible* if, for all polynomials poly and sufficiently large n, we have that $\nu(n) \leq 1/\mathsf{poly}(n)$.

We start by defining the notion of a *policy* and of an *allowed attribute vector* for a policy.

Definition 1. *Fix the number $\ell > 0$ of attributes and alphabet Σ. A policy* $\mathsf{Pol} = \langle \mathsf{Pol}_1, \ldots, \mathsf{Pol}_\ell \rangle \in (2^\Sigma \setminus \emptyset)^\ell$ *is a sequence of ℓ non-empty subsets of Σ. The* *set* $\mathbb{X}_{\mathsf{Pol}}$ *of allowed attribute vectors for policy* Pol *consists of all vectors* $\boldsymbol{x} \in \Sigma^\ell$ *such that for* $i \in [\ell]$ *we have that* $x_i \in \mathsf{Pol}_i$.

Our predicate encryption schemes are for the predicate $\mathsf{Match} : \Sigma^\ell \times \Sigma_\star^\ell \to \{0, 1\}$ defined as follows: $\mathsf{Match}(\boldsymbol{x}, \boldsymbol{y}) = 1$ if and only if $\boldsymbol{x} = \langle x_1, \ldots, x_\ell \rangle$ and $\boldsymbol{y} = \langle y_1, \ldots, y_\ell \rangle$ agree in all positions i for which $y_i \neq \star$. We remark that a predicate encryption scheme for the Match predicate implies efficient constructions for several other predicates (see [BW07] for the descriptions of the reductions).

Definition 2. *A* Predicate Encryption Scheme with Partial Public Keys *for the predicate* Match *consists of five algorithms:*

$\mathsf{Setup}(1^n, 1^\ell)$: *Given the security parameter* n *and the number of attributes* $\ell = \mathsf{poly}(n)$, *procedure* Setup *outputs the* secret key SK.

$\mathsf{PPKeyGen}(\mathsf{SK}, \mathsf{Pol})$: *Given the secret key* SK *and the policy* $\mathsf{Pol} \in (2^\Sigma \setminus \emptyset)^\ell$, *procedure* PPKeyGen *outputs the* partial public key $\mathsf{PPK}_{\mathsf{Pol}}$ *relative to policy* Pol. *We denote by* PK *the public key relative to policy* $\mathsf{Pol} = \Sigma^\ell$.

$\mathsf{Encryption}(\mathsf{PPK}_{\mathsf{Pol}}, \boldsymbol{x})$: *Given the partial public key* $\mathsf{PPK}_{\mathsf{Pol}}$ *relative to policy* Pol *and the attribute vector* $\boldsymbol{x} \in \mathbb{X}_{\mathsf{Pol}}$, *procedure* Encryption *outputs an encrypted attribute vector* $\tilde{X}$.

$\mathsf{GenToken}(\mathsf{SK}, \boldsymbol{y})$: *Given the secret key* SK *and the pattern vector* $\boldsymbol{y} \in \Sigma_\star^\ell$, *procedure* GenToken *outputs token* $T_{\boldsymbol{y}}$.

$\mathsf{Test}(\tilde{X}, T_{\boldsymbol{y}})$: *given the encrypted attribute vector* $\tilde{X}$ *corresponding to attribute vector* $\boldsymbol{x}$ *and the token* $T_{\boldsymbol{y}}$ *corresponding to pattern* $\boldsymbol{y}$, *procedure* Test *returns* $\mathsf{Match}(\boldsymbol{x}, \boldsymbol{y})$ *with overwhelming probability. More precisely, for all* $\ell = \mathsf{poly}(n)$, *all policies* $\mathsf{Pol} \in (2^\Sigma \setminus \emptyset)^\ell$, *all attribute vectors* $\boldsymbol{x} \in \mathbb{X}_{\mathsf{Pol}}$, *and all patterns* $\boldsymbol{y} \in \Sigma_\star^\ell$, *we have that*

$$\mathrm{Prob}[\, \mathsf{SK} \leftarrow \mathsf{Setup}(1^n, 1^\ell); \ \mathsf{PPK}_{\mathsf{Pol}} \leftarrow \mathsf{PPKeyGen}(\mathsf{SK}, \mathsf{Pol}) :$$
$$\mathsf{Test}(\mathsf{Encryption}(\mathsf{PPK}_{\mathsf{Pol}}, \boldsymbol{x}), \mathsf{GenToken}(\mathsf{SK}, \boldsymbol{y})) \neq \mathsf{Match}(\boldsymbol{x}, \boldsymbol{y}) \,]$$

is negligible in n.

Next we state security in the selective attribute model.

2.1   Semantic Security

Semantic security deals with an adversary that tries to learn information from ciphertexts. We define the security requirement by means of an indistinguishability experiment in which the adversary $\mathcal{A}$ selects two challenge attribute vectors $\boldsymbol{z}_0$ and $\boldsymbol{z}_1$ and a policy Pol. The adversary $\mathcal{A}$ then receives the partial public key $\mathsf{PPK}_{\mathsf{Pol}}$ and is allowed to issue token queries for patterns $\boldsymbol{y}$ such that $\mathsf{Match}(\boldsymbol{z}_0, \boldsymbol{y}) = \mathsf{Match}(\boldsymbol{z}_1, \boldsymbol{y}) = 0$. Finally, $\mathcal{A}$ receives encrypted attribute vector $\tilde{X}$ corresponding to a randomly chosen challenge attribute vector $\boldsymbol{z}_\eta$. We require that $\mathcal{A}$ has probability essentially $1/2$ of guessing η.

We model the semantic security property by means of the following game $\mathsf{SemanticExp}_{\mathcal{A}}$ between a challenger $\mathcal{C}$ and adversary $\mathcal{A}$.

SemanticExp$_{\mathcal{A}}(1^n, 1^\ell)$

1. Initialization Phase. The adversary $\mathcal{A}$ announces two challenge attribute vectors $z_0, z_1 \in \Sigma^\ell$ and policy Pol $\in (2^\Sigma \setminus \emptyset)^\ell$.

2. Key-Generation Phase. Challenger $\mathcal{C}$ computes the secret key SK by running the Setup procedure on input $(1^n, 1^\ell)$ and the partial public key PPK$_{\text{Pol}}$ by running PPKeyGen(SK, Pol).
   PPK$_{\text{Pol}}$ is given to $\mathcal{A}$.

3. Query Phase I. $\mathcal{A}$ can make any number of token queries.
   $\mathcal{C}$ answers token query for pattern y as follows. If Match$(z_0, y) =$ Match$(z_1, y) = 0$, then $\mathcal{A}$ receives the output of GenToken(SK, y). Otherwise, $\mathcal{A}$ receives $\bot$.

4. Challenge construction. $\mathcal{C}$ chooses random $\eta \in \{0, 1\}$ and gives the output of Encryption(PK, z_η) to $\mathcal{A}$.

5. Query Phase II. Identical to Query Phase I.

6. Output phase. $\mathcal{A}$ returns η'.
   If $\eta = \eta'$ then the experiments returns 1 else 0.

   Notice that in SemanticExp$_{\mathcal{A}}$ we can assume, without loss of generality, that $\mathcal{A}$ always asks for PK (the public key that allows to encrypt all attribute vectors). We chose the formulation above to keep it similar to the game used to formalize the token security property (see Section 2.2).

Definition 3. *A predicate encryption scheme with partial public keys* (Setup, PPKeyGen, Encryption, GenToken, Test) *is semantically secure, if for all probabilistic polynomial-time adversaries $\mathcal{A}$*

$$\left| \text{Prob}[\text{SemanticExp}_{\mathcal{A}}(1^n, 1^\ell) = 1] - 1/2 \right|$$

is negligible in n for all $\ell = \text{poly}(n)$.

2.2 Token Security

In this section, we present an experiment that models the fact that a token T gives no information on the associated pattern y but the position of the $\star$-entries. We use an indistinguishability experiment in which the adversary $\mathcal{A}$ picks two challenge patterns y_0 and y_1 such that $y_{0,i} = \star$ iff $y_{1,i} = \star$ and a policy Pol such that for all $x \in \mathbb{X}_{\text{Pol}}$ we have that Match$(x, y_0) =$ Match$(x, y_1) = 0$. $\mathcal{A}$ receives the partial public key PPK$_{\text{Pol}}$ associated with Pol and $\mathcal{A}$ is allowed to issue token queries for patterns y of his choice. Finally, $\mathcal{A}$ receives the token associated to a randomly chosen challenge pattern y_η. We require that $\mathcal{A}$ has probability essentially $1/2$ of guessing η.

   We model the token security property by means of the following game TokenExp$_{\mathcal{A}}$ between a challenger $\mathcal{C}$ and adversary $\mathcal{A}$.

TokenExp$_{\mathcal{A}}(1^n, 1^\ell)$

1. Initialization Phase. The adversary $\mathcal{A}$ announces two challenge patterns $y_0, y_1 \in \Sigma_\star^\ell$ and a policy Pol such that for all $x \in \mathbb{X}_{\text{Pol}}$ we have that Match$(x, y_0) =$ Match$(x, y_1) = 0$.

If there exists $i \in [\ell]$ such that $y_{0,i} = \star$ and $y_{1,i} \neq \star$ or if there exists $i \in [\ell]$ such that $y_{1,i} = \star$ and $y_{0,i} \neq \star$ then the experiment returns 0.

2. Key-Generation Phase. The secret key SK is generated by the Setup procedure. The partial public key $\mathsf{PPK_{Pol}}$ relative to policy Pol is generated running procedure $\mathsf{PPKeyGen(SK, Pol)}$. $\mathsf{PPK_{Pol}}$ is given to $\mathcal{A}$.

3. Query Phase I. $\mathcal{A}$ can make any number of token queries that are answered by returning $\mathsf{GenToken(SK, y)}$.

4. Challenge construction. η is chosen at random from $\{0,1\}$ and receives $\mathsf{GenToken(SK, y_\eta)}$.

5. Query Phase II. Identical to Query Phase I.

6. Output phase. $\mathcal{A}$ returns η'.
   If $\eta = \eta'$ then the experiments returns 1 else 0.

Definition 4. *A predicate encryption scheme with partial public keys* (Setup, PPKeyGen, Encryption, GenToken, Test) *is token secure if for all probabilistic polynomial-time adversaries* $\mathcal{A}$,

$$\left| \mathrm{Prob}[\mathsf{TokenExp}_{\mathcal{A}}(1^n, 1^\ell) = 1] - 1/2 \right|$$

is negligible in n *for all* $\ell = \mathsf{poly}(n)$.

Definition 5. *A predicate encryption scheme with partial public keys* (Setup, PPKeyGen, Encryption, GenToken, Test) *is a secure predicate encryption scheme with partial public keys if it is both semantically secure and token secure.*

3 Background and Complexity Assumptions

Linear secret sharing In our assumptions and constructions we use the concept of a (k, n) linear secret sharing scheme (LSSS), for $k \leq n$. A (k, n) LSSS takes as input a *secret* s (typically from a finite field $\mathbb{F}_p$) and returns k *shares* $(s_1, \ldots, s_k)$ with the following properties. Any set of $k - 1$ (or fewer) shares are independent among themselves and are independent from the secret s. In addition, the secret s can be expressed as a linear combination of the shares held by any k participants. More precisely, for any $F \subseteq [n]$ of size k there exist reconstruction coefficients α_i such that $s = \sum_{i \in F} \alpha_i s_i$. For instance, in Shamir's secret sharing scheme [Sha79], the reconstruction coefficients are the Lagrange interpolation coefficients. We stress that the reconstruction coefficients depend only on the set F and not on the actual shares.

The symmetric bilinear setting. We have two multiplicative groups, the *base* group $\mathbb{G}$ and the *target* group $\mathbb{G}_T$ both of prime order p and a non-degenerate bilinear pairing function $\mathsf{e} : \mathbb{G} \times \mathbb{G} \to \mathbb{G}_T$. That is, for all $x \in \mathbb{G}$, $x \neq 1$, we have $\mathsf{e}(x, x) \neq 1$ and for all $x, y \in \mathbb{G}$ and all $a, b \in \mathbb{Z}_p$, we have $\mathsf{e}(x^a, y^b) = \mathsf{e}(x, y)^{ab}$. We denote by g and $\mathsf{e}(g, g)$ generators of $\mathbb{G}$ and $\mathbb{G}_T$. We call a tuple $\mathcal{I} = [p, \mathbb{G}, \mathbb{G}_T, g, \mathsf{e}]$ a *symmetric bilinear* instance and assume that there exists an efficient generation procedure that, on input security parameter 1^n, outputs an instance with $|p| = \Theta(n)$.

We now review and justify the hardness assumptions we will use for proving security of our constructions.

Our first two assumptions posit the hardness of distinguishing whether the exponents relative to given bases of a sequence of $(2\ell-1)$ elements of $\mathbb{G}$ constitute the shares of 0 with respect to an $(\ell, 2\ell - 1)$ LSSS or one of the exponents (the exponent of the *challenge* element, usually denoted by Z in the following) is random. This computational problem is clearly trivial if $\ell-1$ elements share the same base A with the challenge element Z. Indeed, given an ordered ℓ-subset $F = \langle f_1, \ldots, f_\ell \rangle$ of $[2\ell - 1]$, base A, elements A^{s_i} for $i \in \langle f_1, \ldots, f_{\ell-1} \rangle$ and challenge $Z = A^{s_{f_\ell}}$, checking if the exponents s_i constitute ℓ shares of 0 of an $(\ell, 2\ell - 1)$ LSSS is trivial by the linearity of the secret sharing scheme. In a bilinear setting, the problem remains easy in the base group if $(\ell - 1)$ elements share the same base $A \in \mathbb{G}$ even though this is different from the base $B \in \mathbb{G}$ of the challenge element. Specifically, given bases A and B, elements A^{s_i}, for $i \in \langle f_1, \ldots, f_{\ell-1} \rangle$ and challenge $Z = B^r$ it is possible to check whether the s_i's and r constitute ℓ shares of 0 of an $(\ell, 2\ell - 1)$ LSSS in the following way. First, use linearity to compute $A^{s_{f_\ell}}$ and then use bilinearity to check if $r = s_{f_\ell}$ by comparing $\mathsf{e}(A^{s_\ell}, B)$ and $\mathsf{e}(A, B^r)$. If instead less than $\ell - 1$ elements share the same base then the problem seems to be computationally difficult.

The Linear Secret Sharing Assumption (see Section 3.1 below) makes a formal statement of this fact. Specifically, we are given bases $U_1, \ldots, U_{2\ell-1} \in \mathbb{G}$, elements $U_1^{a_1}, \ldots, U_\ell^{a_{2\ell-1}} \in \mathbb{G}$ and index $j \in [2\ell - 1]$ of the challenge element and we have to decide whether $(a_1, \ldots, a_{2\ell-1})$ constitute an $(\ell, 2\ell - 1)$ secret sharing of 0 or the exponent a_j of the challenge element is random. We stress that, for sake of ease of exposition, in stating the Linear Secret Sharing Assumption we have not tried to reduce the number of bases: we have $(2\ell - 1)$ bases for $(2\ell - 1)$ elements. It is not difficult to see that we could have used only 4 bases to formulate an assumption that is sufficient for proving the security of our constructions.

If we consider the same problem in the target group $\mathbb{G}_T$, it seems that it remains difficult even if $\ell-1$ elements share the same base which is different from the base used for the challenge element. Indeed in the target group we are not allowed to use the pairing function e and thus we cannot use the same approach employed for the base group. The F-Linear Secret Sharing Assumption (see Section 3.2 below) makes a formal statement of this fact. By looking ahead, in the F-Linear Secret Sharing Assumption we have ℓ shares corresponding to an ordered subset $F = \langle f_1, \ldots, f_\ell \rangle$ of elements of $[2\ell - 1]$ which appear as exponents of ℓ elements of $\mathbb{G}_T$: $\ell - 1$ of these elements share the same base $\mathsf{e}(g, g)$ (specifically, in the assumption we have $\mathsf{e}(\bar{U}_{f_j}, V_{f_j}) = \mathsf{e}(g, g)^{a_{f_j}}$ for $2 \leq j \leq \ell$) and the challenge element uses a different base (specifically, $\mathsf{e}(U_{f_1}, V_{f_1}) = \mathsf{e}(U_{f_1}, U_{f_1})^{a_{f_1}}$). The task is to decide whether the a_i's for $i \in F$ constitute an $(\ell, 2\ell - 1)$ secret share of 0 or the a_{f_ℓ} is random. We state our assumptions using elements of $\mathbb{G}$ (i.e., the $\bar{U}_j$'s and the V_j's) instead of elements of $\mathbb{G}_T$ (i.e., giving only $\mathsf{e}(\bar{U}_j, V_j)$).

For each of the two above assumptions, we have a *split* version which we call the Split Linear Secret Sharing Assumption (see Section 3.3) and the F-Split Linear Secret Sharing Assumption (see Section 3.4). The split versions of our assumptions are derived by mixing the assumptions based on linear secret sharing with the Decision Linear Assumption (see [BW06]). In the Decision Linear Assumption, the task is to decide, given A, A^r, B, B^s, C, C^z whether $z = r - s$ or z is random. Specifically, in the Split Linear Secret Sharing Assumption we have bases $U_1, \ldots, U_{2\ell-1}$, elements $U_1^{a_1}, \ldots, U_{2\ell-1}^{a_{2\ell-1}}$ and $g^{a_1}, \ldots, g^{a_{2\ell-1}}$ with $(a_1, \ldots, a_{2\ell-1})$ constituting an $(\ell, 2\ell-1)$ LSSS of 0, and $2\ell-2$ related instances of the Decision Linear Assumptions for a randomly chosen $j \in [2\ell-1]$: $U_i^u, U_j^{a_j}, W^s$, with $i \in [2\ell - 1] \setminus \{j\}$ in which we have to decide whether $s = u - a_j$. In addition, we are also given $\hat{U} = W^{u_j}$ where $U_j = g^{u_j}$. The F-Split Linear Secret Sharing Assumption is obtained is a similar way from the F-Linear Secret Sharing Assumption.

3.1   Linear Secret Sharing Assumption

Consider the following game between a challenger $\mathcal{C}$ and an adversary $\mathcal{A}$.

$\mathsf{LSSExp}_{\mathcal{A}}(1^n, 1^\ell)$
01. $\mathcal{C}$ computes shares $a_1, \ldots, a_{2\ell-1}$ of 0 using an $(\ell, 2\ell - 1)$ LSSS;
02. $\mathcal{C}$ chooses instance $\mathcal{I} = [p, \mathbb{G}, \mathbb{G}_T, g, \mathsf{e}]$ with security parameter 1^n;
03. $\mathcal{C}$ chooses random $j \in [2\ell - 1]$;
04. **for** $i \in [2\ell - 1]$
        $\mathcal{C}$ chooses random $u_i \in \mathbb{Z}_p$ and sets $U_i = g^{u_i}$ and $V_i = U_i^{a_i}$;
05. $\mathcal{C}$ chooses random $\eta \in \{0, 1\}$;
06. **if** $\eta = 1$ **then** $\mathcal{C}$ sets $Z = U_j^{a_j}$ **else** $\mathcal{C}$ chooses random $Z \in \mathbb{G}$;
07. $\mathcal{C}$ runs $\mathcal{A}$ on input $[\mathcal{I}, j, (U_i)_{i \in [2\ell-1]}, (V_i)_{i \in [2\ell-1] \setminus \{j\}}, Z]$;
08. Let η' be $\mathcal{A}$'s guess for η;
09. **if** $\eta = \eta'$ **then** return 1 **else** return 0.

Assumption 1 (LSS Assumption). *The Linear Secret Sharing Assumption states that for all probabilistic polynomial-time algorithms $\mathcal{A}$,*
$$\left|\mathrm{Prob}[\mathsf{LSSExp}_{\mathcal{A}}(1^n, 1^\ell) = 1] - 1/2\right| \text{ is negligible in } n \text{ for all } \ell = \mathsf{poly}(n).$$

3.2 F-Linear Secret Sharing Assumption

Let $F = \langle f_1, \ldots, f_\ell \rangle$ be a sequence of ℓ distinct elements from $[2\ell - 1]$. We formalize the F-Linear Secret Sharing Assumption (F-LSS Assumption) by means of the following game between a Challenger $\mathcal{C}$ and an Adversary $\mathcal{A}$.

$F\text{-}\mathsf{LSSExp}_{\mathcal{A}}(1^n, 1^\ell)$
01. $\mathcal{C}$ computes shares $a_1, \ldots, a_{2\ell-1}$ of 0 using an $(\ell, 2\ell - 1)$ LSSS;
02. $\mathcal{C}$ chooses instance $\mathcal{I} = [p, \mathbb{G}, \mathbb{G}_T, g, \mathsf{e}]$ with security parameter 1^n;
03. **for** $i \in F$
        $\mathcal{C}$ chooses random $u_i \in \mathbb{Z}_p$ and sets $U_i = g^{u_i}$, $\bar{U}_i = g^{1/u_i}$, and
        $V_i = U_i^{a_i}$;

04. $\mathcal{C}$ chooses random $\eta \in \{0,1\}$;
05. **if** $\eta = 1$ **then** $\mathcal{C}$ sets $Z = U_{f_\ell}^{a_{f_\ell}}$ **else** $\mathcal{C}$ chooses random $Z \in \mathbb{G}$;
06. $\mathcal{C}$ runs $\mathcal{A}$ on input $[\mathcal{I}, F, (U_i)_{i \in F}, (\bar{U}_i)_{i \in F \setminus \{f_1\}}, (V_i)_{i \in F}, Z]$;
07. Let η' be $\mathcal{A}$'s guess for η;
08. **if** $\eta = \eta'$ **then** return 1 **else** return 0.

Assumption 2 (F-LSS Assumption). *The F-Linear Secret Sharing Assumption states that for all probabilistic polynomial-time algorithms $\mathcal{A}$, $\left|\mathrm{Prob}[F\text{-LSSExp}_\mathcal{A}(1^n, 1^\ell) = 1] - 1/2\right|$ is negligible in n for all $\ell = \mathsf{poly}(n)$.*

The proof of the following theorem is similar to, but simpler than, the proof of Theorem 2. So, we omit it.

Theorem 1. *For any sequences F and K each of ℓ distinct elements from $[2\ell - 1]$, F-LSS implies K-LSS.*

3.3   Split Linear Secret Sharing Assumption

In this section we present the Split Linear Secret Sharing Assumption (the SplitLSS Assumption) which is similar to the Linear Secret Sharing Assumption. The only difference is that whereas in the LSS Assumption the task is to decide whether $V_j = U_j^{a_j}$ or V_j is random, here the task is to decide, whether $Z = W^{u-a_j}$ or Z is a random element of $\mathbb{G}$. We formalize the SplitLSS Assumption by means of the following game between a Challenger $\mathcal{C}$ and an Adversary $\mathcal{A}$.

SplitLSSExp$_\mathcal{A}(1^n, 1^\ell)$
01. $\mathcal{C}$ computes shares $a_1, \ldots, a_{2\ell-1}$ of 0 using an $(\ell, 2\ell - 1)$ LSSS;
02. $\mathcal{C}$ chooses instance $\mathcal{I} = [p, \mathbb{G}, \mathbb{G}_T, g, \mathsf{e}]$ with security parameter 1^n;
03. $\mathcal{C}$ chooses random $u, w \in \mathbb{Z}_p$ and sets $W = g^w$;
04. **for** $i \in [2\ell - 1]$
        $\mathcal{C}$ chooses random $u_i \in \mathbb{Z}_p$ and sets $U_i = g^{u_i}, V_i = U_i^{a_i}$,
        $A_i = g^{a_i}$, and $S_i = U_i^u$;
05 $\mathcal{C}$ picks a random $j \in [2\ell - 1]$ and sets $\hat{U} = U_j^w$;
06. $\mathcal{C}$ chooses random $\eta \in \{0, 1\}$;
07. **if** $\eta = 1$ **then** $\mathcal{C}$ sets $Z = W^{u-a_j}$ **else** $\mathcal{C}$ chooses random $Z \in \mathbb{G}$;
08. $\mathcal{C}$ runs $\mathcal{A}$ on input
        $[\mathcal{I}, j, (U_i)_{i \in [2\ell-1]}, (V_i)_{i \in [2\ell-1]}, (A_i)_{i \in [2\ell-1]}, (S_i)_{i \in [2\ell-1] \setminus \{j\}}, W, \hat{U}, Z]$;
09. Let η' be $\mathcal{A}$'s guess for η;
10. **if** $\eta = \eta'$ **then** return 1 **else** return 0.

Assumption 3 (SplitLSS Assumption). *The Split Linear Secret Sharing Assumption states that for all probabilistic polynomial-time algorithms $\mathcal{A}$, $\left|\mathrm{Prob}[\text{SplitLSSExp}_\mathcal{A}(1^n, 1^\ell) = 1] - 1/2\right|$ is negligible in n for all $\ell = \mathsf{poly}(n)$.*

3.4 F-Split Linear Secret Sharing Assumption

Let $F = \langle f_1, \ldots, f_\ell \rangle$ be a sequence of ℓ distinct elements from $[2\ell - 1]$. We formalize the F-Split Linear Secret Sharing Assumption (F-SplitLSS Assumption) by means of the following game between $\mathcal{C}$ and $\mathcal{A}$.

$F\text{-SplitLSSExp}_{\mathcal{A}}(1^n, 1^\ell)$
01. $\mathcal{C}$ computes shares $a_1, \ldots, a_{2\ell-1}$ of 0 using an $(\ell, 2\ell - 1)$ LSSS;
02. $\mathcal{C}$ chooses instance $\mathcal{I} = [p, \mathbb{G}, \mathbb{G}_T, g, \mathsf{e}]$ with security parameter 1^n;
03. $\mathcal{C}$ chooses random $u \in \mathbb{Z}_p$;
04. **for** $i \in F$,
    $\mathcal{C}$ chooses random $u_i \in \mathbb{Z}_p$ and sets $U_i = g^{u_i}$, $\bar{U}_i = g^{1/u_i}$, $V_i = U_i^{a_i}$,
    and $S_i = U_i^u$;
05. $\mathcal{C}$ chooses random $w \in \mathbb{Z}_p$ and sets $W = g^w$ and $\bar{W} = g^{1/w}$.
06. $\mathcal{C}$ chooses random $\eta \in \{0, 1\}$;
07. **if** $\eta = 1$ **then** $\mathcal{C}$ sets $Z = W^{u - a_{f_\ell}}$ **else** $\mathcal{C}$ chooses random $Z \in \mathbb{G}$;
08. $\mathcal{C}$ runs $\mathcal{A}$ on input
    $[\mathcal{I}, F, (U_i)_{i \in F}, (\bar{U}_i)_{i \in F \setminus \{f_1\}}, (V_i)_{i \in F}, (S_i)_{i \in F}, W, \bar{W}, Z]$;
09. Let η' be $\mathcal{A}$'s guess for η.
10. **if** $\eta = \eta'$ **then** return 1 **else** return 0.

Assumption 4 (F-SplitLSS Assumption)). *The F-Split Linear Secret Sharing Assumption states that for all probabilistic polynomial-time algorithms $\mathcal{A}$ $\left| \mathrm{Prob}[F\text{-SplitLSSExp}_{\mathcal{A}}(1^n, 1^\ell) = 1] - 1/2 \right|$ is negligible in n for all $\ell = \mathsf{poly}(n)$.*

The proof of the next theorem is found in Appendix A.

Theorem 2. *For any two sequences F and K each of ℓ distinct elements from $[2\ell - 1]$, we have that F-SplitLSS implies K-SplitLSS.*

4 The Scheme

In this section, we describe a new proposal for a secure predicate encryption scheme with partial public keys. Our description is for binary alphabets; it is possible to convert our scheme to a scheme for any alphabet by increasing the size of the key, but not the size of ciphertexts and tokens.

The Setup procedure. On input security parameter 1^n and the number of attributes $\ell = \mathsf{poly}(n)$, Setup proceeds as follows.
1. Select a symmetric bilinear instance $\mathcal{I} = [p, \mathbb{G}, \mathbb{G}_T, g, \mathsf{e}]$ with $|p| = \Theta(n)$.
2. For $i \in [2\ell - 1]$, choose random $t_{1,i,0}, t_{2,i,0}, t_{1,i,1}, t_{2,i,1} \in \mathbb{Z}_p$ and set

$$\mathsf{K}_i = \begin{pmatrix} T_{1,i,0} = g^{t_{1,i,0}}, & T_{2,i,0} = g^{t_{2,i,0}} \\ T_{1,i,1} = g^{t_{1,i,1}}, & T_{2,i,1} = g^{t_{2,i,1}} \end{pmatrix} \text{ and}$$

$$\bar{\mathsf{K}}_i = \begin{pmatrix} \bar{T}_{1,i,0} = g^{1/t_{1,i,0}}, & \bar{T}_{2,i,0} = g^{1/t_{2,i,0}} \\ \bar{T}_{1,i,1} = g^{1/t_{1,i,1}}, & \bar{T}_{2,i,1} = g^{1/t_{2,i,1}} \end{pmatrix}.$$

3. Return $\mathsf{SK} = [\mathcal{I}, (\mathsf{K}_i, \bar{\mathsf{K}}_i)_{i \in [2\ell - 1]}]$.

The PPKeyGen procedure. On input SK and policy
$\mathsf{Pol} = \langle \mathsf{Pol}_1, \ldots, \mathsf{Pol}_\ell \rangle \in (2^{\{0,1\}} \setminus \emptyset)^\ell$ of length ℓ, PPKeyGen proceeds as follows.
1. For $i = 1, \ldots, \ell$,
   for every $b \in \mathsf{Pol}_i$, add $T_{1,i,b}$ and $T_{2,i,b}$ to PPK_i.
2. For $i = \ell + 1, \ldots, 2\ell - 1$,
   add $T_{1,i,0}$ and $T_{2,i,0}$ to PPK_i.
3. Return $\mathsf{PPK}_{\mathsf{Pol}} = [(\mathsf{PPK}_i)_{i \in [2\ell-1]}]$.

The Encryption procedure. On input partial public key $\mathsf{PPK}_{\mathsf{Pol}}$ and attribute vector $\boldsymbol{x} = (x_1, \ldots, x_\ell)$ of length ℓ, Encryption proceeds as follows.
1. If $\boldsymbol{x} \notin \mathbb{X}_{\mathsf{Pol}}$ return $\perp$.
2. Extend $\boldsymbol{x}$ to a vector with $2\ell - 1$ entries by appending $(\ell - 1)$ 0-entries.
3. Pick s at random from $\mathbb{Z}_p$.
4. Compute shares $(s_1, \ldots, s_{2\ell-1})$ of 0 using an $(\ell, 2\ell - 1)$ linear secret sharing scheme.
5. For $i = 1, \ldots, 2\ell - 1$,
   set $X_{1,i} = T_{1,i,x_i}^{s-s_i}$ and $X_{2,i} = T_{2,i,x_i}^{-s_i}$.
6. Return the encoded attribute vector $\tilde{X} = [(X_{1,i}, X_{2,i})_{i \in [2\ell-1]}]$.
Notice that if $\boldsymbol{x} \in \mathbb{X}_{\mathsf{Pol}}$, then for every i it holds that $T_{1,i,x_i}, T_{2,i,x_i} \in \mathsf{PPK}_{\mathsf{Pol}}$.
Hence, the Encryption procedure will be able to execute the steps above.
In the following will use sometimes the writing
$\mathsf{Encryption}(\mathsf{PPK}_{\mathsf{Pol}}, \boldsymbol{x}; s, (s_i)_{i \in [2\ell-1]})$ to denote the encoded attribute vector $\tilde{X}$ output by Encryption on input $\mathsf{PPK}_{\mathsf{Pol}}$ and $\boldsymbol{x}$ when using s as random element and $(s_i)_{i \in [2\ell-1]}$ as shares of an $(\ell, 2\ell - 1)$ linear secret sharing scheme for the secret 0.

The GenToken procedure. On input secret key SK and pattern vector $\boldsymbol{y} = (y_1, \ldots, y_\ell)$ of length ℓ, GenToken proceeds as follows.
1. Let h be the number of non-$\star$ entries of $\boldsymbol{y}$. Extend $\boldsymbol{y}$ to a vector with $(2\ell - 1)$ entries by appending $(\ell - h)$ 0-entries and $(h - 1)$ $\star$-entries and denote by $S_{\boldsymbol{y}}$ the indices of the non-$\star$ entries of the extended vector. Notice that $|S_{\boldsymbol{y}}| = \ell$.
2. Compute shares $(r_1, \ldots, r_{2\ell-1})$ of 0 using an $(\ell, 2\ell - 1)$ linear secret sharing scheme.
3. Pick random $r \in \mathbb{Z}_p$.
4. For $i \in S_{\boldsymbol{y}}$,
   set $Y_{1,i} = \bar{T}_{1,i,y_i}^{r_i}$ and $Y_{2,i} = \bar{T}_{2,i,y_i}^{r-r_i}$.
5. Return $T_{\boldsymbol{y}} = [S_{\boldsymbol{y}}, (Y_{1,i}, Y_{2,i})_{i \in S_{\boldsymbol{y}}}]$.
In the following we will sometimes use the writing
$\mathsf{GenToken}(\mathsf{SK}, \boldsymbol{y}; r, (r_i)_{i \in S_{\boldsymbol{y}}})$ to denote the token $T_{\boldsymbol{y}}$ computed by GenToken on input SK and $\boldsymbol{y}$ and using r as random element and $(r_i)_{i \in S_{\boldsymbol{y}}}$ as ℓ shares of an $(\ell, 2\ell - 1)$ LSSS for the secret 0.

The Test procedure. On input token $T_{\boldsymbol{y}} = [S, (Y_{1,j_1}, Y_{2,j_1}, \ldots, Y_{1,j_\ell}, Y_{2,j_\ell})]$ and attribute vector $\tilde{X} = [(X_{1,i}, X_{2,i})_{i \in [2\ell-1]}]$, Test proceeds as follows. Let

$v_{j_1}, \ldots, v_{j_\ell}$ be the reconstruction coefficients for the set $S = \{j_1, \ldots, j_\ell\}$. Then, the Test procedure returns

$$\prod_{i \in [\ell]} [\mathsf{e}(X_{1,j_i}, Y_{1,j_i}) \cdot \mathsf{e}(X_{2,j_i}, Y_{2,j_i})]^{v_{j_i}}.$$

The proof of next theorem is found in Appendix A.

Theorem 3. *The quintuple of algorithms*
(Setup, PPKeyGen, Encryption, GenToken, Test) *specified above is a predicate encryption scheme with partial public keys.*

5   Semantic Security

In this section, we show that, if the Linear Secret Sharing Assumption and the Split Linear Secret Sharing Assumption hold, then the scheme presented in Section 4 is semantically secure. Specifically, we show that, for any attribute vector z and for any policy Pol, the encoded attribute vector output by the Encryption procedure is indistinguishable from a sequence of $2 \cdot (2\ell - 1)$ random elements of $\mathbb{G}$ to a polynomial time adversary $\mathcal{A}$ that has the partial public key associated with Pol and oracle access to GenToken for all pattern vectors y such that $\mathsf{Match}(z, y) = 0$. As it is easily seen, this implies semantic security.

The experiments. We start by describing 3ℓ experiments with a probabilistic polynomial-time adversary $\mathcal{A}$.

Experiment k with $0 \leq k \leq 2\ell - 1$. In this experiment, $\mathcal{A}$ outputs an attribute vector z and a policy Pol, receives the partial public key $\mathsf{PPK}_{\mathsf{Pol}}$ relative to Pol, and has oracle access to GenToken for all pattern vectors y such that $\mathsf{Match}(z, y) = 0$. Then $\mathcal{A}$ receives challenge $\tilde{X} = [(X_{1,i}, X_{2,i})_{i \in [2\ell - 1]}]$ computed as follows and outputs a bit.
1. Extend z to a $2\ell - 1$ vector by appending $(\ell - 1)$ 0-entries.
2. Compute shares $(s_1, \ldots, s_{2\ell - 1})$ of 0 using an $(\ell, 2\ell - 1)$ LSSS.
3. For $i = 1, \ldots, k$, randomly choose $X_{1,i} \in \mathbb{G}$ and set $X_{2,i} = T_{2,i,z_i}^{s_i}$.
4. For $i = k + 1, \ldots, 2\ell - 1$, set $X_{1,i} = T_{1,i,z_i}^{s - s_i}$ and $X_{2,i} = T_{2,i,z_i}^{s_i}$.

Experiment $2\ell + k - 1$ with $k \in [\ell]$. These experiments differ from the previous ones only in the way in which the challenge $\tilde{X}$ is computed. More precisely, $\tilde{X} = [(X_{1,i}, X_{2,i})_{i \in [2\ell - 1]}]$ is computed as follows.
1. Extend z to a $2\ell - 1$ vector by appending $(\ell - 1)$ 0-entries.
2. Compute shares $(s_1, \ldots, s_{2\ell - 1})$ of 0 using an $(\ell, 2\ell - 1)$ LSSS.
3. For $i = 1, \ldots, k$ randomly choose $X_{1,i}, X_{2,i} \in \mathbb{G}$.
4. For $i = k + 1, \ldots, 2\ell - 1$ randomly choose $X_{1,i} \in \mathbb{G}$ and set $X_{2,i} = T_{2,i,z_i}^{s_i}$.

Clearly, in Experiment 0, vector $\tilde{X}$ is a well-formed encryption of z whereas in Experiment $3\ell - 1$ vector $\tilde{X}$ consists instead of randomly chosen elements from $\mathbb{G}$. We denote by $p_k^{\mathcal{A}}$ the probability that $\mathcal{A}$ outputs 1 when playing Experiment k. We start by proving that, under the Split Linear Secret Sharing Assumption, the difference $|p_k^{\mathcal{A}} - p_{k-1}^{\mathcal{A}}|$ is negligible, for $k \in [2\ell - 1]$.

Due to space limit, some proofs are omitted and they can be found in the full version of this paper [BIP10].

Indistiguishability of the first $2\ell - 1$ experiments.

Lemma 1. *Assume the Split Linear Secret Sharing Assumption. Then, for $k \in [2\ell - 1]$, it holds that $|p_k^{\mathcal{A}} - p_{k-1}^{\mathcal{A}}|$ is negligible for all probabilistic polynomial-time adversaries $\mathcal{A}$.*

Indistiguishability of the last ℓ experiments.

Lemma 2. *Assume the Linear Secret Sharing Assumption. Then, for $k \in [\ell]$, it holds that $|p_{2\ell+k-2}^{\mathcal{A}} - p_{2\ell+k-1}^{\mathcal{A}}|$ is negligible for all probabilistic polynomial-time adversaries $\mathcal{A}$.*

Lemma 1 and Lemma 2 imply the following theorem.

Theorem 4. *Assume LSS and SplitLSS. Then, predicate encryption scheme with partial public keys* (Setup, PPKeyGen, Encryption, GenToken, Test) *is semantically secure.*

6   Token Security

In this section, we show that, if the F-Linear Secret Sharing Assumption and the F-Split Linear Secret Sharing Assumption hold, the scheme presented in Section 4 is token secure. Specifically, let z be a pattern and Pol a policy such that $\mathbb{X}_{\mathsf{Pol}}$ does not contain any attribute vector x such that $\mathsf{Match}(x, z) = 1$. Then we show that no probabilistic polynomial-time adversary $\mathcal{A}$ that has oracle access to GenToken and the public key relative to Pol can distinguish a well formed token for pattern z from a sequence of random elements of $\mathbb{G}$. It is straightforward to see that this implies token security.

The experiments. We start by describing 4ℓ experiments with a probabilistic polynomial-time adversary $\mathcal{A}$.

Experiment j with $0 \le j \le 2\ell - 1$. In this experiment, $\mathcal{A}$ outputs a pattern $z \in \{0, 1, \star\}^{\ell}$ and a policy Pol, receives the partial public key $\mathsf{PPK}_{\mathsf{Pol}}$ relative to Pol and has oracle access to GenToken for all pattern vectors y. If there exists an attribute vector $x \in \mathbb{X}_{\mathsf{Pol}}$ such that $\mathsf{Match}(x, z) = 1$ then, $\mathcal{A}$ receives $\bot$; otherwise, $\mathcal{A}$ receives challenge T_z computed as follows. In both cases $\mathcal{A}$ outputs a bit.

1. Let h be the number of non-$\star$ entries of z. Extend z to a vector with $(2\ell-1)$ entries by appending $(\ell - h)$ 0-entries and $(h - 1)$ $\star$-entries. With a slight abuse of notation, we call z the extended vector and denote by S_z the set of indices i such that $z_i \in \{0, 1\}$. Notice that $|S_z| = \ell$.

2. Choose random $r \in \mathbb{Z}_p$ and compute shares $(r_1, \ldots, r_{2\ell-1})$ of 0 using an $(\ell, 2\ell - 1)$ LSSS.

3. For $i \in S_z$ and $i \leq j$, set $Y_{1,i} = g^{r_i/t_{1,i,z_i}}$ and $Y_{2,i} = g^{(r-r_i)/t_{2,i,z_i}}$.

4. For $i \in S_z$ and $i > j$, set $Y_{1,i} = g^{r_i/t_{1,i,z_i}}$ and choose random $Y_{2,i} \in \mathbb{G}$.

5. Set $T_z = [S_z, (Y_{1,i}, Y_{2,i})_{i \in S_z}]$.

Experiment j with $2\ell \leq j \leq 4\ell - 1$. The experiments differ from the previous ones only in the way the challenge T_z is computed. More precisely, the challenge T_z is computed as follows.

1. Let h be the number of non-$\star$ entries of z. Extend z to a vector with $(2\ell-1)$ entries by appending $(\ell - h)$ 0-entries and $(h - 1)$ $\star$-entries. With a slight abuse of notation, we call z the extended vector and denote by S_z the set of indices i such that $z_i \in \{0, 1\}$. Notice that $|S_z| = \ell$.

2. Choose random $r \in \mathbb{Z}_p$ and compute shares $(r_1, \ldots, r_{2\ell-1})$ of 0 using an $(\ell, 2\ell - 1)$ LSSS.

3. For $i \in S_z$, set $Y_{2,i}$ to a random element in $\mathbb{G}$.

4. For $i \in S_z$ and $i \leq j$, set $Y_{1,i} = g^{r_i/t_{1,i,z_i}}$.

5. For $i \in S_z$ and $i > j$, set $Y_{1,i}$ to a random element in $\mathbb{G}$.

6. Set $T_z = [S_z, (Y_{1,i}, Y_{2,i})_{i \in S_z}]$.

Clearly in Experiment 0, T_z is a well formed token for pattern z whereas in Experiment $4\ell - 1$, T_z consists of 2ℓ randomly chosen elements of $\mathbb{G}$. We denote by $p_j^{\mathcal{A}}$ the probability that $\mathcal{A}$ outputs 1 when playing Experiment j. We start by proving that, under the F-Linear Secret Sharing, the difference $|p_j^{\mathcal{A}} - p_{j-1}^{\mathcal{A}}|$ is negligible for $j \in [2\ell - 1]$.

Indistinguishability of the first 2ℓ experiments.

Lemma 3. *Assume F-Split Linear Secret Sharing holds. Then, for $j \in [2\ell-1]$, it holds that $|p_j^{\mathcal{A}} - p_{j-1}^{\mathcal{A}}|$ is negligible for all probabilistic polynomial-time adversary $\mathcal{A}$.*

Indistinguishability of last 2ℓ experiments.

Lemma 4. *Assume F-Linear Secret Sharing holds. Then, for $j = 2\ell, \ldots, 4\ell - 1$, it holds that $|p_j^{\mathcal{A}} - p_{j-1}^{\mathcal{A}}|$ is negligible for all probabilistic polynomial-time adversary $\mathcal{A}$.*

Next theorem holds.

Theorem 5. *Assume F-Linear Secret Sharing and F-Split Linear Secret Sharing. Then predicate encryption*

(Setup, PPKeyGen, Encryption, GenToken, Test) *is token secure.*

Acknowledgments

This work is partially founded by the Italian Ministry of University and Research Project PRIN 2008 *PEPPER: Privacy and Protection of Personal Data* (prot. 2008SY2PH4).

References

[BDOP04] Boneh, D., Di Crescenzo, G., Ostrovsky, R., Persiano, G.: Public key encryption with keyword search. In: Cachin, C., Camenisch, J. (eds.) EUROCRYPT 2004. LNCS, vol. 3027, pp. 506–522. Springer, Heidelberg (2004)
[BIP10] Blundo, C., Iovino, V., Persiano, G.: Predicate encryption with partial public keys. Cryptology ePrint Archive, Report 2010/476 (2010), http://eprint.iacr.org/
[BW06] Boyen, X., Waters, B.: Anonymous Hierarchical Identity-Based Encryption (Without Random Oracles). In: Dwork, C. (ed.) CRYPTO 2006. LNCS, vol. 4117, pp. 290–307. Springer, Heidelberg (2006)
[BW07] Boneh, D., Waters, B.: Conjunctive, subset and range queries on encrypted data. In: Vadhan, S.P. (ed.) TCC 2007. LNCS, vol. 4392, pp. 535–554. Springer, Heidelberg (2007)
[GPSW06] Goyal, V., Pandey, O., Sahai, A., Waters, B.: Attribute-Based Encryption for Fine-Grained Access Control for Encrypted Data. In: ACM CCS 2006: 13th Conference on Computer and Communications Security, Alexandria, VA, USA, October 30-November 3, pp. 89–98. ACM Press, New York (2006)
[IP08] Iovino, V., Persiano, G.: Hidden-vector encryption with groups of prime order. In: Galbraith, S.D., Paterson, K.G. (eds.) Pairing 2008. LNCS, vol. 5209, pp. 75–88. Springer, Heidelberg (2008)
[KSW08] Katz, J., Sahai, A., Waters, B.: Predicate Encryption Supporting Disjunction, Polynomial Equations, and Inner Products. In: Smart, N. (ed.) EUROCRYPT 2008. LNCS, vol. 4965, pp. 146–162. Springer, Heidelberg (2008)
[Sha79] Shamir, A.: How to share a secret. Communications of the Association for Computing Machinery 22(11), 612–613 (1979)
[SSW09] Shen, E., Shi, E., Waters, B.: Predicate privacy in encryption systems. In: Reingold, O. (ed.) TCC 2009. LNCS, vol. 5444, pp. 457–473. Springer, Heidelberg (2009)

A   Appendix

Theorem 2. *For any two sequences F and K each of ℓ distinct elements from $[2\ell - 1]$, we have that F-SplitLSS implies K-SplitLSS.*

Proof. Let $F = \langle f_1, \ldots, f_\ell \rangle$ and $K = \langle k_1, \ldots, k_\ell \rangle$ be sequences of ℓ distinct elements from $[2\ell - 1]$. Given an F-SplitLSS instance

$$[\mathcal{I}, F, (U_j)_{j \in F}, (\bar{U}_j)_{j \in F \setminus \{f_1\}}, (V_j)_{j \in F}, (S_j)_{j \in F}, W, \bar{W}, Z]$$

we show how to get from it a K-SplitLSS instance

$$[\mathcal{I}, K, (U_j')_{j \in K}, (\bar{U}_j')_{j \in K \setminus \{k_1\}}, (V_j')_{j \in K}, (S_j')_{j \in K}, W', \bar{W}', Z'].$$

For $i = 1, \ldots, \ell$, let $\alpha_i = v_{f_i, F}/v_{k_i, K}$, where $v_{f_i, F}$ ($v_{k_i, K}$) is the publicly known value associated to f_i-th (k_i-th) share when participants whose identities are in F (resp., K) collaborate to the reconstruction of the secret in an $(\ell, 2\ell - 1)$ LSSS. Set $U_{k_1}' = U_{f_1}$. For $i = 2, \ldots, \ell$, set

$$U_{k_i}' = U_{f_i} \text{ and } \bar{U}_{k_i}' = \bar{U}_{f_i}.$$

For $i = 1, \ldots, \ell$, set

$$V_{k_i}' = V_{f_i}^{\alpha_i} \text{ and } S_{k_i}' = S_{f_i}^{\alpha_\ell}.$$

Set $W' = W$ and $\bar{W}' = \bar{W}$. Finally, set $Z' = Z^{\alpha_\ell}$. It is immediate to see that the values $(U_j')_{j \in K}, (\bar{U}_j')_{j \in K \setminus \{k_1\}}, (V_j')_{j \in K}, (S_j')_{j \in K}, W', \bar{W}', Z'$ define a K-SplitLSS instance.

Theorem 3. *The quintuple of algorithms*
(Setup, PPKeyGen, Encryption, GenToken, Test) *specified above is a predicate encryption scheme with partial public keys.*

Proof. It is sufficient to verify that the procedure Test returns 1 when Match$(\boldsymbol{x}, \boldsymbol{y}) = 1$. Let $\tilde{X} = [(X_{1,i}, X_{2,i})_{[2\ell-1]}]$ be the output of

Encryption(PPK$_{\mathsf{PoI}}, \boldsymbol{x}; s, (s_i)_{[2\ell-1]})$ and let $T_{\boldsymbol{y}} = [S_{\boldsymbol{y}}, (Y_{1,i}, Y_{2,i})_{i \in S_{\boldsymbol{y}}}]$ be the output of procedure GenToken(SK, $\boldsymbol{y}; r, (r_i)_{i \in S_{\boldsymbol{y}}}$). Let $v_{j_1}, \ldots, v_{j_\ell}$ be the reconstruction coefficients for set $S_{\boldsymbol{y}} = \{j_1, \ldots, j_\ell\}$. We have,

$$\begin{aligned}
\mathsf{Test}(\tilde{X}, T_{\boldsymbol{y}}) &= \prod_{i \in [\ell]} [e(X_{1,j_i}, Y_{1,j_i}) \cdot e(X_{2,j_i}, Y_{2,j_i})]^{v_{j_i}} \\
&= \prod_{i \in [\ell]} e(T_{1,j_i,x_{j_i}}^{s-s_{j_i}}, \bar{T}_{1,j_i,y_{j_i}}^{r_{j_i}})^{v_{j_i}} \cdot e(T_{2,j_i,x_{j_i}}^{-s_{j_i}}, \bar{T}_{2,j_i,y_{j_i}}^{r-r_{j_i}})^{v_{j_i}} \\
&= \qquad (\text{since } x_{j_i} = y_{j_i} \text{ for } i \in [\ell]) \\
&= \prod_{i \in [\ell]} e(g,g)^{r_{j_i} v_{j_i}(s-s_{j_i})} \cdot e(g,g)^{-s_{j_i} v_{j_i}(r-r_{j_i})} \\
&= \prod_{i \in [\ell]} e(g,g)^{sr_{j_i} v_{j_i} - rs_{j_i} v_{j_i}} = e(g,g)^{s \sum_{i \in [\ell]} r_{j_i} v_{j_i}} \cdot \\
&\quad e(g,g)^{-r \sum_{i \in [\ell]} s_{j_i} v_{j_i}} = 1.
\end{aligned}$$

The last equality is satisfied as the r_{j_i}'s and the s_{j_i}'s for $i \in [\ell]$ are ℓ shares of an $(\ell, 2\ell - 1)$ linear secret sharing scheme for the secret 0 and the v_{j_i}'s are the reconstructing coefficient for set $S_{\boldsymbol{y}} = \{j_1, \ldots, j_\ell\}$. Hence, we have that $\sum_{i \in [\ell]} r_{j_i} v_{j_i} = 0$ and $\sum_{i \in [\ell]} s_{j_i} v_{j_i} = 0$.

Anonymous Credential Schemes
with Encrypted Attributes

Jorge Guajardo[1], Bart Mennink[2], and Berry Schoenmakers[3]

[1] Information and System Security Group
Philips Research, Eindhoven, The Netherlands
jorge.guajardo@philips.com
[2] Dept. Electrical Engineering, ESAT/COSIC and IBBT
Katholieke Universiteit Leuven, Belgium
bart.mennink@esat.kuleuven.be
[3] Dept. of Mathematics and Computer Science
Technische Universiteit Eindhoven, The Netherlands
berry@win.tue.nl

Abstract. In anonymous credential schemes, users obtain credentials on certain attributes from an issuer, and later show these credentials to a relying party anonymously and without fully disclosing the attributes. In this paper, we introduce the notion of (anonymous) credential schemes with *encrypted* attributes, in which issuers certify credentials on *encrypted* attributes to users. These schemes allow for the possibility that none of the involved parties, including the user, learns the values of the attributes. In fact, we will treat several variations differing in which parties see which attributes in the clear. We present efficient constructions of these new credential schemes, starting from a credential scheme by Brands, and we show that the security of Brands' original scheme is retained. Finally, we sketch several interesting applications of these novel credential schemes.

1   Introduction

Anonymous credential schemes, credential schemes for short, allow users to obtain credentials on particular attributes from issuers certifying that the users comply with particular conditions. These credentials can subsequently be shown to a relying party (or, verifier) in order to gain access to a service. Credential schemes were introduced by Chaum [14,15], and many efficient constructions are known [16,20,5,7,22,24,10,11,12,2], as well as several variations and extensions [17,3,4,1,9,8] (incl. anonymous cash). Credential schemes can be seen as a refined form of blind signatures, inheriting the unforgeability and unlinkability properties, but adding an extra level of privacy by allowing users to control which private information is disclosed when showing a credential. For example, irrelevant attributes need not be disclosed at all [7,1], or a zero-knowledge proof that the attributes satisfy some given constraints may suffice [6].

In general, a credential scheme consists of a key generation algorithm, an issuance protocol and a verification (or, showing) protocol. The key generation

S.-H. Heng, R.N. Wright, and B.-M. Goi (Eds.): CANS 2010, LNCS 6467, pp. 314–333, 2010.

algorithm supplies an issuer with a key pair, which is used to issue credentials on lists of attributes $x = (x_1, \ldots, x_l)$ to users. Such a credential is of the form $(p, s, \sigma(p))$, where p is the public part authenticated by $\sigma(p)$ and s is a secret part (including x) corresponding to p. To show a credential to a verifier, a user sends the public parts p and $\sigma(p)$, and proves knowledge of the secret part s, possibly revealing some of the attributes in x. Common to all credential schemes in the literature, however, is the property that the focus is on *authentication via credentials* to protect access to services. Naturally, the user knows the attributes in these applications. As we will see below, however, in many applications the user is not allowed to know (some of) the attributes or does not want to know these. In this paper, we therefore introduce credential schemes with encrypted attributes, or *encrypted credential schemes* for short, in which credentials may contain encrypted attributes. In the most general case, all parties involved (issuer, user, and verifier) might have access to the attributes in encrypted form only. We note that these encrypted credential schemes still offer the authentication property, as described above.

APPLICATIONS. Basically, there are two types of applications of encrypted credential schemes: one can think of scenarios where the user *is not allowed* to learn the attributes, as well as cases where the user *does not want* to learn the attributes. As a simple example of the former case, consider the use of confidential letters of recommendation: if a person wants to apply for a job or wants to enter into a graduate program at a university, he can request a letter of recommendation from his former supervisor, which he in turn shows to the potential employer. Since the holder of this letter is not allowed to learn the supervisor's opinions, the letter can be implemented as an encrypted credential. Note that this allows one to apply for jobs anonymously (at first) and without the supervisor knowing of the job applications. For the case where the user does not want to learn the attributes, one can think of people who do not want to learn about (genetic) diseases they suffer, but still need credentials on these data for various purposes.

More generally, encrypted credential schemes provide the missing link between anonymous credentials (where credentials are merely issued on data), and secure multiparty computation (MPC) based on threshold homomorphic cryptosystems (where one can usually trace which outputs are used in subsequent computations). For instance, one can consider a first MPC where the total wealth of a party is computed based on his salary, registered possessions (such as real estate), etc. This computation may involve various database lookups, and in particular, the identity of the party may be known. The computed result representing the total wealth, however, should remain hidden, and will only be given in encrypted form to the credential issuer. A second MPC could then be the millionaires protocol, or a more elaborate computation, on the wealths of a number of parties, where all involved parties should remain anonymous. Here, one needs the encrypted data to be accompanied with an anonymous credential. The parties performing the secure computations will remain oblivious to the links between the inputs and outputs of the secure computations.

CHALLENGES AND TECHNICAL ISSUES. Observe that one cannot solve this problem by naively issuing 'standard' credentials on the ciphertexts (as being the attributes): while the message to be signed can be re-blinded by the user, the attributes themselves *cannot*. As a consequence, the ciphertexts cannot be re-randomized in this solution to the problem. More generally, common credential schemes rely in an essential way on the fact that (at least) the user has access to the attributes in the clear. Instead, to handle encrypted attributes some major changes are needed. Below we highlight some of the issues.

- If users do not know the attributes of a credential, they cannot render zero-knowledge proofs for these attributes as part of the verification protocol. To resolve this issue, verification of a credential with encrypted attributes will involve some type of plaintext equality test, ensuring minimal disclosure of the attributes. The secret key used for verification will in general be distributed among multiple parties;
- Given the use of some type of plaintext equality test in the verification protocol, it must be prevented that a malicious user abuses the verification protocol to gain partial information on the encrypted attributes (which in general need to remain hidden from the user). This contrasts with the verification protocol in common credential schemes, where the verifier only generates a random challenge and has no private inputs;
- To achieve unlinkability for the credential scheme, honest users should be able to blind the encrypted attributes as provided by the issuer, basically by performing random re-encryptions of these attributes. Malicious users, however, should not be able to abuse this mechanism by replacing a particular encrypted attribute with a target encryption. Based on the success or failure in a run of the verification protocol, a malicious user would then be able to find out if the encryption of the attribute and the target encryption contain the same plaintext or not.

OUR CONTRIBUTIONS. We introduce and define the notion of *encrypted credential schemes* as a new concept in the area of anonymous credentials (Sect. 3). Our definition captures the case of an issuer certifying *encrypted* attributes, such that none of the involved participants learns the encryptions. For a variation where the issuer knows the attributes (see the above-mentioned applications), we introduce a concrete construction of an efficient scheme (Sect. 4), but we notice that the scheme can be easily adjusted to a more general case where the issuer does not learn (some of) the attributes (Sect. 6). The security of the scheme is analyzed in Sect. 5. The construction of our scheme is based on an efficient and well-established credential scheme by Brands [7]. By combining the newly introduced schemes with Brands' original scheme, we also obtain schemes in which users learn *some* of the attributes in the clear, and *some* in encrypted form (Sect. 6). Since Brands' schemes are providing single-use credentials, our schemes do so as well[1]. We leave it as an open problem to construct multi-use

---

[1] As observed in [8], Brands' schemes allow for efficient issuance of multiple credentials on the same attribute list. The same remark applies to our schemes.

credential schemes with encrypted attributes (e.g., starting from [11,12]), or to show the impossibility of such construction.

2 Preliminaries

Below, we introduce some well-known cryptographic primitives along with some relevant notation. In particular, we give some background on the credential scheme by Brands on which our constructions are based. Throughout, we use $x \in_R V$ to denote that x is drawn uniformly at random from V, and $\mathcal{H}$ to denote a cryptographic hash function (viewed as a random oracle) with range equal to $\mathbb{Z}_q$. Here, q is a prime of bit-length k, for a security parameter k. Furthermore, we let $\langle g \rangle$ denote a cyclic group of prime order q, representing a suitable discrete log setting.

ELGAMAL CRYPTOSYSTEM. Our scheme uses the additively homomorphic El-Gamal cryptosystem. For cyclic group $\langle g \rangle$, the secret key is a $\lambda \in_R \mathbb{Z}_q$ and the corresponding public key is the group element $f = g^\lambda$. A message $x \in \mathbb{Z}_q$ is encrypted by taking an $r \in_R \mathbb{Z}_q$ and computing $c = (g^r, g^x f^r) =: [\![g^x]\!]$. For efficient decryption, x must be limited to a sufficiently small set (but this is no limitation for our setting). The decryption function is denoted by D. The homomorphic properties ensure that an encryption can be re-blinded by multiplying it with a random zero-encryption $[\![g^0]\!]$.

Σ-PROTOCOLS. Informally, a zero-knowledge proof of knowledge is a two-party protocol for a prover to convince a verifier that he knows something, without leaking any information other than the value of the assertion that is being proved. More specifically, for a relation $R = \{(x; w)\}$ and for an x, common input for the prover and verifier, the prover proves in zero-knowledge that he knows a value w (the witness) such that $(x; w) \in R$. We use the notion of Σ-protocols, cf. Cramer et al. [19]. A Σ-protocol consists of a conversation (a, c, r), where the prover sends a commitment a, the verifier returns a random challenge c and the prover sends a response r. Afterwards, the verifier either accepts or rejects. A Σ-protocol needs to satisfy three properties: *completeness*, *special soundness* and *special honest-verifier zero-knowledge*.

BRANDS' CREDENTIAL SCHEME. We will consider Brands' credential scheme [7] based on the blind Chaum-Pedersen signature scheme [18][2]. Given cyclic group $\langle g \rangle$, the issuer's public key consists of the group elements $h_0, g_1, \ldots, g_l \in_R \langle g \rangle$, and a credential on attribute list $(x_i)_{i=1}^l$ is a tuple $(h', (x_i)_{i=1}^l, \alpha, \sigma(h'))$ satisfying

$$\sigma(h') \text{ is a signature on } h', \text{ and } (g_1^{x_1} \cdots g_l^{x_l} h_0)^\alpha = h'. \tag{1}$$

---

[2] Brands also introduced a more efficient DL-based scheme, but this scheme does not offer the possibility for the issuer to issue a credential without knowing the attributes in the clear, while this is clearly a requirement in encrypted credential schemes.

Upon verification of a credential, the user sends the public part of the credential, $(h', \sigma(h'))$, to the verifier, and executes a Σ-protocol to prove knowledge of $((x_i)_{i=1}^l, \alpha)$ satisfying $(g_1^{x_1} \cdots g_l^{x_l} h_0)^\alpha = h'$. A summary of Brands' scheme is given in App. A.

3   Definition of Encrypted Credential Schemes

In this section, the notion of encrypted credential schemes will be introduced more precisely. As mentioned above, an encrypted credential scheme considers the case of an issuer certifying *encrypted* attributes to users, such that none of the involved participants learns the attributes. The basic ingredients for this type of schemes are three protocols: a key generation protocol for generating public and secret keys, and protocols for issuance and verification of encrypted credentials. Informally stated, the security and privacy of these schemes comprise the following. Security roughly means (1) that credentials are unforgeable, meaning that it is hard for a user to convince the verifier with a forged credential and (2) that no unauthorized participant learns the encrypted attributes. Privacy means that anonymity of the users is guaranteed and executions of the verification protocol cannot be linked. More precisely, we propose the following definition of encrypted credential schemes, taking into account the technical issues mentioned in Sect. 1.

Definition 1. *An* encrypted credential scheme *consists of the following protocols, where $\mathcal{I}$ denotes an issuer, $\mathcal{U}$ denotes a user, and $\mathcal{V}$ denotes a verifier:*

- *A key generation protocol for $\mathcal{I}$ and $\mathcal{V}$, that on input of security parameter k outputs public/secret keys $(pk, sk_\mathcal{I}, sk_\mathcal{V})$, where pk includes the system parameters. It also includes a key pair for an encryption scheme, of which the secret key is owned by $\mathcal{V}$. We write $(pk, sk_\mathcal{I}, sk_\mathcal{V}) \leftarrow \mathsf{keygen}_{\mathcal{I}, \mathcal{V}}(k)$;*
- *An issuance protocol for $\mathcal{I}$ and $\mathcal{U}$, that on input of pk and a list of encrypted attributes C, together with $\mathcal{I}$'s secret key, outputs a credential $(p, s, \sigma(p))$ for the user. This credential satisfies that $\sigma(p)$ is a signature on p, that p is a public key part for which s is a secret key, and that p contains re-encryptions of the encryptions in C. The protocol is denoted by $(p, s, \sigma(p)) \leftarrow \mathsf{issue}_{\mathcal{I}(sk_\mathcal{I}); \mathcal{U}}(pk, C)$;*
- *A verification protocol for $\mathcal{U}$ and $\mathcal{V}$, that on input of pk, $\mathcal{U}$'s input $(p, s, \sigma(p))$ and $\mathcal{V}$'s secret key outputs a bit, representing either acceptance or rejection. We write $\mathsf{verify}_{\mathcal{U}(p, s, \sigma(p)); \mathcal{V}(sk_\mathcal{V})}(pk)$ to denote a run of the protocol, which outputs a bit.*

These protocols satisfy the following properties for any $(pk, sk_\mathcal{I}, sk_\mathcal{V})$ resulting from an execution of the key generation protocol:

- Completeness. *For any honest $\mathcal{I}, \mathcal{U}$ and $\mathcal{V}$, the credential obtained by $\mathcal{U}$ in the execution of the issuance protocol, will be accepted in the verification protocol;*

- Security. *The credentials are unforgeable and no unauthorized party learns the encrypted attributes;*
- Privacy. *The scheme offers unlinkability, and anonymity of the users is guaranteed.*

In practice, the verifier's secret key can be shared among multiple verifiers using threshold cryptography, such that the user can execute the verification protocol with any qualified set of verifiers [25]. The definition is formulated in a general way, but variations are possible as well. The definition can for instance be adjusted to the case where $\mathcal{I}$ learns the attributes, but $\mathcal{U}$ and $\mathcal{V}$ do not. Note that this is precisely the case in the specific applications mentioned in Sect. 1. Furthermore, we notice that Def. 1 does not restrict credentials to be single-use or multi-use. The remainder of the paper, however, concentrates on single-use credentials. In particular, Def. 2 below states completeness, security and privacy properties more concretely for single-use credentials, following Brands' definition of secure credential schemes [7]. Throughout, a (potentially) malicious participant is indicated by an apostrophe, as in $\mathcal{U}'$. A participant is called semi-honest in case he follows the protocol but tries to obtain as much information as possible. For simplicity, we consider semi-honest verifiers only. This can be guaranteed by implementing $\mathcal{V}$ as a set of parties using threshold cryptography (see also Sect. 6).

Definition 2. *A key generation, issuance and verification protocol involving parties $\mathcal{I}, \mathcal{U}$ and $\mathcal{V}$ constitute a secure encrypted credential scheme (cf. Def. 1) if the following properties are satisfied for any $(pk, sk_\mathcal{I}, sk_\mathcal{V})$ resulting from an execution of the key generation protocol:*

- Completeness. *For any attribute list C and honest $\mathcal{I}, \mathcal{U}$ and $\mathcal{V}$, the issuance protocol on input of C results in a valid credential for $\mathcal{U}$. More formally, for any C we have*

$$\Pr\left(\text{verify}_{\mathcal{U}(p,s,\sigma(p));\mathcal{V}(sk_\mathcal{V})}(pk) = 1 \;\middle|\; (pk, sk_\mathcal{I}, sk_\mathcal{V}) \leftarrow \text{keygen}_{\mathcal{I};\mathcal{V}}(k);\right.$$
$$\left.(p, s, \sigma(p)) \leftarrow \text{issue}_{\mathcal{I}(sk_\mathcal{I});\mathcal{U}}(pk, C)\right) = 1;$$

- User privacy. *For any two issued credentials, a malicious $\mathcal{I}'$ cannot distinguish between the public key parts of these credentials. More formally, there exists a negligible $\nu(k)$, such that for any C_0, C_1 we have*

$$\Pr\left(\mathcal{I}'(pk, sk_{\mathcal{I}'}, (p, \sigma(p))_b, (p, \sigma(p))_{1-b}, \text{view}_0, \text{view}_1) = b \;\middle|\;\right.$$
$$(pk, sk_{\mathcal{I}'}, sk_\mathcal{V}) \leftarrow \text{keygen}_{\mathcal{I}';\mathcal{V}}(k); \; b \in_R \{0, 1\};$$
$$\left.(p, s, \sigma(p))_j \leftarrow \text{issue}_{\mathcal{I}'(sk_{\mathcal{I}'});\mathcal{U}}(pk, C_j) \text{ for } j = 0, 1\right) < \frac{1}{2} + \nu(k),$$

where view_j denotes $\mathcal{I}'$'s view on the j-th issuing execution $(j = 0, 1)$, i.e. all values $\mathcal{I}'$ sees during the execution;

- One-more unforgeability. *Suppose that for any $K \geq 0$, malicious $\mathcal{U}'$ can perform K arbitrarily interleaved credential queries on adaptively chosen attribute lists C_j $(j = 1, \ldots, K)$. Then, the probability that $\mathcal{U}'$ outputs $K + 1$ distinct credentials is negligible in k. More formally, there exists a negligible $\nu(k)$, such that for any $K \geq 0$ we have*

$$\Pr\left(\forall_{i=1}^{K+1}\left[\mathsf{verify}_{\mathcal{U}'((p,s,\sigma(p))_i);\mathcal{V}(sk_\mathcal{V})}(pk) = 1\right] \,\middle|\, (pk, sk_\mathcal{I}, sk_\mathcal{V}) \leftarrow \mathsf{keygen}_{\mathcal{I};\mathcal{V}}(k);\right.$$
$$\left.\{(p, s, \sigma(p))_i\}_{i=1}^{K+1} \leftarrow \mathcal{U}'^{\,\mathsf{issue}_{\mathcal{I}(sk_\mathcal{I});\mathcal{U}'}(pk,\cdot)}\right) < \nu(k),$$

  where $\mathcal{U}'$ queries its oracle K times;
- Blinding-invariance unforgeability. *Suppose that for any $K \geq 0$, malicious $\mathcal{U}'$ can perform K arbitrarily interleaved credential queries on adaptively chosen attribute lists C_j $(j = 1, \ldots, K)$, and that $\mathcal{U}'$ outputs L credentials $((p, s, \sigma(p))_i)_{i=1}^L$ for some $L \leq K$. Then, for any of the attribute lists in these L credentials, the number of credentials on this list does not exceed the number of times a credential has been issued on this list. More formally, there exists a negligible $\nu(k)$, such that for any $K \geq L \geq 0$ we have*

$$\Pr\left(\forall_{i=1}^{L}\left[\mathsf{verify}_{\mathcal{U}'((p,s,\sigma(p))_i);\mathcal{V}(sk_\mathcal{V})}(pk) = 1\right] \wedge R \not\subseteq S \,\middle|\right.$$
$$(pk, sk_\mathcal{I}, sk_\mathcal{V}) \leftarrow \mathsf{keygen}_{\mathcal{I};\mathcal{V}}(k);$$
$$R := \{D(\mathsf{inv}(p_i))\}_{i=1}^L \text{ and } S := \{D(C_j)\}_{j=1}^K \text{ multisets;}$$
$$\left.\left(\{(p, s, \sigma(p))_i\}_{i=1}^L, Q\right) \leftarrow \mathcal{U}'^{\,\mathsf{issue}_{\mathcal{I}(sk_\mathcal{I});\mathcal{U}'}(pk,\cdot)}\right) < \nu(k),$$

  where $\mathcal{U}'$ queries its oracle K times, and $Q = \{C_j\}_{j=1}^K$ are the corresponding attribute lists. Here, inv is some non-constant function that, on input of the public key part of a credential, outputs the corresponding list of encrypted attributes (cf. Def. 1), and R and S denote multisets of plaintext (decrypted) attribute lists;
- Secure verification. *For any credential $(p, s, \sigma(p))$, the verification protocol is a secure two-party protocol for proving knowledge of s such that $(p, s, \sigma(p))$ is a valid credential, where $\mathcal{U}$ sent $(p, \sigma(p))$ to $\mathcal{V}$.*

Additionally, no unauthorized party learns the encrypted attributes.

We note that these properties indeed cover the privacy and security requirements informally introduced in the beginning of this section. In particular, the two unforgeability statements encompass any possible forgery: a forger can either construct more credentials than he is issued on, or less but on different attributes[3]. The property that no unauthorized party learns the attributes usually follows directly from the other properties (cf. App. B). In particular, the encrypted attributes do not leak during the verification execution, as the verifier is semi-honest and this protocol is a secure two-party protocol.

---

[3] Even though the idea of encrypted credential schemes is that the user will not learn the attributes in the clear, the unforgeability requirements are defined so as to cover security against malicious users adaptively choosing the attributes. This is done in order to achieve similar security results as in the credential scheme by Brands.

4   An Encrypted One-Show Credential Scheme

In this section, we construct an encrypted credential scheme. The scheme presented below is for the case where the issuer knows the attributes in the clear, while the user and verifier[4] do not learn these. This is one of the most interesting variations (see the applications in Sect. 1). In Sect. 6 we will consider the extension to the case where the issuer does not learn the attributes either. For simplicity, it is assumed that the attributes are binary, i.e., $x_1^*, \ldots, x_{l-1}^* \in \{0,1\}$. The scheme is based on a credential scheme by Brands [7], and in particular our scheme can be combined with Brands' scheme, for instance such that the user learns x_1^*, x_2^*, but does *not* learn $x_3^*, \ldots, x_{l-1}^*$. See also Sect. 6.

The encrypted credential scheme will be introduced from a constructive point of view: at first the ideas of the protocols are described with respect to Brands' scheme, and then the mathematical descriptions of the protocols are given. A general remark is that the issuer will actually certify attributes $x_i = x_i^* + \phi_i$ for some $\phi_i \in_R \mathbb{Z}_q$ unknown to the user (for $i = 1, \ldots, l-1$), and an additional $x_l \in_R \mathbb{Z}_q$. This adjustment turns out to be important for solving the third technical issue of Sect. 1: without this modification, the verification of a credential with an encrypted attribute replaced with a target or faked encryption would succeed with significant probability for attributes from a limited range (e.g. binary attributes). By artificially extending their range to $\mathbb{Z}_q$, such attack succeeds with negligible probability only. Interestingly, it turns out that an issuer can use the same tuple $(\phi_i)_{i=1}^{l-1}$ for all executions of the issuance protocol.

4.1   Key Generation

Essentially, we combine the key generation algorithms of Brands' scheme and of the ElGamal cryptosystem. Additionally, the values $(\phi_i)_{i=1}^{l-1}$ are needed as well, as mentioned above. However, in our scheme the key generation is actually a protocol between issuer and verifier, because the verifier needs secret data as well. The verifier will use its secret data in a plaintext equality test, cf. Eq. (2b), and these values are not needed by the issuer for issuing credentials.

The key generation protocol can now be described as follows. Given a security parameter k, system parameters (q, g), with prime $q > 2^k$, are generated first. Then, public key $(h_0, f, \hat{f}, (g_i)_{i=1}^{l}, (f_i)_{i=1}^{l-1})$ is generated jointly, corresponding to the issuers secret key $x_0, (\phi_i)_{i=1}^{l-1} \in_R \mathbb{Z}_q$ and the verifiers secret key $\lambda, (y_i)_{i=1}^{l} \in_R \mathbb{Z}_q$ satisfying

$$h_0 = g^{x_0}, \quad f = g^\lambda, \quad \hat{f} = f^{x_0} = h_0^\lambda, \quad \forall_{i=1}^{l} : g_i = g^{y_i}, \quad \forall_{i=1}^{l-1} : f_i = g^{\phi_i}.$$

4.2   Credential Issuance

As in Brands' issuance protocol, the user's attributes are signed indirectly via the group element $h = g_1^{x_1} \cdots g_l^{x_l} h_0 \neq 1$. The attributes are provided to the user in

---

[4] Recall that we consider semi-honest verifiers only.

encrypted form only, and are blinded by the users by random re-encryption. To indeed restrict the users to random re-encryptions of the encrypted attributes, the issuer uses the values $(\phi_i)_{i=1}^{l-1}$ when forming h and the encryptions c_i of the attributes, as well as the values $z_i = c_i^{x_0}$ and $e_i = c_i^{w}$. The protocol for issuing a credential on $(x_i^*)_{i=1}^{l-1}$ is given in Fig. 1. It results in a credential for $\mathcal{U}$, which consists of a tuple $(h', (c_i')_{i=1}^{l}, \alpha, z', (z_i')_{i=1}^{l}, c', r')$ satisfying

$$c' = \mathcal{H}([c_i', z_i', (c_i')^{r'}(z_i')^{-c'}]_{i=1}^{l}; h', z', g^{r'}h_0^{-c'}, (h')^{r'}(z')^{-c'}), \qquad (2a)$$

$$\text{and } (D((c_1')^{y_1} \cdots (c_l')^{y_l})h_0)^{\alpha} = h' \neq 1, \qquad (2b)$$

where $c_i' = [\![g^{x_i^*} f_i]\!]$ for $i = 1, \ldots, l-1$, and $c_l' = [\![g^{x_l}]\!]$ for $x_l \in_R \mathbb{Z}_q$.

Note that these credentials mainly differ from Brands' credentials in the second part (2b). By defining $x_i := x_i^* + \phi_i$ for $i = 1, \ldots, l-1$, we have $(c_i')^{y_i} = [\![g^{x_i}]\!]^{y_i} = [\![g^{x_i}]\!]$ for all i. Consequently, (2b) simplifies to $(g_1^{x_1} \cdots g_l^{x_l} h_0)^{\alpha} = h'$, which is the same equation as in Brands' credential scheme, cf. (1). The crucial difference is that the verification of (2b) is done through a plaintext equality test and requires access to a secret key.

4.3   Credential Verification

For verification of a credential $(h', (c_i')_{i=1}^{l}, \alpha, z', (z_i')_{i=1}^{l}, c', r')$, the user sends the public part (all values except for α) to the verifier, and proves knowledge of α such that (2) holds. This protocol is given in Fig. 2. Upon successful verification, the verifier will extract the encrypted attributes $[\![g^{x_i^*}]\!]$ by computing $c_i'[\![f_i]\!]^{-1}$ (for $i = 1, \ldots, l-1$).

The verification protocol can be viewed as a proof of knowledge for relation $\{(h', (c_i')_{i=1}^{l}; \alpha) \mid (h')^{\alpha^{-1}} = D((c_1')^{y_1} \cdots (c_l')^{y_l})h_0 \wedge \alpha \neq 0\}$, except that the verifier uses a secret input as well for the evaluation of a plaintext equality test. For this reason, the protocol is not a Σ-protocol, and an explicit fourth round has been added to inform the user whether verification succeeded.

5   Security Analysis

Using Def. 2, we analyze the security of the above encrypted credential scheme.

Theorem 1. *The protocols introduced in Sect. 4 constitute a secure encrypted credential scheme cf. Def. 2.*

The proof is rather technical, and is included in App. B. It is based on Ass. 2. Intuitively, this assumption states that if a malicious user can succeed showing a credential, then (with overwhelming probability) he has been issued a credential on precisely the same attributes, and he moreover knows the blinding factors $(\delta_i)_{i=1}^{l}$ corresponding to this issuance. It corresponds to the fourth property of Def. 2, and is similar to an assumption Brands needed to prove his scheme secure (Ass. 3). In particular, the level of security of Brands' scheme is retained. We refer to [23, App. A] for a detailed heuristic analysis of Ass. 2.

$$\mathcal{U} \qquad\qquad\qquad \mathcal{I}$$

$$\text{(knows: } (x_i^*)_{i=1}^{l-1} \,; x_0, (\phi_i)_{i=1}^{l-1})$$

$$(r_i)_{i=1}^l \,, x_l \in_R \mathbb{Z}_q$$

$$\left(c_i \leftarrow (g^{r_i}, g^{x_i^*} f_i f^{r_i}) \right)_{i=1}^{l-1}$$

$$c_l \leftarrow (g^{r_l}, g^{x_l} f^{r_l})$$

$$h \leftarrow \prod_{i=1}^{l-1} g_i^{x_i^*+\phi_i} g_l^{x_l} h_0$$

$$z \leftarrow h^{x_0}, \quad (z_i \leftarrow c_i^{x_0})_{i=1}^l$$

$$w \in_R \mathbb{Z}_q, \quad a \leftarrow g^w, \quad b \leftarrow h^w$$

$$\tilde{f} \leftarrow f^w, \quad (e_i \leftarrow c_i^w)_{i=1}^l$$

$$\xleftarrow{\quad h,z,(c_i,z_i)_{i=1}^l;\ a,b,\tilde{f},(e_i)_{i=1}^l \quad}$$

$$\alpha \in_R \mathbb{Z}_q^*, \quad \beta, \gamma \in_R \mathbb{Z}_q$$

$$h' \leftarrow h^\alpha, \quad z' \leftarrow z^\alpha$$

$$a' \leftarrow h_0^\beta g^\gamma a, \quad b' \leftarrow (z')^\beta (h')^\gamma b^\alpha$$

$$\begin{pmatrix} \delta_i \in_R \mathbb{Z}_q, \quad c_i' \leftarrow c_i \cdot (g,f)^{\delta_i} \\ z_i' \leftarrow z_i \cdot (h_0, \hat{f})^{\delta_i} \\ e_i' \leftarrow (z_i')^\beta (c_i')^\gamma e_i \cdot (a, \tilde{f})^{\delta_i} \end{pmatrix}_{i=1}^l$$

$$c' \leftarrow \mathcal{H}([c_i', z_i', e_i']_{i=1}^l; h', z', a', b')$$

$$c \leftarrow c' + \beta \bmod q \xrightarrow{\quad c \quad}$$

$$\xleftarrow{\quad r \quad} r \leftarrow cx_0 + w \bmod q$$

$$a \overset{?}{=} g^r h_0^{-c}, \quad b \overset{?}{=} h^r z^{-c}$$

$$\tilde{f} \overset{?}{=} f^r \hat{f}^{-c}, \quad (e_i \overset{?}{=} c_i^r z_i^{-c})_{i=1}^l$$

$$r' \leftarrow r + \gamma \bmod q$$

Fig. 1. Issuance protocol of the encrypted credential scheme

$$\mathcal{U} \qquad\qquad\qquad \mathcal{V}$$

$$\text{(knows: } h', z', (c_i', z_i')_{i=1}^l, c', r'; \alpha) \qquad \text{(knows: } (y_i)_{i=1}^l, \lambda)$$

$$u \in_R \mathbb{Z}_q, \quad a \leftarrow (h')^u \xrightarrow{\quad a;(c_i',z_i')_{i=1}^l,\ h',z',c',r' \quad}$$

$$\xleftarrow{\quad c \quad} c \in_R \mathbb{Z}_q$$

$$r \leftarrow u + c\alpha^{-1} \bmod q \xrightarrow{\quad r \quad}$$

$$b_1 \leftarrow c' \overset{?}{=} \mathcal{H}([c_i', z_i', (c_i')^r (z_i')^{-c'}]_{i=1}^l;$$

$$h', z', g^r h_0^{-c'}, (h')^r (z')^{-c'})$$

$$b_2 \leftarrow (h')^r \overset{?}{=} a(D((c_1')^{y_1} \cdots (c_l')^{y_l}) h_0)^c$$

$$b \leftarrow [b_1 \wedge b_2]$$

$$\xleftarrow{\quad b \quad}$$

Fig. 2. Verification protocol of the encrypted credential scheme

324   J. Guajardo, B. Mennink, and B. Schoenmakers

Assumption 2. *If $\mathcal{U}'$ produces, after $K \geq 0$ arbitrarily interleaved executions of the protocol in Fig. 1 on adaptively chosen $\left(x_{ji}^*\right)_{i=1}^{l-1}$ $(j = 1, \ldots, K)$ a tuple $(h', (c_i')_{i=1}^l, \alpha, z', (z_i')_{i=1}^l, c', r')$, then this tuple does not satisfy (2), or with overwhelming probability there exists a $j \in \{1, \ldots, K\}$ such that*

$$\mathcal{U}' \text{ knows values } (\delta_i)_{i=1}^l \text{ such that } (c_i')_{i=1}^l = \left(c_{ji}(g,f)^{\delta_i}\right)_{i=1}^l, \tag{3}$$

where $(c_{ji})_{i=1}^l$ is the list of encryptions coming from the first round of the j-th issuance execution. More formally, there exists a p.p.t. extractor $\mathcal{E}$ that may use $\mathcal{U}'$ as a subroutine and also outputs a tuple $(h', (c_i')_{i=1}^l, \alpha, z', (z_i')_{i=1}^l, c', r')$, but additionally outputs the values $(h_j, (c_{ji})_{i=1}^l)_{j=1}^K$ on which the user is issued credentials, and a value $\tau \in \{0, \ldots, K\}$: $\tau = 0$ meaning that (3) is not satisfied for any j (and implying that (2) is not satisfied), and $\tau \neq 0$ meaning that it is satisfied for $j = \tau$, in which case the extractor also outputs a tuple $(\delta_i)_{i=1}^l$ satisfying (3).

6 Variations

It is possible to adjust the scheme of Sect. 4 to the scenario where all parties *only* learn the encryptions $(c_i^*)_{i=1}^{l-1}$. However, it turns out that for the computation of h the issuer then needs secret values $(y_i)_{i=1}^l, \lambda$, and thus we need to identify the role of $\mathcal{I}$ with the role of $\mathcal{V}$. This adjustment is quite simple: in Fig. 1, the issuer now computes $(c_i)_{i=1}^{l-1}$ and h as

$$\left(c_i \leftarrow c_i^* \cdot (g^{r_i}, f_i f^{r_i})\right)_{i=1}^{l-1}, \qquad\qquad h \leftarrow D(c_1^{y_1} \cdots c_l^{y_l})h_0.$$

The remainder of the scheme remains unchanged (this variation is discussed in detail in [23]). Furthermore, it is possible to combine our encrypted credential schemes with Brands' credential scheme [7]: for instance, it is straightforward to construct a scheme for the case that both the user and issuer only know a specific (possibly non-overlapping) subset of the attributes in the clear. For these constructions, the security proofs are similar. Recall that our schemes achieve the same level of security as Brands' schemes.

We notice that the semi-honest behavior of the verifier (as mentioned in Sect. 5) can be achieved by implementing $\mathcal{V}$ as a set of parties using threshold cryptography. Indeed, the secret keys can be threshold shared among the parties using a distributed key generation protocol [21], and the plaintext equality test in the protocol of Fig. 2 can then be securely evaluated. Additionally, the possibility of multiple verifiers can be realized using verifiable secret redistribution, where the verifier redistributes his secret key to other verifiers (possibly also implemented as sets of parties) [25].

7 Conclusions

The notion of *encrypted credential schemes* is introduced and defined as a new concept in the area of anonymous credentials. We have presented and analyzed various efficient constructions of this new type of digital credential schemes, starting from a credential scheme by Brands [7]. Our schemes are comparable in security and efficiency to Brands' schemes, except that the cost grows linearly with the number of encrypted attributes. These new credential schemes have a lot of interesting applications, in particular to scenarios where the user *is not allowed* to learn the attributes in the clear (e.g., letters of recommendation), or where the user *does not want* to learn these data (e.g., medical information about illnesses). The schemes can also be used in the context of secure multiparty computation, where credentials can be issued on the results of a secure computation, which may be used as input in another secure computation, without the parties performing the computations learning anything about the links between these computations, and about the secret data.

The encrypted credential schemes constructed in this paper operate with single-use credentials. It would be interesting to extend existing multi-use credential schemes (such as [11,12]) with the functionality of encrypted attributes. Since our techniques do not readily apply to this case, we leave this as an open problem. Additionally, the construction of an encrypted credential scheme with publicly verifiable credentials remains open.

Acknowledgments. This work has been funded in part by the European Community's Sixth Framework Programme under grant number 034238, SPEED project - Signal Processing in the Encrypted Domain, in part by the IAP Program P6/26 BCRYPT of the Belgian State (Belgian Science Policy), and in part by the European Commission through the ICT program under contract ICT-2007-216676 ECRYPT II. The second author is supported by a Ph.D. Fellowship from the Institute for the Promotion of Innovation through Science and Technology in Flanders (IWT-Vlaanderen).

References

1. Bangerter, E., Camenisch, J., Lysyanskaya, A.: A cryptographic framework for the controlled release of certified data. In: Christianson, B., Crispo, B., Malcolm, J.A., Roe, M. (eds.) Security Protocols 2004. LNCS, vol. 3957, pp. 20–42. Springer, Heidelberg (2004)
2. Belenkiy, M., Chase, M., Kohlweiss, M., Lysyanskaya, A.: P-signatures and noninteractive anonymous credentials. In: Canetti, R. (ed.) TCC 2008. LNCS, vol. 4948, pp. 356–374. Springer, Heidelberg (2008)
3. Brands, S.: Untraceable off-line cash in wallet with observers. In: Stinson, D.R. (ed.) CRYPTO 1993. LNCS, vol. 773, pp. 302–318. Springer, Heidelberg (1993)
4. Brands, S.: Off-line electronic cash based on secret-key certificates. In: Baeza-Yates, R., Poblete, P.V., Goles, E. (eds.) LATIN 1995. LNCS, vol. 911, pp. 131–166. Springer, Heidelberg (1995)
5. Brands, S.: Restrictive blinding of secret-key certificates. In: Guillou, L.C., Quisquater, J.-J. (eds.) EUROCRYPT 1995. LNCS, vol. 921, pp. 231–247. Springer, Heidelberg (1995)
6. Brands, S.: Rapid demonstration of linear relations connected by boolean operators. In: Fumy, W. (ed.) EUROCRYPT 1997. LNCS, vol. 1233, pp. 318–333. Springer, Heidelberg (1997)
7. Brands, S.: Rethinking Public Key Infrastructures and Digital Certificates - Buildin. Privacy. PhD thesis, Eindhoven University of Technology, Eindhoven (1999), http://www.credentica.com/the_mit_pressbook.html

8. Brands, S., Demuynck, L., De Decker, B.: A practical system for globally revoking the unlinkable pseudonyms of unknown users. In: Pieprzyk, J., Ghodosi, H., Dawson, E. (eds.) ACISP 2007. LNCS, vol. 4586, pp. 400–415. Springer, Heidelberg (2007)
9. Camenisch, J., Hohenberger, S., Lysyanskaya, A.: Compact e-cash. In: Cramer, R. (ed.) EUROCRYPT 2005. LNCS, vol. 3494, pp. 302–321. Springer, Heidelberg (2005)
10. Camenisch, J., Lysyanskaya, A.: An efficient system for non-transferable anonymous credentials with optional anonymity revocation. In: Pfitzmann, B. (ed.) EUROCRYPT 2001. LNCS, vol. 2045, pp. 93–118. Springer, Heidelberg (2001)
11. Camenisch, J., Lysyanskaya, A.: A signature scheme with efficient protocols. In: Cimato, S., Galdi, C., Persiano, G. (eds.) SCN 2002. LNCS, vol. 2576, pp. 268–289. Springer, Heidelberg (2002)
12. Camenisch, J., Lysyanskaya, A.: Signature schemes and anonymous credentials from bilinear maps. In: Franklin, M. (ed.) CRYPTO 2004. LNCS, vol. 3152, pp. 56–72. Springer, Heidelberg (2004)
13. Canetti, R.: Security and composition of multi-party cryptographic protocols. Journal of Cryptology 13, 143–202 (2000)
14. Chaum, D.: Blind signatures for untraceable payments. In: CRYPTO 1982. LNCS, pp. 199–203. Plenum Press, New York (1983)
15. Chaum, D.: Security without identification: Transaction systems to make big brother obsolete. Communications of the ACM 28(10), 1030–1044 (1985)
16. Chaum, D., Evertse, J.: A secure and privacy-protecting protocol for transmitting personal information between organizations. In: Odlyzko, A.M. (ed.) CRYPTO 1986. LNCS, vol. 263, pp. 118–167. Springer, Heidelberg (1987)
17. Chaum, D., Fiat, A., Naor, M.: Untraceable electronic cash. In: Goldwasser, S. (ed.) CRYPTO 1988. LNCS, vol. 403, pp. 319–327. Springer, Heidelberg (1990)
18. Chaum, D., Pedersen, T.: Wallet databases with observers. In: Brickell, E.F. (ed.) CRYPTO 1992. LNCS, vol. 740, pp. 89–105. Springer, Heidelberg (1993)
19. Cramer, R.: Modular Design of Secure yet Practical Cryptographic Protocols. PhD thesis, University of Amsterdam, Amsterdam (1997)
20. Damgård, I.: Payment systems and credential mechanisms with provable security against abuse by individuals. In: Goldwasser, S. (ed.) CRYPTO 1988. LNCS, vol. 403, pp. 328–335. Springer, Heidelberg (1990)
21. Gennaro, R., Jarecki, S., Krawczyk, H., Rabin, T.: Secure distributed key generation for discrete-log based cryptosystems. In: Stern, J. (ed.) EUROCRYPT 1999. LNCS, vol. 1592, pp. 295–310. Springer, Heidelberg (1999)
22. Lysyanskaya, A., Rivest, R., Sahai, A., Wolf, S.: Pseudonym systems. In: Heys, H.M., Adams, C.M. (eds.) SAC 1999. LNCS, vol. 1758, pp. 184–199. Springer, Heidelberg (2000)
23. Mennink, B.: Encrypted Certificate Schemes and Their Security and Privacy Analysis. Master's thesis, Eindhoven University of Technology, Eindhoven (2009)
24. Verheul, E.: Self-blindable credential certificates from the weil pairing. In: Boyd, C. (ed.) ASIACRYPT 2001. LNCS, vol. 2248, pp. 533–551. Springer, Heidelberg (2001)
25. Wong, T., Wang, C., Wing, J.: Verifiable secret redistribution for archive system. In: IEEE Security in Storage Workshop, pp. 94–106 (2002)

A   Brands' Credential Scheme

In this appendix, the credential scheme by Brands, as introduced in Sect. 2, is discussed in technical detail.

Key Generation. Given a security parameter k, system parameters (q, g), with prime $q > 2^k$, are generated first, followed by the generation of a secret key $x_0 \in_R \mathbb{Z}_q$. Finally the public key $(h_0, g_1, \ldots, g_l)$ is generated as $h_0 = g^{x_0}$ and $g_1, \ldots, g_l \in_R \langle g \rangle$.

Credential Issuance. Given attribute list $(x_i)_{i=1}^l$, one sets $h = g_1^{x_1} \cdots g_l^{x_l} h_0 \neq 1$. The issuance protocol is given in Fig. 3. It results in a credential for $\mathcal{U}$ on $(x_i)_{i=1}^l$, which consists of a tuple $(h', (x_i)_{i=1}^l, \alpha, z', c', r')$ satisfying

$$c' = \mathcal{H}(h', z', g^{r'} h_0^{-c'}, (h')^{r'}(z')^{-c'}), \text{ and } (g_1^{x_1} \cdots g_l^{x_l} h_0)^\alpha = h' \neq 1. \quad (4)$$

$$
\begin{array}{ll}
\mathcal{U} & \mathcal{I} \\
\text{(knows: } h) & \text{(knows: } h; x_0) \\
 & z \leftarrow h^{x_0}, \quad w \in_R \mathbb{Z}_q \\
 & \xleftarrow{\quad z;a,b \quad} \quad a \leftarrow g^w, \quad b \leftarrow h^w \\
\alpha \in_R \mathbb{Z}_q^*, \quad \beta, \gamma \in_R \mathbb{Z}_q & \\
h' \leftarrow h^\alpha, \quad z' \leftarrow z^\alpha & \\
a' \leftarrow h_0^\beta g^\gamma a, \quad b' \leftarrow (z')^\beta (h')^\gamma b^\alpha & \\
c' \leftarrow \mathcal{H}(h', z', a', b'), \quad c \leftarrow c' + \beta \bmod q & \xrightarrow{\quad c \quad} \\
 & \xleftarrow{\quad r \quad} \quad r \leftarrow cx_0 + w \bmod q \\
a \stackrel{?}{=} g^r h_0^{-c}, \quad b \stackrel{?}{=} h^r z^{-c} & \\
r' \leftarrow r + \gamma \bmod q & \\
\end{array}
$$

Fig. 3. Issuance protocol

Credential Verification. For verification of a credential, the user sends the public part (h', z', c', r') to the verifier, and proves knowledge of $((x_i)_{i=1}^l, \alpha)$ such that (4) holds. This is a Σ-protocol for relation $\{(h'; (x_i)_{i=1}^l, \alpha) \mid h_0 = (h')^{\alpha^{-1}} g_1^{-x_1} \cdots g_l^{-x_l} \wedge \alpha \neq 0\}$ (note that $\alpha \neq 0$ should indeed hold as $h' \neq 1$), and it is given in Fig. 4.

$$
\begin{array}{ll}
\mathcal{U} & \mathcal{V} \\
\text{(knows: } h', z', c', r'; (x_i)_{i=1}^l, \alpha) & \\
u_1, \ldots, u_l, u_\alpha \in_R \mathbb{Z}_q & \\
a \leftarrow (h')^{u_\alpha} g_1^{-u_1} \cdots g_l^{-u_l} & \\
 & \xrightarrow{\quad a; h', z', c', r' \quad} \\
 & \xleftarrow{\quad c \quad} \quad c \in_R \mathbb{Z}_q \\
(r_i \leftarrow u_i + c x_i \bmod q)_{i=1}^l & \\
r_\alpha \leftarrow u_\alpha + c \alpha^{-1} \bmod q & \\
 & \xrightarrow{\quad (r_i)_{i=1}^l, r_\alpha \quad} \\
 & c' \stackrel{?}{=} \mathcal{H}(h', z', g^{r'} h_0^{-c'}, (h')^{r'}(z')^{-c'}) \\
 & (h')^{r_\alpha} g_1^{-r_1} \cdots g_l^{-r_l} \stackrel{?}{=} a h_0^c \\
\end{array}
$$

Fig. 4. Verification protocol

Security Analysis. Security of the above credential scheme is analyzed in [7]. Apart from some standard assumptions, the following specific assumption is needed as well.

Assumption 3. *If $\mathcal{U}'$ produces, after $K \geq 0$ arbitrarily interleaved executions of the protocol in Fig. 3 on adaptively chosen $(x_{ji})_{i=1}^{l}$ ($j = 1, \ldots, K$) a valid tuple $(h', (x_i)_{i=1}^{l}, \alpha, z', c', r')$, then this tuple does not satisfy (4), or with overwhelming probability there exists a $j \in \{1, \ldots, K\}$ such that $(x_i)_{i=1}^{l} = (\alpha x_{ji} \bmod q)_{i=1}^{l}$.*

B   Proof of Thm. 1

In this appendix we prove Thm. 1. The properties to be proven (cf. Def. 2) can be divided in the first four properties concerning the issuance protocol, and the last concerning the verification protocol. These protocols will be proven secure in Sects. B.1 and B.2, respectively. We will consider the five properties for any probabilistic key generation execution, resulting in a tuple $(pk, sk_\mathcal{I}, sk_\mathcal{V})$. We assume that this protocol execution is done properly, i.e. that the system parameters are correctly constructed. Notice that the issuer is the only party who learns the encrypted attributes: the verification protocol is a secure two-party protocol, and in the issuing execution the user only learns perfectly hiding commitments of the encrypted data.

B.1   Correctness of Issuance Protocol

Proposition 1 (Completeness). *If both $\mathcal{U}$ and $\mathcal{I}$ follow the protocol, then for any attribute list $C = (x_i^*)_{i=1}^{l-1}$, the resulting credential of the issuance execution will be accepted upon verification.*

Proof. See [23, Sect. 6.4].   □

Proposition 2 (User privacy). *For any pair of attribute lists C_0, C_1, if $\mathcal{U}$ and $\mathcal{I}'$ engaged in the issuance execution for both lists, obtaining credentials $(p, s, \sigma(p))_0$, $(p, s, \sigma(p))_1$, then it is hard for malicious $\mathcal{I}'$ to guess b correctly, given $(p, \sigma(p))_b$ and $(p, \sigma(p))_{1-b}$ with $b \in_R \{0, 1\}$.*

Proof. The game played by $\mathcal{I}'$ and $\mathcal{U}$ is the following: given any two different attribute lists C_0, C_1, $\mathcal{I}'$ and $\mathcal{U}$ engage in an issuance execution for C_j ($j = 0, 1$), $\mathcal{U}$ takes $b \in_R \{0, 1\}$ and sends the public parts of the b-th and $(1-b)$-th credential to $\mathcal{I}'$ (in that order). $\mathcal{I}'$ wins if he guesses b correctly. Denote by $\Pr(A)$ the success probability of $\mathcal{I}'$ in this game. We slightly change this game, obtaining game B. Now, in each issuance execution $\mathcal{U}$ sets for each $i = 1, \ldots, l$:

$$c_i' \leftarrow (g, f)^{\delta_i}, \qquad z_i' \leftarrow (h_0, \hat{f})^{\delta_i}, \qquad e_i' \leftarrow (z_i')^\beta (c_i')^\gamma (a, \tilde{f})^{\delta_i}, \qquad (5)$$

and executes the remainder as is. (Note that the resulting tuple does not yield a valid credential as $(D((c_1')^{y_1} \cdots (c_l')^{y_l})h_0)^\alpha = h'$ need not be satisfied. However, $\mathcal{I}'$ will not notice as he is p.p.t. and does not have the decryption key.) Denote $\mathcal{I}'$'s success probability in the new game by $\Pr(B)$. Now, the only difference between the games is in the encryptions, and as $\mathcal{I}'$ is p.p.t. and does not have the decryption key, if $\mathcal{I}'$ is able to distinguish between the two games, he is able to distinguish between the constructions of one of the $6l$ encryptions. Hence, the success probabilities in the different games are of negligible difference by the semantic security of the cryptosystem. Formally, there exists a negligible $\nu(k)$ such that

$$|\Pr(A) - \Pr(B)| < \nu(k). \tag{6}$$

We consider the success probability of $\mathcal{I}'$ in game B. We will first prove that for any public part of a credential, and any view on an issuance execution by $\mathcal{I}'$, there is exactly *one* possible secret random tuple $\mathcal{U}$ could have chosen. In particular this means that from $\mathcal{I}'$'s point of view, $(p, \sigma(p))_b$ could have come from the 0-th or 1-th issuance execution with equal probability, and similar for $(p, \sigma(p))_{1-b}$. Then, as $\mathcal{U}$ takes his values uniformly at random, $\mathcal{I}'$ can only succeed in guessing b correctly with probability $\frac{1}{2}$. Hence $\Pr(B) = \frac{1}{2}$, which by (6) implies that the success probability in the original game is upper bounded by $\frac{1}{2} + \nu(k)$ for negligible $\nu(k)$.

So we prove that for any public part of a credential, $(h', z', (c_i', z_i')_{i=1}^l, c', r')$, and all values a malicious $\mathcal{I}'$ sees during the issuance execution of a credential, $(h, z, (c_i, z_i)_{i=1}^l)$ and $(a, b, \tilde{f}, (e_i)_{i=1}^l, c, r)$ satisfying (as $\mathcal{U}$ accepted)

$$a = g^r h_0^{-c}, \qquad b = h^r z^{-c}, \qquad \tilde{f} = f^r \hat{f}^{-c}, \qquad \forall_{i=1}^l : e_i = c_i^r z_i^{-c}, \tag{7}$$

there exists exactly *one* possible combination of random values $\alpha, \beta, \gamma, (\delta_i)_{i=1}^l$ that $\mathcal{U}$ could have chosen to end up with that credential. The values α, β and γ are determined by (h, h'), (c, c') and (r, r'), namely as $\alpha = \log_h h'$, $\beta = c - c' \bmod q$ and $\gamma = r' - r \bmod q$. Furthermore, for each i, δ_i is determined by c_i' as $\delta_i = \log_g(c_i')_1 = \log_f(c_i')_2$. Remains to prove that this choice satisfies $c' = \mathcal{H}([c_i', z_i', e_i']_{i=1}^l; h', z', a', b')$. But the issued credential satisfies $\mathcal{H}([c_i', z_i', (c_i')^{r'}(z_i')^{-c'}]_{i=1}^l; h', z', g^{r'} h_0^{-c'}, (h')^{r'}(z')^{-c'})$, from which the equality follows if $a' = g^{r'} h_0^{-c'}$, $b' = (h')^{r'}(z')^{-c'}$ and $\forall_{i=1}^l : e_i' = (c_i')^{r'}(z_i')^{-c'}$. But the first two equations are easy to check, and for the third we have for all $i = 1, \ldots, l$:

$$
\begin{aligned}
(c_i')^{r'}(z_i')^{-c'} &= (c_i')^\gamma (z_i')^\beta (c_i')^r (z_i')^{-c} && \{\text{setup } r', c'\} \\
&= (c_i')^\gamma (z_i')^\beta (g^r h_0^{-c}, f^r \hat{f}^{-c})^{\delta_i} && \{\text{equation (5)}\} \\
&= (c_i')^\gamma (z_i')^\beta (a, \tilde{f})^{\delta_i} && \{\text{equation (7)}\} \\
&= e_i' && \{\text{equation (5)}\}. \qquad \square
\end{aligned}
$$

Remark 1. For the proof of Prop. 2, $\mathcal{I}'$ may only work in probabilistic polynomial time[5], simply because different issuance executions might involve different encryptions. However, if the two attribute lists are the same, so $C_0 = C_1$, then the changeover to game B is unnecessary. In particular, the issuance executions

---

[5] In case the issuer would know the secret decryption key, e.g. if the issuer plays the role of the verifier as well, we moreover require the issuer to be semi-honest. However, as $\mathcal{V}$ is semi-honest (Sect. 5), this is naturally enforced.

then become unlinkable even for issuers with unlimited resources. This is relevant in case the issuer issues many credentials on the same attribute list.

The proof of one-more unforgeability relies on tightly reducing the credentials to signatures of the blind signature scheme by Chaum and Pedersen [18]. Briefly, the blind Chaum-Pedersen signature scheme considers a cyclic group $\langle g \rangle$ and a public $h \in \langle g \rangle$ corresponding to secret key x, known by the signer. The issuance of a signature on message m happens in four rounds, starting with the user blinding m and sending it to the signer. The protocol results in a signature (m, z, c', r') such that $c' = \mathcal{H}(m, z, g^{r'} h^{-c'}, m^{r'} z^{-c'})$. The reader is referred to [18] for a more detailed discussion of the scheme. In what follows, we assume this scheme to be secure. Note that forging a Chaum-Pedersen signature is just as hard as forging a signature of the form (ξ, m, z, c', r') such that $c' = \mathcal{H}(\xi, m, z, g^{r'} h^{-c'}, m^{r'} z^{-c'})$ for any arbitrary bit string ξ. This is due to the properties of the cryptographic hash function.

Proposition 3 (One-more unforgeability). *Under the assumption that the blind Chaum-Pedersen signature scheme is secure against one-more forgeries, it is impossible for a user $\mathcal{U}'$ to, after $K \geq 0$ arbitrarily interleaved executions of Fig. 1 on adaptively chosen attribute lists $C_j = (x^*_{ji})_{i=1}^{l-1}$ ($j = 1, \ldots, K$), with non-negligible probability output $K + 1$ different credentials satisfying (2).*

Proof. Suppose it is possible, so after K executions of the protocol of Fig. 1, on adaptively chosen $(x^*_{ji})_{i=1}^{l-1}$ for $j = 1, \ldots, K$, $\mathcal{U}'$ can output $K + 1$ different credentials $(h', (c'_i)_{i=1}^{l}, \alpha, z', (z'_i)_{i=1}^{l}, c', r')$ satisfying (2), with non-negligible probability. We construct an interactive polynomial time forger $\mathcal{F}$ that is issued K Chaum-Pedersen signatures by a Chaum-Pedersen signer $\mathcal{S}$, possibly on different messages m for each execution $j = 1, \ldots, K$, and uses $\mathcal{U}'$ to output $K + 1$ different Chaum-Pedersen signatures. By assumption that is impossible, and hence we obtain a contradiction.

Let $\langle g \rangle, h_{CP}$ be the system parameters of the Chaum-Pedersen signature scheme, for which $\mathcal{S}$ knows $x = \log_g h_{CP}$. Now $\mathcal{F}$ simulates the credential issuer for Fig. 1 as follows:

1. *Initialization*: For the encryption scheme, $\mathcal{F}$ takes secret key $\lambda \in_R \mathbb{Z}_q$ and publishes $f = g^{\lambda}$. Furthermore, $\mathcal{F}$ inherits $\mathcal{S}$'s system parameters, and takes moreover $(y_i)_{i=1}^{l}, (\phi_i)_{i=1}^{l-1} \in_R \mathbb{Z}_q$ and publishes $h_0 = h_{CP}$, $f_i = g^{\phi_i}$ ($i = 1, \ldots, l-1$), $g_i = g^{y_i}$ ($i = 1, \ldots, l$), and $\hat{f} = h_0^{\lambda}$;

2. *Issuance*: For each of the K issuance protocol executions, $\mathcal{F}$ operates as follows[6]:

    i. *Commitment part 1*: $\mathcal{F}$ obtains $x^*_i \in \{0, 1\}$ from $\mathcal{U}'$ ($i = 1, \ldots, l-1$).[7] He takes $(r_i)_{i=1}^{l}, x_l \in_R \mathbb{Z}_q$, sets $(x_i \leftarrow x^*_i + \phi_i \bmod q)_{i=1}^{l-1}$, sets

---

[6] For ease of presentation, the first round of the original protocol in Fig. 1, the commitment part, is separated into two phases i and ii. That is, firstly $(h, z, (c_i, z_i)_{i=1}^{l})$ is sent to $\mathcal{U}'$, and then $(a, b, \tilde{f}, (e_i)_{i=1}^{l})$.

[7] Recall that $\mathcal{U}'$ can adaptively choose the attribute list. If $\mathcal{U}'$ would adaptively choose *encrypted* attributes $[\![x^*_i]\!]$ instead (for instance in the variation of the scheme, cf. Sect. 6), $\mathcal{F}$ can still obtain the plaintext attributes by using the decryption key λ.

$$c_i \leftarrow (g^{r_i}, g^{x_i} f^{r_i}) \text{ and } z_i \leftarrow (h_0^{r_i}, h_0^{x_i + \lambda r_i}), \text{ for each } i = 1, \dots, l,$$

and $h \leftarrow g_1^{x_1} \cdots g_l^{x_l} h_0$. For the setup of z, $\mathcal{F}$ sends $\tilde{m} \leftarrow h$ to $\mathcal{S}$, in order to obtain $\tilde{z}$. The forger sends $z \leftarrow \tilde{z}$ to $\mathcal{U}'$;

ii. *Commitment part 2*: $\mathcal{F}$ receives $\tilde{a}, \tilde{b}$ from $\mathcal{S}$, he sets $(a, b) \leftarrow (\tilde{a}, \tilde{b})$ and $\tilde{f} \leftarrow a^\lambda$, and for each $i = 1, \dots, l$ he takes $e_i \leftarrow (a^{r_i}, a^{x_i + \lambda r_i})$. He sends $(a, b, \tilde{f}, (e_i)_{i=1}^l)$ to $\mathcal{U}'$;

iii. *Challenge*: $\mathcal{F}$ receives c from $\mathcal{U}'$ and sends $\tilde{c} \leftarrow c$ to $\mathcal{S}$;

iv. *Response*: $\mathcal{F}$ receives $\tilde{r}$ from $\mathcal{S}$ and sends $r \leftarrow \tilde{r}$ to $\mathcal{U}'$;

3. *Signature forging*: Now $\mathcal{U}'$ outputs, with non-negligible probability, $K + 1$ distinct credentials $(h', (c_i')_{i=1}^l, \alpha, z', (z_i')_{i=1}^l, c', r')$. For each of these credentials $\mathcal{F}$ computes Chaum-Pedersen forgery

$$(\overline{\xi}, \overline{z}, \overline{c}, \overline{r}, \overline{m}) \leftarrow ([c_i', z_i', (c_i')^{r'} (z_i')^{-c'}]_{i=1}^l, z', c', r', h'), \tag{8}$$

and he outputs these $K + 1$ Chaum-Pedersen signatures.

The proof that this reduction works can be found in [23, Sect. 6.4]. □

The proof of blinding-invariance unforgeability relies on Ass. 2. However, this assumption is not sufficient: it essentially says that a malicious user cannot with non-negligible probability output any credential on a *different* plaintext attribute list than he is issued credentials on, while blinding-invariance unforgeability more generally requires that for any attribute list the user cannot output more credentials on it than he is issued. So similar to Brands' scheme [7, Ass. 4.4.5], blinding-invariance unforgeability of our scheme is slightly more general than the corresponding assumption. Therefore, it is stated without proof.

Proposition 4 (Blinding-invariance unforgeability). *If $\mathcal{U}'$ comes, after $K \geq 0$ arbitrarily interleaved executions of Fig. 1 on adaptively chosen attribute lists $C_j = (x_{ji}^*)_{i=1}^{l-1}$ ($j = 1, \dots, K$), with L different credentials satisfying (2), Then, for any of the attribute lists in these L credentials, the number of credentials on this list does not exceed the number of j's such that this attribute list equals C_j.*

B.2   Correctness of Verification Protocol

We need to prove that the verification protocol in Fig. 2 is a secure two-party protocol for proving knowledge of α such that $(h', (c_i')_{i=1}^l, \alpha, z', (z_i')_{i=1}^l, c', r')$ is a valid credential. The equality $c' \stackrel{?}{=} \mathcal{H}(\cdot)$ can be checked publicly and is therefore assumed to hold. Consequently, the verification protocol of Fig. 2 simplifies to Fig. 5, where $\mathcal{U}$ can be an active attacker, but $\mathcal{V}$ can only be passive.

So, we need to prove that the protocol in Fig. 5 is a secure two-party protocol for $\mathcal{U}$ to prove knowledge of α such that $(h')^{\alpha^{-1}} = D((c_1')^{y_1} \cdots (c_l')^{y_l}) h_0$. The protocol should be a secure proof of knowledge for relation

$$R = \{(h', (c_i')_{i=1}^l; \alpha) \mid (h')^{\alpha^{-1}} = D((c_1')^{y_1} \cdots (c_l')^{y_l}) h_0 \wedge \alpha \neq 0\}.$$

$$\mathcal{U} \qquad\qquad\qquad \mathcal{V}$$

$$\text{(knows: } h', (c_i')_{i=1}^l ; \alpha) \qquad\qquad \text{(knows: } h', (c_i')_{i=1}^l ; (y_i)_{i=1}^l , \lambda)$$

$$u \in_R \mathbb{Z}_q, \quad a \leftarrow (h')^u \qquad \xrightarrow{\quad a \quad}$$

$$\xleftarrow{\quad c \quad} \qquad c \in_R \mathbb{Z}_q$$

$$r \leftarrow u + c\alpha^{-1} \bmod q \qquad \xrightarrow{\quad r \quad}$$

$$b \leftarrow \left[(h')^r \stackrel{?}{=} a(D((c_1')^{y_1} \cdots (c_l')^{y_l}) h_0)^c \right]$$

$$\xleftarrow{\quad b \quad}$$

Fig. 5. Simplified verification protocol of the encrypted credential scheme

We need to prove that the protocol in Fig. 5 is a proof of knowledge, and that it is secure. Demonstrating that it is a proof of knowledge is captured by proving 'completeness' and 'special soundness' (cf. Sect. 2) for relation R. For security, using the multiparty computation model of [13], we need to prove that the adversarial view on the protocol can be simulated for any allowed adversary structure: $\mathcal{V}$ being semi-honest or $\mathcal{U}$ being malicious. Therefore, we construct two simulators that may both use the adversarial party as a subroutine, and that simulate the conversations of the corrupted party with an honest participant in an indistinguishable way, on any common input $(h', (c_i')_{i=1}^l)$.

Proposition 5. *The protocol in Fig. 5 is complete and special sound.*

Proof. See [23, Sect. 6.4]. □

Proposition 6. *For any common input $(h', (c_i')_{i=1}^l)$, the protocol in Fig. 5 can be simulated in a perfectly indistinguishable way, for any semi-honest $\mathcal{V}'$.*

Proof. Given a common input $(h', (c_i')_{i=1}^l)$. For any honest prover $\mathcal{U}$ and semi-honest verifier $\mathcal{V}'$ following the protocol, the real conversations satisfy the following distribution[8]:

$$\{(a, c, r, b) \mid u, c \in_R \mathbb{Z}_q; a \leftarrow (h')^u; r \leftarrow u + c\alpha^{-1} \bmod q;$$
$$b \leftarrow \left[(h')^r \stackrel{?}{=} a(D((c_1')^{y_1} \cdots (c_l')^{y_l}) h_0)^c \right] \}.$$

This distribution is perfectly simulated by:

$$\{(a, c, r, b) \mid c, r \in_R \mathbb{Z}_q; a \leftarrow (h')^r (D((c_1')^{y_1} \cdots (c_l')^{y_l}) h_0)^{-c};$$
$$b \leftarrow \left[(h')^r \stackrel{?}{=} a(D((c_1')^{y_1} \cdots (c_l')^{y_l}) h_0)^c \right] \}.$$

Note that the simulator knows the values $((y_i)_{i=1}^l , \lambda)$ as he may use $\mathcal{V}'$ as subroutine, and therefore he can compute $D((c_1')^{y_1} \cdots (c_l')^{y_l})$, where D is the decryption function. □

---

[8] Effectively, $b = 1$ by construction. To keep the simulation clear, it is however denoted in full.

The construction of a simulator for the view of a malicious prover $\mathcal{U}'$ on the protocol relies on Ass. 2. We can assume that $\mathcal{U}'$ did $K \geq 0$ credential issuance queries, and output a tuple $(h', (c_i')_{i=1}^l, \alpha, z', (z_i')_{i=1}^l, c', r')$. We recall that the equation $c' = \mathcal{H}(\cdot)$ of (2) is assumed to hold.

Proposition 7. *For any common input $(h', (c_i')_{i=1}^l)$, the protocol in Fig. 5 can be simulated in a perfectly indistinguishable way, for any malicious $\mathcal{U}'$.*

Proof. Given a common input $(h', (c_i')_{i=1}^l)$. For any prover $\mathcal{U}'$ and honest verifier $\mathcal{V}$, the real conversations are as follows:

- Receive a from $\mathcal{U}'$, send $c \in_R \mathbb{Z}_q$ to $\mathcal{U}'$, and receive r from $\mathcal{U}'$;
- Set $b \leftarrow \left[(h')^r \overset{?}{=} a(D((c_1')^{y_1} \cdots (c_l')^{y_l})h_0)^c \right]$, and output (a, c, r, b).

We construct a simulator that also has input $(h', (c_i')_{i=1}^l)$ and may use $\mathcal{U}'$ as a subroutine:

- Receive a from $\mathcal{U}'$, send $c \in_R \mathbb{Z}_q$ to $\mathcal{U}'$, and receive r from $\mathcal{U}'$;
- Use the extractor $\mathcal{E}$ of Ass. 2 to obtain $(h_j, (c_{ji})_{i=1}^l)_{j=1}^K$ and $\tau \in \{0, \ldots, K\}$;
- Set $b \leftarrow \begin{cases} 1, & \text{if } \tau \neq 0 \text{ and } (h')^r = ah_\tau^c, \\ 0, & \text{if } \tau = 0 \text{ or } (h')^r \neq ah_\tau^c; \end{cases}$
- Output (a, c, r, b).

Remains to prove that these two distributions are indistinguishable, given any common input $(h', (c_i')_{i=1}^l)$. But the values (a, c, r) are constructed the same in both conversations, remains to show that b is distributed the same in both sets.

Suppose that in the real conversation $b = 1$. By the special soundness property (Prop. 5), with overwhelming probability $\mathcal{U}'$ knows an α such that $(h')^{\alpha^{-1}} = D((c_1')^{y_1} \cdots (c_l')^{y_l})h_0$. By Ass. 2 (and as $c' = \mathcal{H}(\cdot)$ holds), this implies that with overwhelming probability there exists a j such that (3) holds, which by definition means that $\tau \neq 0$. It moreover implies that:

$$
\begin{aligned}
(h')^r &= a(D((c_1')^{y_1} \cdots (c_l')^{y_l})h_0)^c & \{\text{since } b = 1\} \\
&= a(D((c_{\tau 1})^{y_1} \cdots (c_{\tau l})^{y_l})h_0)^c & \{\text{equation (3)}\} \\
&= ah_\tau^c & \{\text{by construction}\}.
\end{aligned}
$$

So by construction the simulator sets $b = 1$ as well.

Conversely, suppose that in the simulated conversation $b = 1$. By construction this means that $\tau \neq 0$ and $(h')^r = ah_\tau^c$. By definition, $\tau \neq 0$ implies that (3) is satisfied with $j = \tau$. Now:

$$
\begin{aligned}
(h')^r &= ah_\tau^c & \{\text{since } b = 1\} \\
&= a(D((c_{\tau 1})^{y_1} \cdots (c_{\tau l})^{y_l})h_0)^c & \{\text{by construction}\} \\
&= a(D((c_1')^{y_1} \cdots (c_l')^{y_l})h_0)^c & \{\text{equation (3)}\},
\end{aligned}
$$

which implies that also in the real execution $b = 1$. Concluding, with overwhelming probability b is computed the same in both conversations, and hence the real and simulated conversations are perfectly indistinguishable. □

One Time Anonymous Certificate: X.509 Supporting Anonymity*

Aymen Abed[1] and Sébastien Canard[2]

[1] Logica IT Service - 17 place des Reflets - 92097 Paris la Défense Cedex - France
[2] Orange Labs - 42 rue des Coutures - BP6234 - F-14066 Caen Cedex - France

Abstract. It is widely admitted that group signatures are today one of the most important cryptographic tool regarding privacy enhancing technologies. As evidence, the ISO organization has began a subject on authentication mechanisms supporting anonymity, in which group signatures are largely studied. However, it seems difficult to embed group signatures into other standards designed for classical authentication and signature mechanisms, such as the PKI X.509 certification. In fact, X.509 public key certificates are today widely used but not designed to support anonymity. One attempt has been done by Benjumea *et al.* but with the drawback that (i) the solution loses the principle of one certification per signer, (ii) revocation cannot be performed efficiently and (iii) the proposed architecture can not be applied to anonymous credentials, a concept close to group signature and today implemented by IBM or Microsoft. This paper presents a new approach which permits to use the X.509 standard to group signature schemes and anonymous credentials in a more standard and efficient way than related work.

1 Introduction

Anonymity is today considered as one possible base of individual privacy. In this case, the customer is anonymous when she is accessing a specific service. A lot of theoretical work has been done to design new signature schemes which provide such anonymity of the signer. Among them, one of the most popular is the concept of group signature scheme, introduced by Chaum and van Heyst [17] and currently under discussion to be standardized at the ISO organization.

Several constructions of group signature schemes today exist, either secure in the random oracle model [1,5,18] or in the standard model [21]. Many variants have also been proposed, such as Direct Anonymous Attestation [16,7,15,30] or anonymous credential systems [12,2].

The latter, a.k.a. "need to know approach", is an emerging concept which is currently under development in concrete systems such as IBM Idemix [23] or

---

* This work has been financially supported by the French Agence Nationale de la Recherche under the PACE project and by the European Commission's Seventh Framework Programme (FP7) under contract number ICT-2007-216676 ECRYPT II, while 1st author was working at Orange Labs.

S.-H. Heng, R.N. Wright, and B.-M. Goi (Eds.): CANS 2010, LNCS 6467, pp. 334–353, 2010.

Microsoft Credentica UProve [25]. This concept permits users to access services that are conditioned to some attributes such as the age, the address or the nationality, while revealing the minimum of information about them. In fact, this is not necessary to reveal the information on all the attributes (age, nationality, etc.) or indeed the attribute itself. For example, one needs not to reveal her date of birth to prove that she is more than 65 years old.

However, there is not enough work to apply these theoretical concepts in current existing infrastructures. As one of the main example, X.509 public key certificates are widely used but not designed to support anonymity. A Public Key Infrastructure (PKI) serves to identify the holder of a private key in a standard fashion and has been used in many types of transactions and communications over Internet. It seems however that it is not possible to use a X.509 certificate with e.g. group signatures, for the following reasons.

1. In the *Subject Name* field of a X.509 certificate, one can find the true identity of the user to whom it was issued.
2. In the *Subject Public Key Information* field, the Certification Authority (CA) puts the public key of the signer.
3. A certificate is published by the CA, for example in a directory system, which may be widely accessible.

Recently at CANS 2007, Benjumea, Choi, Lopez and Yung [4] have proposed Anonymity 2.0 as a way to use X.509 certificates in the group signature scheme setting (as well as in the case of traceable and ring signatures). In a nutshell, they consider a unique X.509 certificate for the whole group, which one is valuable to all group members. In the proposed X.509 certificate, the *Serial Number* and the *Validity Period* are fixed, the *Subject Name* field is the identity Id_G of the whole group and the *Subject Public Key Information* field contains the public key gpk of the whole group.

However, there are three main issues to solve regarding the use of X.509 certificates for group signatures.

1. Using Anonymity 2.0 [4], the "real identity" of the user is lost since the X.509 certificate is no more related to one unique signer.
2. X.509 standard includes the way to revoke one particular signer. Regarding Anonymity 2.0, either each revocation in the group modifies the group public key gpk and thus the corresponding X.509 certificate, which is a very expensive solution, or the X.509 certificate is not used for the revocation purpose, which makes Anonymity 2.0 not in accordance with the X.509 standard.
3. Anonymity 2.0 can not be applied in the context of anonymous credentials.

It also exist some work on the way to use X.509 certificates while protecting the privacy of the owner of the certificate. For example, the RFC 5636 on traceable anonymous certificate [28] defines a practical architecture and protocols such that the X.509 certificate contains a pseudonym, while still retaining the ability to map such a certificate to the real user who requested it. The architecture separates the authority responsible for the verification of the ownership of a private key and the one who validate the contents of a certificate. In [29], Persiano and Visconti propose the concept of Secure and Private Socket Layer protocol, that is an extension of the SSL/TLS protocol that allows the server to present a list of certificates and the client to prove that she owns the private key related to at least one of those certificates, without revealing which one. But none of these two approaches focuses on group signatures nor anonymous credentials, which is the case in this paper.

In fact, we propose a new approach, called One-Time Anonymous Certificate (OTAC), which permits to better use group signatures with X.509 certification. Our solution also solve all the above issues and, this way, permits the user to become the center of the whole system.

The paper is organized as follows. Section 2 presents the concept of group signature schemes, and describes some existing constructions. In Section 3, we recall X.509 certification and give some words on Anonymity 2.0. Section 4 describes our work and compares it with other proposed solutions to make X.509 supporting anonymity. The Section 5 finally studies the application of OTAC to anonymous credentials.

2   Group Signature Schemes

We first focus on group signature schemes which have been introduced in [17]. In such scheme, any member of the group can sign messages on behalf of the group. Such signature remains anonymous and unlinkable for anyone except a designated authority who has the ability to identify the signer. This entity is sometimes called the Opening Manager. The group is typically controlled by a Group Manager $\mathcal{GM}$ that handles enrollment of members.

2.1   Concept of Group Signatures

In the following, we consider a group which is publicly identified by a unique identifier denoted Id_G. This group can be dynamic in the sense that group members can enter or leave the group at any time during the life cycle of this group. Following [3], a group signature scheme is a digital signature scheme comprised of the following procedures.

- SETUP is a probabilistic algorithm which on input a security parameter 1^λ, outputs the initial group public key gpk, the secret key gmsk for the group manager and the opening secret key osk.
- USERKG is a user key generation which on input a user i and the group public key gpk, outputs a personal public and private key pair (upk[i], usk[i]). We assume that upk[i] is public.
- JOIN is a protocol between the group manager, on input gmsk and a user i on input (upk[i], usk[i]) that results in the user becoming a new group member. The user's output is a membership secret gsk[i], which may include (upk[i], usk[i]). $\mathcal{GM}$ makes an entry reg[i] in the registration table which is included into gpk.

- SIGN is a probabilistic algorithm that on input a group public key gpk, a membership secret gsk[i] and a message m outputs group signature σ of m.
- VERIFY is an algorithm for establishing the validity of a group signature σ of a message m with respect to the group public key gpk.
- OPEN is an algorithm that, given a message m, a valid group signature σ on it, the group public key gpk and an opening secret key osk, determines the identity i of the signer, together with a proof τ that the opening has been correctly done. This algorithm output i = 0 in case no group member has produced this signature.
- JUDGE takes on input the group public key gpk, an integer j ≥ 1, the public key upk[j] of the user j, a message m, a valid signature σ of m, and a proof τ. Its aim is to check that τ is a proof that user j has truly produced σ.

As described in [3], a secure group signature scheme must satisfy the following properties. Note that we do not give the formal definitions since our aim in this paper is not to study in detail the security properties of some schemes.

- **Correctness:** a signature which is produced by the group member i using SIGN must be accepted by VERIFY. Moreover, in case of opening, the OPEN procedure should output i together with a proof τ which is accepted by the JUDGE algorithm.
- **Anonymity:** an adversary should not be able to decide whether one given signature has been produce by two known group members of her choice. The adversary has chosen both members and may know their respective secret keys. The adversary has also access to both group manager's secret keys gmsk and can ask for the opening of group signatures (except for the given signature).
- **Traceability:** this property describes that an adversary is not able to produce a signature such that either the honest opener declares herself unable to identify the origin of the signature (that is OPEN outputs i = 0), or she has find the signer but is unable to produce a correct proof τ of its claim.
- **Non-frameability:** the adversary should not be able to create a proof, accepted by the JUDGE algorithm, that an honest user produced a certain valid signature unless this user really did produce this signature.

2.2   Constructions of Group Signature Schemes

Group signatures have been the subject of many research papers over the past years. Currently proposed group signatures have been proved secure in the random oracle model [1,5,18], or in the standard one [21].

Zero-knowledge proof of knowledge. Roughly speaking, a zero knowledge proof of knowledge is an interactive protocol during which an entity proves to a verifier that he knows a set of secret values $\alpha_1, \ldots, \alpha_q$ verifying a given relation R without revealing anything else. These protocols are also used to prove that some public values are well-formed from secret ones known by the prover. In the sequel, we denote by $\text{SOK}(\alpha_1, \ldots, \alpha_q : R(\alpha_1, \ldots, \alpha_q))(m)$ a signature of knowledge based on a proof of knowledge of the secrets $\alpha_1, \ldots, \alpha_q$ verifying the relation R, and where m is the message to be signed. As shown in [14,9], it is today possible to prove discrete-logarithm based predicates (e.g. representation, equality of discrete logarithms, belonging to a public interval).

Constructions in the random oracle model. Group signature scheme constructions that are secure in the random oracle model [1,8,5,20,18] are based on the use of signatures of knowledge [13], that is zero-knowledge proofs of knowledge transformed into signatures using the Fiat-Shamir heuristic [19].

In a nutshell, such group signature schemes are based on the same structure. Namely, during the JOIN procedure, the new group member obtains from the group manager a (Camenisch-Lysyanskaya [11] type) signature on one secret computed by both entities but only known by the member. The group signature is next the proof of knowledge of the signature from $\mathcal{GM}$ on the secret, without revealing the signature nor the secret, to ensure anonymity. We recall below the ACJT group signature scheme [1] and give some words on the BBS one [5].

The ACJT group signature scheme. In this scheme[1], the underlying signature scheme [11] is based on the Flexible RSA assumption (a.k.a. Strong RSA assumption). The secret key, denoted x is signed by the group manager and the resulting signature is the couple (A, e) such that $A^e = a_0 a^x \pmod{n}$ where n is a safe RSA modulus and a_0 and a are publicly known random elements of $QR(n)$, the group of quadratic residues modulo n. The secret key of the group manager is the factorization of n, which permits to choose at random e and to compute the corresponding A in the signature.

The group signature next consists in first encrypting one part of the obtained signature using the public key related to the opening secret one. In the ACJT case, the encryption is done by using the El Gamal encryption scheme, as

$$T_1 = Ay^w, T_2 = g^w, T_3 = g^e h^w$$

where y is the public key corresponding to the opening secret key θ (that is, $y = g^\theta$). The couple (T_1, T_2) is the El Gamal encryption of A while T_3 corresponds to a commitment on e. Note that these values, denoted in the following $K_g = (T_1, T_2, T_3)$, do not depend on the message m to be signed. The second part of the group signature, denoted S_m in the following, is the signature of knowledge on the message m, which can be written as

$$U = \text{SOK}\left[x, e, w, ew : a_0 = T_1^e/(a^x y^w e) \wedge T_2 = g^w \wedge 1 = T_2^e/g^{we} \wedge T_3 = g^e h^w\right](m)$$

with the message m on input. The group signature σ on m is finally (T_1, T_2, T_3, U), which correspond to the couple (K_g, S_m).

---

[1] One better solution in terms of efficiency and security has afterward been proposed by Camenisch and Groth in [8] but we do not need to detailed it in this paper, since both solutions can be used in our result and the ACJT description is easier to give.

The BBS group signature scheme. The BBS group signature scheme [5] and its variants [20,18] are based on the q-SDH assumption. For example in [20,18], the secret key, denoted y, is signed by the group manager and the resulting signature is the couple (A, x) such that $A^{\gamma+x} = g_0 h^y$ in a group of prime order. γ is the secret key of the group manager and g_0 and h are publicly known generators.

Next, the group signature is composed of a ciphertext and a signature of knowledge. For example, in the XSGS variant [18] of the BBS group signature scheme, the encryption is done by using the double El Gamal encryption scheme. Next, the first part of the group signature scheme consists in

$$T_1 = A y_1^\alpha, T_2 = g^\alpha, T_3 = A y_2^\beta, T_4 = g^\beta,$$

where y_1 and y_2 are the public keys corresponding to the opening secret keys ζ_1 (that is, $y_1 = g^{\zeta_1}$) and ζ_2 (that is, $y_2 = g^{\zeta_2}$). The tuple (T_1, T_2, T_3, T_4) is the double El Gamal encryption of A and does not depend on the message m. The second part of the group signature is the signature of knowledge on the message m, which can be written as

$$U = \text{SOK}\big[\alpha, \beta, x, z : T_2 = g^\alpha \wedge T_4 = g^\beta \wedge T_1/T_3 = y_1^\alpha/y_2^\beta \wedge$$
$$e(T_1, g_2)^x e(h, w)^{-\alpha} e(h, g_2)^z = e(g_1, g_2)/e(T_1, w)\big](m)$$

where g_1 and g_2 are random generators, $z = x\alpha + y$, $w = g_2^\gamma$ and with the message m on input. The resulting group signature is composed of the (double) El Gamal or linear encryption $K_g = (T_1, T_2, T_3, T_4)$ and the signature of knowledge $S_m = U$ on the message m. The group signature is again a couple of the form (K_g, S_m) where K_g does not depend on the message, while S_m does.

Constructions in the standard model. Groth Sahai NIWI proofs [22] permit to prove to a third party that some given values are well-formed (they lie in the correct given language) but do not permit to prove the knowledge of these values. As it is sometimes necessary to prove the knowledge of some secret values related to the group membership (e.g. in group signature schemes), the use of Groth-Sahai technique is not enough.

In [21], Groth proposes to use certified signatures. During the JOIN protocol, each group member obtains a signature v_i on a user secret key (x_i, a_i, b_i). During the signature procedure, the signer chooses at random a new key pair $(\text{vk}_{sots}, \text{sk}_{sots})$ for a one-time signature, produces a certificate σ and two intermediary values (a, b) (one a which is revealed and the other, b, which is kept secret by the signer) of the corresponding public key using her certified secret key. The Groth signature also encrypts the value v_i to open the group signature if needed. She next produces proofs (π, ψ) that all the above values are well-formed. Finally, she signs the message m and the values $\text{vk}_{sots}, a, \pi, y, \psi)$, using the one-time secret key, obtaining σ_{sots}. The final group signature is $(\text{vk}_{sots}, a, \pi, y, \psi, \sigma_{sots})$, which is again of the form (K_g, S_m) with $K_g = (\text{vk}_{sots}, a, \pi, y, \psi)$ (not depending on m) and $S_m = \sigma_{sots}$ (depending on m).

2.3   Group Member Revocation

It may be necessary, in some cases, to handle the situation in which a group member wants to leave a group or is excluded by the Group Manager. In both cases, it is necessary to set up a mechanism to prevent the possibility for a revoked member to produce a valid group signature. It exists today two different methods to revoke a group member in a group signature scheme, that is, make it infeasible for a membership secret gsk[i] to be used to produce a group signature which will be accepted by the VERIFY algorithm: the use of accumulators or the verifier-local revocation approach.

It is moreover necessary to modify the above security model. In fact, it currently exists two different models dealing with group member revocation, one for each underlying solution: accumulator technique [10], or verifier local revocation [6,24]. We do not detail these models as it is not really necessary to the understanding of our paper.

The accumulator technique. This technique has been introduced by Camenisch and Lysyanskaya [10] and is based on the use of dynamic accumulators [10,27]. In a nutshell, $\mathcal{GM}$ publishes a single value v which accumulates a group member secret key per valid group member. The group signature should next include a zero-knowledge proof that the member knows a secret value and a corresponding witness that this value is truly accumulated in v. It should be hard for users outside the group to forge such proof (by finding an appropriate witness). The accumulator is next updated each time a new user becomes a group member (a new value is added into the accumulator) and after each revocation (the value of the revoked group member is deleted from the public accumulator).

The main problem with this method is that it implies for each group member to update her secret data (more precisely a witness that her value is truly accumulated) after each modification (addition and deletion) in the group.

The Verifier Local Revocation (VLR) technique. This technique has been proposed by Boneh and Shacham [6]. The idea behind is to manage a revocation list with a data for each revoked group member. During the SIGN process, the group member produces an extra value which is used by the verifier who run through the revocation list to test, for each entry, if this is related to the value used to produce the signature.

This gives a solution without any update for group members. However, the group public key gpk needs to be regularly updated to avoid backward linkability (see *e.g.* [26]). More importantly, the time complexity from the verifier's side is linear in the number of entries in the revocation list.

Using one of these solutions, a revocation mechanism can thus be used for group signatures with the X.509 principles. The side effect is that each group member needs to produce two different group signatures, which makes our solution less efficient than the Anonymity 2.0 one [4]. However, our solution is more efficient considering the revocation mechanism since the use of Anonymity 2.0 implies the creation of a new certificate at each modification within the group.

2.4   Anonymous Credential

In the context of anonymous credentials, users have to show a kind of tokens to prove statements about themselves. For this purpose, each user has one or several credentials which are issued by some organizations that ascertain the authenticity of the information and can be provided to check things on demand. The certified attributes can be *e.g.* a name, an address, an age, etc. Users may also be required to prove a predicate on the attributes encoded in their credentials, such as for example that her age is greater than a fixed value, revealing neither their age nor other attributes.

The main requirements an anonymous credential systems are (i) *unforgeability* which states that the user can not prove the validity of forged credentials or predicates that are encoded on her issued credential and (ii) *privacy* which states that the verifier should not be able to learn any information about the user's credentials (e.g. other attributes) beyond what can be logically inferred from the status of the proven predicate.

It is possible to construct an anonymous credential system using the same techniques as group signature schemes. For example [12], it is possible to construct such system based on the ACJT group signature scheme as follows. The credential (A, e) is of the form $\mathsf{A}^\mathsf{e} = \mathsf{a}_0 \mathsf{a}_1^{c_1} \cdots \mathsf{a}_\ell^{c_\ell} \mathsf{b}^\mathsf{x} \pmod{\mathsf{n}}$ where n is a safe RSA modulus, $\mathsf{b}, \mathsf{a}_0, \cdots, \mathsf{a}_\ell$ are publicly known random elements of $QR(\mathsf{n})$, the c_i's are the certified attributes (for example c_1 represents the nationality, c_2 the address, c_3 the date of birth, etc.) and x is a secret value only known by the user, but jointly computed.

The proof of possession of a credential is done similarly as for a group signature. For example, if one user wants to prove that her first attribute (her nationality) is the value c_1, she has first to compute

$$\mathsf{T}_1 = \mathsf{A}\mathsf{y}^\mathsf{w}, \mathsf{T}_2 = \mathsf{g}^\mathsf{w}, \mathsf{T}_3 = \mathsf{g}^\mathsf{e}\mathsf{h}^\mathsf{w}$$

where w is a random value and next to produce the signature of knowledge

$$\mathsf{U} = \mathrm{SOK}\Big[\mathsf{x}, \mathsf{e}, \mathsf{w}, \mathsf{ew}, c_2, \cdots, c_\ell : \mathsf{a}_0 \mathsf{a}_1^{c_1} = \mathsf{T}_1^\mathsf{e}/(\mathsf{a}_2^{c_2} \cdots \mathsf{a}_\ell^{c_\ell} \mathsf{b}^\mathsf{x} \mathsf{y}^\mathsf{we}) \wedge$$
$$\mathsf{T}_2 = \mathsf{g}^\mathsf{w} \wedge 1 = \mathsf{T}_2^\mathsf{e}/\mathsf{g}^\mathsf{we} \wedge \mathsf{T}_3 = \mathsf{g}^\mathsf{e}\mathsf{h}^\mathsf{w}\Big](m)$$

with the message m on input. The user sends $(c_1, \mathsf{T}_1, \mathsf{T}_2, \mathsf{T}_3, \mathsf{U})$ to the verifier who verifies the signature of knowledge to be convinced that the credential embed the certified attribute c_1, as expected.

3   X.509 Certification and Anonymity

In this section, we recall X.509 certification principles and we describe the paper from Benjumea *et al.* [4], which is the first to propose the use of X.509 certification for signature schemes with anonymity (group, traceable and ring signatures).

3.1   X.509 Certification

X.509 is an ITU-T standard for a public key infrastructure (PKI). Its aim is to make the link between a public key and an entity, since public key cryptography, for example in the signature setting, only permits to know that one signature has been produced by this particular public key but not by this particular entity. In a nutshell, a X.509 certificate is the certification by a trusted authority called the Certification Authority (CA) that the public key in the certificate belongs to the identity in this certificate.

Thus, when sending a signed message, one has to give the message m, the signature σ produced using her secret key, and the X.509 certificate on the corresponding public key. The verification step consists next in verifying the validity of the certificate, extracting the verification public key and using it to verify σ on m. More precisely, all X.509 certificates have the following data, in addition to the signature from the CA on all these fields.

- *Version*: this identifies which version of the X.509 standard applies to this certificate, which affects what information can be specified in it. We do not detail this field as it is not really important in our study, except that we should use X.509 Version 3, which is the most recent one.
- *Serial Number*: the entity that created the certificate is responsible for assigning it a serial number to distinguish it from other certificates it issues. This information is used in numerous ways, for example when a certificate is revoked its serial number is placed in a Certificate Revocation List (CRL).
- *Signature Algorithm Identifier*: this identifies the algorithm used by the CA to sign the certificate.
- *Issuer Name*: the X.500 name of the entity that signed the certificate. This is normally a CA. Using this certificate implies trusting the entity that signed this certificate[2].
- *Validity Period*: each certificate is valid only for a limited amount of time. This period is described by a start date and time and an end date and time, and can be as short as a few seconds or almost as long as a century. The chosen validity period depends on a number of factors, such as the strength of the private key used to sign the certificate or the amount one is willing to pay for a certificate. This is the expected period that entities can rely on the public value, if the associated private key has not been compromised.
- *Subject Name*: the name of the entity whose public key is embedded into the certificate. This name uses the X.500 standard, so it is intended to be unique across the Internet. This is the Distinguished Name (DN) of the entity.
- *Subject Public Key Information*: this is the public key of the entity being named, together with an algorithm identifier which specifies which public key cryptosystem this key belongs to and any associated key parameters.
- *Extensions*: there are today some common extensions in use. *KeyUsage* limits the use of the keys to particular purposes such as "signing-only", *AlternativeNames* allows other identities to also be associated with this public key,

---

[2] Note that in some cases, such as root or top-level CA certificates, the issuer signs its own certificate.

e.g. DNS names, Email addresses, IP addresses. Extensions can be marked critical to indicate that the extension should be checked and enforced/used. For example, if a certificate has the KeyUsage extension marked critical and set to "keyCertSign" then if this certificate is presented during SSL communication, it should be rejected, as the certificate extension indicates that the associated private key should only be used for signing certificates and not for SSL use.

3.2   Anonymity 2.0

X.509 public key certificates were designed to support the concept of one public key, corresponding to a unique private key, for one identity, referring to the one who has the private key. As a consequence, this does not support the anonymity of the signer, by construction. If one want to use such X.509 certificate structure for group signature, one has to face to the problem that the group member key reg[i] (see group signatures above) can not be transcript in the certificate, as it reveals the identity of the group member.

Benjmumea *et al.* propose in [4] to consider one single standard X.509 certificate for the whole group. The elegant model which is given in [4] is based on adding some semantic extensions, keeping the same X.509 structure. More precisely, they define a X.509 public key certificate with extended semantic where the public key is not bound to a single entity but it is bound to a concept (see the Appendix A in [4] for the specification of extension fields in ASN.1). Considering the above structure of an X.509 certificate, they propose the following modifications.

1. Fix the *Serial Number*.
2. Fix the *Validity Period*, that is the start date/time and the end date/time.
3. Put the identity of the group Id_G in the *Subject Name* field.
4. Put the public key of the group gpk in the *Subject Public Key Information* field.

In fact, this concept stated that members belonging to the same group possess the same X.509 certificate. Thanks to their solution we get the desired anonymity and their solution can next be applied for group signatures, but also for traceable and ring signatures, as shown in [4].

3.3   Remaining Issues

However, there remains three main issues to solve regarding the use of X.509 certificates for group signatures.

1. One aim of X.509 certificates is to make the link between one public key and one identity. However, using Anonymity 2.0 [4], the "real identity" binding the X.509 certificate is lost since it is no more related to one unique signer, but to the whole group. From this point of view, Anonymity 2.0 is not totally in accordance with the X.509 standard principles.

2. X.509 standard includes the way to revoke one particular signer. Using for example a revocation list, this is possible to put the certificate corresponding to a revoked signing secret key onto this revocation list so that, in case of fraud, a signature produced by this private key is no more accepted. Thus, in this case, each time a signature and a certificate is received, the verifier needs to verify whether this certificate belongs to the revocation list. Regarding group signature schemes, this is not completely possible using Anonymity 2.0. In fact, there are two ways to consider revocation using Anonymity 2.0.

   (a) Each revocation in the group modifies the group public key gpk and thus the corresponding X.509 certificate. As a consequence, if a revoked member uses the wrong group public key to be accepted, then the X.509 will be refused and the whole group signature rejected. This is thus necessary to put each time the previous X.509 certificate onto the revocation list and next to recreate a new one for the whole group. This is therefore a very expensive solution.

   (b) The X.509 certificate is not used for the revocation purpose. This way, the revocation is only done using the simple group signature and usual techniques (see Section 2.3). The X.509 certificate is here only used to make the link between the group public key and the group in which belong the signer. With such solution, Anonymity 2.0 is again not in accordance with the X.509 standard.

3. Anonymity 2.0 can not be applied in the context of anonymous credentials (see Section 2.4), which can be described as a way to manage several groups (group of people who live in the same town, group of people being more than 65 years old, group of people being a student, etc.) in an efficient and compact way. Using Anonymity 2.0 in such context, having one X.509 certificate per group, implies as many certificates as there are possibilities of attributes, which makes it unusable in practice.

In the following, we design a new way to consider X.509 certificates in the context of group signature schemes and anonymous credential systems in such a way that the proposed solution is totally in accordance with the X.509 standard, without the multiplication of revoked and/or issued certificates.

4   One Time Anonymous Certificate

In this section, we present our result on X.509 certification for group signature schemes. We name it OTAC for One-Time Anonymous Certificate and we here give a general overview and next detail our system.

4.1   Overview of Our Solution

We want to add to a message m and its group signature σ a X.509 certificate which

1. can be used by the verifier to verify the group signature, as a standard X.509 certificate;

2. does not compromise the security aspects of group signature schemes (see Section 2.1);
3. is unique for a given group member and
4. directly permits the management of revocation of group members.

The idea is to put on the *Subject Public Key Information* field a cryptographic key which is at the same time unique for a given group member and different from one signature to another (to obtain the anonymity property). This "public key" should be generated by the group member as many time as she wants and should not depend on the message to be signed (since it is included into the certificate).

If we examine different group signature schemes (See Section 2.2), we remark that for each of them, the final group signature is divided into two parts. The first one, denoted K_g does not depend on the message to be signed and is related to the identity of the group member. The second one, denoted S_m, depends on the message m. Moreover, the first part K_g includes the way for the Opening Manager to open the group signature (see (T_1, T_2) for the ACJT group signature [1] or y for the Groth one [21]). In Figure 1, we resume the way current group signatures are divided in that way.

Group signature	K_g	S_m
ACJT [1]	T_1, T_2, T_3	U
BBS [5,18]	T_1, T_2, T_3, T_4	U
Groth [21]	$vk_{sots}, a, \pi, y, \psi$	σ_{sots}

Fig. 1. The K_g and S_m variables for some group signatures

As a consequence, regarding the X.509 certificate as described in Section 3.1, we can consider that the key K_g corresponds to the *Subject Public Key Information*, while S_m is a true signature (of knowledge) on the message m, using, in some senses, the "public key" K_g. We thus obtain our "one-time" X.509 certificate.

Focusing on the X.509 certificate signature, it is obvious that we cannot ask for the certification authority to perform this task each time a group member produces a group signature, for two obvious reasons. First, this implies that this authority is always on-line, which comes against the principle of certification. Second, this goes against the anonymity of the user.

In this paper, we adopt a new approach which consists in delegating to each group member the power to sign a certificate. As it is necessary for the signer to be anonymous, we introduce the concept of "major group" in which each user is thus able to produce a major group signature on the above one-time X.509 certificate. Consequently, each group member lies into two (or more) groups, the "major" one for signing a X.509 certificate and the "minor" one(s) for the purpose of the initial group.

4.2 Detailed Description

In the following, we consider a group, called the *minor group*, identified by Id_G, and related to a group signature scheme. This minor group is managed by a Group Manager and an Opening Manager and is composed of several group members. Each of them has a membership secret gsk[i] for this minor group and is thus capable of producing group signatures σ on messages m, on behalf of this minor group.

We now describe how to create a One Time Anonymous Certificate (OTAC) for group signatures.

Creation of an OTAC. On input a message m, the group member executes SIGN(gpk, gsk[i], m) and obtains the group signature σ. We here remember that we use one of the group signature that has been described in Section 2.2. Thus, the obtained signature is of the form (K_g, S_m) (see Section 2.2 and Figure 1). The user can now create the one-time certificate, based on the X.509 structure, with the following fields.

- *Serial Number*: as we consider that this certificate is one-time, and according to the standard X.509 certificate, the serial number should be different from one OTAC to another. It may for example be the (collision-resistant) hash value of K_g;
- *Signature Algorithm Identifier*: the signature algorithmID must be redefined to be the major group signature. We detail this step just below;
- *Issuer Name*: the issuer name will be the CA;
- *Validity Period*: according to [4] the validity period needs to be fixed by *e.g.* the CA and is consequently common to all group members;
- *Subject Name*: the subject name should not give any information about the identity of the group member. It may for example be the hash value of $K_g \| Id_G$;
- *Subject Public Key Information*: this field is defined by the user to be the value K_g outputted by the execution of SIGN(gpk, gsk[i], m) = σ = (K_g, S_m).

Remark 1. As this is the user herself who fill in these fields, one can argue that the verifier may not be convinced that those values have been correctly computed. In fact, for most of them (serial number and subject name), it is in the user's interest to fill in them as described above. For the other ones (Signature Algorithm ID, Issuer Name, Validity Period), it is possible for the verifier to check them using a parent certificate.

Note that this certificate is necessarily one-time since it includes the value K_g, which is specific to this signature, for unlinkability purpose. It now remains to sign this certificate. As said above, this is not possible to ask the certification authority to sign this message, since it goes against the essence of X.509 certification (the CA should be off-line) and compromises the user anonymity.

Signature of the OTAC. To cope with the above problem, our solution consists in enabling the user to sign certificates on behalf of the CA. As the user should be anonymous, we thus use again a group signature scheme. For this purpose, the user should also belong to a *major group*, managed by the Certification Authority himself. Thus, each above user should interact with the CA using the JOIN protocol of the major group signature scheme, to obtain a membership secret here denoted $GSK[i]$. She can now produce a group signature Σ on some message M.

After having creating a certificate and fulfilled the different fields as described above, the generated certificate should be hashed in order to get m_{cert} that plays the role of a generic message m. The group member next generates a "major" group signature Σ which play the role of the signature of the X.509 certificate (performed by the CA in the standard certification process).

Finally, the complete certificate OTAC is composed of the above fields, including the subject public key with K_m, and the complete major group signature Σ.

General scenario. After having created the certificate, the user can send all the information to the verifier, that is (i) the message m, (ii) the signature S_m from the minor group signature and (iii) the above signed OTAC. The verifier next first verifies the validity of the certificate by using the VERIFY algorithm for the major group on input the signature Σ and the message m_{cert} (that is, the hash value of all the fields in the X.509 certificate). Next, the verifier gets back from the OTAC the value K_g and uses it with the signature S_m on the message m on input of the VERIFY procedure for the minor group signature. If both verifications succeed, then the signature is accepted.

Remark 2. In the standard PKI setting, a fraudulent user may give her X.509 certificate and the corresponding secret key to another user so that the verification is falsely said to be correct. We have the same problem with our above description since one fraudulent user can give her major group membership secret key to another user (which is not necessarily a group member in this major group). There are two ways to deal with this problem. First, we can assume that the Group Manager for the minor group does not accept a user which is not a group member in the major group. This also implies that this Group Manager and the Certificate Authority communicate one to each other in case of revocation of a user. The second solution consists in checking that the two valid group signatures $\sigma = (K_g, S_m)$ and Σ are based on the same initial secret. In fact, this is easily possible in the group signature schemes described in Section 2.2, as shown in Section 5.4.

4.3   Conclusion on OTAC for Group Signature Schemes

As a first conclusion, our OTAC proposal for group signature schemes solves all the issues considered in Section 3.3, except the case of anonymous credentials which is considered in the next section. More precisely, we have the following points.

Unicity. With OTAC, it is clear that the resulting X.509 certificate is unique for a given group member and is related to the identity of her. In fact, the certificate is related to K_g which includes, as described in Section 2.2, the encryption of a value which is used by the Opening Manager to open the group signature. OTAC is here totally in accordance with the X.509 standard.

Revocation. In Section 2.3, we have described the two ways to deal with the revocation of a group member in group signature schemes. It is relatively obvious that our OTAC solution can easily include both mechanisms. In fact, we only need to add a revocation mechanism to the minor group (even if this is also possible for the major one), which makes the revocation totally in accordance with the revocation principles in the X.509 standard.

1. **Accumulator:** this revocation solution includes the modification of the group signature, that is the addition of some T_i's (see *e.g.* [10] for details in the ACJT case) which are included into K_g and the modification of the signature of knowledge U accordingly. If a group member is revoked, she will not be able to produce a valid U.
2. **VLR:** similarly, this mechanism add some values to K_g and implies the modification of the signature of knowledge U, accordingly. This time, the revocation is next based on the principle of a certificate revocation list (CRL) which is used by the verifier to check whether this certificate is related to a revoked user. This way, OTAC is in accordance with the X.509 standard.

In some cases, the Group Manager can inform the Certification Authority to revoke a particular user in the major group.

Security considerations. We should be careful that the addition of an OTAC does not compromise the security of the initial group signature scheme. As a consequence, the whole protocol (including the OTAC) should verify the correctness, anonymity, traceability and non-frameability properties, as described in Section 2 and in [3]. We here do not give a formal security proof but assuming that both chosen group signature schemes (minor and major) are secure in the BSZ sense [3] and according to the fact that the filled fields in OTAC does not compromise any security property, then it is clear that the global scheme is also secure in the BSZ sense. The latter argument can be validated by the fact that the fields depending on K_g (*Serial Number*, *Subject Name* and *Subject Public Key Information*) do not reveal any extra information that is not initially revealed. Finally, the other OTAC fields do not give any useful information to the adversary.

5   The Case of Anonymous Credentials

In this section, we show how our OTAC solution can be applied to anonymous credentials [2]. As described in Section 2.4, anonymous credentials permits users to give evidence to service providers that they have the right attribute (age, address, nationality, etc.) while revealing the minimum of information on the certified attributes. We here first recall the two ways to treat attributes with the X.509 standard. (X.509 Attribute Certificate or the addition of an extension to X.509 traditional certificates) and we next give our solution.

5.1   X.509 Attribute Certificate

An attribute certificate is a digital document that describes a written permission from the issuer to use a service or a resource that the issuer controls or has access to use. It is also known as an authorization certificate. The most recent ITU-T X.509 recommendation standardizes this concept of attribute certificate and describes how the attribute certificate should be used with a standard PKI X.509 certificate.

As shown in Figure 2, the Attribute Certificate is close to the identity certificate described in Section 3.1. The main differences come from the field *Holder* which contains the serial number of the related PKI X.509 certificate, and the *Attribute* one which contains the attributes of the related user. This certificate is signed by an authority designated to be the Attribute Authority (AA). The revocation process is defined using the concept of Attribute Certificate Revocation List, which is used in the same way as a Certification Revocation Lists in the PKI case.

Version Number	Holder
Serial Number	Attributes
Signature Algorithm	Issuer Unique Identifier
Issuer	Extensions
Validity Period	AA Signature

Fig. 2. The Attribute Certificate

5.2   An Alternative to X.509 Attribute Certificate

Most of recent applications do not use attribute certificate to carry permission or authorization information. User attributes (date of birth, address, nationality, etc.) are rather put inside a X.509 PKI certificate, using an extension field which contain a sequence of one or more attributes. This extension field, namely *Subject Directory Attributes*, is defined in ASN.1 as follows.

```
subjectDirectoryAttributes Extension::=  {
    SYNTAX AttributesSyntax
    IDENTIFIED BY id-ce-subjectDirectoryAttributes }
AttributesSyntax ::= SEQUENCE SIZE (1..MAX) OF Attribute
```

5.3   Using OTAC with the Extension Field

As described in Section 2.4, we suppose that a user obtains from an organization a credential which contains several user certified attributes $c_1, \cdots, c_\ell$. As we use the extension field, we can consider that the credential system is the *minor group* in our OTAC solution. Thus, following the description given in Section 2.4, the proof of possession of a credential is composed of a first part $K_g = (\mathsf{T}_1, \mathsf{T}_2, \mathsf{T}_3)$ and a second part $S_m = \mathsf{U}$. We can now apply our technique described in Section 4. As a first consequence, this user (as any other in the same case) belongs to the major group managed by the Certification Authority (CA). She can thus produce group signatures Σ on some message M, using her membership secret $\mathsf{GSK}[i]$.

When this user wants to prove that her first attribute c_1 is for example equal to the correct nationality (see our example in Section 2.4), she has to create a one-time certificate OTAC as described in Section 4, with the following modified fields:

- *Subject Public Key Information*: the public key contains $K_g = (\mathsf{T}_1, \mathsf{T}_2, \mathsf{T}_3)$;
- *Subject Directory Attributes*: this extension field contains the value of the revealed attribute c_1.

The signature of the OTAC is done by the user as a member of the major group, and is thus equal to Σ_M, that is the group signature on the hash of all the fields of the OTAC. The verification is done similarly as for the simple OTAC described in Section 4 and is not repeated again.

5.4   Using OTAC with the Attribute Certificate

We now describe how to use the attribute certificate with our OTAC solution. As described above, the Attribute Certificate is sent together with a standard X.509 PKI certificate. In our case, the idea is to replace the latter by our OTAC during the proof of possession. Before describing our solution, we first focus on the link between a group signature and an anonymous credential, based on Remark 2 (see Section 4.2).

Link between a group signature and an anonymous credential. In the following, we need to make a link between a group membership secret and a certified credential. For this purpose, we take the example of the ACJT based construction but, in fact, we need to slightly modify both. According to Section 2 and based on the result in [11], we remark that the group membership secret can also be taken as (A, e, v, x) such that $A^e = a_0 a^v b^x \pmod{n}$, where x is only known by the group member but computed by both her and the Group Manager, while v is randomly chosen by the group member. Moreover, this group member may have a credential on her attributes $(c_1, \cdots, c_\ell)$ of the form $(\mathsf{A}, \mathsf{e}, v, \mathsf{x})$ such that $\mathsf{A}^{\mathsf{e}} = \mathsf{a}_0 \mathsf{a}_1^{c_1} \cdots \mathsf{a}_\ell^{c_\ell} \mathsf{a}^v \mathsf{b}^{\mathsf{x}} \pmod{\mathsf{n}}$, with the same v, which does not compromise the security, as show in [11].

From these two related tuples, (A, e, v, x) and $(\mathsf{A}, \mathsf{e}, v, \mathsf{x})$, this becomes possible for the group member to provide a proof that both tuples are truly related. Thus, from

$$T_1 = Ay^w, T_2 = g^w, T_3 = g^e h^w \, and \, \mathsf{T}_1 = \mathsf{A}\mathsf{y}^w, \mathsf{T}_2 = \mathsf{g}^w, \mathsf{T}_3 = \mathsf{g}^\mathsf{e}\mathsf{h}^w,$$

which are computed at each group signature and proof of possession (see Sections 2.2 and 2.4), the group member can provide the following signature of knowledge

$$V = \mathrm{SOK}\big[v, x, e, w, we, \mathsf{x}, \mathsf{e}, \mathsf{w}, \mathsf{we}, c_2, \cdots, c_\ell : a_0 = T_1^e/(a^v b^x y^{we}) \wedge$$
$$\mathsf{a}_0 \mathsf{a}_1^{c_1} = \mathsf{T}_1^\mathsf{e}/(\mathsf{a}_2^{c_2} \cdots \mathsf{a}_\ell^{c_\ell} \mathsf{a}^v \mathsf{b}^\mathsf{x} \mathsf{y}^\mathsf{we})\big](m),$$

which in particular proves that the same v is used in both equations. Given that, we are now able to describe our solution.

Our solution. We consider that a user is the member of a (minor) group, identified by Id_G, and can therefore produce group signatures on behalf of this group. Again, this user also belongs to a major group which permits her to sign X.509 based group signature OTAC on behalf of the Certification Authority, as explained in Section 4. Finally, the organization in the anonymous credential plays the role of the Attribute Authority AA.

According to the X.509 Attribute Certificate principle (see Section 5.1), the user has to send a message, a signature, a PKI X.509 certificate and a X.509 attribute certificate. In our solution, the PKI X.509 certificate is the OTAC as described in Section 4. Most of the fields (e.g. *Serial Number*, *Validity Period*) are unchanged from the initial description of OTAC, which includes the two following ones.

- *Subject Public Key Information*: this field is defined by the user to be the value $K_g = (T_1, T_2, T_3)$ outputted by the execution of $\mathrm{SIGN}(\mathsf{gpk}, \mathsf{gsk}[i], m) = \sigma = (K_g, S_m)$.
- *OTAC Signature*: this field is the "major" group signature Σ, using the membership secret $\mathsf{GSK}[i]$.

The Attribute Certificate is next constructed as follows.

- *Holder*: we put in this field the signature of knowledge V described above, which permits to make the link between the key embed in the OTAC and the anonymous credential.
- *Attribute*: this field contains the revealed attribute c_1. In case no attribute needs to be given (if the user has for example to prove that she is more than 65 years old), this field is set to the empty string.
- *AA Signature*: the Attribute Certificate signature is the whole proof of possession of an anonymous credential $(\mathsf{T}_1, \mathsf{T}_2, \mathsf{T}_3, \mathsf{U})$ as described in Section 2.4.

Remark 3. Another solution can be to have only one group which permits to sign both the message and the OTAC. This is more efficient but implies the delegation of the group management to a single authority.

References

1. Ateniese, G., Camenisch, J., Joye, M., Tsudik, G.: A practical and provably secure coalition-resistant group signature scheme. In: Bellare, M. (ed.) CRYPTO 2000. LNCS, vol. 1880, pp. 255–270. Springer, Heidelberg (2000)
2. Belenkiy, M., Camenisch, J., Chase, M., Kohlweiss, M., Lysyanskaya, A., Shacham, H.: Randomizable proofs and delegatable anonymous credentials. In: Halevi, S. (ed.) CRYPTO 2009. LNCS, vol. 5677, pp. 108–125. Springer, Heidelberg (2009)
3. Bellare, M., Shi, H., Zhang, C.: Foundations of group signatures: The case of dynamic groups. In: Menezes, A. (ed.) CT-RSA 2005. LNCS, vol. 3376, pp. 136–153. Springer, Heidelberg (2005)
4. Benjumea, V., Choi, S.G., Lopez, J., Yung, M.: Anonymity 2.0 - x.509 extensions supporting privacy-friendly authentication. In: Bao, F., Ling, S., Okamoto, T., Wang, H., Xing, C. (eds.) CANS 2007. LNCS, vol. 4856, pp. 265–281. Springer, Heidelberg (2007)
5. Boneh, D., Boyen, X., Shacham, H.: Short group signatures. In: Franklin, M. (ed.) CRYPTO 2004. LNCS, vol. 3152, pp. 41–55. Springer, Heidelberg (2004)
6. Boneh, D., Shacham, H.: Group signatures with verifier-local revocation. In: ACM Conference on Computer and Communications Security 2004, pp. 168–177. ACM, New York (2004)
7. Brickell, E.F., Camenisch, J., Chen, L.: Direct anonymous attestation. In: ACM Conference on Computer and Communications Security 2004, pp. 132–145. ACM, New York (2004)
8. Camenisch, J., Groth, J.: Group signatures: Better efficiency and new theoretical aspects. In: Blundo, C., Cimato, S. (eds.) SCN 2004. LNCS, vol. 3352, pp. 120–133. Springer, Heidelberg (2004)
9. Camenisch, J., Kiayias, A., Yung, M.: On the portability of generalized schnorr proofs. In: Joux, A. (ed.) EUROCRYPT 2009. LNCS, vol. 5479, pp. 425–442. Springer, Heidelberg (2009)
10. Camenisch, J., Lysyanskaya, A.: Dynamic accumulators and application to efficient revocation of anonymous credentials. In: Yung, M. (ed.) CRYPTO 2002. LNCS, vol. 2442, pp. 61–76. Springer, Heidelberg (2002)
11. Camenisch, J., Lysyanskaya, A.: A signature scheme with efficient protocols. In: Cimato, S., Galdi, C., Persiano, G. (eds.) SCN 2002. LNCS, vol. 2576, pp. 268–289. Springer, Heidelberg (2002)
12. Camenisch, J., Lysyanskaya, A.: Signature schemes and anonymous credentials from bilinear maps. In: Franklin, M. (ed.) CRYPTO 2004. LNCS, vol. 3152, pp. 56–72. Springer, Heidelberg (2004)
13. Camenisch, J., Stadler, M.: Efficient group signature schemes for large groups (extended abstract). In: Kaliski Jr., B.S. (ed.) CRYPTO 1997. LNCS, vol. 1294, pp. 410–424. Springer, Heidelberg (1997)
14. Canard, S., Coisel, I., Traoré, J.: Complex zero-knowledge proofs of knowledge are easy to use. In: Susilo, W., Liu, J.K., Mu, Y. (eds.) ProvSec 2007. LNCS, vol. 4784, pp. 122–137. Springer, Heidelberg (2007)
15. Canard, S., Schoenmakers, B., Stam, M., Traoré, J.: List signature schemes. Discrete Applied Mathematics 154(2), 189–201 (2006)
16. Canard, S., Traoré, J.: List signature schemes and application to electronic voting. In: Proceedings of Workshop on Coding and Cryptography (WCC 2003), pp. 81–90 (2003)

17. Chaum, D., van Heyst, E.: Group signatures. In: Davies, D.W. (ed.) EUROCRYPT 1991. LNCS, vol. 547, pp. 257–265. Springer, Heidelberg (1991)
18. Delerablée, C., Pointcheval, D.: Dynamic fully anonymous short group signatures. In: Nguyên, P.Q. (ed.) VIETCRYPT 2006. LNCS, vol. 4341, pp. 193–210. Springer, Heidelberg (2006)
19. Fiat, A., Shamir, A.: How to prove yourself: Practical solutions to identification and signature problems. In: Odlyzko, A.M. (ed.) CRYPTO 1986. LNCS, vol. 263, pp. 186–194. Springer, Heidelberg (1986)
20. Furukawa, J., Imai, H.: An efficient group signature scheme from bilinear maps. In: Boyd, C., González Nieto, J.M. (eds.) ACISP 2005. LNCS, vol. 3574, pp. 455–467. Springer, Heidelberg (2005)
21. Groth, J.: Fully anonymous group signatures without random oracles. In: Kurosawa, K. (ed.) ASIACRYPT 2007. LNCS, vol. 4833, pp. 164–180. Springer, Heidelberg (2007)
22. Groth, J., Sahai, A.: Efficient non-interactive proof systems for bilinear groups. In: Smart, N.P. (ed.) EUROCRYPT 2008. LNCS, vol. 4965, pp. 415–432. Springer, Heidelberg (2008)
23. IBM. Idemix - identity mixer (2004),
    http://www.zurich.ibm.com/security/idemix/
24. Libert, B., Vergnaud, D.: Group signatures with verifier-local revocation and backward unlinkability in the standard model. In: Garay, J.A., Miyaji, A., Otsuka, A. (eds.) CANS 2009. LNCS, vol. 5888, pp. 498–517. Springer, Heidelberg (2009)
25. Microsoft. Microsoft U-Prove CTP (2010),
    https://connect.microsoft.com/content/
    content.aspx?contentid=12505&siteid=642
26. Nakanishi, T., Funabiki, N.: Verifier-local revocation group signature schemes with backward unlinkability from bilinear maps. In: Roy, B. (ed.) ASIACRYPT 2005. LNCS, vol. 3788, pp. 533–548. Springer, Heidelberg (2005)
27. Nguyen, L.: Accumulators from bilinear pairings and applications. In: Menezes, A. (ed.) CT-RSA 2005. LNCS, vol. 3376, pp. 275–292. Springer, Heidelberg (2005)
28. Park, S., Park, H., Won, Y., Lee, J., Kent, S.: Traceable Anonymous Certificate. RFC 5636 (Experimental) (August 2009)
29. Persiano, P., Visconti, I.: User privacy issues regarding certificates and the tls protocol: the design and implementation of the spsl protocol. In: ACM Conference on Computer and Communications Security, pp. 53–62 (2000)
30. Trusted Computing Group. Direct Anonymous Attestation (2004),
    http://www.zurich.ibm.com/security/daa/

Author Index

GPSR Compliance

The European Union's (EU) General Product Safety Regulation (GPSR) is a set of rules that requires consumer products to be safe and our obligations to ensure this.

If you have any concerns about our products, you can contact us on ProductSafety@springernature.com

In case Publisher is established outside the EU, the EU authorized representative is:

Springer Nature Customer Service Center GmbH
Europaplatz 3
69115 Heidelberg, Germany

Batch number: 09474011

Printed by Printforce, the Netherlands